ASSAY DEVELOPMENT

ASSAY DEVELOPMENT

Fundamentals and Practices

GE WU

WILEY

A JOHN WILEY & SONS, INC., PUBLICATION

Published by John Wiley & Sons, Inc., Hoboken, New Jersey
Published simultaneously in Canada

Library of Congress Cataloging-in-Publication Data:

Wu, Ge, 1962–
Assay development: fundamentals and practices / Ge Wu.
p. cm.
Includes index.
ISBN 978-0-470-19115-6 (cloth)
1. Biological assay. I. Title.
QH90.57.B5W84 2010
612′.015—dc22

2009031386

Printed in the United States of America
10 9 8 7 6 5 4 3 2 1

CONTENTS

FOREWORD

As the first decade of the 21st century draws to a close, profound changes have come to many of the traditional biomedical practices that provided productivity for decades. A transformation has begun in the pharmaceutical sector, where well-known, established companies are now witnessing much of their expertise and R&D efforts spilling into the hands of the academic community. Today, we see initiatives by universities, research institutions, and state and federal government, both in the US and abroad, enabling so-called academic screening centers; the mission of these centers is to discover new classes of chemicals for development into chemical probes or drug leads using high throughput screening (HTS) and related discovery paradigms. The research conducted in these HTS laboratories necessitates databases, many of which are constructed in an open format to enable information sharing. These databases warehouse the unavoidable stream of data generated by the automation of experimentation. Along with HTS and databases come laboratories that specialize in medicinal chemistry; these laboratories fine-tune an initial 'hit' into a chemical series with pharmacological merit.

This changing landscape offers both new opportunities and responsibilities. Accepting a greater role in the search for and development of new chemical entities requires a shift toward interdisciplinary collaboration, an acceptance of team-based science, and for the first time, an effort to stem suffering from rare and neglected diseases; in the past, shareholder-funded companies had not focused on these types of diseases due to low prevalence or low profile, high cost of development, and the lack of appropriate business models. Integration of disciplines involving biology, chemistry, physics, engineering, and informatics is critical to our successful treatment of diseases that have plagued humanity for millennia, and for emergent diseases, as nature inevitably evolves to circumvent our current defenses. At the heart of any program to discover, design and evaluate new therapeutics for our ills is a means to measure, or 'assay', biological processes central to the pathology under investigation. The level of sophistication of that bioassay is partially dependent upon our knowledge of the pathophysiology itself, as well as the technologies that can enable the measurement.

Thus, the practice of assay development is the linchpin in an area referred to by various terms, such as translational research or bench-to-bedside science, aimed at constructing a path leading from basic research to clinical trials; proportionally greater lengths of this path now sit within an academic environment. The development of assays suitable for use in HTS, or in the follow-up optimization of new lead entities, can be a challenging undertaking and paramount to the success of any program in a directed drug discovery effort. In this volume, Dr. Ge Wu provides a guide to navigating the assay development process. Written with not only the practical knowledge that comes from his years in the pharmaceutical and biotech sectors, but with a scholarly approach to the subject matter, Dr. Wu's book provides an important treatise for 21st century drug discovery pioneers.

James Inglese, Ph.D.

Bethesda, MD

PREFACE

Bioassay development is a multidisciplinary science that utilizes the knowledge of biology, analytical chemistry, instrumentation, and statistics. Bioassay development plays a pivotal role in both the pharmaceutical industry and the diagnostic industry. The recent advances in high-throughput screening (HTS) further demand high-quality assays for a variety of novel biological targets. To meet these demands, many novel assay technologies have emerged in recent years. There are a few books and book chapters dealing with bioassay development. However, these books tend to focus either on specific assay technologies or on specific assay targets. There is no book that treats the subject of assay development systematically from the fundamental principle behind the assay to the final practice of assay development. This book fills the gap by discussing the fundamental principles on assay development and how the principles are applied to the practice of assay development. Because assay development is applied to many biological problems, it is impossible and it is not the intent of this book to discuss specific biology subjects in depth. Instead, a few commonly encountered biological targets in assay development (e.g., proteases, kinases, ion channels, and G protein-coupled receptors) are selected to help illustrate the art of assay development. Just enough biological background is discussed for these biological targets so that the readers can follow the logic in assay development. On the other hand, this book tries to introduce as many of the most widely used assay development technologies as possible, though it is impossible to cover them all in one book. This book's primary audiences are scientists in pharmaceutical/biotechnology companies and graduate students who plan to do drug discovery research after graduation. There is a trend that more and more academic institutions are setting up HTS capabilities. This book will be very helpful for students developing assays for HTS in academic settings as well.

ORGANIZATION OF BOOK

This book is organized into four parts. The first part of the book, comprised of Chapters 1 and 2, discusses the basics of assay development. Chapter 1 discusses the concept and application of bioassays in drug discovery, as well as the general approach to assay development. Chapter 2 discusses instrumental methods that are commonly used in bioassays with emphasis on optical methods. Chapters 1 and 2 together build the basic foundation required to carry out bioassay development. Bioassays can be generally divided into two broad categories: biochemical assays and cell-based assays. Biochemical assays employ isolated proteins as assay systems and cell-based assays employ live cells as the assay systems. The second part of the book, comprised of Chapters 3 through 7, discusses biochemical assays with a focus on the assay strategies for protein binding and enzymatic activity. In Chapter 3, the fundamental principles for both binding and enzymatic assays are introduced. In Chapter 4, the separation techniques that are very important in heterogeneous assays are discussed. In Chapter 5, general approaches for binding assays are discussed. Since all binding events follow the same principle, the techniques discussed here can be applied to all protein targets. Enzymatic assays on the other hand vary greatly, depending on the enzyme's function in the cell. It is impossible to cover all enzymatic assays in this book, instead, I selected protease assay as an example to discuss single-substrate enzymes (Chapter 6) and kinase assay as an example to discuss bi-substrate enzymes (Chapter 7). The third part of the book, comprised of Chapters 8 to 12, discusses cell-based assays. Chapter 8 discusses the fundamental principles and practices of cell-based assays. Cell-based assays can be performed based on either the measurement of specific components in specific cellular pathways or the measurement of global properties of the target cells as an integrated signal from many individual cellular component changes. Assay methods for specific pathways are target-dependent, which makes it impossible to cover them all in this book. Thus, I chose to discuss the two most popular cell-based assay targets: ion channels (Chapter 9) and G protein-coupled receptors (Chapter 10). The assay methods for integrated cellular changes are target-independent and are discussed in Chapter 11. High-content screening has emerged as an indispensible assay method in recent years and is discussed in Chapter 12. In the last part of the book, I selected two topics that have general interest for assay developers. Chapter 13 discusses high-throughput screening, for which assay development found most uses in recent years. Chapter 14 discusses an actual assay development case: the development process of the now popular Caliper off-chip kinase assay.

At the end of each chapter, publications that can help the readers understand the topics are listed. These references are chosen not based upon their scientific significance, but instead upon whether they can help the readers further understand the topics. In addition, the websites of many databases and vendors are listed to help the readers locate the databases, specific technologies, or instruments mentioned in the associated chapter. I have no financial arrangement with any of the sources listed while writing this book. I had welcome suggestions and recommendations with regards to the topics covered in this book, and you may contact me by e-mail at wuge22@gmail.com.

ACKNOWLEDGMENTS

This book is written based on my first-hand experiences in assay development while working in several organizations over the last two decades. I would like to thank my colleagues and supervisors at UCLA, Harvard Medical School, Cornell Medical College, Merck and Co., Caliper Technologies, Amphora Discoveries, and Five Prime Therapeutics. The majority of the manuscript was reviewed by friends and colleagues. I would like to thank Dr. Anthony Chao, Dr. Ning Lei, Dr. Yanlong Li, Dr. Yi Liu, Dr. Pengguang Wu, and Dr. Hongbing Zhang for proofreading and valuable suggestions. I would also like to thank Ernestine Franco of Pern Editorial for editorial assistance. I would like to thank Anita Lekhwani, Rebekah Amos, and Christine Punzo at Wiley for administrative assistance. Part of the book was taught at Peking University in the summer of 2008 to graduate students and postdoctoral candidates from both Peking University and the Chinese Academy of Science. This experience helped to make the book more understandable to young scientists. I am grateful to my wife, Liping Fan, and my daughters, Lily and Cora, for their patience and support while I devoted most of my time over the past couple of years researching for and writing this book instead of spending time with them.

GE WU

La Canada Flintridge, California

ABBREVIATIONS

ACE	Angiotensin Converting Enzyme
ADE	Acoustic Droplet Ejection
ADME	Absorption, Distribution, Metabolism, and Excretion
AEBSF	4-[2-Aminoethyl]benzenesulfonyl Fluoride
ADP	Adenosine Diphosphate
AMC	7-Amino-4-Methyl Coumarin
AMP	Adenosine Monophosphate
AMPA receptor	α-Amino-3-hydroxy-5-methyl-4-isoxazolepropionate receptor
AP	Alkaline Phosphatase
APC	Allophycocyanin
API	Active Pharmaceutical Ingredient
ATP	Adenosine 5'-TriPhosphate
BCR	Breakpoint Cluster Region
BLA	Biological License Application
BLyS	B-Lymphocyte Stimulator
BODIPY	Boron-Dipyrromethene
BrdU	Bromodeoxyuridine
BRET	Bioluminescence Resonance Energy Transfer
BSA	Bovine Serum Albumin
cAMP	Cyclic Adenosine Monophosphate
CaMK	Calmodulin-dependent Protein Kinase
CBER	Center for Biologics Evaluation and Research
CCD	Charge-Coupled Device
CDER	Center for Drug Evaluation and Research
CDK	Cyclin-Dependent Kinase
CEC	Capillary ElectroChromatography
CFR	Code of Federal Regulation
CK	Casein Kinase
CML	Chronic Myelogenous Leukemia
CRE	cAMP Response Element
CV	Coefficient of Variation
DABCYL	4-(4-Dimethylaminophenyl) Diazenylbenzoic Acid
DAG	Diacylglycerol
DAPI	4′,6-Diamidino-2-phenylindole
DFP	Diisopropylphosphofluoridate
DMSO	Dimethyl Sulfoxide
DNA	DeoxyriboNucleic Acid
DPBS	Dulbecco's Phosphate-Buffered Saline

DTT	Dithiothreitol
EC	Enzyme Commission
EC_{50}	The concentration of an effective molecule at which half of the maximum signal is elicited in a test system.
EDANS	5-(2-Aminoethyl)aminonaphthalene-1-sulfonic Acid
EDC	1-Ethyl-3-(3-dimethylaminopropyl)-carbodiimide
EDTA	Ethylenediaminetetraacetic Acid
EFC	Enzyme Fragment Complementation
EGFR	Epidermal Growth Factor Receptor
EIA	Enzyme Immunoassay
ELISA	Enzyme-Linked Immunosorbent Assay
EPO	Erythropoietin
EOF	Electroosmotic Flow
ERK	Extracellular Signal-Regulated Kinases
FAT	Factory Acceptance Test
FDA	Food and Drug Administration of the United States
FGF	Fibroblast Growth Factor
FGFR	Fibroblast Growth Factor Receptor
FP	Fluorescence Polarization
FRET	Fluorescence Resonance Energy Transfer
GABA	γ-Aminobutyric Acid
GC	Gas Chromatography
GCP	Good Clinical research Practice
GDP	Guanosine-5'-Diphosphate
GEF	Guanine Nucleotide Exchange Factor
GFP	Green Fluorescent Protein
GLP	Good Laboratory Practice
GMP	Good Manufacturing Practice
GPCR	G Protein-Coupled Receptor
Grb2	Growth Factor Receptor-Bound Protein 2
GRK	G Protein Receptor Kinase
GST	Glutathione S-Transferase
GTP	Guanosine-5'-Triphosphate
hERG	Human Ether-à-go-go Related Gene
HIV	Human Immunodeficiency Virus
HPLC	High Performance Liquid Chromatography
HRP	Horse Radish Peroxidase
HTRF	Homogenous Time-Resolved Fluorescence
HTS	High Throughput Screening
IC_{50}	Inhibitory concentration at which half (50%) of the maximal activity is produced.
ICH	The International Conference on Harmonisation of Technical Requirements for Registration of Pharmaceuticals for Human Use
ICP	Instrument Control Program

IND	Investigational New Drug
InsR	Insulin Receptor
IP_1	Inositol Monophosphate
IP_3	Inositol Triphosphate
IRS	Insulin Receptor Substrate
IUBMB	International Union of Biochemistry and Molecular Biology
IUPAC	International Union of Pure and Applied Chemists
IUPHAR	International Union of Basic and Clinical Pharmacology
LC	Liquid Chromatography
MAP	Mitogen-Activated Protein
MAPK	Mitogen-Activated Protein Kinase
MBP	Myelin Basic Protein
MWCO	Molecular Weight Cut-Off
nAcChR	Nicotinic Acetylcholine Receptor
NAD^+	Nicotinamide Adenine Dinucleotide
NDA	New Drug Application
NFAT	Nuclear Factor of Activated T cells
NMDA	N-Methyl D-aspartate
NMR	Nuclear Magnetic Resonance
PBS	Phosphate-Buffered Saline
PCA	Protein Fragment Complementation Assay
PCR	Polymerase Chain Reaction
PDE	Phosphodiesterase
PDGFR	Platelet Derived Growth Factor Receptor
PEG	Polyethylene Glycol
PGE	Prostaglandin E
PI-3 kinase	Phosphoinositide-3 Kinases
PIP_2	Phosphatidylinositol 4,5-bisphosphate
PKA	Protein Kinase A
PLC	Phospholipase C
PMSF	Phenylmethylsulphonyl Fluoride
PMT	Photomultiplier
PS	Phosphatidylserine
PSSCLs	Positional Scanning Synthetic Combinatorial Libraries
PVT	Polyvinyl Toluene
QC	Quality Control
RFU	Relative Fluorescence Unit
RIA	Radioimmunoassay
RTK	Receptor Tyrosine Kinase
SAR	Structure-Activity Relationship
SAT	Site Acceptance Test
SDS	Sodium Dodecyl Sulfate
SDS-PAGE	SDS-Polyacrylamide Gel Electrophoresis
SNARF-1	Seminaphtharhodafluor-1
SOP	Standard Operating Procedure

SPA	Scintillation Proximity Assay
STM	Single-pass Transmembrane
TAFI	Thrombin-Activable Fibrinolysis Inhibitor
TCA	Trichloroacetic Acid
TIMPs	Tissue Inhibitors of Metalloproteases
TLC	Thin Layer Chromatography
TMR	Tetramethylrhodamine
TNFα	Tumor necrosis Factor-Alpha
TPA	Tripropylamine
TPCK	Tosylphenoalanine Chloromethyl Ketone
TR-FRET	Time-Resolved Fluorescence Resonance Energy Transfer
TRF	Time-Resolved Fluorescence
TRP channel	Transient Receptor Potential channel
TUNEL	TdT-mediated dUTP Nick End Labeling
UPLC	Ultra Performance Liquid Chromatography
WGA	Wheat Germ Agglutinin
YSi	Yttrium Silicate

CHAPTER 1

INTRODUCTION TO ASSAY DEVELOPMENT

IN THIS chapter, we first give definitions of the assay and the bioassay. Drug discovery and development processes are then reviewed to show the role bioassay plays in this process. Because drug development is performed in regulated environments, brief discussion of regulations is given. In pharmaceutical research, a drug substance's characterization involves physiochemical characterization and bioassays. Bioassays also play a significant role in screening for potential drug candidates. While physiochemical characterization is a direct measurement, bioassay is an indirect measurement. Because bioassay is indirect, the relevance of the assay to its intended purpose is a significant factor for bioassay development. Finally, common bioassay categories are discussed.

1.1 ASSAY AND BIOASSAY

1.1.1 Definitions

An assay is a well-defined analytical method that contains the measurement procedure and how the measurement should be interpreted to obtain the properties of a system or object. Assays are very important tools in the pharmaceutical industry and in the medical diagnostics industry. A bioassay is defined as an assay that measures biological activity of a substance based on the response of a biological test system to the test substance. In the pharmaceutical industry, bioassays are commonly applied to characterize a substance's biological properties, to study a biological process, to detect the presence and quantity of a substance in a sample, and to screen for active molecules from a library of molecules. Before a substance is approved for human use, it has to be fully characterized. The characterization of a substance requires the determination of its physiochemical properties by physiochemical assays (characterization) and the determination of its biological activities by bioassays. The physiochemical properties of a drug substances include its chemical composition, chemical structure, solubility, particle size, crystal property, purity, and the like. With biological drug substances (such as proteins), additional physiochemical properties, such as

Assay Development: Fundamentals and Practices. By Ge Wu

amino acid sequence, modifications (phosphorylation, glycolation, etc.), and tertiary or higher structure may need to be determined as well. Physiochemical properties of a substance can be directly assayed by studying the drug substance alone using well-established physical and chemical techniques, such as high-performance liquid chromatography (HPLC), mass spectrum, nuclear magnetic resonance (NMR), X-ray crystallography, amino acid sequencing, and so forth. In contrast, the biological activity of a substance by definition is the effect of the substance on a biological test system. Thus, a substance's biological activity cannot be measured by studying the drug substance alone. A biological test system is required for a bioassay. The biological test system can be (1) biochemical, such as the activity of an enzyme or the ability to bind to a predefined protein; (2) cell based, such as isolated primary cells or transformed cell lines; (3) tissue or organ based; and (4) animal based. Due to the diversity of the choices of the biological test system, a variety of a bioassay can be developed for a given project. The most common bioassays are biochemical assays with isolated proteins and cell-based assays. To study a substance, several bioassays can be configured in either biochemical assay format or cell-based format. In biochemical assay, a substance's binding to a protein or the substance's effects on the protein's enzymatic activity can be measured. In cell-based assays, the substance's effects on cell morphology, cell cycle, total number of cells, modification and localization of intracellular proteins, the identity and the quantity of proteins secreted by the cells, transcription activity, or the beating rate and strength of isolated cardiomyocytes can be measured. It is challenging to pick a bioassay system that is best suited for a particular study from many potential bioassays.

1.1.2 Comparison of Physiochemical Measurement and Bioassay

Figure 1.1 shows the difference in direct physiochemical measurements and the indirect bioassays. Physiochemical properties of a substance can be measured by directly

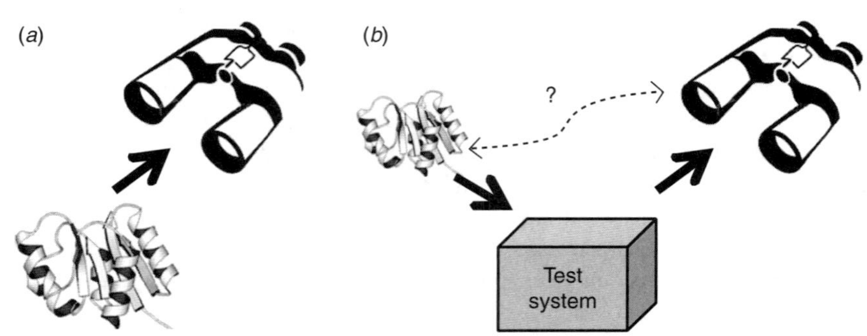

Figure 1.1 Comparison between direct physiochemical measurement and indirect bioassay. (*a*) A substance's physiochemical properties can be directly measured using HPLC, mass spectrum, NMR, IR, etc. (*b*) A substance's bioactivity can only be indirectly measured with a predefined biological test system. In bioassays, observations are made to the test system instead of to the substance. The biological activity of the substance is inferred from the observed changes in the test system based on prior knowledge about the test system.

analyzing the substance with one or more analytical instruments. An expert in a particular analytical method usually can make accurate conclusions on the aspect of the particular physiochemical measurement. For example, the composition of an organic substance can be readily obtained using well-established elemental analysis for carbon, hydrogen, nitrogen, and oxygen. A crystallographer armed with well-established analysis software can obtain three-dimensional structure of a substance with high-quality crystals. Physiochemical measurement is usually performed on a well-defined substance either in its pure form or in a formulation. It is an absolute measurement and a reference material usually is not required. Physiochemical measurements of a substance are commonly used in the quality control of drug manufacturing process and in the monitoring of drug distribution in the body. In contrast to physiochemical measurement, bioassay measures the responses of a test system to an external stimulation. A substance's bioactivity can only be indirectly measured by its effect on a predefined biological test system. The observations are made to the test system instead of to the substance. The biological activity of the substance is then inferred from the observed changes in the test system based on prior knowledge about the test system. Bioassay is commonly used in screening unknown molecules to discover active molecules and in accessing the known biological activity of the drug. To develop a bioassay, the assay developer has to first establish a test system and decide which response from the test system is relevant and how to interpret the response. Because the values of biological responses are relative, a control, which is a substance known to have an effect on the system, is required in bioassays. One characteristic of bioassay is the high sensitivity that substance at subnanomolar concentration can be readily detected with many well-designed bioassays. This is due to the fact that (1) the assay designer can choose a measurement among many responses in a bioassay system; (2) the test system can be manipulated to employ chemical/biological amplification (different from electronic amplification that raise the signal and background with the same amplitude); and (3) the most sensitive detection technology can be used. In comparison, the measurement of a substance's physiochemical properties in physiochemical assay usually requires the substance at a concentration of more than tens of micromolar (with the exception of mass spectrum technology).

1.1.3 Biological Relevance versus Experimental Control, Complexity, and Data Quality

For any type of measurement, it is most desirable to have full control of the system so that the system can be set at any predefined experimental conditions. A response from a system can only be measured after perturbations to the system are made. To obtain reliable interpretation about the responses from a system, bioassay scientists usually only vary one parameter at a time. All the responses from individual perturbation of the system are then recorded and synthesized to reach a conclusion. If more than two conditions are varied simultaneously, the interpretation of the experimental results will be difficult. Full control of a test system is not always possible. Physical scientists, such as physicists and chemists, have the luxury to vary experimental conditions in their system almost at will (only limited by existing technology and imagination). In contrast, sociology is at the other extreme: No changes can be made to a society in order to study a hypothesis. Sociologists can only observe the existing society or study its past to draw conclusions. Between the two extremes is life science in terms of experimental control. We can pick a system over which we have almost full control (purified stable proteins), a system over which we have some control (live cells and animals), or a system over which we have almost no control (human). The "biological relevance" limits full control over cells or animals used in the biological assays. Biological relevance means how much the test system resembles the system in its native state. Dramatically exerting controls over live entities (such as cells or animals) may make the measurement irrelevant. Protein kinase A (PKA) is used below as an example to illustrate different test systems. The hypothetical goal is to find a molecule that will inhibit PKA in humans.

PKA is a tetramer that comprises two regulatory subunits that are regulated by adenosine cyclic 3′,5′-phosphate (cAMP) and two catalytic subunits that catalyze the transfer of the phosphate group from adenosine 5′-triphosphate (ATP) to its protein (or peptide) substrate. The tetramer is inactive. The regulatory subunits dissociate from the catalytic subunits upon cAMP binding. The free catalytic subunit is active. For the purpose of finding an inhibitor for PKA, the simplest study system is the isolated catalytic subunits of PKA. The catalytic subunit of PKA has two binding pockets, one for ATP and one for a protein substrate. The simplest assay for the catalytic subunit of PKA is the competitive binding assay. The assay can be designed to screen a compound library for compounds that bind to the catalytic subunit resulting in interfering with the binding of ATP or the substrate protein to PKA. The assay developer can exert a great level of controls on this simple assay system: The assay can be carried out in the presence or absence of ATP and the substrate; the assay can be performed at any concentrations of ATP or the substrate if they are present; and the assay can be done at any temperature, pH, and any salt concentration (before it denatures). With this much control, the assay is easy to perform. However, this is the least biologically relevant system, and the resulting inhibitors will be of less value compared to the other systems. Only competitive inhibitors that bind to the binding sites of either ATP or the protein substrate can be identified in this assay. Other types of PKA inhibitors will be missed in this assay (see Chapter 3). The functional assay for PKA's ability to phosphorylate a peptide substrate is a step closer to biological relevance. The transfer of the phosphate group from ATP to a peptide substrate is a reaction with a net loss of free energy. This means as long as there is sufficient ATP and peptide substrate, the active kinase will continue to turn the peptide substrate into a phosphorylated product. With one step closer to biological relevance, we lose some controls of our experimental conditions in this assay. To keep the kinase functional, the experiment has to be done in conditions with the following constraints: a small temperature range (~20–37°C); the presence of Mg^{2+} or Mn^{2+} in a small concentration range (~1–20 mM); the presence of ATP and peptide substrate; and a narrow pH range for PKA to be active. With this functional assay, all inhibitors that can affect the kinetics of PKA-catalyzed phosphorylation of the peptide substrate can be detected. Though the functional phosphorylation assay system is a step closer to native biological system than the binding assay, it is still far less biologically relevant because the peptide substrate is not the native substrate(s) for PKA in cells and the isolated catalytic subunits of PKA is not properly regulation. In addition, the ATP concentration in the artificial functional assay may be different from what it is in the native system (the cells).

The next step closer to physiological conditions is the cell-based assay in which the activities of PKA in live cells are measured by the phosphorylation of PKA's native substrates. In assays using cells as test systems, more experimental controls are lost. Cells can only grow in conditions with a very tight range of pH, temperature, and oxygen. In addition, the phosphorylation of the native protein substrates inside the cells can only occur with fixed ion species, at fixed ionic and ATP concentration maintained by the cells. Varying these conditions will destroy the assay system (the cells). Furthermore, test compounds have to be able to pass cell membrane to exert direct effect on PKA. After the test compounds pass the cell membrane, they may act on some other proteins that indirectly influence the PKA's activity. The interpretation of experimental data is more difficult in complicated uncontrolled biological systems. One of the major challenges in cell-based assays due to the loss of control is that many cell types require undefined media containing serum to survive. In this case, uncontrollable unknown matrix contain hundreds of thousands of substances (the serum), which must be present to maintain the survival of the test systems, can cause significant variations between assays at different times. However, the results obtained from these cell-based assays are more biologically relevant because the proteins are in a native state at native concentrations and are properly regulated by other interacting proteins.

The relationship among physiological relevance, assay complexity, controllability of the assay system, throughput, quantitation, and data quality for bioassays performed with different assay systems are summarized in Figure 1.2. Isolated proteins were the most used bioassay systems. Binding assays and enzymatic activity assays using proteins as the test systems are the simplest bioassay, and they were the

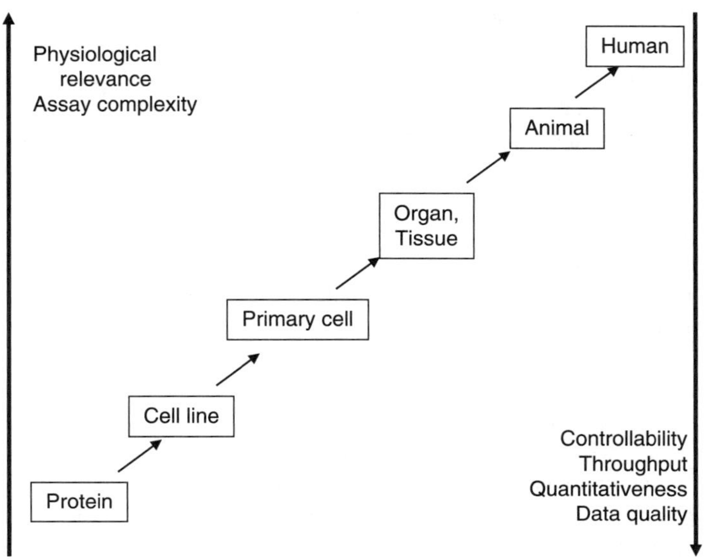

Figure 1.2 Relationship among physiological relevance, assay complexity, controllability, throughput, quantitation, and data quality for bioassays performed with different biological assay system. From Protein to Human, the physiological relevance and the assay complexity increase while the experimental control, assay throughput, ability to obtain quantitative data, and the quality of the acquired data decreases.

predominant assay format in the early day of bioassays. In modern drug discovery research, a drug target (a protein) is first identified and isolated. Molecules that can interact with the drug target are obtained by bioassays. Even with these simple protein assay systems, only a small portion of proteins (enzymes) is fully characterized before they are used as assay systems. The uncertainty in the test system will cause uncertainty in the assay outcome. Bioassays with transformed cell lines as assay systems are becoming more common recently because it is a step closer to biological relevance than isolated proteins. Transformed cell lines can be obtained in large quantities with homogeneous populations for large-scale screening, and they are easy to maintain. One advantage of using cell lines as assay systems is that many controls can be exerted on cell lines. Their genome can be readily manipulated to change the cells' characteristics to fit specific assay needs. With the advance of modern molecular biology, specific genes can be inserted or deleted from cells to make the cells gain or lose specific functions. However, the transformed cells may loss many properties of the primary cells from which they were derived. This makes the assay less biologically relevant compared with primary cells. Primary cells, especially human primary cells, are difficult to obtain in large quantity and cannot be easily manipulated. In addition, primary cells vary a lot among donors, which can cause large variations in assays. Because of these issues, primary cells are rarely used in primary screening. Cell-based assays are much more complicated than protein-based assays. Due to the complexity of the cell, it is safe to say that no cell is fully characterized. In addition, which response of the cell is relevant to the study and how the response is related to the study are not trivial to decide in a study. Further up the biological relevance ladder is to use tissue as a bioassay system in assays. Tissues preserve the interaction of cells with their native matrix environment, which is disrupted when cells are isolated. In some cases the isolated cells will not function well, and tissue or cells in artificial cell matrix should be used in bioassays. Beta cells in islet and chondrocytes in three-dimensional matrices are examples of such situations. Preserving these cells' native function is very challenging. Minimum control of the experimental condition can be exerted on these assay systems.

The next level of assay system based on biological relevance is the organ. There are a few examples of bioassays using organs as assay systems. For example, isolated animal heart has been used to perform bioassays. Whole animals are rarely used in initial bioassay, though they have been used extensively in preclinical research to investigate the toxicity and efficacy of drug candidates. Limited control can be imposed on these animals, though modern technology has allowed the creation of transgenic and knockout species. There are some attempts to use whole zebra fish as a bioassay system in initial compound screening. This approach still remains to be judged when more data is available. The advantage of using animals as test systems is that the results will be physiologically relevant. The disadvantage is that the results will be less quantitative and there is less control of the assay system. It is difficult to set up a predefined condition in an animal in order to measure the changes caused by an outside perturbation.

Because of less experimental control and the high cost associated with assay systems higher than tissues in terms of the biological relevance, these complicated bioassay systems are reserved for late-stage studies with a handful of compounds that have passed the hurdles in early-stage studies. The relatively simple biochemical assays with isolated proteins and cell-based assays are the most common assays in early drug discovery research. Because many bioassays may exist for a given project, picking the right bioassay system is very important. For example, in the drug discovery phase, it is more important to find a chemical entity that is physiologically relevant than to be quantitatively accurate about its potency. Serious consideration may be given to more complicated bioassay systems depending on the stage to which the molecule has advanced. In the drug development phase, especially when the study is under regulated environments, quality of data is very important. A simple bioassay system is preferred whenever possible if it is appropriate for the study. Biochemical assays are more quantitative and are amenable to accessing the quality of a product and product-related variants. Since a majority of bioassays in pharmaceutical research and development are biochemical and cell-based assays, this book will focus on these two bioassay test systems. Furthermore, the terms assay and bioassay will be used interchangeably from here since most of the assays we will discuss are bioassays.

1.2 DRUG DISCOVERY PROCESS AND ROLE OF ASSAYS IN THE PROCESS

To understand the role assays play in the pharmaceutical industry, it is important to first understand the process to obtain approved drugs. This process can be divided into the drug discovery phase and the drug development phase.

1.2.1 Drug Discovery Phase

The first phase in the drug life cycle is the drug discovery phase. In this phase, one or more drug candidates for a particular disease are identified. In the early days, most drugs were discovered by chance or luck when scientists in academic research laboratories investigating particular diseases accidentally found the drugs. This model of drug discovery still exists, but it only accounts for a small fraction of drug candidates moving into the preclinical development stage. Most pharmaceutical and biotechnology companies follow more systematic methods to obtain drug candidates. The companies first identify the disease area they want to pursue based on a combination of factors, such as potential market size of the disease, expertise in the disease area, and tractable target for the disease. This is followed by thorough investigation of the disease by studying the literature and performing some key experiments to identify one or more disease targets. The disease targets can be an enzyme that is over- or less reactive, a receptor that is over- or less responsive to its ligand, a ligand for a cell receptor that is at too high or too low a concentration, a component molecule in a signal transduction pathway that can be modified to balance the abnormal signal coming from upstream diseased proteins, and the like. Some examples of drug targets are discussed below.

Gleevec (imatinib), the first approved drug for chronic myelogenous leukemia (CML) based on a kinase inhibitor, is a good example of selecting an overactive enzyme as the drug target. It was found that 95% of people with CML have a chromosomal abnormality called Philadelphia translocation. In this case, part of the *BCR* (breakpoint cluster region) gene from chromosome 22 (region q11) is fused with part of the *ABL* gene on chromosome 9 (region q34). The fused *bcr-abl* gene codes for a protein that has tyrosine kinase activity. The *bcr-abl* transcript is constitutively active. It activates a number of cell cycle-controlling proteins and inhibits DNA (deoxyribonucleic acid) repairs. Thus, it is reasonable to hypothesize that inhibiting the kinase activity of BCR-ABL may offer a cure for CML. Thus, BCR-ABL was identified as the disease target for CML.

Tumor necrosis factor-alpha (TNFα) and its membrane-associated receptors offer a good example for drugs targeting ligand or its receptor. TNFα is a cytokine produced by monocytes and macrophages. TNF receptors are found on the surface of virtually all nucleated cells. TNFα mediates the immune response by increasing the transport of white blood cells to sites of inflammation, and through additional molecular mechanisms that initiate and amplify inflammation. Thus, it is reasonable to propose that interrupting the binding between TNFα and its membrane-bound receptor may offer a way to treat inflammatory disease, such as rheumatoid arthritis. Currently, there are three approved drugs for rheumatoid arthritis that block the activity of TNFα ligand: adalimumab (Humira), etanercept (Enbrel), and infliximab (Remicade). There is no drug targeting TNF receptor yet. This may be due to the fact that TNF receptors are expressed in too many cell types and its inhibition may lead to unwanted side effects.

In another scenario, if the disease target at the protein level cannot be identified, the cells involved in the diseases can serve as the disease target in the initial drug discovery effort. For example, scientists have tried to find drugs that stop B cells from causing inflammation. B cells cause joint inflammation in people with rheumatoid arthritis, though the detailed mechanism is not clear. Thus, it is reasonable to hypothesize that reducing the number of B cells in the body may reduce inflammation. A recently approved drug, rituximab, intercepts B cells and stops them from completing their tasks. Several other approaches to stopping B cells are under investigation. One investigational drug, belimumab, is a fully human monoclonal antibody that specifically recognizes and inhibits the biological activity of B-lymphocyte stimulator (BlyS). In this case, the protein target (BlyS) is known because BlyS is necessary for the maturation of B lymphocytes. In another example, isolated cancer cells can be used as the disease target for screening compounds that can selectively kill the cancer cells.

After identification of the tractable disease target, the next task is to identify molecules that can interact with the disease target. There are two prevailing strategies in the current drug discovery paradigm. One strategy is to physically test a large library of compounds against the disease target with one or more predefined bioassay. This process is referred to as screening. The molecules that produce signals that meet the assay criteria are called "primary hit" or "hit." Since the late 1980s, most pharmaceutical companies have built a collection of molecules (or "library") that is tested against the disease targets with predefined assays. The compound library may range from a few hundred thousands to several million compounds. When the assay throughput is high enough, the process is called high-throughput screening (HTS), which will be discussed in Chapter 13. Molecules that interact with proteins or with diseased cells can be obtained through HTS. Many newly discovered drugs were the results of significant contributions from HTS, such as Gleevec to treat CML. The other strategy is to use computational methods (or virtual screening) to identify molecules that interact with the disease target. This strategy requires the identification of the protein as the disease target and the knowledge of the target protein's structure. Many drugs were discovered successfully using this method, especially in the discovery of HIV (human immunodeficiency virus) protease inhibitors and reverse transcriptase inhibitors. The discovery of the HIV protease inhibitor darunavir (Prezista) is a good example of this approach. However, there are many limitations for computational methods. One is the availability of the crystal structure of the target protein. Further, a crystal structure does not always accurately depict how a molecule will behave in vivo. In addition, medicinal chemists often found it difficult to develop new structures for the "rational" approach. Computational methods have limited use in finding molecules that act on diseased cells. HTS remains the only way to find molecules that act on diseased cells. The advantage of the computation method is the large number of potential small molecules it can test and the resource and money needed to test the molecules. It is estimated that there are $>10^{30}$ conceivable compounds in the chemical space with molecular weight less than 500. It is not practical to physically screen every disease target with a library approaching 10 million compounds. Even 10 million is only a small number compared with the enormous chemical space. One strategy is to limit the potential pool of molecules going to HTS by first filtering out unlikely drug molecules through computation. Another approach is to build targeted screening libraries for specific drug targets.

After a collection of "hit" molecules is identified, the structure–activity relationship (SAR) is evaluated if the drug target is a known protein. This includes the determination of IC_{50} values of a series of inhibitors and the characteristics of the inhibition (reversibility, binding kinetics, inhibition mechanism, etc.). In addition to studying the hit molecule's interaction with the disease target, other properties of the hit molecule are further evaluated in vitro to determine whether it meets initial criteria as a drug. This includes the molecule's solubility in aqueous solution, octanol–water partition (log *P*), permeability to biomembranes, and so forth. Since a majority of clinical drug candidates failed because of their cardiotoxicity and hepatotoxicity, initial in vitro toxicity evaluations are sometimes performed at this stage to avoid costly failure in late stage. For example, the interaction of a hit molecule with hERG potassium channels may hint at cardiotoxicity, and the interaction of the hit molecule with different forms of cytochrome C isozymes may indicate potential hepatotoxicity. Furthermore, the compounds may be tested with tissues or intact organs in drug discovery phase. Only a limited number of the compounds survived this phase. The survivors are further moved to the preclinical stage. The drug discovery phase is very dynamic and new ways of doing research are constantly evolving. It is expected that with the rapid advancement of understanding of human biology, new approaches to drug discovery will certainly emerge in the future.

1.2.2 Drug Development Phase and Regulations

The drug discovery phase only shows that a substance can interact with the disease target protein or diseased cells. There is no information about whether the substance can be successfully delivered to the target protein in vivo, how long the substance remains in the body after administration to exert its effect, whether there is a disease modifying effect or not, and whether there are toxic side effects on the whole body. These questions are addressed in the drug development phase. To protect human safety, it is required by regulations that necessary safety information be collected from animal studies before any new substance is introduced to humans. Thus, drug development is divided into preclinical (animal study) and clinical (human) phases. The studies in the drug development phase must meet the regulatory requirements and they must be carried out by closely following regulatory rules and guidelines. The regulations have evolved slowly compared with the rapid advance of science and technology. This is understandable since the regulations' primary goal is to protect public health. However, sometimes it is difficult to obtain clear regulatory guidance when working with cutting edge technologies. In such cases, frequent communications with the regulatory authority is important.

Unlike in the drug discovery phase, there is less freedom to the approaches a researcher can adopt in the drug development phase. The drug development phase follows a process that is guided by government regulations and international standards. In the following discussion of preclinical and clinical development, related regulatory articles in the United States and international standards are cited so that the reader can further study the rules if in-depth knowledge is desired. In the United States, the Code of Federal Regulation (CFR) Title 21 deals with food and drug regulations. The Food and Drug Administration (FDA) in the United States publishes guidelines to further explain and clarify the regulations. Because drugs are marketed in different countries, there is a need for internationally recognized standard and harmonized regulations. The International Conference on Harmonisation of Technical Requirements for Registration of Pharmaceuticals for Human Use (ICH) is a project that brings together the regulatory authorities of Europe, Japan, and the United States and experts from the pharmaceutical industry in the three regions to discuss scientific and technical aspects of product registration. The objectives of ICH are the more economical use of human, animal, and material resources, the elimination of unnecessary delay in the global development and availability of new medicines while maintaining safeguards on quality, safety, and efficacy, and regulatory obligations to protect public health. ICH publishes many guidelines that are divided into four major topics: Quality (Q), Safety (S), Efficacy (E), and Multidisciplinary (M).

Good laboratory practice (GLP), good clinical research practice (GCP), and good manufacturing practice (GMP) are managerial and regulation concepts dealing with the process and conditions under which laboratory studies, clinical research studies, and manufacturing are planned, performed, recorded, and reported. The basic philosophy for GxP (GLP, GCP, and GMP) is that the laboratory or the manufacturing facility should design and perform studies or manufacturing processes carefully. All activities should be performed according to predefined standard operating procedure (SOP) and documented in such a way that studies can be reconstructed in the future. At different drug development stages, different GxPs are applied as shown in Figure 1.3. GLP is applied in safety assessment of a drug's pharmacology, pharmacokinetics, and toxicity in animal studies. GCP is applied in all three phases of clinical studies to protect human subjects. GMP is applied to both the manufacturing

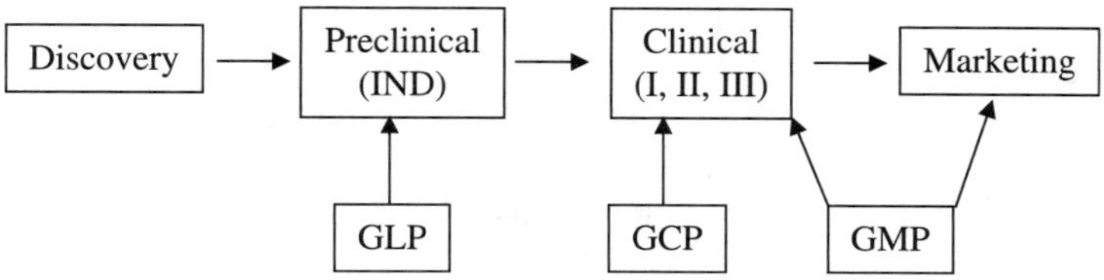

Figure 1.3 Required applications of GxP at different phases of drug development. There is no regulation in the drug discovery phase. GLP is applied in safety assessment of a drug's pharmacology, pharmacokinetics, and toxicity in animal studies. GCP is applied in all three phases of clinical studies to protect human subjects. GMP is applied to both the manufacturing of the reagents used in clinical testing and the manufacturing of the final marketed product after the approval of the drug.

of the reagents used in clinical testing and the manufacturing of the final marketed product after the approval of the drug.

During the *preclinical development* phase the drug candidate's initial safety and activity profile is investigated to support the investigational new drug (IND) application filing. The studies in this phase will determine the route of injection, duration, and total exposure in clinical patients based on pharmacological and toxicological evaluations. Biological activity of the substance may be evaluated using in vitro assays to determine which effects of the substance may be related to clinical activity. In vitro assays with cell lines derived from mammalian cells can also be used to predict specific aspects of in vivo activity and to assess quantitatively the relative sensitivity of various species. This study can assist in the selection of an appropriate animal species for further in vivo pharmacology and toxicology studies. The primary goals of preclinical safety evaluations are: (1) to identify an initial safe dose and subsequent dose escalation schemes in humans; (2) to identify potential target organs for toxicity and for the study of whether such toxicity is reversible; and (3) to identify safety parameters for clinical monitoring. Regulations require a set of tests being performed on animals to collect necessary information before the substance is introduced into the human body in clinical trials. At present, these studies have to be done with animals, though the industry and regulatory body have been working hard to find substitute methodology. In the United States, federal regulations on IND can be found in the Code of Federal Regulation Title 21 Part 312 (21 CFR 312). The IND submission for Phase I clinical study is required to contain the following sections:

A. Cover sheet

B. Table of contents

C. Introductory statement and general investigational plan

D. Investigator's brochure

E. Protocols

F. Chemistry, manufacturing, and control (CMC) information

G. Pharmacology and toxicology information

H. Previous human experience with the investigational drug

The most important section of the IND filing is the substance's pharmacology and toxicity profile. FDA guideline on pharmacology specifically indicates that "this section should contain, if known, (1) a description of the pharmacologic effects and mechanism of actions of the drug in animal, and (2) information on the absorption, distribution, metabolism, and excretions of the drug." The first part of pharmacology is a pharmacodynamic study that deals with how the drug interacts with its target cell or organ and how it exerts its effects or side effects. The second part of pharmacology involves pharmacokinetic studies that gather the data on the substance's absorption, distribution, metabolism, and excretion (ADME). The study of a substance's pharmacological effect (efficacy) and the mechanism of action in the preclinical stage may be important to address safety issues and may assist in the evaluation of toxicity data. However, the lack of efficacy information in IND filing will not cause a Phase I clinical hold. In vivo efficacy study of a substance requires the availability of the animal model for a particular disease. The lack of good animal model for a given disease is a major obstacle for new drug discovery and development. In contrast, in vivo pharmacokinetics and toxicity studies of a substance can be investigated with normal animals.

Toxicity studies gather information that is used to determine the safety range of dosing for phase I clinical studies in humans. ICH guideline M3 titled *Non-clinical Safety Studies for the Conduct of Human Clinical Trials for Pharmaceuticals* recommends the following basic required tests for preclinical development:

> The non-clinical safety study recommendations for the marketing approval of a pharmaceutical usually include single and repeated dose toxicity studies, reproduction toxicity studies, genotoxicity studies, local tolerance studies and for drugs that have special cause for concern or are intended for a long duration of use, an assessment of carcinogenic potential. Other non-clinical studies include pharmacology studies for safety assessment (safety pharmacology) and pharmacokinetic (ADME) studies.

Toxicity and pharmacokinetic studies are usually performed simultaneously with normal health animals, typically with at least two different species. Rodents are used for early studies and more expensive nonhuman primates are used in later studies when large amounts of data have been already collected from studying rodents. Acute and chronic toxicity studies are usually performed. With acute toxicity studies, animals are administrated with a single does of substance with different doses for different groups of animals. The animals are sacrificed a few days after administration and the organs are analyzed. In the same experiment, pharmacokinetic data can be collected too. In a chronic toxicity study, one or more administration of the substance to the animal per day is done, and the dosing lasts several weeks to several months with different doses for different dosing groups. At the end of the dosing schedule, the animals are sacrificed and the organs are analyzed. Again, pharmacokinetic data can be obtained in the same experiments. The IND-enabling toxicity studies should be performed in compliance with GLP. GLP is a quality system concerned with the organizational process and the conditions under which nonclinical health and environmental safety studies are planned, performed, monitored, recorded, archived, and reported. GLP regulations for the pharmaceutical industry is covered by Code of Federal Regulation Title 21 Part 58 (21 CFR 58). Though an unaudited draft report is accepted for IND filing, fully quality-assured documents must be available to the FDA, upon request, within 120 days of the start of human studies.

Preclinical development of biotechnology-derived pharmaceuticals (biopharmaceuticals) follows the general guidelines discussed above. However, special properties of biopharmaceuticals require different treatment in some areas from their small-molecule counterpart. ICH guideline S6 titled *Preclinical Safety Evaluation of Biotechnology-Derived Pharmaceuticals* provides general principles for designing scientifically acceptable preclinical safety evaluation programs. The biological activity together with species and/or tissue specificity of many biopharmaceuticals often preclude standard toxicity testing designs in commonly used species (e.g., rats and dogs) for small-molecule drug development. It is important to select relevant animal species for toxicity testing. A relevant species is one in which the test material is pharmacologically active due to the expression of the receptor or an epitope of monoclonal antibodies. Biopharmaceuticals intended for humans are often derived from the source with human origins. Thus, they may be immunogenic in animals used in preclinical testing. The antibody detected in animals associated with the administration of biopharmaceuticals may complicate the toxicology studies. Some of the standard toxicity tests that are routinely performed with small-molecule substances are not applicable to biopharmaceuticals and are not needed, such as metabolic study of the substance's biotransformation, genotoxicity studies, and carcinogenicity studies. There is a significant difference in the physical properties between biopharmaceuticals and small molecules. The safety concerns may arise from the impurities and contaminants in the biopharmaceuticals because they are derived from host cells, such as bacteria, yeast, insect, plant, and mammalian cells. Thus, the product that is used in the IND-enabling pharmacology and toxicology studies should be comparable to the product proposed for the initial clinical studies. Though IND-enabling toxicity studies should be performed in compliance with GLP, some non-GLP compliant specialized tests needed for biopharmaceuticals are acceptable to regulatory agency. Areas of noncompliance should be identified and their significance evaluated relative to the overall safety assessment.

Another important part in IND filing is the CMC section. The emphasis in Phase I CMC submission is placed on providing information to assure the proper identification, quality, purity, and strength of the investigational drug that will allow evaluation of the safety of subjects in the proposed study. For preclinical studies to be useful in assuring the safety of human studies, the drug product being proposed for use in a clinical study must be able to relate to the drug product used in the animal toxicity studies. The quantity of the drug candidates required for preclinical study is in several orders of magnitude more than what is required in the drug discovery phase. In the drug discovery phase, the quantity of a drug candidate used in studies, usually for physiochemical and biological characterization, is relatively small (on the order of a few grams). A bench chemist can make the neccessary quantity of small molecules in a flask and a biologist can make such quantities of proteins with several cell culture flasks. In preclinical studies, a gram or more of the drug may be administered to one animal in a single does. Multiply this by repeated doses, the hundreds of animals in one study, and the numbers of different studies, several hundred grams or even several kilograms of the drug may be needed in the preclinical study. Thus, the substance used in preclinical testing may come from different batches of production. At this stage, the production group also constantly improves the process aimed at more efficient production processes and even larger production quantities for future clinical testing. Production process modification may result in large changes in biological activity between different batches of the substance, even though there is no detectable change in physiochemical properties. The problem is more prevalent for protein production than for small molecules. For protein production, many potential changes, such as degree of denaturization, different stable forms, posttranslational modification (phosphorylation, glycosylation, acylation, prenylation, methylation, etc.), can affect biological activity. For small molecules, the major activity changes may come from the different solid forms of the substance (polymorphs, amorphous, solvate, salt, and co-crystal). It is crucial to have good communication between material production groups and the preclinical testing groups so that any changes in the production process are taken into account in the interpretation of unexpected experimental results. The production process developed in preclinical phase should be quite stable or be similar to the substance production process for clinical testing in order to obtain consistent results and protect the safety of human subjects in clinical testing. Though different formulations of a drug may be used during drug development phases, links between formulations must be established by bioequivalence studies to allow interpretation of the preclinical and clinical study results. The substance used in the preclinical stage is not required by regulation to be manufactured by a GMP-approved facility. However, a GMP-produced substance is required at the clinical stage by FDA regulation. It will be of advantage if the substance is produced in a GMP facility in the late-stage preclinical testing to smooth the progress to clinical testing. The FDA guidelines suggest the following information being provided for review of the manufacturing procedures for drug products used in Phase I clinical studies.

1. Chemistry and manufacturing introduction
    a. Potential risk from the chemistry or the manufacturing of the drug substance or the drug product
    b. Chemistry and manufacturing difference between drug product for clinical use and the drug product used in the animal toxicity study
2. Drug substance
    a. Description of the drug substance (physical, chemical, biological)
    b. Manufacturer of the clinical drug substance
    c. Method of preparation of the drug substance
    d. Analytical methods to assure identity, strength, quality, and purity of the drug substance and acceptable limits
    e. Stability of the drug substance during the toxicology studies and proposed clinical studies
3. Drug product: Similar requirement as outlined in guideline 2 above
4. A brief general description of the composition, manufacture, and control of any placebo to be used in the proposed clinical trial

5. A copy of all labels and labeling to be provided to each investigator
6. A claim for categorical exclusion from or submission of an environmental assessment

The *drug substance* mentioned above means an active pharmaceutical ingredient (API) that is intended to furnish pharmacological activity or other direct effect in the diagnosis, cure, mitigation, treatment, or prevention of disease or to affect the structure or any function of the human body, but does not include intermediates used in the synthesis of such ingredient. The *drug product* means a finished dosage form that contains a drug substance in association with one or more other ingredients (tablet, capsule, or solution).

The *clinical development* phase is aimed at determining the drug candidate's dosage range, safety profile, ADME, clinical end point, and efficacy. Clinical development is generally divided into three consecutive phases (Phase I, II, and III) to support the filing of a new drug to the regulatory agencies for marketing. In addition, the FDA Amendments Act of 2007, which went into effect on October 1, gives FDA the power to require drug makers to do postmarketing clinical trials (Phase IV). The phase concept is a description but not a set of requirement. The logic behind serial studies is that the emerging data from prior studies will guide the planning of later studies. This will also minimize the risk for trial subjects. Each phase of the clinical studies has a general objective: Phase I, human pharmacology; Phase II, exploratory therapeutics; and Phase III, confirmatory therapeutics. There are many individual studies in each phase of the clinical trials to answer different questions. Each individual study should contain objectives, design, conduct, analysis, and report. Dose–response information of the drug should be obtained at all stages of the development. In the United States, an NDA is filed with the Center for Drug Evaluation and Research (CDER) and a Biological License Application (BLA) is filed with Center for Biologics Evaluation and Research (CBER). Detailed format and contents requirements on NDA filing can be found in the Code of Federal Regulation Title 21 Part 314 (21 CFR 314) and BLA filing can be found in the Code of Federal Regulation Title 21 Part 601 (21 CFR 601). The drug used in the clinical phase is required by regulation to be manufactured by a GMP-approved facility. GMP is a part of quality assurance, which ensures that products are consistently produced and controlled to the quality standards appropriate to their intended use. Regulations for pharmaceutical GMP can be found in 21 CFR 210–226. ICH also published Q7: "Good Manufacturing Practice Guide for Active Pharmaceutical Ingredients." In addition, the conduct of clinical research is highly regulated by federal regulations in the United States and throughout the world. ICH published guidelines for clinical trials in publication E8: "General Considerations for Clinical Trials," which describe the principle and practice in the conduct of both individual clinical trials and overall development strategy for an investigational drug. In clinical development, GCP must be followed. GCP is an international ethical and scientific quality standard for the design, conduct, performance, monitoring, auditing, recording, analyses, and reporting of clinical trials that provides assurance that the data and reported results are credible and accurate, and that the rights, integrity, and confidentiality of trial subjects are protected. GCP guideline is published in ICH publications E6: "Good Clinical Practice: Consolidated Guideline."

In the United States, the Phase I clinical trial is initiated 30 days after the submission of IND to FDA unless the FDA puts a clinical hold because of toxicology concerns or questions about the study design. The objectives of Phase I clinical development are typically nontherapeutic but are aimed at establishing the pharmacological and the toxicological properties of the drug candidate in humans and to determine the tolerability of the dose range to be used for later clinical studies. The emphasis here is to establish the drug candidate's safety characteristics. A small number of healthy volunteers or patients (20–100) are tested in this phase. The studies typically include both single and multiple dose administrations. Dose range and dose scheduling are commonly studied. For safety reasons, doses below the proposed treatment level are usually tested first and the dose is increased over time. Studies in this phase can be open, baseline controlled, or double blinded. The duration of clinical Phase I studies on average takes up to 1 year.

The primary objective of Phase II clinical study is the initial exploring of the therapeutic efficacy in patients. Usually 100–300 patients are enrolled in the study. The patients are selected by relatively narrow criteria with a relatively homogeneous population. Initial therapeutic exploratory studies may use a variety of study designs, including concurrent controls and comparisons with baseline status. Subsequent trials are usually randomized and concurrently controlled to evaluate the efficacy of the drug and its safety for a particular therapeutic indication. An important goal for this phase is to determine the dose and regimen for Phase III trials. Doses used in Phase II are usually less than the highest doses used in Phase I. Additional objectives of clinical trials conducted in Phase II may include evaluation of potential study end points, therapeutic regimens, and target populations. These objectives may be served by exploratory analyses, examining subsets of data and by including multiple end points in trials. The duration of clinical Phase II studies on average takes up to 2 years. The number of patients enrolled in Phase II trials is not large enough to obtain unambiguous statistical information to prove efficacy and safety.

Phase III clinical development is conducted with a large number of patients, usually in the range from 1000 to 3000. The objectives of Phase III studies are to gather enough statistically important evidence to confirm the efficacy and safety of the drug candidate for its intended indication in the targeted patient population. The data collected from Phase III clinical studies form the basis for market approval. Studies in Phase III may also further explore the dose–response relationship or explore the drug's use in wider populations, in different stages of disease, or in combination with another drug. Long-term effects of the drug candidate are studied for drugs that are intended for long-term use. The duration of clinical Phase III studies on average takes up to 3 years.

1.2.3 Role of Assays in Drug Discovery and Development

From the drug discovery and development processes described above, it is clear that assays play very important roles from the initial drug discovery phase all the way to the clinical phases. In the drug discovery phase, assays are performed for testing hypothesis of unknown biological processes, for detecting specific biological pathways, and for testing the effect of a substance on proteins, cells, tissues, or organs.

Bioassay is a major part of high-throughput screening. Bioassays also play a vital role in the computational drug discovery approach because the lead molecules obtained from computational methods must be confirmed experimentally. In drug development phases, assays play important roles in pharmacological and toxicity testing, substance characterization, manufacturing process development and validation, formulation development, manufacturing quality control, evaluating long-term and short-term (accelerated) stability, establishing comparability between batches of manufactured drug substance and drug product. As discussed before, physiochemical and biological characterization of a drug substance is required in the specification part of the document for regulatory submission to access the substance's potency. The assay here is not to identify an unknown species but to make sure a known molecule is detected over and over with the same accuracy. The biological characterization is a measurement of the substance's ability to elicit a biological response or potency. While physiochemical characterization can be performed by physical/chemical techniques, the biological characterization can only be accessed by bioassay.

When the drug is a small molecule, physical/chemical measurement alone usually can establish the equivalency between different batches and between different manufacturing processes because small molecules usually can be precisely characterized with a set of well-established techniques, such as gas chromatography (GC), HPLC, NMR, mass spectrum, ultraviolet (UV), infrared (IR), and the like. There is a direct correlation between a substance's structure and activity for small molecules. Thus, assay for biological activity may not be required for each batch of small molecules. For protein-derived drugs, the identity, purity, and quantity can be estimated by sodium dodecyl sulfate–polyacrylamide gel electrophoresis (SDS–PAGE), HPLC, amino acid sequencing, sugar analysis, and the like. However, these physical/chemical methods may not be able to fully characterize the drug; or the physical/chemical properties do not correlated with the biological activity due to post-translational modification, protein folding, media effect on protein, and so forth. Bioassay is especially important in protein-based drug development. One specific example is to detect neutralizing antibodies for a given protein drug that can only be measured with bioassays.

1.3 BIOASSAY DEVELOPMENT

Bioassay development is the process to obtain a final assay system that is appropriate for its intended use and can be reliably performed repeatedly. The final bioassay protocol is developed through careful evaluation of all potential parameters that may affect the assay. The bioassay development process can be divided into following parts:

- **A.** Carefully studying the biological target to determine what biological parameters should be determined to answer specific questions
- **B.** Setting up a bioassay system in which some of its components, having relationships with the predefined biological parameter, can be directly measured with well-established methods

C. Understanding the selected measurement's application range (boundary conditions) and making sure the intended uses are within the boundary

D. Obtaining a control substance known to generate the intended responses in the test system

E. Building algorithms that can mathematically relate the final detected signal to the intrinsic biological function

Part B can be divided into two stages: (1) a series of manipulation of the test system to generate a detectable signal and (2) the measurement of the signal. When developing an assay, the first step is to develop the detection system before putting much effort to manipulate the test system. Bioassay usually involves several steps of manipulation of the test system to reach the final stage that a detectable signal is generated. The intermediate stages usually cannot be detected. When an assay does not give an expected signal, it is very difficult to tell which step among a series of steps leads to the wrong signal. In some cases, the signal is not detected not because of malfunction of the test system but because of the wrong detection system. The first step in developing an assay is to artificially generate the detectable signal and then test whether the signal can be properly measured. This can be achieved by obtaining the substance that gives off the detectable signal or by establishing an artificial test system mimicking the final stage of the assay to generate the detectable signal. When unexpected results happen, a diagnostic procedure should be performed from the last step closest to the signal generation backward stepwise to the beginning of the assay. The reason for going from back to front is that the signal can only be detected at the last step.

Here is a real case example to demonstrate that the proper sequence of experiments can save a lot time and effort: An inexperienced postdoc was working on a project to generate bacteria strings that secreted the most surfactant. He found in the literature a method using measurement of surface tension as the final assay reading because the more surfactant secreted the more changes in the surface tension. After he generated many strings of bacteria and tested them with his detection system, he could not see changes in surface tension among all the bacteria tested. He suspected that he might not generate a good string of bacteria and continued generating more bacteria. After all these efforts, he learned that he should first separate the bacteria from detection. He artificially applied different concentrations of the same surfactant secreted by the bacteria to a test solution to establish an artificial system mimicking the final stage of the assay. He then tried to detect the change using the detection method described in the literature. To his surprise, there was no change at all for any concentration of the surfactant he tested. Now he realized that he could not detect any change in surface tension with the method described in the literature even if there is a change. He then went back to the beginning of the assay development: to obtain a reliable detection system.

An example of bioassay development is shown to illustrate each step in the process of assay development (see Fig. 1.4). The background of this study is based on a real case scenario. A growth factor (L) is a potential therapeutics found in drug discovery phase and it is currently under development. It is known that L may denature in

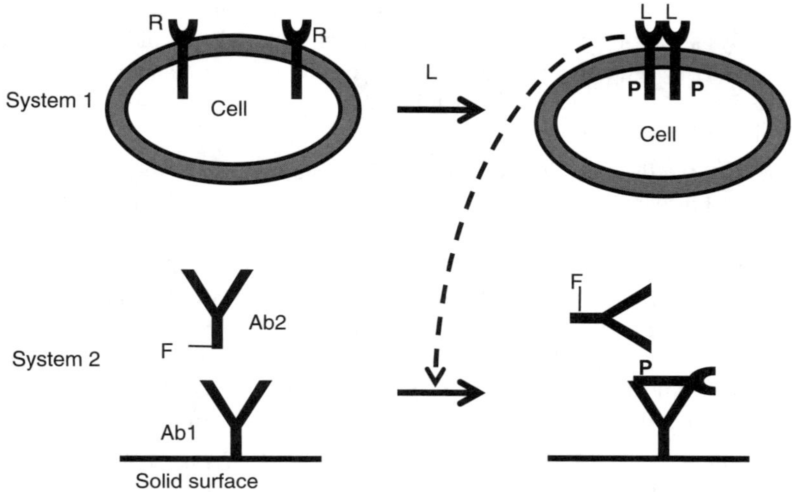

Figure 1.4 Illustration of bioassay for active ligand (L) of a growth factor receptor (R). The first biological test system (System 1) is a cell line that expresses the receptor on its surface. The R receptors dimerize and phosphorylate each other upon binding to ligand L. The second biological test system (System 2) is an ELISA system with the first antibody (Abl) attached to a solid surface that captures solubilized receptor R. The fluorescence-labeled second antibody (Ab2) binds only to phosphorylated R. Upon removal of the unbound Ab2, the fluorescent signal can be measured with a fluorimeter. The intensity of the fluorescence is proportional to the quantity of the phosphorylated receptor R that in turn is proportional to the quantity of the active ligand L.

some formulation resulting in lose of activity. An assay for L is needed to support the preclinical testing in mice.

A. *Define the Goal of the Assay* The goal here is to determine the active substance in each batch of L to guide the administration of correct amount of L into mouse. In this case, physiochemical characterization of each batches of L in formulation is not good enough to determine the quantity of the active molecules administrated into mouse. A functional bioassay is required.

B. *Design an Assay* It is known that L binds to a receptor (R) to elicit the downstream biological effects that form the basis for the intended therapeutic use of L. Thus, a biological test system that contains receptor R can be constructed. The response from the test system can be directly measured when L binds to receptor R. It is also known that L's biological activity started with its binding to the cell surface receptor R. The binding of L initiated the dimerization of receptor R on the cell surface. This is followed by autophosphorylation of receptor R that is followed by many other biological responses. From this information, a test system based on cells that express receptor R at their surface can be established. With this test system, the biological activity of L can be measured by the binding between L and R, the dimerization of R, the phosphorylation of R, or other downstream biological responses. After detailed analysis with these different readouts, the phosphorylation of R is chosen as the final assay readout. After a literature search, a stable cell line that expresses receptor R on its surface is found readily available, and it will be used as the biological assay system.

C. *Understanding the Measurement* There are many bioassays to measure the phosphorylation state of membrane receptors. Here we choose to measure the phosphorylation of receptor R with a sandwiched enzyme-linked immunosorbent assay (ELISA) system. In this assay, one antibody attached to the bottom surface of a microplate captures solubilized receptor R. Another fluorescently labeled antibody recognizing the phosphate group is used for detection. It is important to establish the boundary conditions for the ELISA system and to find the most sensitive part in the detection window (linear range). Many assay conditions should be tested, such as pH, buffer, temperature, and duration.

D. *Obtaining a Control Substance Known to Elicit the Desired Response from the Test Systems* The proposed assay contains two consecutive assays with two test systems. The first test system is the cell line that expresses receptor R and is responsive to active L. The second test system is the ELISA, which is responsive to the phosphorylated R. A preferred control to qualify the ELISA bioassay system is phosphorylated receptor R with a known amount of phosphorylation. In the absence of this control, a ligand known to induce phosphorylation of receptor R in the cell line can be used as a control to test the two assay systems together. The biological activity of this control ligand must be known. The same commercially available growth factor L with known activity is used as a control here.

E. *Relating Measurement to Intrinsic Biological Function* The raw data obtained from the above bioassay is the relative fluorescence unit (RFU). Unlike absorbance measurement that is an absolute measure, RFU changes between fluorimeters and within a given fluorimeter when instrument parameters (such as gain, slit width, voltage on photomultiplier) change. Thus, the measurement of RFU at one experimental condition should be converted to an absolute measurement. In this case, the RFU is converted to units of control ligand L (either activity unit or concentration because there is a correlation between the two in the control) based on the response generated. The first attempted experiment generated the initial standard dose–response curve (Figure 1.5). These data indicate that the assay is sensitive in the region between 1500 and 9500 RFU, which corresponds to the biological response of the test system to 0.05 and 1 nM of the control ligand L. Further experiments should be performed to obtain more data points in this region to plot a higher quality dose–response curve. These data also help setting the boundary condition as to how much test sample containing phosphorylated R from test system 1 can be applied in test system 2 (ELISA system).

A well-developed assay is a finished product that has many intrinsic properties and performance characters. It is important to bear in mind the performance

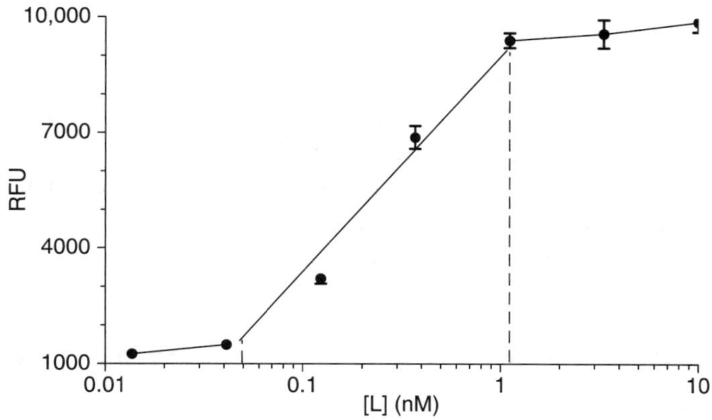

Figure 1.5 Standard dose–response curve of control ligand L known to cause receptor dimerization and autophosphorylation. The sensitive region for this assay is between 1500 to 9500 fluorescence count that corresponds to a dynamic range between 0.05 and 1 nM of the known ligand L.

characters when developing bioassays. Documentation of the performance characters is important for the transfer of the assay and for satisfying the regulatory requirement if the assay is intended for IND, NDA, or BLA filing. Following are important assay performance characteristics that should be investigated in the assay development process.

1. *Accuracy* The closeness of the *mean* test results obtained by the assay to the true value or the accepted reference value of the analyte. For many bioassays, it is very hard to obtain the true value. Accepted reference value is commonly used as a standard for accuracy measurement.
2. *Precision* The closeness of *individual* measurements of an analyte when the same assay procedure is applied repeatedly to multiple aliquots of a single homogenous sample. Precision measurement is obtained in an assay with a particular concentration of the analyte without referencing to a standard sample. Precision of an assay at a particular analyte concentration is commonly expressed as the coefficient of variation (CV). Precision can be further divided into: within-run precision (also called repeatability) and between-run precision (also called intermediate precision).
3. *Sensitivity* The ability of the assay to discriminate between small differences in analyte concentration (detailed discussion can be found in Chapter 2).
4. *Specificity* The ability of the assay to differentiate and quantify the intended analyte in the presence of other components expected to be present in the sample. Specificity is especially important in multiplex assay in which the assay is designed to analyze more than one analyte simultaneously. Specificity is also a very important factor in cell-based assays in the presence of potentially interfering serum.

5. *Detection Limits* The lowest concentration of the analyte in the sample that produces an assay signal that can be distinguished from the assay background (the assay signal in the absence of the analyte).
6. *Lower Quantitation Limits* The lowest concentration of the analyte in a sample that can be quantitatively determined with predefined precision and accuracy. Quantitation limits are always higher than the detection limits.
7. *Upper Quantitation Limits* The concentration of the analyte at which and beyond that cause the assay to fail to obtain quantitative results [e.g., deviate from known biology, oversaturating the detection instruments, oversaturating one of the assay components, unreasonable phenomenon such as "hook effect" (a phenomenon in an assay in which the response initially going up with higher concentration of analyte and then turns lower with even higher concentration producing a hooklike dose–response curve)].
8. *Linearity* The phenomenon in an assay by which the measured test results are directly proportional to the concentration of the analyte in the sample within a range of the analyte concentration.
9. *Dynamic Range* The concentration of the analyte between lower quantitation limits and the upper quantitation limits.
10. *Robustness* The measurement of an assay's tolerance to small perturbations in one or more components in the assay system.
11. *Boundary Conditions* The range of the assay components beyond which the assay is not valid (e.g., pH, temperature, buffer components and their concentration, enzyme concentration, substrate concentration, cell number).
12. *Reproducibility* The ability to carry out the assay and obtain specified results with a combination of any of the following: a different scientist, a different time, a different location, a different batch of assay components, and different instruments, etc.
13. *Scalability* The ability of the assay to perform in different formats (test tubes, different microtiter plates: 12-well, 24-well, 48-well, 96-well, 384-well, 1536-well) and in different scales (screen a few sample per run vs. screen a few hundred thousand sample per run). The scalability is especially important for high-throughput screening operations and for large-scale clinical lab testing.

1.4 BIOASSAY CLASSIFICATIONS

There is no unified systematic classification of bioassays. Some commonly used bioassay classification and associated nomenclatures are listed below.

1. *Classification According to Test System Used in Bioassay* Isolated protein-based assay (or biochemical assay), cell-based assay, tissue-based assay, organ-based assay, and animal-based assay. This book is organized using this classification method by first discussing isolated protein-based assays

(Chapters 3 to 7) that is followed by discussing cell-based assays (Chapters 8 to 12).

2. *Classification According to Assay Target Class* Protein-binding, protease, kinase, GPCR, ion channel, metabolite transporter, and so forth.
3. *Classification According to Whether Employing Separation Methods in the Assay* With this method, all the assays are divided into homogeneous assay and heterogeneous assay. Homogenous assays sometimes are also referred to as "mix and read" assays. There is no separation step to remove the interfering species in the assay systems from the analyte in homogenous assays. Higher background as the result of the signal from the interfering species in the assay system is the major issue affecting homogeneous assay. Opposite to homogenous assay is heterogeneous assay, which involves separation steps, such as washing, filtration, and centrifugation, to physically remove the interfering components from the analyte (the techniques are discussed in Chapter 4). Separation steps usually are tedious to perform and can result in higher variations in the assay. In addition, heterogeneous assays are more difficult to implement in HTS operations compared with homogeneous assays. Evolving technologies have made this classification system based on separation difficult to apply. For example, the "off-chip" kinase assay marketed by Caliper Technologies is a homogenous assay from operation point of view. However, physical separation is incorporated into the detection. It is a homogeneous assay but with high signal-to-background ratio comparable to heterogeneous assay (see Chapter 14).
4. *Classification According to Whether a Label is Introduced in the Assay System* Most traditional bioassays employ a foreign tag (or label) attached to one or more components in the assay system. The foreign label can be a small molecule (fluorescent molecules, biotin, etc.), a small peptide (epitope peptide, peptide substrate for biotin attachment, etc.), or a large protein (GST, streptavidin, etc.). This scheme allows the assay detection to focus only on the label (intensity or distribution), while other changes in the test system are invisible unless they indirectly cause changes in the label intensity or distribution. The downside of this scheme is that the foreign tag may interfere with the native biological system and make the assay less biological relevant. Label-free technologies have emerged in recent years that do not use a label in the test systems. The systemwide changes in the whole test system are monitored instead of just monitoring the label. By introducing specific detection techniques, the technologies may be able to detect a specific signal in the background of all the other signals. Marketed label-free detection technologies can detect the mass changes after binding events occurred (surface plasmon resonance technology from GE Health Life Sciences, formerly Biacore Life Sciences; bio-layer interferometry technology from Fortebio; EPIC from Corning), the impedance change between cells and the electrode that they stick to (RT-CES from ACEA Bioscience and Cellkey from MDS-Sciex), and extracellular microenvironment changes (XF96 Extracellular Flux Analyzer from Seahorse Bioscience).

5. *Classification According to the Format or Specialized Technology* ELISA, SPA (scintillation proximity assay, from GE Health Life Sciences, formerly Amersham Biosciences), AlphaScreen (amplified luminescent proximity homogeneous assay, from PerkinElmer), HTRF (homogeneous time-resolved fluorescence, from Cisbio International), microfluidic (Caliper Life Sciences & others), EFC (enzyme fragment complementation assay from DiscoverX), Branched DNA (QuantiGene, from Panomics), ECL (electrochemiluminescence assay from Meso Scale Discovery), and the like.

Useful Websites

http://www.accessdata.fda.gov/scripts/cdrh/cfdocs/cfCFR/CFRSearch.cfm

http://www.fda.gov/oc/gcp/

http://www.fda.gov/cdrh/comp/gmp.html

http://www.ich.org/cache/compo/276-254-1.html

http://www.oecd.org/document/63/0,2340,en_2649_34381_2346175_1_1_1_1,00.html

http://www.who.int/medicines/areas/quality_safety/quality_assurance/production/en/

http://www.emea.europa.eu/Inspections/GMPhome.html

http://wwwn.cdc.gov/clia/regs/toc.aspx

BIBLIOGRAPHY

Achyuthan, K. E. and Whitten, D. G. (2007) Design considerations for high-throughput screening and in vitro diagnostic assays. *Comb. Chem. High Throughput Screen.* **10**, 399–412.

Bleicher, K. H., Bohm, H.-J., Muller, K., and Alanine, A. I. (2003) Hit and lead generation: Beyond high-throughput screening. *Nat. Rev. Drug Discov.* **2**, 369–378.

Burtis, C. A., Ashwood, E. R., and Bruns, D. E. (2007) *Tietz Fundamentals of Clinical Chemistry*, 6th ed. Saunders, New York.

Chorghade, M. S. (ed.) (2006) *Drug Discovery and Development, Drug Discovery*, Vol. 1. Wiley, Hoboken, NJ.

Corey, M. J. (2009) *Coupled Bioluminescent Assays: Methods, Evaluations, and Applications*. Wiley, Hoboken, NJ.

Crowther, J. R. (2001) *The ELISA Guidebook*. Humana, Totowa, NJ.

Ding, C. (2008) Belimumab, an anti-BLyS human monoclonal antibody for potential treatment of inflammatory autoimmune diseases. *Expert Opin. Biol. Ther.* **8**, 1805–1814.

Druker, B. J. (2008) Translation of the Philadelphia chromosome into therapy for CML. *Blood* **112**, 4808–4817.

Ermer, J. and Miller, J. H. M. (eds.) (2005) *Method Validation in Pharmaceutical Analysis: A Guide to Best Practice*. Wiley-VCH, Weinheim.

Ghosh, A. K., Chapsal, B. D., Weber, I. T., and Mitsuya, H. (2008) Design of HIV protease inhibitors targeting protein backbone: An effective strategy for combating drug resistance. *Acc. Chem. Res.* **41**, 78–86.

Kambach, C. (2007) Pipelines, robots, crystals and biology: What use high throughput solving structures of challenging targets? *Curr. Protein Pept. Sci.* **8**, 205–217.

Kaplan, L. A., Pesce, A., and Kazmierczak, S. (2003) *Clinical Chemistry: Theory, Analysis, Correlation*, 4th ed. Mosby, Philadelphia.

Kitano, H. (2007) A robustness-based approach to systems-oriented drug design. *Nat. Rev. Drug. Discov.* **6**, 202–210.

Knowles, J. and Gromo, G. (2003) Target selection in drug discovery. *Nat. Rev. Drug. Discov.* **2**, 63–69.

Lee, D. C. and Webb, M. (eds.) (2003) *Pharmaceutical Analysis (Sheffield Analytical Chemistry).* Blackwell, Oxford.

Lindsay, M. A. (2005) Finding new drug targets in the 21st century. *Drug Discov. Today* **10**, 1684–1687.

Manly, C. J., Chandrasekhar, J., Ochterski, J. W., Hammer, J. D., and Warfield, B. B. (2008) Strategies and tactics for optimizing the Hit-to-Lead process and beyond—A computational chemistry perspective. *Drug Discov. Today* **13**, 99–109.

McInnes, C. (2007) Virtual screening strategies in drug discovery. *Curr. Opin. Chem. Biol.* **11**, 494–502.

Mdinor, L. (ed.) (2006) *Handbook of Assay Development in Drug Discovery.* CRC, Boca Raton, FL.

Ng, R. (2009) *Drugs: From Discovery to Approval*, 2nd ed. Wiley-Blackwell, Hoboken, NJ.

Pollok, B. (2005) Assay development: An increasingly creative endeavour. *Nat. Rev. Drug Discov.* **4**, 956–957.

Prabhakar, U. and Kelley, M. (eds.) (2008) *Validation of Cell-Based Assays in the GLP Setting: A Practical Guide.* Wiley, Hoboken, NJ.

Pritchard, J. F., et al. (2003) Making better drugs: Decision gates in non-clinical drug development. *Nat. Rev. Drug Discov.* **2**, 542–553.

Rees, D. C., Congreve, M., Murray, C. W., and Carr, R. (2004) Fragment-based lead discovery. *Nat. Rev. Drug Discov.* **3**, 660–672.

Rishton, G. M. (2003) Nonleadlikeness and leadlikeness in biochemical screening. *Drug Discov. Today* **8**, 86–96.

Rishton, G. M. (2005) Failure and success in modern drug discovery: Guiding principles in the establishment of high probability of success drug discovery organizations. *Med. Chem.* **1**, 519–527.

Seiler, J. P. (2005) *Good Laboratory Practice: The Why and the How*, 2nd ed. Springer, Berlin.

Silverman, G. J. and Weisman, S. (2003) Rituximab therapy and autoimmune disorders: Prospects for anti-B cell therapy. *Arthritis Rheum.* **48**, 1484–1492.

Smith, C. G. and O'Donnell, J. T. (eds.) (2006) *The Process of New Drug Discovery and Development*, 2nd ed. Informa HealthCare, New York.

Spilker, B. (2009) *Guide to Drug Development: A Comprehensive Review and Assessment.* Lippincott Williams and Wilkins, Philadelphia.

Stevens, J. L. and Baker, T. K. (2009) The future of drug safety testing: Expanding the view and narrowing the focus. *Drug Discov. Today* **14**, 162–167.

Stewart, K. K. and Ebel, R. E. (2000) *Chemical Measurements in Biological Systems.* Wiley, New York.

Tracey, D., Klareskog, L., Sasso, E. H., Salfeld, J. G., and Tak, P. P. (2008) Tumor necrosis factor antagonist mechanisms of action: A comprehensive review. *Pharmacol Ther.* **117**, 244–279.

Vogel, H. (ed.) (2007) *Drug Discovery and Evaluation: Pharmacological Assays*, 3rd ed. Springer, New York.

Weinberg, S. (ed.) (2007) *Good Laboratory Practice Regulations*, Vol. 168, 4th ed. Informa HealthCare, New York.

Yang, Y., Adelstein, S. J., and Kassis, A. I. (2009) Target discovery from data mining approaches. *Drug Discov. Today* **14**, 147–154.

CHAPTER 2

MEASUREMENT AND INSTRUMENTATION

WE DISCUSSED in Chapter 1 that a bioassay involved a test system that contained many components. After the biological and chemical manipulation of the test system, the bioassay reaches the final measurement stage when one or more signals from the test system are detected. The primary detection method in modern chemical and biological sciences is instrumental analysis. The instruments employed in bioassays usually record electronic signals, such as electric current, voltage, and impedance. If the final assay signal is not electric, it is usually converted into an electric signal by a converter. For example, a light signal is commonly converted into an electric signal by a photomultiplier, a photodiode, or a charge-coupled device (CCD). While it is impossible to cover all instrumental analysis, which is a separate discipline, we will discuss in this chapter the most commonly used bioassay-relevant analytical methods so that biologists practicing bioassays will have a good understanding of the underlying analytical principles. It is very important for biologists to know the applicability and the limit of the technology that are employed in bioassays. Lack of understanding often leads to the failure of internal bioassay development and/or the improper use of commercial bioassay kits.

2.1 MEASUREMENT AND PERTURBATION

The majority of measurements involve a stimulation (perturbation) of the system under study that is followed by monitoring the responses of the system to the stimulation. Observing a large object under daylight with unaided eyes is analogues to scientific measurement. The large object is the system under study, the daylight is the stimulus, and the reflected light from the object is the response to the stimulation from the system. The response from the system is observed with our eyes, which convert the reflected light signal into cellular responses in the retina at the back of the eyes where the signal is amplified. The signal is then transmitted to the brain for further processing. In this case, the object under study is hardly perturbed by the stimulus (visible light) and what we observed truly represents the object. The same process is employed

Assay Development: Fundamentals and Practices. By Ge Wu

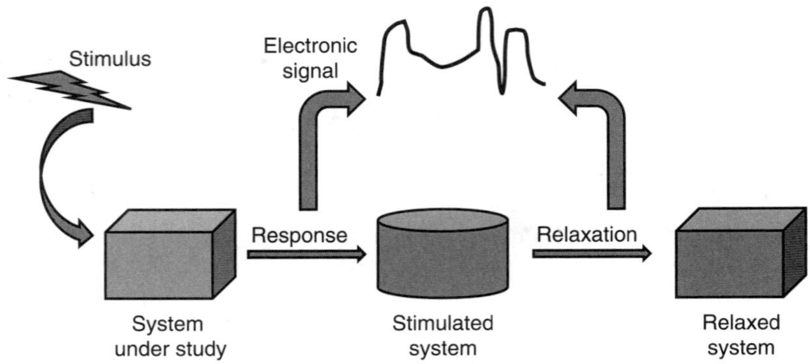

Figure 2.1 Process of instrumental analysis. A stimulus (usually energy such as radiation, electrical force, mechanical force, heat) is applied to the system under study. In response to the stimulus, the system will change and the signal can be recorded. The stimulated system will relax back to the original state (nondestructive stimulus) or change to a different stage (destructive stimulus). This process also emits signal that can be recorded.

by instrumental analysis: A stimulus is applied to a test system and the response signal from the system to the stimulus is captured. The signal is then converted into recordable electrical signal (Fig. 2.1). Depending on the analytical technique, the stimulus can be radiation, electric forces, or mechanical forces. In any measurement, it is desirable to exert minimum perturbation to the native system under study. Otherwise, it is difficult to relate what was observed to what really happened in the native systems because the perturbation to the system may change the properties of the original system. In bioassays, the perturbations to the system by the external stimulus are usually small and may be negligible. The largest perturbations to the native biological system are usually from the exogenous probes introduced into the native system to aid detection and from the special sample preparation procedure to allow a particular detection.

Visible light (human eye responsive region: approximately 400 to 700 nm) stimulation exerts minimal perturbation on the system under study. It is nondestructive and the system under investigation usually can be studied in the native state. It is also more receptive because the object under study can be directly viewed with human eyes. As the saying goes: Seeing is believing. However, unaided eyes cannot resolve anything under the millimeter scale. The next thing closest to direct observation with the naked eye is light microscope, and it is the most desirable method for biological study. The simplest microscopic technique is bright field microscopy. With this technique, the object is illuminated from below with white light and the transmitted light is observed from above. The visibility of light microscope depends on two factors: contrast and resolution. The contrast is the difference in light intensity between an object and its background (or surroundings render the object distinct). The contrast from colorless native biological sample is rarely high enough to obtain a high-quality image, though a variety of techniques have been used (phase contrast, dark field, oblique illumination, reflection interference, polarized light, and differential interference contrast). Staining of cell membranes and subcellular components or labeling specific cellular proteins with fluorescent probes allows us to obtain very high contrast when the staining is specific. However, these treatments significantly perturb the native biological system. The resolution of a conventional optical microscope is limited by the diffraction of light that is governed by the Ernst Abbe equation:

$$d = \frac{0.612\lambda}{n \sin \alpha} \tag{2.1}$$

where d is the distance between two points that can be resolved, λ is the radiation wavelength, n is the refractive index of the medium between specimen and the objective lens, and α is the half-angle of the cone of radiation from the specimen plane to the surface of the lens (the product of $n \sin \alpha$ are also referred as numerical aperture or N.A.). The resolution of a conventional optical microscope is thus limited by the following physical limitations: α cannot be greater than 90° (with practical limit at about 70°), there is no media having n more than 1.6, and human eyes cannot see wavelength shorter than 400 nm. On top of these theoretical limitations, conventional optical microscopes encounter other practical limitations and in practice can only resolve two points more than 200 nm apart. Figure 2.2 shows the sizes of commonly encountered objects in bioassays. Cells, subcellular components, and bacteria can be observed with conventional light microscope. However, viruses and objects smaller than viruses cannot be observed with a conventional optical microscope. Recent developments in light microscope design have pushed the resolution of light microscope close to or even broken the diffraction limits. These new light microscopes are referred to as supermicroscopes or nanoscopes. By using multiphoton excitation and two objectives, resolution down to 80 nm has been achieved. By employing the scanning method and other innovations, single-digit nanometer resolution has been achieved too. Supermicroscopes or nanoscopes are beyond the scope of this book and will not be discussed here. Interested readers can study the listed bibliography.

Quantum theories indicate that electron beams can be in waveforms and they can substitute light in a microscope. This substitution results in electron microscopes

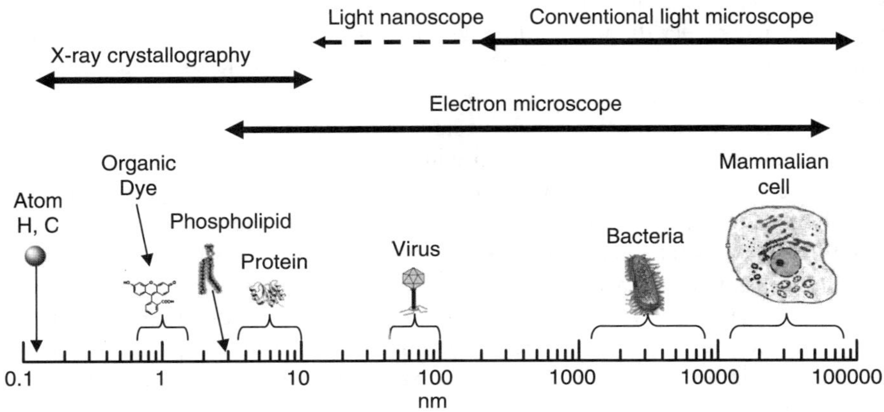

Figure 2.2 Sizes of objects commonly encountered in bioassay and the methods with proper spatial resolutions to study them.

that are built with electrical components instead of optical components. The wavelength of the electron beam can go 5 orders of magnitude shorter than visible wavelength when the electrons are accelerated to high velocity (higher velocity, shorter wavelength) with high voltage. The resolution of electron microscope is also governed by Eq. (2.1). With electron microscope, $n = 1$ because the chamber is under vacuum; and $\sin \alpha = \alpha$ because α is extremely small (about 0.01 rad). If the electrons are accelerated to 100,000 V, the theoretical limit of resolution in electron microscope will be 0.24 nm. However, practical limitations, mostly the biological specimen preparation limitations, limited the electron microscope to a resolution at about 20 nm. Viruses, ribosomes, large proteins, and biomembranes can be observed with electron microscopes though not at atomic resolution. Electron microscope requires special treatment of the biological samples to obtain thin specimen ($\sim$100 nm thick) so that electron beams can penetrate without losing much energy or being absorbed. Such treatment results in substantial perturbation of the native system. Thus, there is a risk that the observation may not be relevant to the native system.

To directly observe a biomolecule at atomic level resolution (1 to 2 Å), the wavelength of the radiation source has to be on the same order or less. X-ray emission occurs when the electrons in the outer shells of metal atoms return to the inner shell after these electrons are knocked out from the inner shell by high-energy electrons. The X-ray radiation wavelength is between 0.1 and 100 Å, which is an ideal radiation source for atomic resolution. Biomolecules in crystals will diffract X rays upon radiation. The diffraction pattern reveals the structure of the biomolecule at atomic resolution. This technique is called X-ray crystallography. The X-ray crystallography studies require that the biomolecules under study must be extracted from their native environment and be purified. Furthermore, the technique requires that the purified molecules be in a crystal form, which may differ from the structure of the molecules in solution. Thus, X-ray crystallography studies of biomolecules significantly perturb the native system. However, there is no other substituting technology that can directly observe biomolecules at atomic level in their native environment.

For most bioassays, labels (or tags) are essential to allow measurement of the system. However, the labels may perturb the native system so much that the measurement may not reflect the true biology and defeat the purpose of the measurement. The perturbations of the native system by foreign labels mostly come from the label's molecular size, charge, and the amount of the label added to the native system. A general rule for introducing a label is that the label should be significantly smaller than the volume of the system under study and the label should be neutral if possible. Based on the dimension of the biological objects shown in Figure 2.2, it is fine to use organic dye, fluorescent protein (such as green fluorescent protein, or GFP), or dye-labeled antibody to study the global properties of mammalian cells and bacteria purely from molecular size consideration. However, the use of dye-labeled lipids to study membrane bilayer's property may be questionable since the dye molecules are approaching the size of the lipids and may significantly perturb the microenvironment that the dye is sensing in the lipid bilayers. In such a study, isotope labeling, which is a substitution at atomic level, will provided information more closely to the native environment. In addition to the molecular property of the label itself, the total amount of labels applied to a system should be as little as possible to reduce perturbation to the native system. As a rule of thumb, the amount of labeled species should be kept at less than 1% of the total native molecules they are mimicking in the native system. For example, the number of labeled lipids should not exceed the total number of lipids in the biomembrane they are mimicking.

Isotope atomic substitution to a molecule under study is a commonly used method in bioassay. Stable isotope substitutions are usually used in NMR studies and in mass spectrum studies. Unstable (or radioactive) isotope substitutions are commonly used to quantify a molecule or to follow a molecule in its distribution and transformation. Isotope substitutions have minimum perturbation to the native molecule they substitute. However, the considerations of the environment impact due to the production and the disposal of radioactive isotopes have significantly reduced the use of the radioactive labeling method in recent years. Nonradioactive labeling methods have replaced radioactive labeling methods in many applications in recent years. However, radioisotope labeling is still widely used in situations where radiolabels are irreplaceable (e.g., distribution and metabolism of drug substances).

Due to the complexity of the cells, dramatic perturbation or even destruction of the cells under study often occurs in cell-based assays. For example, cells are often lysed to expose their interior component proteins to enable examination of the changes in the protein quantities or protein modifications as a result of the cells' responses to a test molecule. Another example of dramatic perturbation to the cells is the fixation of the cells after they are exposed to a test molecule. It is speculated that many changes in the cell test system may occur during this process and the final measurement may not represent what happened in the native cells. Experimental results obtained from these studies should be examined carefully. However, these alternations are necessary in many situations due to the lack of an alternative less destructive technology.

In addition to matching the size of the probe with the system to be studied (spatial resolution), it is very important to match the speed of the measurement to the speed of the change in the biological system (temporal resolution). Mismatch in the speed of measurement, both too slow and too fast, can significantly affect the performance of an assay. For example, the protein binding process (reaching equilibrium, discussed in Chapter 3) usually happens in the milliseconds scale, and the ion flux through ligand-gated ion channels also happen in the milliseconds scale. Thus, stopped-flow and quenched-flow devices (described in Chapter 3) with temporal resolution in the millisecond scale should be used to study the two events. Without such instruments, the kinetics of the protein binding process cannot be studied. Common instrumentation methods are too slow, which only allow obtaining the equilibrium binding data. On the other hand, measurement that is too fast compared with the biological events will obtain information only in a small window of a long process and yield incomplete information about the system. For example, if an active enzyme that converts substrate to product in the scale of several hours, measurement of the process within a few seconds will lead to the conclusion that the enzyme is inactive because there is no detectable substrate turnover in the small time window of measurement. Temporal resolution mismatch also can have huge consequence that leads to product failure (discussed in Chapter 14).

2.2 COMMON INSTRUMENTAL METHODS AND INSTRUMENT COMPONENTS

The majority of the methods employed in bioassays involve the measurement of components in the test system at the molecular level (e.g., molecular absorption and luminescence). A remaining portion of the assays involve the measurement of components in the test system at atomic level (e.g., atomic absorption), the whole test system (e.g., differential thermal analysis, refractometry), and the interface between the substrate in analytical device and parts of the test system (e.g., electric properties such as current and impedance). The common analytical methods and instruments used in bioassays are listed in Table 2.1. The focus of the discussion in this section will be on optical molecular spectroscopy.

Optical instruments detect light transmitted/emitted from a test system. A typical optical instrument design is shown in Figure 2.3. The instruments usually contain energy source, wavelength selector, sample holder, signal detector that converts optical signal to electrical signal, and signal processor. Depending on the measurement mode, the design of a particular optical instrument may vary. The absorption spectrophotometer may not have the filter in front of the sample holder. While fluorescence and phosphorescence use radiation as the energy source, the electrochemiluminescence measurement uses voltage instead of radiation as the energy source.

TABLE 2.1 Analytical Methods and Instruments Commonly Used in Bioassays

Characteristic Properties	Instruments	Applications
Absorption of radiation	Spectrophotometer	UV, visible, IR, NMR, ESR
Emission of radiation	Fluorimeter	Detect labeled tags that emit fluorescence or phosphorescence upon absorption of radiation
	Luminometer	Detect tags that catalyze the chemical reaction that emits luminescence
Rotation of radiation	Polarimeter	Fluorescence polarization
	CD spectroscope	Circular dichroism
Electrical current	Amperometer	Patch clamp
Mass-to-charge ratio	Mass spectrometer	Measure molecular weight and molecular fragmentation
Radioactivity	Liquid scintillation analyzers	Measure radiolabeled molecules

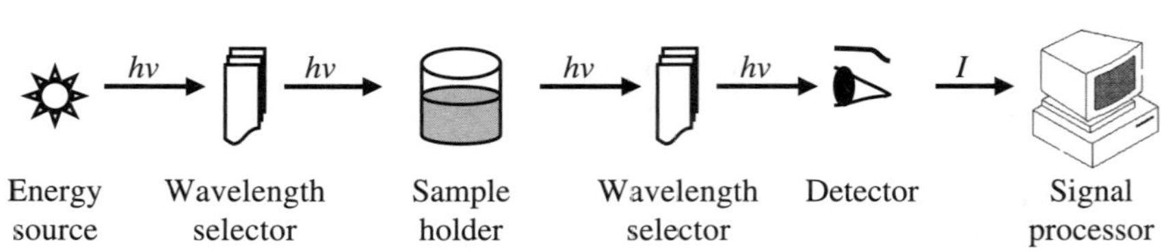

Figure 2.3 Typical design of optical instruments. The instruments contain the following major components: energy source (normally radiation), filter, sample holder, detector, and signal processor (normally a microcomputer).

Chemiluminescence measurement does not need the radiation source and may not need the filters at all. Scintillation counting of radioactivity uses the radiation from the decay of nucleus as energy source and thus external energy source is not needed. Some critical components in the optical instruments are discussed in this section. Understanding these components is important for the assay developer to develop optimal assays and to make wise instrument purchasing decisions.

2.2.1 Energy Sources

Radiation is the most commonly used energy source in optical instruments. The key requirements for radiation source are sufficiently strong power, stable output, and suitable wavelengths that match the absorption wavelength of the molecule to be excited. Energy absorbing molecules employed in most bioassays have their absorption of radiation energy with wavelength between 210 nm (UV) and 2500 nm (IR). There are two types of radiation sources: continuum source and line source. Continuum sources of radiation have relatively constant radiation intensity over a wide range of wavelengths. Xenon lamp and tungsten lamp are the most commonly used continuum radiation sources in the optical instruments for bioassays because both of them cover a wide range of wavelength from UV to near IR. The difference between the two sources is that the xenon lamp covers more UV range while the tungsten lamp covers more IR range. The light output intensity at different wavelengths for xenon lamp is shown in

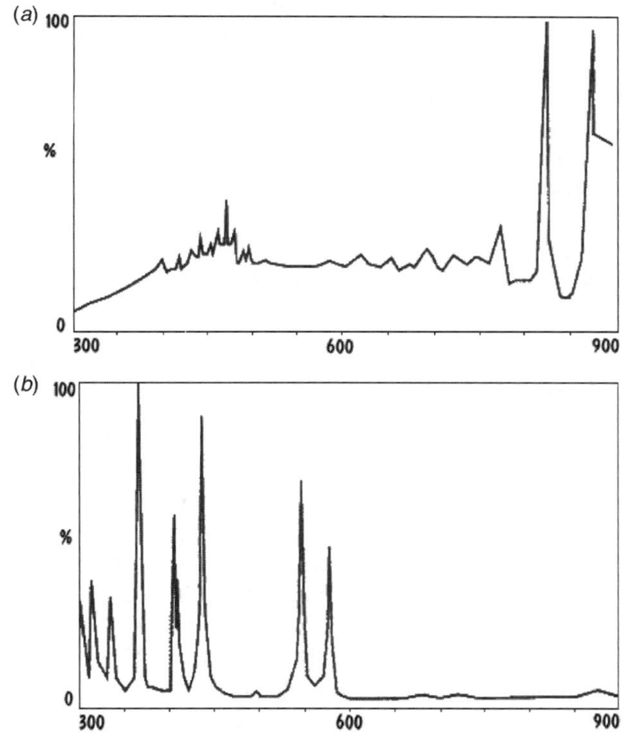

Figure 2.4 Comparison of (*a*) xenon lamp and (*b*) mercury lamp spectra. The spectra are scanned from 300 nm to 900 nm.

Figure 2.4*a*. Line source radiation has a limited number of radiations; each of them span a narrow range of wavelengths. The mercury lamp, sodium lamp, and laser source belong to this type of radiation. The light output intensity at different wavelengths for a mercury lamp is shown in Figure 2.4*b*. It can be seen that a xenon lamp has a more constant light output from the far UV region into the visible region while the mercury lamp has several peaks at specific wavelengths. A continuum radiation lamp is the lamp of choice for any spectra acquisition while the line radiation lamp offers strong excitation when the absorption of the molecule under study matches the "peaks" of the lamp. Line source radiation is the method of choice for kinetic experiments because the excitation wavelengths are fixed in these studies and line source radiation offers very strong excitation. The advantages of using xenon and tungsten lamps as radiation sources are that they are relatively inexpensive compared with laser and they cover wide wavelength spectra. The disadvantages of using these continuum light sources are that they have to be used together with a wavelength selector to increase the signal specificity, and the excitation power is not as strong at a selected wavelength as compared with laser source.

Lasers as a light source have many favorable properties including coherence, collimination, and compatibility with scanning systems. Commonly used laser radiation sources in optical instruments in bioassays are blue diode laser (405 nm), helium–cadmium laser (325 and 442 nm), argon-ion laser (457, 488, and 514 nm), Nd:YAG laser (532 nm), helium–neon laser (543, 594, and 633 nm), and the krypton-ion laser (568 and 647 nm). Lasers are preferred light sources for many bioassay applications because they have strong excitation power and are highly specific due to the narrow band of laser (<1 nm). The disadvantages of using lasers as radiation sources are the high cost of laser producing instrument and the limitation that only a few wavelengths are available.

There are many advances in newer light sources. High brightness light-emitting diodes have been shown to emit consistent light that covers a wide spectra from UV to IR and with power matching that of the xenon or mercury lamp. Another new development is the super continuum lasers that are obtained as consequences of nonlinear optical processes when short laser pulse propagates through specially designed optical fibers.

2.2.2 Wavelength Selector

There are two types of wavelength selectors: the monochromator and the optical filter. A monochromator is an optical instrument designed to separate polychromatic white light into monochromatic light. Monochromators have a dispersing element (diffracting grate or prism) that spectrally decomposes the light passing through an entrance slit and having a beam-deflecting element for selecting the wavelength. The light-dispersing element in a monochromator rotates by a driving mechanism so that the selected wavelength can past through a fixed exit slit. Monochromators can continuously select a wavelength from 180 to 2500 nm. The ideal transfer function of a monochromator is a triangular shape with the selected wavelength at the peak of the triangle. The intensity of the nearby wavelength decreases linearly on both sides of the peak wavelength until a cutoff value is reached, where the intensity stops decreasing. For a typical monochromator, the cutoff level is about one thousandth of the peak value, or 0.1%. An important character for a wavelength selector is the spectral bandwidth that is defined as full width at half maximum (FWHM). For a monochromator, spectral bandwidth is the width of the triangle at the points where the light has reached half the maximum value. A typical spectral bandwidth is about 1 nm. Together with a continuum light source such as xenon lamps, monochromators are commonly found in spectrophotometers to scan the absorption and emission spectra of a molecule.

Optical filters physically appear as a piece of round glass with color or a shining mirror. Optical filters can be divided into two types, interference filters and absorption filters, based on the underlying mechanism of how they selectively pass light with desired wavelengths. Interference filters rely on optical interference to pass a narrow band of light. Absorption filters use colored dye on the glass to absorb unwanted light and pass selected light. Interference filters transmit more light than absorption filters. In addition, interference filters have narrower bandwidths (usually <5 nm) than absorption filters (usually >20 nm). However, interference filters are more expensive to make than absorption filters. In practice, optical filters are commonly classified by their intended use. Commonly used filters in bioassays are listed below:

Long-pass cut-on filters allow light above certain wavelengths to pass while blocking (reflecting) the lower wavelength light. Long-pass filter properties are measured by the edge, which is the wavelength at 50% of the maximum transmission and the slope, which is the sharpness of the transition from transmission to reflection.

Short-pass cut-off filters allow light below certain wavelengths to pass while blocking (reflecting) the higher wavelength light. Short-pass filter properties are measured by the edge and the slope in the same way as the long-pass filter.

Bandpass filters transmit light only within a defined spectral band. Bandpass filter properties are measured by center wavelength (CWL), which is defined as the wavelength at the center of the pass band, FWHM, which is defined as the bandwidth at 50% of the maximum transmission, and peak transmission (T), which is defined as the wavelength of maximum transmission.

Dichroic beamsplitters are long-pass and short-pass filters that are used at an angle (usually at 45°) to the incident light. They highly reflect one specified spectral region while optimally transmitting another. Dichroics can normally reflect and transmit 90% incident light, which make them superior to the normal beamsplitters that only allow 50% of light to be reflected and transmitted, respectively.

Neutral density filters do not select wavelength, but they uniformly attenuate the intensity of light over a broad spectral range. Attenuation is accomplished by either absorption using light-absorbing glass or by a combination of absorption and reflection using a thin-film metal coating. They are commonly used in optical instruments to modulate light intensity for desired applications.

Transmittance (T) and optical density (OD) are the two parameters commonly used to evaluate wavelength selectors. T is defined as the percentage of the total radiation at a given wavelength that passes through the wavelength selector. The transmission value is always between 0 and 100%. OD is defined as $\text{OD} = -\log 10\, T$

(*T* and OD are also discussed in Section 2.3.1 for absorption in solution). When describing the transmitting performance (usually when throughput is more than 1%), the preferred expression is *T*. When describing the attenuating performance of a filter (usually when throughput is less than 1%), the preferred expression is OD. This is because in a linear plot such as *T*, 1% or below is visually indistinguishable when the full scale of the plot is more than 50%. OD is in logarithm and value lower than 1% can be clearly displayed in a plot. However, small fluctuations that are visible in the linear plot of *T* will be hard to see in logarithm plot of OD.

While monochromators found most use in spectrophotometers that study processes that involve broad wavelength and to scan spectrum of a molecule, optical filters are used in a wide variety of applications where fixed spectral isolation is required. Optical filters have many more advantages than monochromators. Optical filters occupy small physical space and are cost effective to be integrated into optical instruments. They transmit more light than monochromators because the grating mechanism in monochromators causes significant loss of radiation. Optical filters normally can reject more than 6 OD of background light to achieved a high signal-to-noise ratio that is crucial for many bioassays when the signal is very low. In comparison, monochromators can only reject about 3 OD of background light due to "stray" light. To achieve the same signal-to-noise ratio, two monochromators can be used in tandem. However, such arrangement will further reduce the radiation power reaching the sample since monochromators cause more loss of radiation compared with optical filters. In addition, the instruments with more monochromators will be more bulky as well. Because a monochromator typically has a narrow bandwidth, it can only excite a small portion of the excitable dyes. In comparison, bandpass filters typically have broader bandwidths and allow excitation of a large portion of the excitable dyes. Thus, fluorimeters with bandpass filter-based wavelength selectors normally can detect lower concentrations of fluorophore than the fluorimeters with monochromator-based wavelength selectors. This argument is also true at the emission end of the instruments. This is why monochromator-based fluorimeters are mostly used to acquire spectrum and filter-based fluorimeters are mostly used in heavy-duty high-throughput screening mode, especially when the concentration of the fluorophores are closer to 1 nM or below.

2.2.3 Sample Holder

Sample holders position the test samples in place for radiation and detection. Radiation light and emission light must pass through the material that makes the sample holders if the sample holders are in the light path. The optical properties of the commonly used materials for sample holders are shown in Figure 2.5. KBr and NaCl crystalline have the best optical property and are transparent to a wide range of wavelengths from UV all the way to IR. However, they are incompatible with aqueous solution and are sensitive to moisture in the air. The weak chemical and mechanical properties of KBr and NaCl prevent them from being used in broad applications. They are mostly used to obtain IR spectra of organic molecules. When acquiring IR spectra, liquid organic molecules are usually placed between two pieces of NaCl crystalline; and solid organic molecules are usually ground together with KBr, and

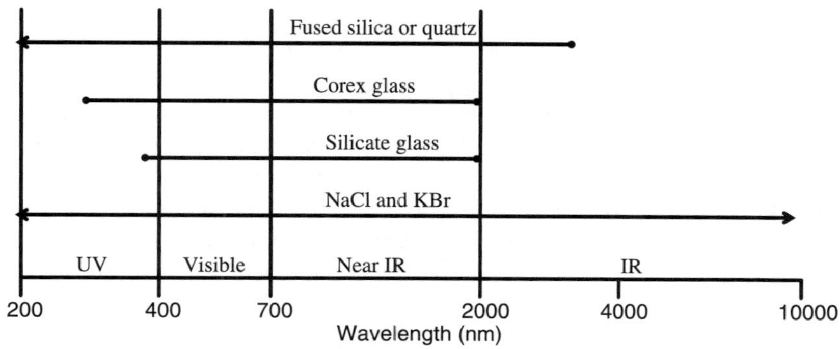

Figure 2.5 Optical properties of commonly used materials as sample holder.

the powers are pressed to make a thin solid film that is then placed on the light path for IR spectra collection. For most bioassay applications, materials that are transparent to UV, visible, and near IR up to 1000 nm are sufficient. Fused quartz and fused silica are types of glass containing primarily silica in amorphous (noncrystalline) form. Fused quartz is manufactured by melting naturally occurring quartz crystals of high purity at approximately 2000°C. Fused quartz can also form naturally. Fused silica is produced using high-purity silica sand as the feedstock. Fused silica and quartz are the best materials for bioassay applications because of their superior mechanical strength, strong resistance to chemicals, and optical transmittance covering broad wavelengths. However, fused silica and quartz are expensive and difficult to machine. When transparency in the UV region is not required, glass and even plastics may be good enough for the applications and they are cost effective. Plastics are more cost effective compared with glass because of the simple manufacturing process to produce the sample holders. Polystyrene is the most common plastic used as sample holders. It is usually used when transmission wavelength of more than 350 nm is required. Polymethyl methacrylate has much better transmission at lower wavelength than polystyrene and may be used down to 290 nm. Most bioassays use disposable plastic sample holders except with microscopic applications where glass slides are preferred because of the requirement for flatness and low background emission.

Microplates are currently the most common sample holders employed in bioassays. Optical instruments specifically designed to interface with microplates are required to measure the samples in the microplates. Commonly used microplates are in 12-well, 24-well, 48-well, 96-well, 384-well, and 1536-well format. Before the wide adoption of microplates, cuvettes with 1-cm paths were traditionally used as the sample holders for absorption and fluorescence measurement. A larger volume of samples is required when cuvettes are used as the sample holders (from 100 μL for microcuvettes to 2 mL for standard cuvettes). In comparison, a much smaller volume of samples are required when microplates are used as the sample holders (from 6 μL for 1536-well microplates to 100 μL for 96-well plates). With the transition from cuvette to microplate wells, the detection limit is reduced because of the shorter light path in the microplate. In addition, care must be taken to avoid air bubbles when pipetting solutions into the microplate because air bubbles will interfere with top reading the microplate readers. In comparison, air bubbles floating on top of solution is not a problem with cuvette because the light is detected/transmitted from the side. Microplates are commonly made with plastics with different shaped wells designed for different applications. Absorption measurements with microplate are usually made with transparent polystyrene flat bottoms though sometimes the bottom is made with glass to allow transmission of UV wavelengths. Microplates with glass bottoms are also used for microscopic applications because of the flatness of the bottom. For fluorescence and luminescence applications with radiation and detection from the top, microplates are usually made with solid low autofluorescent plastics. The excitation source and the detector are positioned on the top of the microplate and the excitation light and emission light does not pass the microplate at all. For some fluorescence applications that require the excitation and detection from the bottom of the microplates, the microplates with transparent bottom and solid sidewalls are employed.

2.2.4 Detector

Modern radiation detectors are devices that convert radiation energy into electric signals. There are two types of radiation: photon (UV to near IR) and heat (near IR to far IR) that are detected with two different types of detector, respectively. Since photons are detected in most bioassays, we will focus the discussion on photon detectors in this section. The ideal detectors should have high sensitivity, high signal-to-noise ratio, constant response to a wide range of wavelengths, fast response time, and low dark current. The most commonly encountered photon detectors in bioassays are photomultiplier (PMT), photodiode, and charge-coupled device (CCD).

The *PMT* is constructed from a glass vacuum tube that houses a photocathode, several dynodes, and an anode. Incident photons strike the photocathode material, which is present as a thin deposit on the entry window of the device, to produce electrons. These electrons are directed by the focusing electrode toward the electron multiplier, where the electrons are multiplied by the process of secondary emission. PMTs are highly sensitive to UV and visible light. PMT tubes typically require 1000 to 2000 V for proper operation. Special care must be taken when using PMTs because they are very sensitive and can be damaged when the light intensity is too high. When powered, PMTs must be shielded from ambient light to prevent their destruction through overexcitation. The sensitivity of PMTs is limited by dark current that is mostly from thermal emission. PMTs are often cooled to -30°C to eliminate thermal dark current.

A *photodiode* is a semiconductor with *pn* junction. When a photon of sufficient energy strikes the diode, it excites an electron, thereby creating a mobile electron and a positively charged electron hole. If the absorption occurs in the *pn* junction's depletion region, or one diffusion length away from it, these carriers are swept from the junction by the built-in field of the depletion region, producing a photocurrent. Photodiodes have excellent linearity of output current as a function of incident light with spectral response from 190 to 1100 nm (silicon photodiode). They are inexpensive, compact, rugged, and can last for a long time. They operate at low voltage and have high quantum efficiency ($\sim$80%) with low noise. However, photodiodes cannot match the performance of PMTs in terms of sensitivity, dynamic range, and signal-to-noise ratio.

A *CCD* is comprised of a two-dimensional array of individual photosensing units (imaging pixels) and a matching array of storage pixels that are coupled together.

After the imaging array is exposed to light, each photosensing unit records the intensity of light as a variable charge. The charges are then transferred to the storage array. The storage array is read a row at a time to the analog-to-digital converters (A/D converters) that transform the charges into binary data to be processed. Because of their sensitivity to infrared, the CCDs are usually cooled in applications requiring high performance. In addition, cooling also reduces the array's dark current, improving the sensitivity of the CCD to low light intensities, even for ultraviolet and visible wavelengths. The performance of CCDs is close to that of PMTs with the added advantage of two-dimensional sensors that make CCDs the ideal choice for applications in imaging cells, imaging microarrays, and simultaneously detecting signals from high-density microplates.

2.3 MOLECULAR ABSORPTION MEASUREMENTS

2.3.1 Absorption Measurement Principles and Instruments

Molecular absorption measures the transmittance (T) or absorbance (A) of a solution containing molecules that absorb the incident light in a transparent sample holder. Absorption in UV and visible wavelength range are mostly used in bioassays and will be the focus of discussion here. Transmittance is defined as the ratio of the exiting light intensity (I) to the incident light intensity (I_0) when a light beam passes a test solution (Fig. 2.6):

$$T = \frac{I}{I_0} \tag{2.2}$$

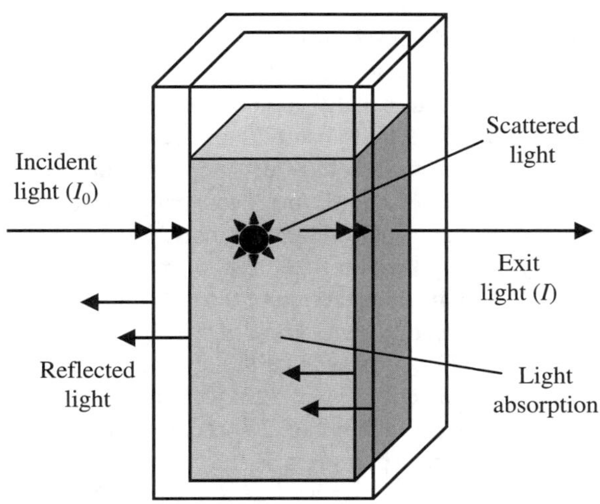

Figure 2.6 Absorbance measurement in a cuvette. In addition to the desired uniform absorption by the test sample, reflection and scattering are two unwanted phenomena that interfere with absorption measurement.

Absorbance is defined as

$$A = -\log T \tag{2.3}$$

If the exit light intensity (I) reduction from the incident light is caused by the molecular absorption of the test sample alone, the transmittance and absorbance is the measurement of the pure absorption process, and the following equation (Beer's law) holds when a single wavelength of light is employed as the incident light:

$$A = \varepsilon bc \tag{2.4}$$

where ε is the molecular extinction coefficient, b is the path length of light across the test solution, and c is the concentration of the test solution.

In practice, a sample holder or other interface between the light and the sample is always required to measure absorption, which introduces the reflection of the incident light due to the refractive index difference between different materials (Fig. 2.6). The reflection is lowest when the incident light is perpendicular to the interface. Even with perpendicular light, reflection can still cause $\sim 8\%$ light loss with a glass cuvette (refractive index $= 1.5$). In addition, scattering of the incident light inside the test solution occurs that will affect the absorbance measurements. The unavoidable reflection and scattering of light in absorbance measurements can be reduced to a negligible level by employing good experimental methods. First, care must be taken to remove large particles that randomly scatter light in the test solution and make sure the sample is placed perpendicular to the incident light to reduce reflection. Second, a blank sample that matches the characteristic of the test sample and without the light-absorbing test molecule should always be used to collect a blank read of the light in the same sample holder. When the value of the blank read is subtracted from the absorption value obtained from the test sample, the reflection and scattering in the two reads are canceled out and thus eliminated.

After applying the experimental methods mentioned above, we can now obtain the true absorbance value of a test sample and apply Beer's law to the measurement to obtain the concentration of the light-absorbing substance. In Beer's law, ε is a property of the light-absorbing molecule and is constant. When the path length of the light crossing the test sample is fixed, usually at 1 cm, the absorbance is directly proportional to the concentration of the light-absorbing molecule. However, there is a limitation for Beer's law that deviation from direct proportionality between absorbance and analyte concentration that occurs at high concentration of the light-absorbing analyte. This deviation has two major effects: incorrect calculated analyte concentration and less sensitive assay. To illustrate these effects, a hypothetical situation is shown in Figure 2.7. At a high concentration of 0.6 mM hypothetical light-absorbing analyte, the directly measured absorbance (point D in Fig. 2.7) will be lower than the theoretical value (point F in Fig. 2.7). When diluting the sample by a factor of 3 to 0.2 mM, the absorbance measurement falls to the linear range where Beer's law holds. The correct absorbance value (point E in Fig. 2.7) is obtained experimentally. The correct absorbance value of 0.6 mM (point F in Fig. 2.7) can be obtained by factoring in the threefold dilution into the experimentally measured absorbance value at point E in Figure 2.7. It is obvious that using the absorbance value at point D instead of the value at point F will result in underestimating the concentration of the analyte. In addition to

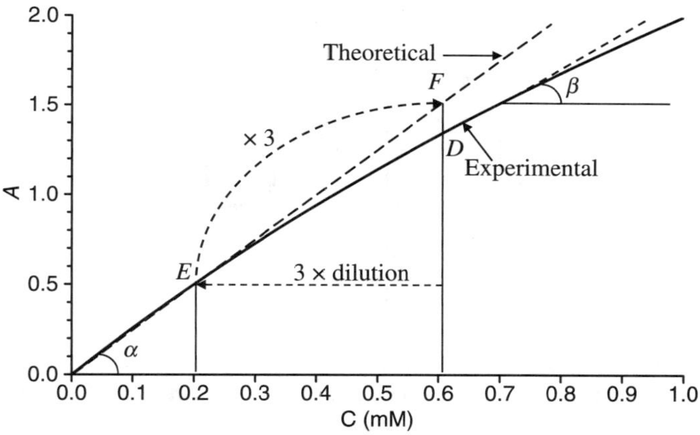

Figure 2.7 Limitations of Beer's law in absorbance measurement. Experimentally collected data deviate from Beer's law at higher analyte concentration (solid curved line) compared with theoretical value (dotted straight line). At hypothetical high concentration of 0.6 mM, direct experimental measurement resulted in incorrect value of *D* that is lower than the theoretical value of *F*. After the sample is diluted 3 times to 0.2 mM in the linear range, a measurement can be made. Correct value can be obtained by multiply the measured value by 3. In addition to deviation from Beer's law, direct measurement at high analyte concentration also results in less sensitive assay as shown that the slope (or angle, α) at linear range is bigger than the slope (or angle, β), at the curved range.

underestimating the analyte concentration, direct measurement at high analyte concentration also results in a less sensitive assay because the slope at the linear range at lower analyte concentration is larger than the slope at the curved range at high analyte concentration. Thus, when developing bioassays with absorption measurement, it is always a good idea to make sure the final absorption measurement is in the linear range. If a batch of samples gives high absorbance values, the whole batch of samples should be diluted. A rule of thumb is that the absorbance value should be always kept at less than 2 for most bioassays and less than 1 for applications demanding high accuracy.

2.3.2 Application of Molecular Absorption Measurement in Bioassays

All organic compounds are capable of absorbing electromagnetic radiation because they contain valence electrons that can be excited to a higher energy level. However, only some molecules with specific functional groups or type of bonds absorb in the UV and visible wavelengths that is detected in many bioassays. For organic molecules, there are two different types of bonds (σ and π) between the atoms making up the molecules. For example, there is a σ bond and a π bond between the two carbon atoms in the ethylene molecule forming the double bond. The electrons forming the chemical bonds are classified as σ electron and π electron when they are in the bonding state. After absorption of energy, the electrons in the bonding

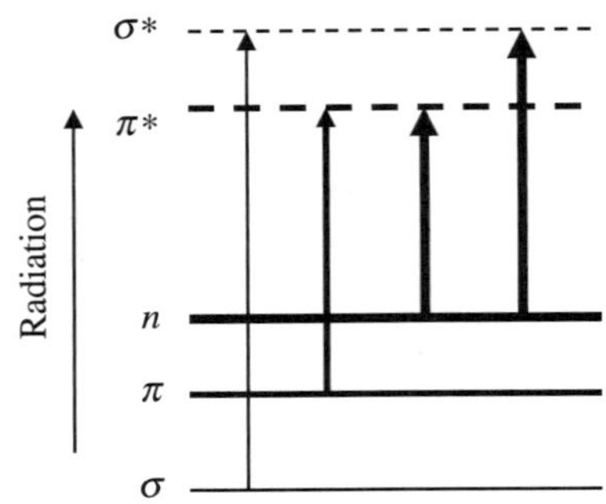

Figure 2.8 Partial electron energy diagram for molecular absorption of organic molecules. The electrons in a molecule can initially reside in either σ, π, or n orbits. Radiation can change the electron state to σ^* or π^*. The lowest transition energy is from n to π^*, followed by n to σ^* and π to π^*. These transitions absorb UV or visible wavelengths depending on the molecular structure. The σ to σ^* transition requires the highest energy that is usually shorter than UV wavelength.

state can jump to antibonding states (σ^* and π^*). In addition to carbon and hydrogen atoms that form either σ or π bonds, there are oxygen, nitrogen, and sulfur atoms in some organic molecules. These atoms possess extra nonbonding electrons (n). Figure 2.8 shows the energy diagram of electrons in bonding, nonbonding, and antibonding states in a typical organic molecule. The absorption of radiation energy can make the electrons transition from the lower energy state to the higher energy state. The transitions from σ to π^* and from π to σ^* are forbidden according to quantum theory. However, the transition from n to either σ^* or π^* is allowed. The electrons in organic molecules can initially reside in σ, π, or n orbits. Radiation can change the electrons from their initial states to σ^* or π^* state. From Figure 2.8, the lowest transition energy required is the transition from n to π^*, which is followed by the transitions from n to σ^* and from π to π^*. These transitions absorb in the UV or visible wavelength region depending on the structure of the organic molecules. The electron transition from σ state to σ^* state requires the highest energy, which is usually shorter than UV wavelength. When the transition from the σ state to the σ^* state occurs with high-energy input, the bond is usually destructed resulting in fragmentation of the organic molecule. This is one of the reasons why there are hardly any applications based on the absorption from the σ state to the σ^* state transition. Most UV and visible absorptions observed in bioassays are from the molecules with double bonds, and preferably with oxygen, nitrogen, and sulfur participating in the double-bond formation. Conjugated double bonds further lower the energy of π^* orbits and shift the absorption spectrum to the visible wavelength. Here are some examples of the absorption wavelengths for molecules involved in bioassays: amido group in protein (214 nm), tryptophan in protein (278 nm), nucleotide in DNA (280 nm), reduced nicotinamide adenine dinucleotide (NADH) (340 nm), and retinal in rhodopsin (500 nm).

Compared with other optical measurement methods that will be discussed next, absorption measurements have several advantages and disadvantages. One advantage is that many biological molecules absorb in the UV and visible region by themselves

(DNA, protein, ATP, NADH, retinal, hemoglobin, etc.) and thus eliminate the need to introduce foreign labels. The experimentally measured absorbance value is absolute and is instrument-independent, which makes the collected data comparable between different instruments. The major disadvantage of absorbance measurement is the high limit of detection (for a definition, see Section 2.10). Absorption measurement requires high concentration of analyte, usually in the range between 0.01 mM to several millimolars. Coupled with relatively large volumes required to fill the light path (1 cm for cuvette), relatively large quantity of analytes are required for absorbance measurement. Fortunately, absorbance measurement is nondestructive and the samples can be recovered in some applications. In chromogentic bioassays, where the detections are based on the generation of light-absorbing molecules by enzymatic cleavage of a nonabsorption substrate, the requirement to generate minimal detectable submillimolar product means a large number of substrate turnover. In comparison, less than 1000-fold of the enzymatic products are required to be detectable with fluorescence or luminescence. Thus, when everything else is equal, chromogentic assays will require either a thousand times more concentrated enzymes or a thousand times longer reaction time to produce detectable product signal compared with fluorescence methods or luminescence methods. Another disadvantage of absorption measurement is that absorbance is a function of light pass length. It is not a problem with standard cuvettes as the sample holders where the light pass length is fixed at 1 cm. For samples in a microplate, the precision in liquid dispensing and the uniformity of the wells within a microplate and between microplates will affect the light pass length and cause variations in the absorbance readings between samples. In comparison, fluorescence measurement in such situations will be independent on the liquid volume and the uniformity of the wells in the microplates when the liquid volume is large enough to cover the whole excitation and emission portion of the solution.

2.4 MOLECULAR LUMINESCENCE MEASUREMENTS

Molecular luminescence refers to the phenomenon for some molecules that lights are emitted when the molecule in excited states returns to ground states. Depending on how the molecules are excited, molecular luminescence can be subdivided into photoluminescence when they are excited by photons, chemiluminescence when they are excited by chemical energy, and electrochemiluminescence when electricity is the initial energy input that is then converted to chemical energy.

2.4.1 Photoluminescence

Photoluminescence can be further divided into fluorescence when the emission is from the excited molecule in a singlet state and phosphorescence when the emission is from the excited molecules in a triplet state. Figure 2.9 shows the partial energy diagram of the single-photon absorption processes that produce fluorescence and phosphorescence. The molecules started at the lowest vibrational level of the ground state (S_0). After absorption of radiation energy that matches the energy between any vibrational levels of ground state (S_0) and excited state (S_1), the molecules are excited

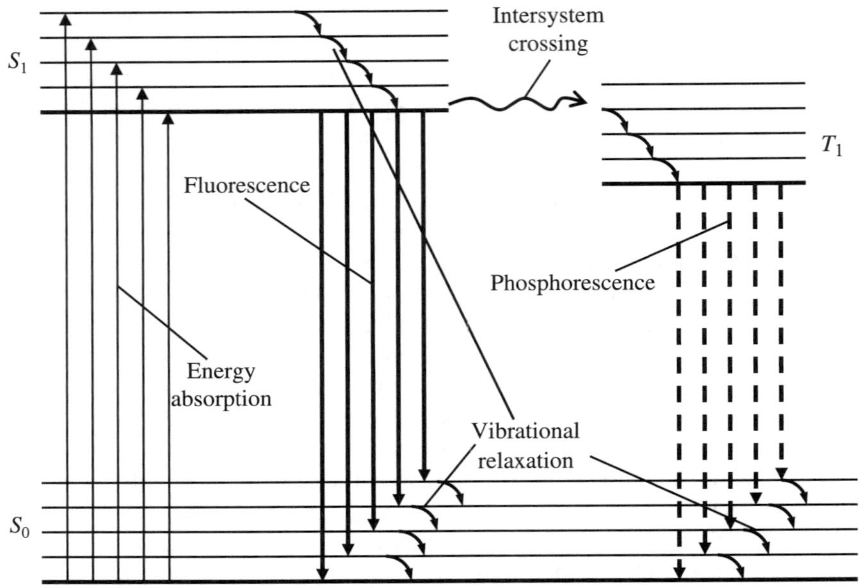

Figure 2.9 Partial energy diagram for fluorescence and phosphorescence. The molecules started at the lowest vibrational level of the ground state (S_0). After absorption of radiation energy that matches the energy between any vibrational levels of ground state (S_0) and excited state (S_1), the molecules are excited to the excited state S_1. Vibrational relaxation will change the excited molecules at higher vibrational levels to the lowest vibrational level in the excited state (S_1). Three processes can happen from here: (1) The molecules return to the ground state by emitting fluorescence; (2) the molecules cross the system to triplet states and then relax to lowest energy level of the triplet state. From there, the molecules return to the ground state by emitting phosphorescence; (3) the molecules return to ground state without emitting light through internal and external conversion (not shown on the graph).

to excited state S_1. Vibrational relaxation will change the excited molecules at higher vibrational levels to the lowest vibrational level in the excited state (S_1). Three processes can happen from here: (1) The molecules return to the ground state by emitting fluorescence; (2) the molecules cross the system to triplet states and then relax to the lowest energy level of the triplet state. From there, the molecules return to the ground state by emitting phosphorescence; and (3) the molecules return to the ground state without emitting light through internal and external conversion (not shown on the graph). The energy diagram shows that the emitted energy will always be lower than the absorbed energy if there is no other source of energy input (such as chemical energy). Thus, the emission wavelength will always be longer than the excitation wavelength. This phenomenon is called "Stokes shift." Photon absorption by fluorophore in solution is instantaneous according to the Franck–Condon principle. After photon absorption, most fluorescence happens in the nanosecond to microsecond scale. In comparison, phosphorescence happens in the millisecond to second or longer scale.

Phosphorescence is the result of the intersystem crossing of electrons from excited singlet state to triplet state, which is followed by the "forbidden" transition from higher energetic triplet state back to the ground singlet (Fig. 2.9). The intersystem crossing requires some overlapping of the vibrational levels of the triplet and singlet states. The phosphorescence can only happen when the electrons return to the lower energy state through the π^* to σ transition, which is the quantum mechanically "forbidden" transitions. Thus, this kinetically unfavored transition progresses at significantly slower time scales. The phenomenon of phosphorescence is rare in solution at room temperature because there exist many other competing deactivation processes though some molecules containing heavy atoms, such as bromine and iodine, may favor phosphorescence. We will focus the discussion on fluorescence here.

What Type of Molecule Emits Fluorescence When Excited? The most commonly used fluorescence molecules in bioassay are organic fluorescent dyes. We have shown in Figure 2.8 that radiation can cause the electronic transition in organic molecules from the σ to σ^* orbit, from the π to π^* orbit, from the n to σ^* orbit, and from the n to π^* orbit. The electronic transition from the σ to σ^* orbit requires high energy that may dissociate the bond. Thus, fluorescence rarely happens with excitation at less than 250 nm. Experimentally, it is found that fluorescence favors molecules having lower transition energy from the π to π^* orbit than from the n to π^* orbit. In addition, molecules with rigid structures are favored to emit fluorescence. Thus, molecules with fused aromatic rings and molecules with highly conjugated double bonds tend to give fluorescence (e.g., tryptophan, NADH, fluorescein, rhodamine, and cyanine). Most organic molecules have fluorescence half-life in the nanosecond scale. Because of the low energy level required for the π to π^* transition, most organic fluorescence molecules absorb in the UV to visible range and emit fluorescence in the visible range.

Chelates of lanthanides are commonly used in bioassays to emit fluorescent signals. Lanthanides in aqueous solution are weakly fluorescent. However, lanthanide chelates are highly fluorescent with a wide spectrum separation between the excitation and the emission wavelengths. In addition, their fluorescence lifetimes are in the submillisecond scale, which makes them ideal for time-resolved fluorescence bioassays (e.g., europium chelates are used in Delfia, Lance, and HTRF assay formats).

Quantum dots are nanometer-sized crystalline clusters (usually between 1 and 10 nm in diameter) made from semiconductor materials. They have broad absorption spectra and narrow symmetric emission bands (25 to 35 nm) that span a wide spectrum from UV to IR with different diameter dots. The excitation at one band of wavelength can simultaneously excite many different sized quantum dots, which is very useful for many bioassays. In addition, quantum dots possess excellent photostability and brightness (quantum yield can reach 90%). Because of these properties, quantum dots have rapidly emerged as potential new fluorescent probes for imaging of biological samples since their discovery in recent years. Quantum dots have been successfully applied in cellular imaging, Western blot, ELISA, and so on. However, some difficulties have been encountered in applying quantum dots in bioassays, such as compatibility in bioassay solution and cellular toxicity.

The green fluorescent protein (GFP), comprised of 238 amino acids (26.9 kDa), was originally isolated from the jellyfish *Aequorea victoria*, which fluoresces green when exposed to blue light. The fluorophore of GFP is formed by the chemical reaction among Ser, Tyr, and Gly residues in the middle of GFP sequences. An enhanced version of GFP and many derivatives with different colors were made in the 1990s by cloning and subsequent mutation of the original GFP. The availability of GFP and its derivatives has revolutionized fluorescence microscopy and the way it is used in cell biology and other biological disciplines. GFPs can be readily introduced into live cells through cloning and transfection. GFPs are not toxic to cells when they are illuminated in living cells. This property enables observing live cells over a long period of time. In comparison, it is difficult to introduce the other types of fluorescent dyes into live cells, and the introduced extrinsic dyes are often toxic to the cells. Cloning of GFPs into specific proteins allows intracellular imaging of protein distribution and trafficking. Another application of GFPs in bioassay is the fragment complementation assay. With this assay, a GFP is split into two parts and each part is covalently linked to one of a pair of binding proteins. The GFP fragments are not fluorescent when they are separated. When the pair of proteins binds to each other, which brings the two fragments of GFP together, a functional GFP is formed and it fluoresces. This is a very useful assay to study protein–protein interaction. In another application, fluorescent proteins cloned into selected transparent organisms have opened a new era to image whole organisms.

Application of Fluorescence Measurement Unlike assays based on absorption measurement that usually measure the absorption of light by native biomolecules directly, few bioassays based on fluorescence measurement directly detect the intrinsic fluorophores present in the native biological system. This is because either the native biological system under study does not contain naturally occurring fluorophores or the naturally occurring fluorophores do not have the fluorescence properties suitable for the measurement. Extrinsic fluorescent molecules are usually used in a bioassay because the excitation and emission range can be selected to avoid interference and because they fluoresce intensely. Extrinsic fluorophores are used as labels that are either covalently linked to the biomolecule under study or are used to stain cellular substructure. In comparison, GFP and their derivatives are usually cloned into the target protein genes so that the target protein and the GFP tags are co-expressed as one fusion protein. The detection limit in fluorescence measurement is several orders lower than that in absorption measurement. In addition, fluorescence measurements have a broad dynamic range of detection. These properties make fluorescence measurement ideal for detecting small changes in biological systems. When used to stain subcellular components in microscopic imaging applications, fluorescent molecules can help improve significantly the contrast between signal and background. Another significant attribute of fluorescence measurement in this application is the multicolor staining that makes possible simultaneous monitoring of many biological processes. When studying enzyme function with fluorogenic assays (see Chapter 6), the fluorescence measurement enables detection of enzymes at a 1000-fold less concentration than the corresponding assay in the chromogenic assay format. In ELISA assays, fluorescent tags or fluorescent substrates play an important role too.

When performing fluorescence measurement, some cautions should be made:

1. Most fluorescent molecules may be photobleached that can cause continuous decrease in fluorescent signals with repeated illumination. However, some fluorophores are more photoresistant than others. Photobleaching can be minimized

by employing less photolabile fluorophores and by reducing local excitation intensity that can be achieved by increasing the area of lumination to achieve the same amount of excitation.

2. Dissolved oxygen in the solution often reduces the fluorescence intensity. When performing biochemical assays when the oxygen is not a required component in the assay, buffer solutions should be purged with argon first to remove oxygen.
3. Some dyes are pH sensitive. For example, fluorescein is a pH-dependent fluorescent molecule and it should be used at a pH higher than 8 to obtain the maximal signal.
4. Background fluorescence can come from both samples and sample holders. Employing a sample holder with less autofluorescence and employing a fluorophore with a longer wavelength can reduce the assay background since few molecules in native biological samples and in compound library fluoresce at longer than 600 nm wavelength.
5. Fluorescence intensity is only linearly proportional to the concentration of fluorophore in dilute solution. If the absorbance of the sample under study is too high due to high fluorophore concentration, the fluorescence measurements can be distorted by artifacts, such as self-absorption and the inner-filter effect. This phenomenon may cause a nonlinear concentration–intensity relationship.

2.4.2 Chemiluminescence

Chemiluminescence is produced when a chemical reaction is the energy source to yield molecules in the excited state that emit light when returning to the ground state. In bioassay applications, oxygen is usually the initial energy donor. Through enzymatic actions, oxygen converts an enzyme's substrate from ground state into excited state (e.g., luciferin in the presence of luciferase) or generates stoichiometrically hydrogen peroxide (e.g., the reaction of glucose and oxygen in the presence of glucose oxidase). The peroxide generated in the reaction can oxidize other molecules (such as luminol) in the presence of peroxidase to generate a product in the excited state. Figure 2.10 shows two common chemical reaction schemes that are employed to generate chemiluminescence: Luciferin reacts with oxygen and luminol reacts with peroxide. Chemiluminescence has the unique properties that no excitation radiation and no filters are required for the detection of light generated in the process. All the lights are collected in chemiluminescent measurement because the lights only come from the specific species under study. The lack of excitation source and filters enables a much simpler luminometer instrumentation that can be built with only a detector, such as PMT, in a dark enclosure. The elimination of photoexcitation also makes the background very low in chemiluminescent assays. Because of the overlapping of the excitation and emission spectra in fluorescence detection and the possible presence of other fluorescent substances in the sample or in the sample holder, wavelength selectors (filters or monochromators) must be used to reject the reflected excitation light and other unwanted emitting lights. In this process, the wavelength selector rejects a significant amount of emitted lights from the substance under study together with the unwanted lights. Even with reduction of light, fluorescence

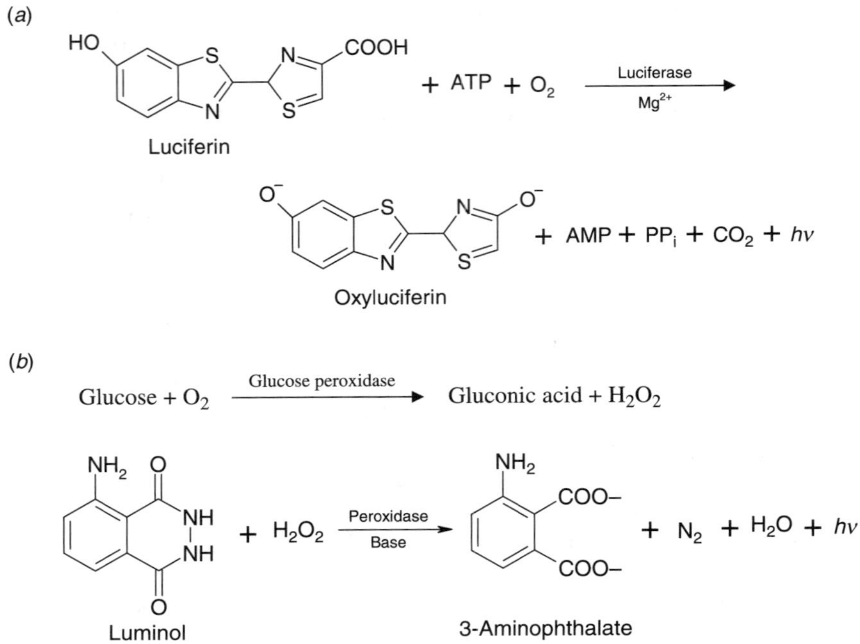

Figure 2.10 Chemoluminescence detection of (*a*) ATP and (*b*) glucose. (*a*) Firefly luciferase converts luciferin to oxyluciferin and emits luminescence in the presence of oxygen and ATP. This is a widely used bioassay to detect the presence and quantity of ATP. (*b*) Glucose peroxidase converts glucose to gluconic acid in the presence of oxygen and generate hydrogen peroxide. The peroxide then oxidizes luminol to generate luminescence. This is a widely used bioassay for detecting many biomolecules, such as uric acid, cholesterol, and lactate.

measurements usually have much higher signal than the chemiluminescence measurement. However, fluorescence measurements usually have much higher background due to the leakage of excitation lights into the emission channel and the autofluorescence transparent to the wavelength selector from other substances in the excitation light path. Chemiluminescence does not enjoy the broad applications in bioassays as fluorescence because of the limited number of biological systems that can produce chemiluminescence.

Chemiluminescence measurement finds most use in the following bioassays: (1) Quantitation of ATP (see Fig. 2.10*a*). In this reaction, firefly luciferase catalyzes the conversion of luciferin to oxyluciferin in the presence of oxygen and ATP. Luminescence is generated in the process and its intensity is proportional to the concentration of ATP. (2) Quantitation of a molecule that can produce stoichiometrically hydrogen peroxide through enzymatic reaction in the presence of oxygen. The quantity of the hydrogen peroxide generated in the reaction can be measured by the intensity of the luminescence produced in a reaction that involves luminol (or other substrate) in the presence of peroxidase (see Fig. 2.10*b*). Thus, the intensity of luminescence is proportional to the quantity of the substance to be measured. This bioassay is widely used to detect many biomolecules, such as glucose, uric acid, cholesterol, and lactate. (3) Quantitation of peroxidase that is covalently linked to the last antibody in ELISA. In ELISA, the quantity of peroxidase is proportional to the quantity of the analyte captured on the solid surface. In the presence of an excess amount of luminol and oxygen, the luminescence intensity is proportional to the quantity of the analyte.

Most early chemiluminescence measurements were carried out with the rapid mixing of all the components and then immediately measure the flash luminescence that rapidly decays with half-life of less than a minute. This assay scheme requires liquid handling devices to be built inside the luminometer to enable immediate measurement after luminesce is generated. Recent bioassay design has significantly reduced the chemical reaction rate in the luminescence generation and a steady emission of luminescence can be obtained over several hours (e.g., the ATP measurement kits from Promega, Perkin-Elmer, and many other vendors). Because flash luminescence has a burst of signal in a very short period, the flash luminescence gives a much stronger signal than the steady luminescence that emits light over a long period of time.

2.4.3 Electrochemiluminescence

Electrochemiluminescence (ECL) is a form of chemiluminescence in which the light-emitting chemiluminescent reaction is preceded by an electrochemical reaction. The first ECL was observed when luminol was oxidized in the presence of dissolved oxygen on the surface of anodes and emitted blue light. Later, it was found that substituting oxygen with hydrogen peroxide results in more intense luminescence on the anodes. This chemiluminescence occurs in the absence of peroxidase that is required in chemiluminescence. In addition to luminol, it was found that several other systems could emit luminescence in the presence of electric current, such as ruthenium chelates and acridan esters. Electrochemical oxidation of an acridan ester in the presence of the peroxide anion also results in blue light emission. Tris(2,2′-bipyridyl) ruthenium (II) [$Ru(bpy)_3^{2+}$] and its derivatives are currently widely used ECL reagents in clinical diagnostics and in drug discovery bioassays. The ECL mechanism of [$Ru(bpy)_3^{2+}$] is shown in Figure 2.11. Tripropylamine (TPA) is oxidized on the electrode resulting in the formation of TPA^+, which spontaneously loses a proton and forms a neutral TPA radical. At the same time, [$Ru(bpy)_3^{2+}$] is oxidized on the electrode to generate $Ru(bpy)_3^{3+}$. $Ru(bpy)_3^{3+}$ is reduced by the TPA radical to generate excited $Ru(bpy)_3^{2+*}$, which emits red light at about 620 nm when it relaxes to the ground state.

Applications of ECL ECL has all the advantages of chemiluminescence, such as no need for excitation radiation and filters and low assay background. In addition, ECL offers precise control of the timing and the location where the luminescent light emits because electrical energy is used to initiate the luminescence and the luminescence only happens on top of the electrode. This precise temporal and spatial control of ECL improves detection limits over chemiluminescence by increasing the

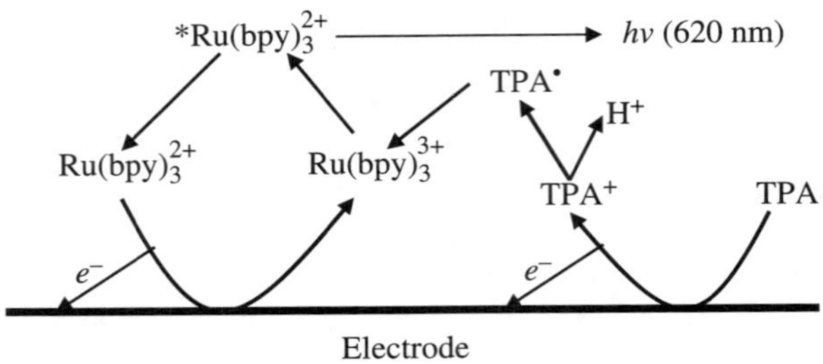

Figure 2.11 Mechanisms of ECL with Tris(2,2′-bipyridyl) ruthenium (II) [Ru(bpy)$_3^{2+}$] and tripropylamine (TPA). TPA is oxidized on the electrode resulting in the formation of TPA$^+$, which spontaneously loses a proton and forms a neutral TPA radical. A$^+$ the same time, Ru(bpy)$_3^{2+}$ is oxidized on the electrode to generate Ru(bpy)$_3^{3+}$. Ru(bpy)$_3^{3+}$ is reduced by TPA radical to generate excited Ru(bpy)$_3^{2+*}$, which emits red light at about 620 nm when it relaxes to the ground state.

ratio of signal to noise. ECL is the most sensitive photon detection technology in bioassays. Substance with concentrations at subpicomolar can be detected with ECL. In addition, spatial control also allows multiplexing that several reactions proceed in a well of microplates. The signals from individual reactions at different locations at the bottom of the well where different electrodes are located can be detected simultaneously by scanning or imaging.

Ru(bpy)$_3^{2+}$/TPA ECL has been extensively applied to many bioassays in microplate format by Meso Scale Discovery. In this technology, functionalized ruthenium chelates (Fig. 2.12) are used to label biomolecules. There are four sulfonate groups added to the molecule to increase its water solubility. The molecule is functionalized with succinamide, which reacts with many biomolecules having primary

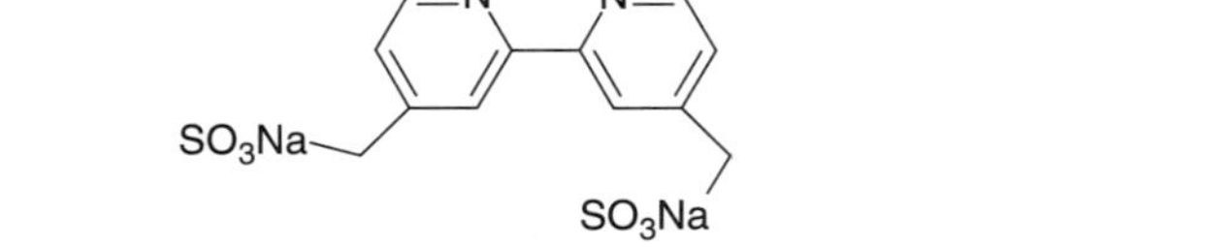

Figure 2.12 Functionalized Tris(2,2′-bipyridyl) ruthenium (II) [Ru(bpy)$_3^{2+}$] used in Meso Scale Discovery platform. The four sulfonate groups are added to the molecule to increase its water solubility. The molecule is functionalized with succinamide that will react with primary and secondary amines present in many biomolecules.

and secondary amines. After the labeled biomolecules bind to the surface of an electrode, ECL can be initiated by turning on the electrical circuits. The ruthenium chelates are stable and the emitted red light has less chance of being interfered by colored organic chemicals in an organic compound library. One precaution when using $Ru(bpy)_3^{2+}$ ECL is that the reaction generates protons, and therefore a relatively strong buffer is required to maintain the pH.

2.4.4 Molecular Luminescence Instruments

Molecular luminescence instruments that detect fluorescence, chemiluminescence, and ECL share similar design in the light detection part. The major differences among the instruments that detect the three different modes are in the light generation and sample holders.

Light Detection Fluorescence detection in bioassays usually involves many different dyes with broad wavelength coverage ranging from UV to visible light. In addition, the emission light output is usually very strong. Thus, most fluorimeters use photodiodes as the detector because of their broad relatively steady response over a wide wavelength range, though photodiodes are not as sensitive as PMT and CCDs. Photodiodes are more tolerable to abuse by accidental overexposure that may cause severe damage to PMTs. Filters must be used in front of the detector to prevent the reflected incident light and light from other fluorescence species in the sample from reaching the detector in fluorescence detection. Sometimes monochromators are used instead of filters, especially in the applications to scan the emission spectra of an unknown molecule.

Chemiluminescence detection deals with light from only one source. Filters are not required. Because the signals are relatively low in chemiluminescence, PMTs are usually used for their detection. PMTs are more sensitive in the blue region, which makes PMTs ideal detectors since blue light is emitted in most chemiluminescence (except ruthenium chelates). In flash chemiluminescence measurement, liquid handling devices are usually integrated into the detection module. In ECL, electrodes must be integrated into the detection module. To take advantage of the positional resolution of ECL, CCDs are usually used as the detectors because CCDs are very sensitive, especially to red wavelength.

Sample Holders Traditional fluorescence was measured with cuvettes as sample holders. To prevent the excitation light from reaching the detector, the detector is placed at 90° from the incident light. This arrangement requires the cuvette being transparent on both sides to allow the pass of the incident light and the emission light. However, modern bioassays are mostly performed in microplates. In this situation, the light excitation and light detection module is placed either on top of the microplate (when solid microplates are used) or beneath the microplate (when bottom-transparent microplates are used). This design allows the maximal throughput for sample detection. However, the signal-to-background ratio is significantly compromised because this design relies totally on the excitation filters and the emission filters to reject the background light. To reduce the intensity of reflected light, black microplates are used in fluorescence measurement. The microplates used in fluorescence measurements are usually made with polypropylene or polystyrene with low binding surface. However, some labeled peptides at low concentrations (in the nanomolar range) in the microplate may gradually lose fluorescent signal. In practice, labeled peptides should be stored at high concentration in aliquots and only be diluted to nanomolar concentrations in microplates before measurement. For chemiluminescence measurement, white microplates are used to gather as much light as possible to the detector since chemiluminescence measurements detect light from only one source and the background emission almost does not exist.

Excitation Light Among the different modes of molecular luminescence discussed so far, only fluorescence measurement needs a radiation source. Xenon lamps are commonly used in fluorimeters because of their relatively steady light output over a wide wavelength range. A bandpass filter that matches the excitation wavelength is usually present to reduce the background light at other wavelengths. Monochromators are used sometimes instead of filters, especially when the application is scanning the excitation spectra of unknown molecules. There is a trend recently that monochromators and filters are used in tandem to take advantage of both types of wavelength selector and in the meantime to increase the flexibility for the fluorimeter to handle different applications.

When studying the fluorescence properties of a new molecule, it is a challenge since we need to fix the excitation wavelength at the region where the molecule absorbs light and then scan the emission spectrum. Alternatively, we need to fix the emission wavelength at the region of emission and then scan the excitation spectrum. However, we know neither the excitation region nor the emission region with an unknown molecule. In this situation, we have to obtain the unknown molecule's absorption spectrum in a solution first with a spectrophotometer. A high concentration of the molecule must be used to obtain its absorption spectrum. The solution is then diluted more than 100-fold so that the fluorescence measurement can be performed. The absorption wavelength is used as the fixed excitation wavelength and the emission spectrum is scanned. The obtained emission wavelength is then used in scanning the excitation spectrum.

The measured value of fluorescence intensity is a relative value that is dependent on individual instruments and the parameters of the instrument when the value is measured. Thus, it is difficult to compare experimental results obtained from two instruments or obtained within one instrument but at different times with different instrument settings. A standard fluorescent sample is essential to compare the fluorescence intensity obtained from different instruments or from the same instrument with different configurations. Because of photobleaching, the standard sample must tolerate light so that it does not change over a long period of use. Fluorescence molecules in solution are usually more photomobile than fluorescence molecules in solid state. High-quality fluorescence standards in solid state are commercially available from vendors such as PerkinElmer.

2.5 LUMINESCENCE LIFETIME MEASUREMENT AND TIME-RESOLVED FLUORESCENCE MEASUREMENT

The measurement of luminescence lifetime was initially performed with phosphorescence due to its long half-life of decay. Because of technology limitations, early fluorescence measurements were made possible by simultaneously exciting the sample and collecting the emitted light. The advance of laser technology allows radiation pulse in picoseconds. Combined with improvements in other instrument components, fluorescence lifetime can be measured after the initial pulse of lumination. It was found that different fluorescent molecules have different half-lives. Most organic fluorescent molecules have half-lives in the nanosecond scale (e.g., fluorescein has a half-life at about 1 to 3 ns). In comparison, lanthanide chelates have long half-lives reaching several hundred microseconds. Fluorescence half-life is not an absolute property of the fluorescent molecule. Factors such as ionic strength, hydrophobicity, oxygen concentration, binding to macromolecules, and the proximity of molecules that can deplete the excited state by resonance energy transfer can all modify the lifetime of a fluorophore. Measurements of lifetimes can therefore be used as indicators of the microenvironment surrounding the fluorophore. Different from fluorescence intensity measurement that is instrument-dependent, fluorescence lifetime is independent of the concentration of the fluorophore and the instruments. The temporal resolution in fluorescence lifetime measurement results in higher signal-to-background ratio. In addition, fluorescence lifetime measurements allow the analysis of a mixture of fluorescent molecules in the same sample because of the temporal resolution. Measurement of fluorescence lifetime has found increased applications in bioassays in recent years. However, the instrumentation required to make fluorescence lifetime measurement is quite expensive, which prevents its wide adoption in bioassays.

Time-resolved fluorescence (TRF) is a technique to measure the delayed fluorescence emission from a fluorophore with a relatively long lifetime after the sample under study is excited with a short pulse of excitation light. The scheme of employing TRF to detect the fluorescence signal after a delay of a hundred microseconds is shown in Figure 2.13. We have discussed that common fluorescence measurement (simultaneous collection of emission signal when the sample is excited) is compromised by high background signals that originate from endogenous sample constituents and the material that makes up the sample holder (autofluorescence) and reflected excitation light. In common fluorescence measurement schemes, the background fluorescence is usually minimized by employing narrowband filters and by selecting probes that absorb and emit at longer wavelengths away from the wavelength of possible interfering fluorescent components in the excitation light pass. However, significantly high background still compromises the common fluorescence measurement. By taking advantage of some fluorophore's (such as lanthanide chelates') long half-life, TRF measurements can significantly reduce the background in fluorescence measurement. In the TRF scheme, the fluorescence measurement was made more than 100 μs after the sample is excited. This delay is long enough that all reflected light and autofluorescent light (in the nanosecond scale) have disappeared before the measurement of the signal from the molecules with long lifetime (such as lanthanide chelates).

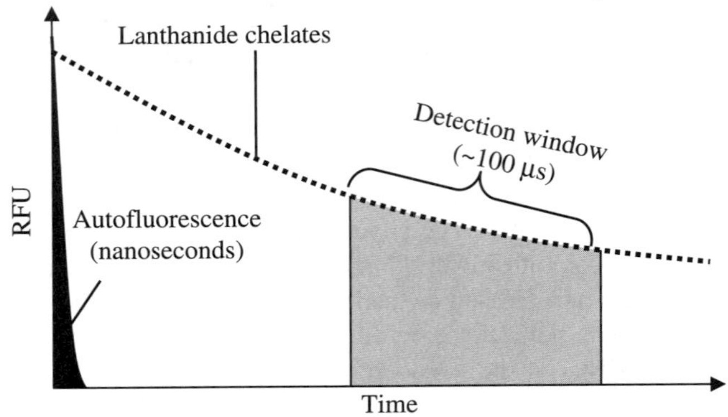

Figure 2.13 Illustration of how time-resolved fluorescence (TRF) significantly reduced fluorescence background. A pulse of excitation radiation is applied to the test sample. Within nanoseconds, autofluorescence from contaminant organic fluorophores and from sample holders will emit and then disappear. Detection of emission from long lifetime fluorophores, such as lanthanide chelates, only starts a few microseconds after the excitation and the detection lasted for a predetermined time in microsecond scales.

DELFIA assays (marketed by PerkinElmer) are widely used TFR-based assays. In DELFIA assays, lanthanides (typically, europium) are used as tracers attached to the molecule of interest, the same way as radiolabels. After completing most of the assay procedures, the molecule of interest with lanthanide tracer attached is separated from the other assay components (usually achieved by attaching the analyte to the surface of the microplate). The lanthanides are practically nonfluorescent during the early part of the assay. The fluorescence signal is developed by the addition of either enhancement solution or DELFIA inducer. The low pH of these formulations efficiently dissociates the europium from the labeled molecule within a few minutes. The free Eu^{3+} rapidly forms a new, highly fluorescent chelate inside a protective micelle with components of the enhancement solution or DELFIA inducer. The fluorescence of the lanthanide chelate is amplified 1 to 10 million times by the enhancement technique. The sensitivity of the DELFIA assay is close to radioactivity assays. Many applications using radiolabels can potentially be adapted in DELFIA format except in situations that the exact original molecule is required and in situations that the lanthanide chelate interferes with the molecular interactions.

2.6 FLUORESCENCE RESONANCE ENERGY TRANSFER (FRET) AND TIME-RESOLVED-FRET

Fluorescence resonance energy transfer is the transfer of the excited state energy from one fluorescent moiety (donor) to another fluorescent moiety (acceptor) when the emission spectrum of the donor overlaps the excitation spectrum of the acceptor and the distance between the donor and the acceptor is within a certain range. The donor transfers its energy to the acceptor through dipole–dipole interactions.

When energy is transferred, no light is emitted from the donor molecules and only the acceptor will emit lights. The rate of energy transfer k_{FRET} is inversely proportional to the sixth power of the donor–acceptor distance as described in the following equation:

$$k_{\mathrm{FRET}} = \frac{1}{\tau_D}\left(\frac{R_0}{r}\right)^6 \qquad (2.5)$$

where τ_{D} is the decay time of the donor in the absence of the acceptor, R_0 is the distance at which FRET is 50% efficient (also called Förster distance at which donor emission is reduced by half of its intensity in the absence of acceptor), and r is the distance between donor and acceptor. Förster distances are in the range of 20 to 90 Å depending on the pairs of donor and acceptor. In bioassays, the Förster values of commonly encountered donor–acceptor pairs are listed in Table 2.2. Because the FRET transfer rate is inversely proportional to the sixth power of the distance between the donor and the acceptor, the transfer efficiency is quickly reduced when the distance increases. For example, if the distance between europium chelates and allophycocyanin (APC) shifted from 90 to 117 Å, the FRET efficiency will drop from 50 to 10% as calculated from Eq. (2.5). For any pairs of donor–acceptor, the transfer efficiency will reduce to only 1.6% if their distance is twice the Förster distance.

Early applications of FRET in biological research were to measure the distance between two amino acid residues in a protein and to measure biomembrane properties using organic dyes. Most current FRET in bioassay development use lanthanide as the donor. The use of the lanthanide enables measurement of FRET in longer distance between donor and acceptor and allowed time-resolved FRET (TR-FRET) measurement that significantly increased the signal-to-background ratio of the assay. TR-FRET was commonly used in bioassays to measure the association of two binding proteins and to detect enzymatic reaction product. HTRF (marketed by CisBio) and Lance (marketed by PerkinElmer) are the most widely used TR-FRET-based commercial assay formats. The two formats share many similar properties but with some differences in terms of reagent choices. HTRF are fixed with europium cryptate as the donor and APC as the acceptor. Lance uses a collection of many different lanthanide chelates and many acceptors. TR-FRET can be adapted to assay many different biological targets including kinase assays, which will be discussed in Chapter 7 in detail. Figure 2.14 shows the application of HTFR assay format to measure protein–protein (or peptide) interactions. Here HTRF format is used to measure the

TABLE 2.2 Förster Distance of Some Donor–Acceptor Pairs

Donor	Acceptor	R_0 (Å)
Dansyl	Fluorescein	37
Fluorescein	Tetramethylrhodamine	51
Terbium	Rhodamine	65
Europium	Cy5	70
Europium	APC	90

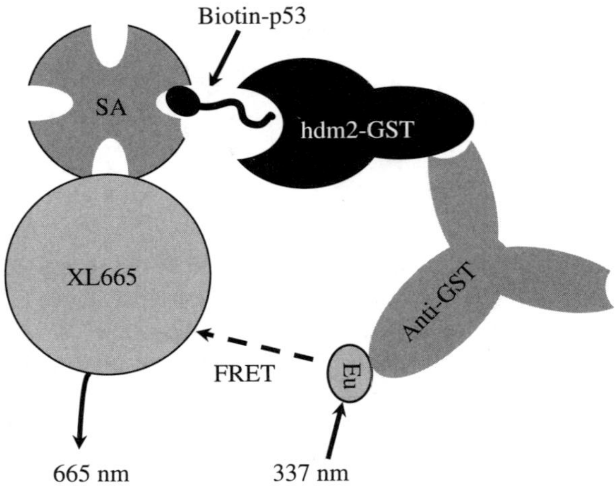

Figure 2.14 Example of TR-FRET-based bioassay in HTRF format to measure the binding between hdm2 and a peptide derived from P53 protein in the binding region. Hdm2 is labeled with GST by co-expression. The binding peptide is labeled with biotin. Acceptor XL665 is labeled by streptavidin and anti-GST antibody is labeled with europium cryptate. When all these reagents are mixed together, a complex is formed that brings the europium chelate close to XL665. When the europium is excited at 337 nm with a laser pulse, the energy is transferred through FRET to XL665. The emission at 665 nm can be measured after a few hundred microseconds.

binding between hdm2 and a peptide derived from P53 protein in the binding region. Hdm2 is labeled with GST by co-expression. The binding peptide is labeled with biotin. Acceptor XL665 is labeled by streptavidin and anti-GST antibody is labeled with europium cryptate. When all these reagents are mixed together, a complex is formed and brings europium close to XL665. When europium cryptate is excited with a pulse of light at 337 nm, the energy is transferred through FRET to XL665. The emission from XL665 at 665 nm is measured a few hundred microseconds after the excitation. The detail of this assay is discussed in Chapter 5.

When designing a bioassay in FRET format, care must be taken to make sure that the FRET distance is within range. When multiple components are present in the complex as in the case shown in Figure 2.14, the assay developer must make sure that the components can adopt a geometry to bring the donor and acceptor within 90 Å. To put this in perspective, the hemoglobin tetramers have a diameter longer than 100 Å. Another complication is that the FRET process is also dependent on the orientation of the donor and the acceptor molecules if they are immobile. In this situation, FRET does not only depend on the distance between the donor and the acceptor. In bioassay applications, we must make sure that the motion of the donor and the acceptor is not restricted so that the fluidic motions will eliminate the dipole–dipole orientation factor. Another factor affecting FRET is the extent of the overlapping between the donor's emission spectrum and the acceptor's excitation spectrum. Only a limited number of donor and acceptor pairs can be found to give good enough overlapping spectra. One disadvantage of FRET format is that it requires double labeling, which may be impossible in some bioassay situations.

The FRET acceptor can be fluorescent or nonfluorescent. When a nonfluorescent molecule is used as the acceptor, the net effect is the disappearance of the donor fluorescence that is sometimes referred to as "fluorescence quenching," which is different from the fluorescence quenching that will be discussed in the next section. Many protease assays use a peptide substrate with a quencher (nonfluorescent FRET acceptor) and a fluorophore at the N-terminus and C-terminus, respectively. This arrangement results in the quenching of the fluorescence due to close proximity between the donor and the acceptor, which allows efficient FRET. After the cleavage of the substrate by the protease, the quencher and the fluorophore are separated far away from each other, resulting in strong fluorescence. Molecular assembly or fragmentation processes such as membrane fusion can be measured with FRET as well when the fluorophore and the quencher are initially loaded in different parts of the assay system.

2.7 FLUORESCENCE QUENCHING

Fluorescence quenching refers to any molecular interactions with the excited fluorophore that cause a decrease of fluorescence intensity. Fluorescence quenching is a bimolecular process that results from collision of quencher with transient excited state or from the formation of nonfluorescent ground-state species between fluorophore and the quencher. Common fluorescence quenchers are dissolved oxygen, molecules containing heavy atoms (iodine, bromine, chlorine), and amine. At high concentration, fluorophore exhibits self-quenching phenomenon that the intensity of the fluorescence is reduced with increasing fluorophore concentration. In addition to bimolecular quenching process, many other factors, such as solvent polarity and pH, also affect fluorescence intensity. The fluorescence quenching from FRET process discussed in the previous section is different from the fluorescence quenching process discussed here. The decrease of the fluorescence intensity of the donor by FRET is through space and does not involve molecular contact between the donor and the acceptor.

Many bioassays are based on fluorescence quenching (or intensity change). One such application is the "molecular beacon" technology that is employed in monitoring DNA hybridization. A DNA probe is designed to adopt a hairpin structure with a fluorophore at the 5′ end and a quencher (DABCYL) at the 3′ end. The hairpin structure brings the fluorophore and DABCYL together and causes efficient quenching of all the fluorescence from the fluorophore. The fluorescence quenching mechanism in this system is not well understood. The DABCYL in this system can quench a wide range of fluorophores with some of them having emission spectra that do not overlap with the DABCYL absorption spectrum at all. Thus, the FRET process cannot explain the quenching since the overlapping of spectra is required for FRET to happen. It is possible that the fluorescence quenching in this system is due to the collision or association between DABCYL and the fluorophores. After hybridization with a target DNA, the DNA probe that originally adopted the hairpin structure now adopts a linear structure that separates the quencher and the fluorophore in space by the relatively rigid linear DNA double helix. No collision or association between the quencher and the fluorophore is possible in the linear structure, which leads to strong fluorescence.

2.8 FLUORESCENCE POLARIZATION (FP)

Polarized Light and Instrumentation Ordinary light consists of a bundle of light waves that travels in different planes along the path of the light. The collection of lights in all planes can be presented by lights in two perpendicular planes. Removal of the lights from one plane by a polarizer (media that selectively absorb or block light in one plane) results in plane polarized light. In fluorescence polarization measurement, monochromatic ordinary light is used as the light source. The light first passes through a polarizer to remove the light from one plane. Only lights from the other plane are allowed to reach the sample. Behind the sample, another polarizer that can rotate 90° is used to let the light in parallel plane (using the incident light as reference) and vertical plane to pass in succession so that the intensities of the light in both planes are collected individually.

Principles and Applications of FP After instant absorption of polarized light, the fluorophores are excited. If the fluorophores are all fixed on the same orientation and motionless, the emitting fluorescence will be completely polarized in the same plane. When the molecule is tumbling freely in solution, the emitting fluorescence will be in all different planes. This is called depolarization. The extent of depolarization can be quantified by a parameter called fluorescence polarization that is defined as

$$P = \frac{I_{\parallel} - I_{\perp}}{I_{\parallel} + I_{\perp}} \tag{2.6}$$

where $I_{\parallel}$ is the fluorescence intensity in the parallel plane to the plane of the polarized excitation light and $I_{\perp}$ is the fluorescence intensity in the plane perpendicular to the plane of the polarized excitation light.

The same phenomenon can also be quantified by fluorescence anisotropy that is defined as

$$A = \frac{I_{\parallel} - I_{\perp}}{I_{\parallel} + 2I_{\perp}} \tag{2.7}$$

Though both P and A describe the same phenomenon, A is preferred to P in practice because it makes the mathematical expression of related equations much simpler. The relationship between A and P is

$$P = \frac{3A}{2 + A} \tag{2.8}$$

$$A = \frac{2P}{3 - P} \tag{2.9}$$

For complete polarization, $I_{\perp}$ is zero and $A = P = 1$. For complete depolarization, $I_{\parallel} = I_{\perp}$ and $A = P = 0$. With homogeneous unorientated sample, the maximum anisotropy value is 0.4 because of photoselection.

Rotation of a fluorophore is primarily responsible for the fluorescence depolarization. The rate of the angular movement of a molecule in solution is described by rotational correlation time (θ). For a spherical molecule,

$$\theta = \frac{\eta V}{RT} \tag{2.10}$$

where η is the viscosity, T is the temperature in Kelvin, R is the gas constant, and V is the volume of the spherical molecule. In typical assay conditions, the experimentally observed rotational correlation time for trypsin (25 kDa) is 13 ns and for albumin (66 kDa) is 42 ns.

The extent of depolarization by a rotating spherical fluorophore is described by the following Perrin equation:

$$\frac{A_0}{A} = 1 + \frac{\tau}{\theta} \tag{2.11}$$

where A_0 is the fundamental anisotropy, which is the observed anisotropy in the absence of the depolarization process such as rotational diffusion, A is the measured anisotropy, τ is the fluorescence lifetime, and θ is the rotational correlation time. We discussed previously that fluorescence lifetime for most organic fluorescent molecules is in the nanosecond scale. The rotational correlation time for a protein larger than 25 kDa at 25°C and in normal bioassay buffer (viscosity close to 1 cP) is much longer than fluorescence lifetime. This leads to $A \approx A_0$ (a larger value) when the fluorescent molecule is larger than 25 kDa. When the fluorescent molecule is smaller than 1 kDa, the rotational correlation time will be shorter than the fluorescence lifetime and that leads to $A < A_0$ (a smaller value). Thus, a free small fluorophore in solution has a small A. After the binding of the small fluorophore to a large protein, a larger A is obtained. This is the fundamental principle of measuring the association between a small fluorophore and a large molecule.

With a mixed population of fluorophores, the final anisotropy is the sum of the product of the anisotropy and the fractional light intensity contributed from each species as described by the following equation:

$$A = \sum_{i=1}^{n} f_i \times A_i \tag{2.12}$$

where f_i is the fraction contribution of light intensity from species i and A_i is the anisotropy of species i. This is a very useful equation for bioassays. For example, when measuring the binding of a small peptide labeled with a fluorophore to a large protein, two labeled species (free peptide and bound peptide) will be present in the solution. The measured anisotropy in the solution is

$$A = f_f A_f + f_b A_b = A_f + f_b(A_b - A_f) \tag{2.13}$$

where f_f is the fraction of the total light intensity contributed from the free peptide and A_f is the anisotropy of the free peptide; and f_b is the fraction of the total light intensity contributed by the protein-bound peptide and A_b is the anisotropy of the protein-bound peptide. A_f and A_b are constant for a given peptide and its binding partner in a solution with constant viscosity and temperature (usually $A_f \approx 0.05$ and $A_b \approx 0.3$). Rearranging Eq. (2.13) leads to Eq. (2.14) in that the fraction of the bound peptide is expressed in terms of light intensity (be aware that f_b here is different from the molar fraction of the bound species):

$$f_b = \frac{A - A_f}{A_b - A_f} \tag{2.14}$$

In applications to study the binding between a fluorescently labeled ligand and a large protein, it is common to assume that the change in fluorescent intensity of the ligand after its binding to the protein is negligible. With this assumption, the fraction of the bound peptides expressed in terms of light intensity is the same as the molar fraction of the bound peptide. If a large increase or decrease of the fluorescence intensity is observed after the ligand binding, a correction factor that takes account of the differences between the two species has to be included when using Eqs. (2.13) and (2.14).

When performing a fluorescence polarization experiment to measure binding between a small fluorescent molecule and a large protein, the concentration of the labeled small molecule should be fixed at a concentration that is at the lower end of the fluorescence detection dynamic range of the instrument. The concentration of the larger molecule is varied over a concentration range around the EC_{50} value. The EC_{50} value by definition is the concentration of the large molecule at which 50% of the labeled small molecules are bound to the large molecule. Conventionally, the fraction of the bound small molecule f_b versus the concentrations of the large molecule is plotted. Because of the linear relationship between A and f_b, a plot of A versus the concentrations of the large molecule should give the same EC_{50} value. However, a plot of P versus the concentrations of the large molecule should not be used to obtain EC_{50} because there is no linear relationship between P and f_b. Figure 2.15 makes a comparison of the EC_{50} when calculation is made using data from polarization (P), anisotropy (A), and fractional bound (f_b). In this theoretical treatment, anisotropy for the unbound small fluorophore is set at 0.05 and the anisotropy for the fluorophore bound to the large binding partner is set at 0.3. The disassociation constant between the small fluorophore and the large binding partner is set at 10 nM. The results showed that the EC_{50} value is about 11% lower than the true value if P instead of A is used to calculate the EC_{50} value. Though the EC_{50} value obtained by directly using polarization value (P) is close to the true EC_{50} obtained with the other two plots, directly plotting P versus the concentrations of the large molecule is theoretically meaningless. This is a common mistake in many published FP assays.

Fluorescence polarization offers the unique advantage of its ratiometric measurement that will tolerate volume variation and other external perturbations. In addition, the anisotropy values obtained are instrument independent and can be easily transferred from lab to lab. However, fluorescence polarization measurement usually has a smaller dynamic range as compared with other fluorescence methods.

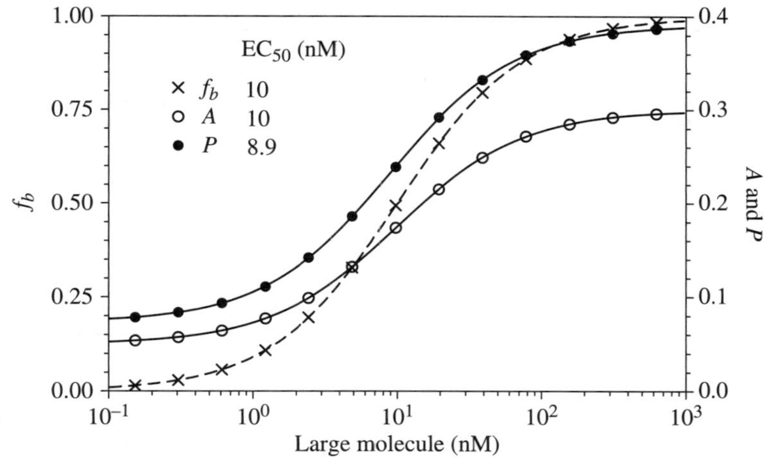

Figure 2.15 Comparison of the EC_{50} value when calculation is made using data from polarization (P), anisotropy (A), and fractional bound (f_b). In this theoretical treatment, anisotropy for the small unbound fluorophore is set at 0.05, and the anisotropy from the bound fluorophore to a large binding partner is set at 0.3. The disassociation constant between the small fluorophore and the large binding partner is set at 10 nM. The results show that the EC_{50} value obtained by directly using polarization data is about 11% lower than the true value.

In addition, its application in bioassay is limited by the requirements that the two binding species must have large molecule weight difference and the lifetime of the fluorophore must fit the temporal resolution dictated by the rotational correlation time of the two binding species. An application of FP in studying the binding between a p53-derived peptide and hdm2 is discussed in Chapter 5.

2.9 RADIOACTIVITY MEASUREMENT

We discussed in previous sections that the radiation emission in molecular luminescence is from the bonding energy released by excited molecules. In this section we will discuss the radiation emission from the energy released by unstable radioisotope during nuclear decay. There are two major forms of nuclear radiation commonly encountered in bioassays, β and γ radiations. Whether the radiation energy can reach a detector determines whether the radiation can be directly detected. The distance the radiation can travel depends on the type of radiation and the radiation energy. The β ray is the emission of negatively charged electrons. Because of its negative charge, the β particle is easily blocked by the nuclei of matter on the travel path of the β ray and thus it does not travel very far in the media. The γ ray is the emission of neutral high-energy waves. It can travel a long distance in the media and may penetrate thick metals. Another important property of an unstable isotope is its half-life, which determines its usage in bioassays. An unstable isotope must have a relatively long half-life to allow some measurements to be made before it disintegrates to an undetectable level. Table 2.3 lists the properties of radioisotopes commonly used in bioassays.

TABLE 2.3 Properties of Commonly Used Radioisotopes

Radioisotopes	Type of Emission	Intensity (keV)	Half-life
^{3}H	β	18.6	12.6 years
^{14}C	β	156	5730 years
^{32}P	β	1710	14 days
^{33}P	β	249	25.4 days
^{35}S	β	167	88 days
^{45}Ca	β	252	165 days
^{125}I	γ ray	28, 31	60 days

Except ^{125}I, which emits γ rays, all other commonly used radioisotopes emit β particles. The emission from low-energy β emitters, such as ^{15}C, ^{35}S, and especially ^{3}H, cannot even pass most detector's windows to allow effective detection. Liquid scintillation counting is a far more efficient way to detect β particles and is the dominant method used in bioassays to detect β radiation.

Liquid scintillation counting is an analytical method that uniformly mixes radiolabeled analyte with a liquid scintillant that produces light flashes when it absorbs the energy of particulate radioactive decay. A liquid scintillation counter usually contains a sample chamber with two PMTs placed in opposite directions relative to the sample to collect light produced from scintillation. The whole system is enclosed in a dark environment. The liquid scintillant is a cocktail containing solvent, emulsifier, and fluor. When nuclear disintegration occurs in the relatively dense liquid, the β particle travels only short distances within a few nanoseconds before all of its kinetic energy is dissipated. The energy is absorbed by the medium in three forms: heat, ionization, and excitation. Some β energy absorbed by solvent molecules makes them excited. When the excited solvent returns to the ground state, it emits UV light. The UV light is absorbed by the fluor molecules that emit blue light flashes upon return to the ground state. The blue light is then detected by the PMTs. Only when the flashes detected by both PMTs simultaneously, within approximately 20 ns, the flashes are counted as true events in coincidence counting to reject background signal caused by chemiluminescence or other events. Nuclear decay events produce approximately 10 photons/keV of energy in liquid scintillation counting. The intensity of the light is proportional to the β particle's initial energy. Thus, PMTs detects two events simultaneously: number of flushes and the intensity of the lights. If there are mixed radioisotopes in one sample and they release sufficiently different energy upon decay, such as tritium and ^{32}P, a detection energy window can be set to measure only the nuclear decay events caused by one isotope. The rate of nuclear decay events is measured by the number of flashes in unit time, usually counts per minute, or CPM. Because not all nuclear decays result in solvent excitation, measured CPM is not the same as the actual nuclear disintegration per minutes, or DPM. A standard sample of a particular isotope with known DPM must be used to calibrate the counting efficiency to enable the conversion from CPM to DPM. The counting efficiency for tritium is only about 40%. In comparison, it is relatively easy to detect γ rays emitted from isotopes such as ^{125}I in that the counting efficiencies are usually over 90%.

Application of Radioisotopes in Drug Discovery Radioisotopes were widely used in tracking the metabolism and distribution of radiolabeled drug substance in the human body in positron emission tomography (PET), in radioimmunoassays (RIA), and recently in scintillation proximity assays (SPA). To reduce environmental impact, RIAs are rarely practiced now, and they are substituted by assays in which the radioisotopes are substituted with luminescence labels. However, some radioisotope applications are very difficult or impossible to substitute with current technologies. Purchased radioisotopes are commonly shipped in a plastic vial labeled with specific radioactivity in mCi/mmol, the concentration, and the total volume. When using the radioisotope in the bioassay, the unit of mCi is usually converted to CPM to estimate the usage of the isotope so that enough signals can be counted: 1 mCi equals to 2.22×10^9 DPM. Because the decay of radioisotope is a random event, CPM measured at a different time with the same sample will not be exactly the same. Longer counting time is required for samples with low count to obtain statistically significant values.

The SPA is a relatively new and powerful bioassay technology. The principle of SPA is shown in Figure 2.16. SPA is a bead-based technology with SPA bead (2 to 10 μm in diameter) coated by scintillant on its surface. In addition, the bead surface is also functionalized with attached capturing molecules. The capturing molecules can be antibodies, protein A, GST, His-tag, wheat germ agglutinin, glutathione, streptavidin, and the like. There are two types of SPA beads, polyvinyl toluene (PVT) and yttrium silicate (YSi). PVT beads emit light that is sensitive to PMT detection. YSi beads emit light that is sensitive to CCD detection. The PVT beads are lighter than normal assay buffer and will float on top of the buffer if the solution in the reservoir is not stirred. The YSi beads are denser than normal assay buffer and will precipitate if the solution is not stirred. When radiolabeled tracers bind to the capture molecules on the surface of the SPA beads, the radioisotope is brought to close proximity to the

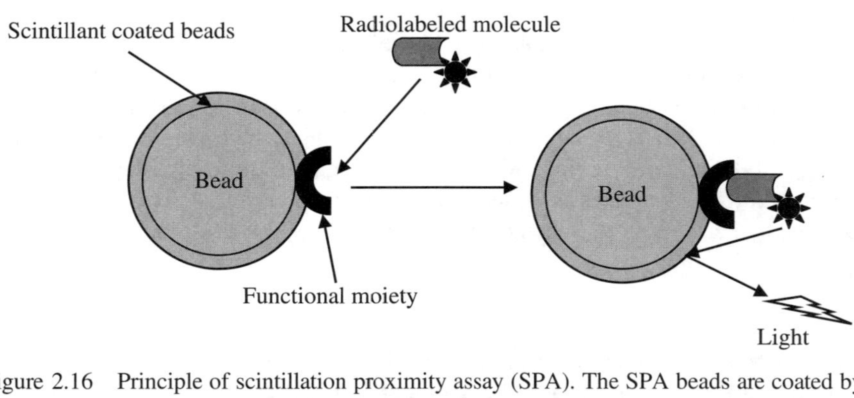

Figure 2.16 Principle of scintillation proximity assay (SPA). The SPA beads are coated by scintillant on their surfaces and a functional moiety is covalently linked to the beads. The functional moiety can be a binding protein or a substrate of an enzyme. The SPA beads have similar density to bioassay buffer so that they can be suspended in the assay solution. Upon the addition of radiolabeled binding protein counterpart or a labeled second substrate of an enzyme, the radiolabel will be brought to close proximity to the scintillant. When the radioisotope decays, the energy hits the scintillant coated on the beads and light will emit.

TABLE 2.4 Commonly Used Radioisotopes in SPA Assays

Isotopes	Mean Path Length (μm)	Recommended Specific Activity (mCi/mmol)
^{3}H	1.5	1–85
^{125}I	17	600–2000
^{14}C	50	0.001–0.1
^{35}S	65	1000
^{33}P	125	1000–3000
$^{32}P^{a}$	2100	Not recommended

^{a32}P is listed for comparison.

scintillant on the beads. When the radiation from radioisotope decays strikes the scintillant, light will emit and can be detected. SPA depends on the distance the radioisotope can travel in the assay media. The mean path length of commonly used radioisotopes and their recommended specific activity are listed in Table 2.4. Isotopes with low-energy β radiation, such as ^{3}H, ^{14}C, or ^{35}S, are preferred for SPA because only radioactive decay close to the bead can cause light emission. Isotopes with high-energy β radiation, such as ^{32}P decay, can travel long distance and may cause adjacent SPA beads to emit light even if the beads do not bind to the decaying isotope. This nonspecific excitation of SPA beads can cause large assay background. A typical SPA uses between 0.2 and 2 mg SPA beads per well in a typical 96-well microplate. SPA has been successfully applied in many bioassays for HTS applications.

FlashPlate (marketed by PerkinElmer) technology is based on similar principles as the SPA technology. Here the scintillant is coated on the surface at the bottom of a microplate instead of on the surface of SPA beads. FlashPlate shares many properties with SPA. The advantage of FlashPlate over SPA is that the unbound radioactivity can be easily removed by washing if the background signal is too high when higher energy radiation isotope, such as P^{33}-labeled ATP in kinase assay, is used. The disadvantage of FlashPlate compared with SPA is the limited capacity on the bottom surface of a microplate for binding. In contrast, the binding capacity with SPA technology is dependent on the number of beads that can be placed in a well in a microplate. With both technologies, the liquid scintillation counting is not perfomed with a regular liquid scintillation counter. Instead, specially designed liquid scintillation counting instruments for microplates are used, such as TopCount and MicroBeta (both are marketed by PerkinElmer). Specific applications of SPA and FlashPlate are discussed in Chapters 5 and 7.

2.10 EVALUATING AND SELECTING AN INSTRUMENTAL METHOD FOR BIOASSAY

We have discussed the most commonly used instruments and methods in bioassays. Here we will discuss the evaluation and selection of instruments and methods for bioassays. In Chapter 1, we introduced the performance characters for bioassays.

Some of the performance characters are also used in evaluating instrumental methods. In addition, those performance parameters are useful in comparing instruments for purchasing decisions.

Instrumental Noise and Precision Any measurement is made up of two components: signal and noise. Signal carries information about the object of interest and noise carries no information about the object. The noise affects the precision, accuracy, and the detection limit of an instrument. Instrumental noise can come from many components that made up the analytical instruments. The noise are unavoidable but can be minimized through good instrument design. Mathematically, the noise function is usually expressed as a signal-to-noise ratio (S/N), which is the reverse expression of relative standard deviation (RSD; see Chapter 14). It is difficult to separate instrumental precision with assay method precision because an assay method is always required to evaluate an instrument. Figure 2.17 shows an example of instrumental noise in fluorescence polarization assay method. This measurement has quite large noise as compared to signal and the collected data set has a large standard deviation. However, because this is a continuous measurement and a large number of data were

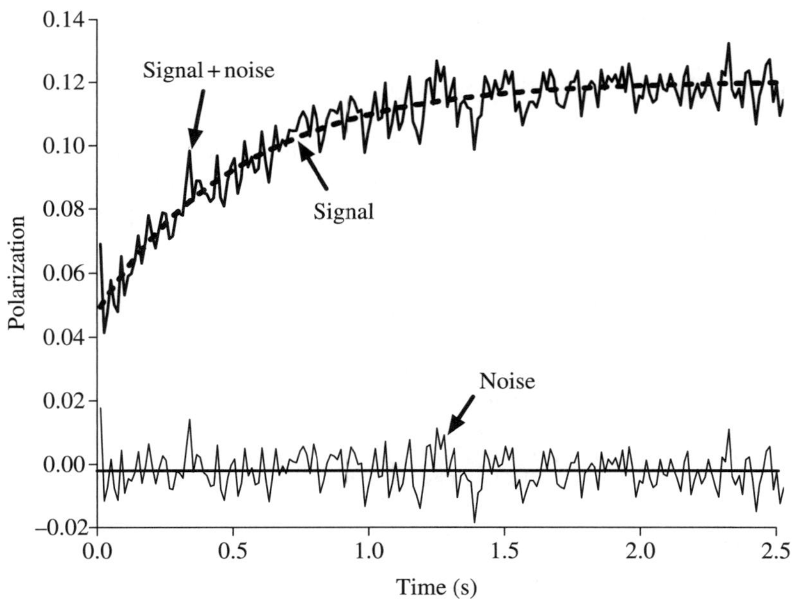

Figure 2.17 Instrument noise as illustrated with the stopped-flow fluorescence polarization measurement of a fluorescein-labeled peptide binding to a protein. The peptide has an initial anisotropy value of ~0.05. After the binding of the peptide to a larger protein, the measured anisotropy value changed to ~0.12 at equilibrium. The top trace is measured experimentally that contains both signal and noise. The dotted trace is the fit to the experimental data that represents the "true" signal. The bottom trace is the noise from the instrument that is calculated by subtracting the "true" signal from experimental signal.

collected in short intervals, the standard deviation of the mean is very small and the true kinetics can be obtained with great precision (this topic is discussed in detail in Chapter 14).

Sensitivity of Detection Sensitivity is the ability of the method to discriminate between small differences in analyte concentration. The noise of the measurement and the slope of the calibration curve affect detection sensitivity. If the signal difference between two data points is close to the sum of the noise of the two measurements, the method is insensitive to discriminate the two concentrations. If a measurements calibration curve is described as

$$S = k * [\mathrm{C}] + \mathrm{b} \tag{2.15}$$

where S is the detection signal, k is the slope, $[C]$ is analyte concentration, and b is the detection background of the analyte, then the *analytical sensitivity* is k/σ_b, where σ_b is the standard deviation of the measurement. It should be noted that the International Union of Pure and Applied Chemists (IUPAC) uses the slope k as the *calibration sensitivity* measurement that does not take into account the noise sensitivity measurement. In both definitions, a higher slope between the two concentration points indicates a more sensitive measurement. A good example of the analytical method affected by sensitivity change is the absorption measurement shown in Figure 2.7 where the sensitivity of the assay decreases with increasing level of absorption because of reduced slope.

Detection Limit of Instrument The detection limit of an instrument is the minimal concentration of the analyte that gives a signal that is larger than a multiple (m) of the variation in the blank on top of the blank signal. Generally accepted m value is 3. The detectable signal limit is mathematically expressed as

$$S_{\min} = b + 3\sigma_b \tag{2.16}$$

Thus, the limit of detection is the concentration at which the signal amplitude of $S_{\min}$ is detected. If $S_{\min}$ falls in the linear dynamic range of detection, then

$$[C]_{\min} = \frac{S_{\min} - b}{k} = \frac{3\sigma_b}{k} \tag{2.17}$$

Dynamic Range The dynamic range is the analyte concentration span from a lower quantitation limit to upper quantitation limits (see Chapter 1). An instrument's lower quantitation limit is generally taken as analyte concentration at which the signal is $10\sigma_b$. The larger the dynamic range of an instrument, the more useful of the instrument to handle samples with large concentration variations. Analytical instruments should have a minimum dynamic range of at least two orders of magnitude.

The performance evaluation of four commercial fluorescence microplate readers is shown in Figure 2.18. In this experiment, fluorescein was serially diluted in a pH 8.0 buffer and was placed in a 96-well solid black microplate. The samples in the same

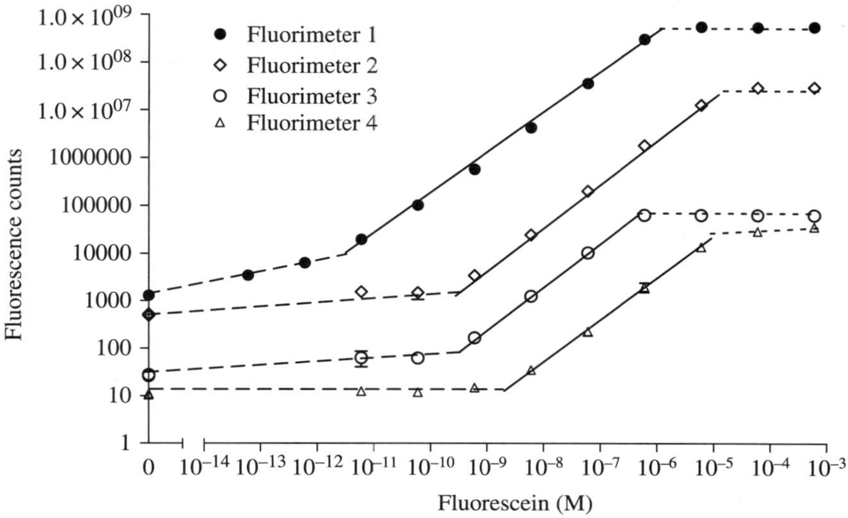

Figure 2.18 Performance evaluation of four commercial fluorescence microplate readers. Fluorescein is serially diluted in pH 8.0 buffer and placed in 96-well solid black microplate. The same plate is read three times with the four fluorimeters. The original data is plotted in double-log scales to cover broad fluorescein concentration range. The dynamic ranges are shown in solid lines and the lower quantitation limits are at about the interception points of the lower dotted lines and the solid lines. The upper quantitation limits are at about the interception points of the upper dotted lines and the solid lines.

microplate were read three times with the four fluorimeters to be evaluated. The original data were plotted in double-log scales to cover a broad fluorescein concentration range. The dynamic ranges for the four fluorimeters were shown in solid lines and the lower quantitation limits were at about the interception points of the lower dotted lines and the solid lines. The upper quantitation limits were at about the interception points of the upper dotted lines and the solid lines. The detection limits for each of the four fluorimeters could not be shown in this graph because the standard deviations were too small to be plotted in the same scale here. The detection limits for all the fluorimeters are close to the lower quantitation limits except instrument 1, which has two linear regions (referred to as higher and lower linear regions here). The detection limit at the lower linear region for instrument 1 is lower than the lowest concentration of fluorescein tested here. Instrument 1 has the largest dynamic range in the higher linear region, which covers more than 5 orders of magnitude. In comparison, instrument 3 only has 3 orders of magnitude of dynamic range. Instrument 1 has the lowest lower quantitation limit that made it very useful to measure fluorophore at subnanomolar concentrations. In comparison, instrument 4 has the highest lower quantitation limit that made it only useful for detecting high concentrations of fluorophore. All the four fluorimeters have similar sensitivity because they have similar slope and standard deviations. Overall, instrument 1 performed the best, but it also cost more than twice the other instruments at the time of testing.

Useful Websites

http://www.newport.com/Lasers/361887/1033/catalog.aspx

http://www.chroma.com/

https://www.omegafilters.com/

http://www.sbsonline.org/msdc/index.php

http://www.mesoscale.com/CatalogSystemWeb/WebRoot/

http://www.perkinelmer.com/

BIBLIOGRAPHY

Alfano, R. R. (2006) The ultimate white light. *Sci. Am.* **295**, 86–93.

Alivisatos, A. P., Gu, W., and Larabell, C. (2005) Quantum dots as cellular probes. *Annu. Rev. Biomed. Eng.* **7**, 55–76.

Bullen, A. (2008) Microscopic imaging techniques for drug discovery. *Nat. Rev. Drug Discov.* **7**, 54–67.

Cazes, J. (ed.) (2005) *Ewing's Analytical Instrumentation Handbook*, 3rd ed., Marcel Dekker, New York.

Corey, M. J. (2008) *Coupled Bioluminescent Assays: Methods, Evaluations, and Applications*. Wiley, Hoboken, NJ.

Glickman, J. F., Schmid, A., and Ferrand, S. (2008) Scintillation proximity assays in high-throughput screening. *Assay Drug Dev. Technol.* **6**, 433–455.

Gore, M. G. (ed.) (2000) *Spectrophotometry and Spectrofluorimetry: A Practical Approach*, 2nd ed. Oxford University Press, Oxford.

Hell, S. W. (2007) Far-field optical nanoscopy. *Science* **316**, 1153–1158.

Hof, M., Hutterer, R., and Fidler, V. (eds.) (2005) *Fluorescence Spectroscopy in Biology: Advanced Methods and Their Applications to Membranes, Proteins, DNA, and Cells*. Springer, Berlin.

Holme, D. and Peck, H. (1998) *Analytical Biochemistry*, 3rd ed. Prentice Hall, Essex.

Issad, T., Blanquart, C., and Gonzalez-Yanes, C. (2007) The use of bioluminescence resonance energy transfer for the study of therapeutic targets: Application to tyrosine kinase receptors. *Expert Opin. Ther. Targets* **11**, 541–556.

L'Annunziata, M. F. (2003) *Handbook of Radioactivity Analysis*, 2nd ed. Elsevier, San Diego.

Lakowicz, J. R. (2006) *Principles of Fluorescence Spectroscopy*, 3rd ed. Springer, New York.

Owicki, J. C. (2000) Fluorescence polarization and anisotropy in high throughput screening: Perspectives and primer. *J. Biomol. Screen.* **5**, 297–306.

Rodriguez-Diaz, R., Wehr, T., and Tuck, S. (eds.) (2005) *Analytical Techniques for Biopharmaceutical Development*. Mercel Dekker, New York.

Rouessac, F., and Rouessac, A. (2007) *Chemical Analysis: Modern Instrumentation Methods and Techniques*, 2nd ed. Wiley, Hoboken, NJ.

Rydberg, J., Cox, M., Musikas, C., and Choppin, G. R. (eds.) (2004) *Solvent Extraction Principles and Practice*, 2nd ed. Marcel Dekker, New York.

Shaner, N. C., Patterson, G. H., and Davidson, M. W. (2007) Advances in fluorescent protein technology. *J. Cell Sci.* **120**, 4247–4260.

Skoog, D. A., Holler, F. J., and Nieman, T. A. (1998) *Principles of Instrumental Analysis*, 5th ed. Brooks/Cole, Toronto, Canada.

Snyder, L. R. and Kirkland, J. J. (1979) *Introduction to Modern Liquid Chromatography*. Wiley, New York.

Snyder, L. R., Kirkland, J. J., and Glajch, J. L. (1997) *Practical HPLC Method Development*, 2nd ed. Wiley, New York.

Stewart, K. K. and Ebel, R. E. (2000) *Chemical Measurements in Biological Systems*. Wiley, New York.

Sutherland, K. (2008) *Filters and Filtration Handbook*, 5th ed. Elsevier Science, Oxford.

Tsien, R. Y. (2003) Imagining imaging's future. *Nat. Rev. Mol. Cell Biol.* **4**, SS16–SS21.

CHAPTER 3

FUNDAMENTAL PRINCIPLES OF ASSAY DEVELOPMENT WITH ISOLATED PROTEINS

PROTEINS PARTICIPATE in every process within and between cells in an organism, making them essential parts of an organism. Proteins can function as enzymes that catalyze biochemical reactions; function in cell signaling, immune responses, cell adhesion, and the cell cycle; and function as structural or mechanical building block. Due to the complexity of live organisms or cells, it is very difficult to study proteins in their native intact environment. To reduce the complexity, proteins are often studied in isolation after they are removed from their native organisms or cells. In most cases, the isolated proteins retain their properties and can still perform many of the same functions as they do in vivo. Thus, isolated proteins can be assayed in vitro, and the assay results are much simpler to interpret because of the simplified assay system. Common in vitro assays for isolated proteins are based on (1) their interaction (binding) with another molecule, (2) their function to facilitate a process (catalyze a chemical reaction or enable ion flux), and (3) their function to induce changes in test cells. Isolated proteins that function as ion channels will be discussed in Chapter 9 and their interaction with cells will be discussed in Chapter 8. The binding of proteins with other molecules and their function as enzymes will be discussed in this chapter. Protein binding is universal to any proteins no matter what functions they perform. A binding assay can thus be applied to any proteins. In addition to binding assays, functional assays can be performed with proteins possessing special functions, such as enzymes that catalyze chemical reactions. If the protein's functions are preserved in isolation, functional assays can be superior to the universal binding assays when the function is well understood. Because enzyme kinetics is a well-established discipline, enzymes are usually assayed with functional enzymatic assays instead of with binding assays. The binding process and enzyme-catalyzed reactions are different in many aspects but the two processes have many similarities. We will use chemical thermodynamics to analyze these two processes here.

Assay Development: Fundamentals and Practices. By Ge Wu

3.1 CHEMICAL POTENTIAL, EQUILIBRIUM, AND KINETICS

Remember from physical chemistry classes that any chemical substance possesses chemical potential that is defined as

$$\mu = \mu^0 + RT \ln a \tag{3.1}$$

where μ^0 is the standard chemical potential, R is the gas constant, T is temperature in the absolute scale (Kelvin), and a is a dimensionless quantity that is proportional to the substance's concentration. For ease of discussion, we will use the concentration of the substance to substitute a. Thus, μ^0 is the chemical potential at 1 M substance concentration. When the concentration of the substance becomes lower, the chemical potential becomes lower too. The chemical potential at concentration below 1 M is lower than its standard chemical potential because the $RT \ln a$ part of Eq. (3.1) is a negative value at a concentration lower than 1 M.

In ligand–protein binding studies (see Fig. 3.1*a*), there is no barrier for the transition from free ligand and free protein to the bound complex and vice versa. The spontaneous transition direction is determined by the free energy difference (ΔG) between the sum of the chemical potential for the free ligand and the free protein and the chemical potential of the bound complex. The substances initially present in the system will be forced to lower their chemical potentials if their total chemical potentials are higher than the total chemical potentials at equilibrium (downhill forces). Lowering their chemical potentials is achieved by lowering their concentrations as dictated in Eq. (3.1). The mass cannot simply disappear when the concentration is reduced. The reduced concentration in one substance gives rise to the increased concentration of the counterpart substance resulting in chemical transformation

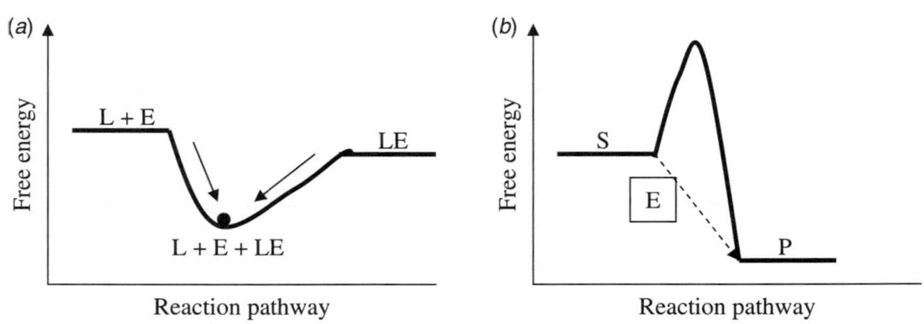

Figure 3.1 Comparison of binding assay and enzymatic kinetic assay from thermodynamic point of view. (*a*) In the binding assay, the protein is part of the chemical reaction system. At equilibrium between the reactants and products, the free energy is zero that is the lowest free energy point. Most binding studies are done in equilibrium. After a system has reached equilibrium, the timing of the study is not a factor. (*b*) In enzymatic kinetic assay, chemical reaction is downhill from reactants to products. The enzymes only serve as the agents to lower the transition barrier but not serve as reactants nor products. The concentrations of reactants and products are constantly changing with time. Thus time is a very important factor in kinetic studies.

(association or dissociation of the binding complex). The chemical transformation is stopped when the system reaches the energy minimum (equilibrium). At equilibrium, $\Delta G = 0$. Thus, if starting with a ligand and a protein alone, a portion of them will associate to form the binding complex. On the other hand, if starting with a pure binding complex, a portion of it will dissociate into free ligand and free protein. The extent of the association and dissociation is governed by the equilibrium constant that we will discuss later. A central theme of equilibrium studies is to determine the equilibrium constant of a system by measuring the concentrations of the bound complex and the free ligand. Most binding studies are performed when the system is at equilibrium. However, binding studies can also be studied in nonequilibrium states by using stopped-flow and quenched-flow techniques that are fast enough to study the kinetics of the chemical transformation before equilibrium is established. An example of stopped-flow study will be discussed in Chapter 5. Nonequilibrium studies can obtain more valuable information about the system (kinetic rate constants) but require more expensive instruments and specialized expertise.

In enzyme-catalyzed reactions, the substrate (S) has a higher chemical potential than the product (P). Thermodynamically, a downhill force will spontaneously transform S to P (see Fig. 3.1*b*). However, this transition does not happen in the absence of an enzyme because of the high transition state barrier (thick solid line). The enzyme serves as an intermediate agent to eliminate the transition barrier (thin dotted line). The enzyme does not become part of the products, and it can be reused over and over to facilitate the chemical transformation. Because the product can only be generated from a bound enzyme–substrate complex, the analysis of the product formed can reveal the binding process between the substrate and the enzyme. Thus, enzyme kinetic studies can be thought of as a special case of a protein binding study that the bound complex is indirectly measured through the analysis of the rate of substrate turnover. The indirect measurement by the rate of substrate turnover has the advantage of signal amplification because the same enzymes are used repeatedly to carry many rounds of reactions. Thus, only a small amount of enzyme is required in enzyme kinetic studies to obtain the binding information between the substrate and the enzyme. In contrast, if the substrate–enzyme complex is analyzed with normal binding methods, a large amount of enzyme is required to obtain measurable signal because there is no signal amplification. Enzyme kinetic studies have two processes built into one system: binding and generation of product. Thus, the analysis of the assay system is more complicated than the simple binding assay system. Two parameters, the binding constant and the substrate turnover rate, are obtained in enzyme kinetic studies versus just one parameter in binding studies.

The processes of binding studies and enzyme kinetic studies are fundamentally different. The binding process is usually studied in equilibrium (at the end of chemical transformation when the energy minimum of the system containing protein and ligand is reached) while the enzyme-catalyzed reactions are studied in the kinetic mode (by studying the initial speed of chemical transformation driven by one-way downhill energy forces toward the energy minimum of the system). Another major difference between the two processes is that proteins participate in the chemical transformation and are part of the final product in binding studies while proteins are only involved as a catalyst but do not become part of the final product in kinetic studies. Enzyme kinetic studies do not directly measure the interaction between enzyme and substrate.

Instead, such interaction is indirectly inferred by analyzing the measured product formation rate at different substrate concentrations. Because enzyme kinetic studies deal with nonequilibrium situations with regard to the substrate and product, time-dependent measurement is an important part of enzymatic kinetic studies. In contrast, binding studies are usually time-independent after the equilibrium is reached.

3.2 PROTEIN BINDING STUDIES AT EQUILIBRIUM

3.2.1 Introduction to Equilibrium and Equilibrium Constant

Let's consider the binding between ligand L and protein E leading to a bound complex LE with one binding site on the protein as shown in Scheme 3.1:

$$\mathrm{L} + \mathrm{E} \rightleftarrows \mathrm{LE}$$

3.1

As discussed in Section 3.1, the chemical potentials of ligand, protein, and the bound complexes are

$$\mu_{\mathrm{L}} = \mu_{\mathrm{L}}^0 + RT\ln[\mathrm{L}] \tag{3.2}$$

$$\mu_{\mathrm{E}} = \mu_{\mathrm{E}}^0 + RT\ln[\mathrm{E}] \tag{3.3}$$

$$\mu_{\mathrm{LE}} = \mu_{\mathrm{LE}}^0 + RT\ln[\mathrm{LE}] \tag{3.4}$$

The free energy difference between the reactants and the product is

$$\Delta G = \mu_{\mathrm{LE}} - (\mu_{\mathrm{L}} + \mu_{\mathrm{E}}) \tag{3.5}$$

Since there is no barrier to prevent the conversion between the product and the reactants, the chemical reaction will go either way if the chemical potential of either side is higher than the other. The chemical reaction will consume the substance resulting in lower concentration and hence lower chemical potential. Finally, the system reaches a state that the free energy difference ΔG is zero and equilibrium is achieved (Fig. 3.1*a*). By substituting Eqs. (3.2), (3.3), and (3.4) into Eq. (3.5) and making $\Delta G = 0$, the following equation is obtained:

$$\Delta G^0 = \Delta H^0 - T\,\Delta S^0 = \mu_{\mathrm{LE}}^0 - (\mu_{\mathrm{L}}^0 + \mu_{\mathrm{E}}^0) = -RT\ln K_a \tag{3.6}$$

where G is the Gibbs function (or free energy), H is enthalpy, S is entropy, and K_a is the association constant that equals to

$$K_a = \frac{1}{K_d} = \frac{[\mathrm{LE}]}{[\mathrm{L}]\cdot[\mathrm{E}]} \tag{3.7}$$

In practice, dissociation constant, $K_d = 1/K_a$, are often used as well. Both of them are called equilibrium constants. Because association (binding) reduces the entropy,

Eq. (3.6) shows that higher temperature favors the dissociation of the binding complex because of the increase in the value of $T\,\Delta S^0$. After the association constant or the dissociation constant is obtained experimentally, the free energy ΔG^0 can be calculated. Further, by measuring ΔG^0 at different temperatures, enthalpy (ΔH^0) and entropy (ΔS^0) can be obtained, which offers insightful information regarding the formation of the bound complex. The central theme in studying the binding between a protein and a ligand at equilibrium is to determine the association or dissociation constant.

3.2.2 Determination of the K_a or K_d Value

From Eq. (3.7), the K_d value can be calculated if the concentration of the ligand, protein, and the bound complex can be measured at equilibrium. However, the concentrations of these species at equilibrium are not readily determinable unless in rare situations where each of them gives a distinct signal that can be detected without disrupting the equilibrium. Most studies to determine K_d are based on the measurement of the concentration of the bound complex together with the known initial concentrations of ligand ($[\mathrm{L}]_0$) and protein ($[\mathrm{E}]_0$) that were placed in the system. The concentrations of the ligand and the protein at equilibrium can be calculated from their initial concentrations and the concentration of bound complex based on mass transformation:

$$[\mathrm{E}] + [\mathrm{LE}] = [\mathrm{E}]_0 \tag{3.8}$$

$$[\mathrm{L}] + [\mathrm{LE}] = [\mathrm{L}]_0 \tag{3.9}$$

Combining Eqs. (3.7), (3.8), and (3.9), we have

$$K_d = \frac{([\mathrm{L}]_0 - [\mathrm{LE}])([\mathrm{E}]_0 - [\mathrm{LE}])}{[\mathrm{LE}]} \tag{3.10}$$

At any given initial concentrations of ligand and protein, a discrete K_d value can be calculated if the bound complex concentration at equilibrium is determined. If we directly use Eq. (3.10) to calculate K_d, a collection of scattered K_d values will be obtained at different initial concentrations of ligand and protein. Some of the calculated K_d value will not be accurate because of the error associated with the measurement and the poor choice of the initial concentration of ligand and protein. Thus, it is difficult to extract the K_d value with precision. Scientifically, Eq. (3.10) must be transformed to an equation so that K_d can be expressed as a parameter in a function that correlates the independent variables and the output. Such treatment enables all sets of the independent variables and the corresponding output to be used to obtain a single best-fit parameter of K_d. Solving Eq. (3.10) leads to

$$[\mathrm{LE}] = \frac{([\mathrm{L}]_0 + [\mathrm{E}]_0 + K_d) - \{([\mathrm{L}]_0 + [\mathrm{E}]_0 + K_d)^2 - 4[\mathrm{L}]_0[\mathrm{E}]_0\}^{1/2}}{2} \tag{3.11}$$

Thus, the K_d value can be obtained by fitting the data into Eq. (3.11) with the measured bound complex concentration at different initial concentrations of ligand and protein.

This equation is universally applicable to all the protein and ligand binding studies. However, this equation is too complicated, and it may not be necessary to use this equation in most situations. As will be discussed next, much simpler equations can be derived when the experiment is designed so that either the ligand or the protein concentration is set at far below K_d. However, the simplified situations are only applicable when K_d is larger than 10 nM. When K_d is very low (less than 1 nM) in tight binding situations, the complicated Eq. (3.11) has to be used because it is difficult to carry the experiment with ligand or protein concentrations much lower than 1 nM.

In practice, the equilibrium constant is obtained using the following simplified approximation method. From Eqs. (3.7), (3.8), and (3.9), we can obtain

$$[\mathrm{LE}] = \frac{[\mathrm{E}]_0[\mathrm{L}]}{K_d + [\mathrm{L}]} = \frac{[\mathrm{L}]_0[\mathrm{E}]}{K_d + [\mathrm{E}]} \tag{3.12}$$

These are the most important relationships in binding studies to derive the following simplified approximation equations. The K_d value can be determined with this equation if the concentration of the bound complex and the ligand or the protein at equilibrium can be measured. In practice, the experiments can be designed to set the starting protein concentration much lower than both the K_d and the starting ligand concentration. Thus, the ligand concentration barely changes when the bound complex is formed. This leads to the fact that the ligand concentration at equilibrium is approximately equal to its initial concentration. Because of this approximation and other deviations from the theoretically correct Eq. (3.12), EC_{50} is used to substitute K_d in Eq. (3.12) and thus we have

$$[\mathrm{LE}] = \frac{[\mathrm{E}]_0[\mathrm{L}]_0}{\mathrm{EC}_{50} + [\mathrm{L}]_0} \tag{3.13}$$

The EC_{50} value is the ligand concentration at which half of the maximum concentration of the bound complex is formed. In ideal situation, EC_{50} is approximately equal to K_d. To use this equation in practice, a series of different concentrations of ligand is mixed with a fixed concentration of protein (at much lower concentration compared with ligand concentration and K_d). The concentration of the bound complex [LE] is measured experimentally. A plot of [LE] vs. $[L]_0$ is generated. EC_{50} is then obtained from fitting the data to Eq. (3.13). When designing experiments based on Eq. (3.13), care must be taken to make sure that the equation is only valid when the boundary condition is met, that is, concentrations of the ligand must be much higher than the protein. Further, this boundary condition implies that the protein concentration must be much lower than the EC_{50} value.

In some situations the ligand must be set at lower concentration than the protein concentration because of experimental constraints. In these situations, Eq. (3.12) can be rewritten as

$$[\mathrm{LE}] = \frac{[\mathrm{L}]_0[\mathrm{E}]_0}{\mathrm{EC}_{50} + [\mathrm{E}]_0} \tag{3.14}$$

When using Eq. (3.14), a series of concentrations of bound complex is measured at different concentrations of protein. This equation is only valid when the boundary condition is met, that is, concentrations of the protein must be much higher than the ligand. Further, this boundary condition implies that the ligand concentration must be much lower than the EC_{50} value. Studying binding between protein and labeled ligand with the fluorescence polarization assay is a good example for this situation, which will be discussed in Chapter 5.

All the discussions above are with proteins having only one binding site for the ligand. In many situations, a protein may possess more than one binding sites for a ligand. The binding sites can be independent of each other or cooperative. A more generalized equation can be used to empirically describe the situation:

$$[\mathrm{LE}] = \sum_i \frac{[\mathrm{E}]_0[\mathrm{L}]_0^n}{\mathrm{EC}_{50}^i + [\mathrm{L}]_0^n} \tag{3.15}$$

where i is the number of independent binding sites and n is number of cooperative binding sites. This model is used to fit up to two independent binding sites (i is equal or less than 2). Increasing the binding sites will fit the data better but may not be very meaningful. In contrast, the value of n can be any number if the system it describes exists. For example, the hemoglobin has four cooperative binding sites for oxygen but has no other independent binding sites. After measuring the bound oxygen–hemoglobin complex at varying oxygen concentrations, the experimental data can be fitted to Eq. (3.15), which gives n a value of 3.8, indicating four cooperative binding sites on the hemoglobin for oxygen. Equation (3.15) can be used to describe any form of the protein binding process. When studying the binding between agonists and receptors, Eq. (3.15) is called the Hill equation where the term n is referred to as the Hill coefficient.

Equation (3.15) can be graphically illustrated in semilog plot as shown in Figure 3.2. Here the concentration of ligand is expressed using the EC_{50} value as the unit. The curves with a higher Hill coefficient (n) value give a steeper slope. The lines at 90% binding and 10% binding are drawn that intercept with each of the three curves to obtain the EC_{90} values and EC_{10} values in the three situations. The ratio of EC_{90} to EC_{10} is used here to describe the span of the ligand concentrations that cause the change from 10% binding to 90% binding. The results are summarized in Table 3.1. To increase the binding from 10 to 90%, 81-fold changes in the ligand concentration is required in the model with one binding site, 9-fold changes in the ligand concentration is required in the model with two binding sites, and only 3-fold changes in the concentration is required in the model with four binding sites. By visually inspecting the concentration–response graph, one can estimate the number of cooperative ligand binding sites on the protein.

The method described above to obtain the binding constant value in a binding process is based on chemical measurement of the concentration of the bound complex. In addition to the chemical measurement method, the binding constant can also be obtained with physical measurement of the thermodynamic of the whole system, that is, the free energy changes (ΔG^0). The K_d value can then be obtained based on

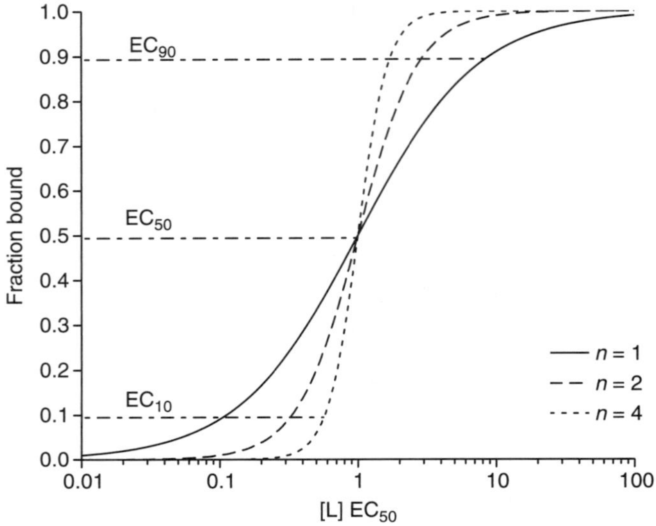

Figure 3.2 Graphical illustration of Hill equations in semilog plot. Here the concentration of ligand is expressed using EC_{50} value as the unit. The lines at 90% binding and 10% binding is drawn that intercept with each of the three curves to obtain the EC_{90} values and EC_{10} values in the three situations. To increase the binding from 10 to 90%, 81-fold changes in the ligand concentration is required for one binding site model and 9-fold changes in the ligand concentration is required for two binding site model and only 3-fold changes in the concentration is required for four binding site model.

TABLE 3.1 Span between EC_{90} and EC_{10} When Hill Coefficient *n* is Varied[a]

n	EC_{90}	EC_{10}	EC_{90}/EC_{10}
1	8.97	0.11	81
2	3.00	0.33	9.1
4	1.73	0.57	3.0

[a] EC_{90} and EC_{10} values expressed in EC_{50} as the units.

the relationship between K_d and ΔG^0 as shown in Eq. (3.6). Further manipulation of this equation leads to two physical measurement methods to obtain the binding constant, isothermal titration calorimetry and differential scanning calorimery. These methods measure the global thermal property of the system containing the binding partners. The disadvantage of these methods is that the binding system used in the measurement must be very pure because all binding processes, both specific binding and nonspecific binding, will contribute to the measured thermodynamic changes. These methods are very slow compared with chemical measurements and are not scalable. Since these methods are mostly used by biophysicists and rarely used by biologists, we will not discuss the details about these techniques here.

3.3 KINETIC STUDIES OF THE PROTEIN BINDING PROCESS

3.3.1 Introduction to Chemical Kinetics

So far we discussed ligand and protein binding in the equilibrium state. We will now discuss the kinetic process of reaching the equilibrium state from initial nonequilibrium state. The study of the rate at which a chemical reaction proceeds is called *chemical kinetics*. The rate of a chemical reaction is defined as the rate of the product formation. Here we consider the chemical reaction (shown in Scheme 3.2) with a single reactant (S) converting to a product (P).

$$S \xrightarrow{k} P$$

3.2

The *reaction rate* is defined as

$$\frac{d[\mathrm{P}]}{dt} = -\frac{d[\mathrm{S}]}{dt} = -k[\mathrm{S}]^n \tag{3.16}$$

where k is the *rate constant* and n is the *reaction order*. When $n = 0$, the reaction is zero order and the rate of reaction is constant at any time in the reaction process. When $n = 1$, the reaction is first-order and the rate of reaction is proportional to the reactant's concentration. This is the most encountered case. Most fundamental principles of chemical reaction and enzyme kinetics are derived from first-order reactions. When $n = 2$, the reaction is second order and the rate of reaction is proportional to the square of the concentration of the reactants. Reaction order higher than 2 can happen but is rare. Chemical reactions may involve two different reactants. If the rate of the chemical reaction involving two different reactants is proportional to the product of the concentration of both reactants, it is a second-order reaction too. Experimentally, the concentration of one reactant is often made much higher than the other so that the kinetics of the reaction resemble a first-order reaction. This situation is called a pseudo-first-order reaction and many equations obtained from first-order kinetic analysis are applicable in this case.

Reaction Progressive Curve and Initial Velocity Reaction progressive curve is a plot with the total product generated (or the product concentration when the reaction volume is fixed) at a given time as the dependent variable and the time as the independent variable. Figure 3.3 shows the reaction progressive curve for the first-order reaction shown in Scheme 3.2. The rate of the reaction is

$$\frac{d[\mathrm{P}]}{dt} = -\frac{d[\mathrm{S}]}{dt} = -k[\mathrm{S}] \tag{3.17}$$

The rate of the first-order reaction is proportional to the concentration of the reactant [S]. When the reaction proceeds, the concentration of the reactant decreases due to the depletion of the reactant. Hence the rate of the reaction decreases. The rate constant can be calculated when the rate of reaction and the concentration of the reactant is

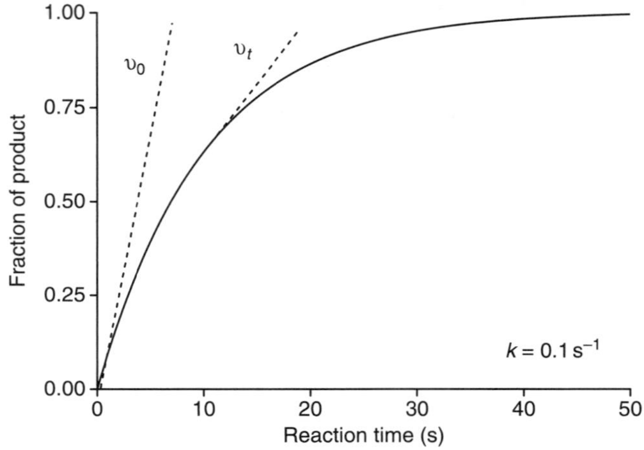

Figure 3.3 First-order reaction progressive curve. The rate constant is set at 0.1 per second. The first-order reaction contains an exponential function. The rate of reaction is the highest at initial velocity (v_0) and gradually decreases over time (v_t). The half-life of the reaction is equal to ln $2/k$ that is 6.9 s in this case.

determined. Because the initial concentration of the reactant is known, the rate constant in a reaction can be obtained by simply measuring the initial rate of the reaction (or *initial velocity*) without the need to measure the concentration of the reactant. The initial rate of a reaction can be obtained by measuring the whole reaction progressive curve and then fitting the data into a known equation. In practice, scientists often do not measure the whole reaction progressive curve, which consumes both time and resources. Instead, a reaction rate at less than 10% substrate turnover is measured, which is approximately equal to the theoretical initial velocity for first-order reactions.

By integrating Eq. (3.17), we obtain

$$[\mathrm{P}] = [\mathrm{S}]_0(1 - e^{-kt}) \tag{3.18}$$

where $[\mathrm{S}]_0$ is the initial concentration of the reactant. When the unit of time is in seconds, the rate constant unit is s^{-1}. Figure 3.3 shows a first-order reaction progressive curve plotted using Eq. (3.18). The rate constant is set at 0.1 per second in this plot. The rate of the reaction is the highest at initial velocity (v_0) and it gradually decreases over time (v_t). The half-life of a first-order reaction is equal to ln $2/k$, which is 6.9 s when the rate constant is 0.1 s^{-1}.

The reaction progressive curve is a very important plot from which the reaction rate at any time in the reaction can be obtained by taking the first derivatives. By definition, the dependent variable in this plot should be the product concentration. In practice, it is fine to use something that is proportional to the product concentration as the dependent variable to extract kinetic parameters. However, novice scientists sometimes mistakenly plot the reaction progressive curve by directly using whatever readout from an instrument that is not proportional to the product concentration. A quick search of articles published in scientific journals reveals that authors often erroneously plot the measured fluorescence polarization values as an independent variable that is not proportional to the concentration of a reaction product. The fluorescence polarization values should be converted to something proportional to the concentration of the products (such as anisotropy) to be meaningful as the dependent variable in a reaction progressive curve (see Chapter 2 for details).

Relationship Between Equilibrium Constant and Kinetic Parameters The same equilibrium scheme outlined in Scheme 3.1 is considered here in terms of kinetics:

$$\mathrm{L + E} \underset{k_{-1}}{\overset{k_1}{\rightleftarrows}} \mathrm{LE}$$

3.3

where k_1 is the rate constant of forward reaction and k_{-1} is the rate constant of backward reaction. The forward and backward reaction rate in the above reaction is

$$\frac{d[\mathrm{LE}]}{dt} = k_1[\mathrm{L}][\mathrm{E}] \tag{3.19}$$

and

$$-\frac{d[\mathrm{LE}]}{dt} = k_{-1}[\mathrm{LE}] \tag{3.20}$$

At equilibrium, the forward rate and backward rate of the reaction are equal. Thus,

$$k_1[\mathrm{L}][\mathrm{E}] = k_{-1}[\mathrm{LE}] \tag{3.21}$$

This leads to

$$\frac{[\mathrm{L}][\mathrm{E}]}{[\mathrm{LE}]} = \frac{k_{-1}}{k_1} = K_d \tag{3.22}$$

This equation establishes the relationship between the equilibrium constant (K_d) and the kinetic constants (k_1 and k_{-1}). This relationship can help obtain the kinetic constants in situations when one of the kinetic constants cannot be directly determined experimentally. For example, k_1 cannot be directly measured if its value is higher than the diffusion limits ($>10^8\,\mathrm{M}^{-1}\,\mathrm{s}^{-1}$), and k_{-1} cannot be accurately measured directly if it is too slow (slower than the time period in which the instruments can maintain its stability). In these situations, one of the kinetic constants that can be experimentally measured is obtained. The equilibrium constant is obtained from a separate experiment. The other kinetic constant that cannot be directly measured experimentally is then calculated using Eq. (3.22).

3.3.2 Determination of Kinetic Parameters in the Protein Binding Process

Quenched-Flow and Stopped-Flow Apparatuses The time it takes to reach equilibrium from an initially nonequilibrium system depends on kinetic parameters.

In order to obtain the kinetic parameters of a binding process that leads to equilibrium, the temporal resolution of the measurement must be faster than the time it takes for the system to reach equilibrium. Stopped-flow and quenched-flow techniques (see Fig. 3.4) are commonly used to study binding kinetics with temporal resolution at millisecond time scales. These apparatuses use pneumatic pressure to drive the push bars that in turn drive the plungers of syringes containing solutions to be mixed. The whole system is airtight and it can sustain high pressures up to a few thousands of pounds per square inch (psi). The pressure is exerted throughout the system by

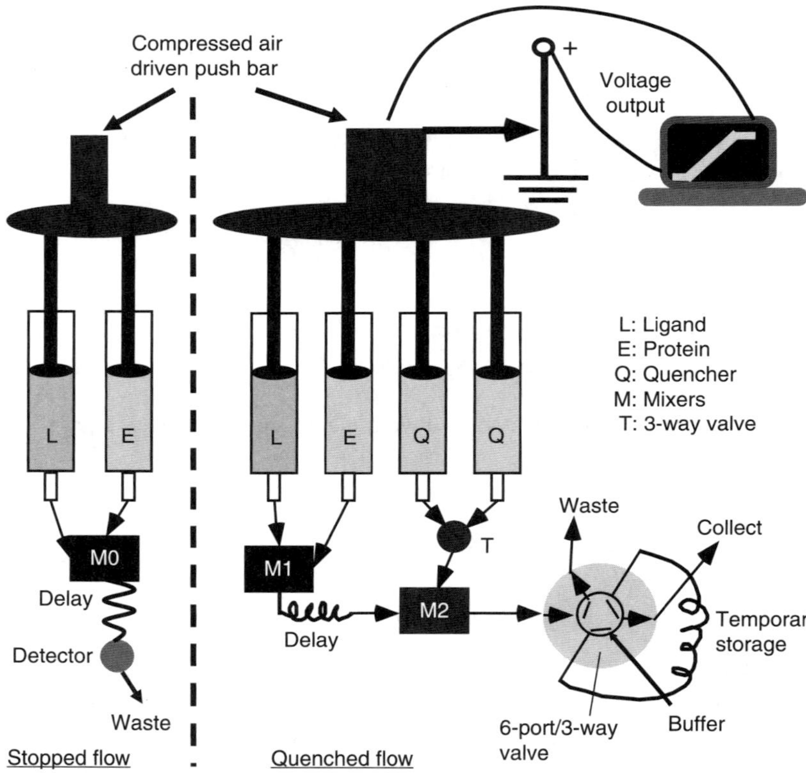

Figure 3.4 Stopped-flow and quenched-flow apparatuses. Pneumatic pressure is used to drive the push bars that in turn drive syringes containing solutions to be mixed. The solutions are first pushed to mixers that mix the solution together. In the stopped-flow apparatus, online detector is placed downstream of the mixer to monitor the reaction on the fly. In quenched-flow apparatus, the reaction in the mixed solutions is stopped when they move to a second mixer where they mix with a reagent that freezes the reaction. The quenched mixtures move downstream to a 6-port/3-way valve that directs the mixtures going to either waste or to a storage loop that is then pushed to collection vials for further analysis offline. The push bar movement is monitored in real time electronically that is used to calculate the reaction time. The time it takes to move the liquid from the mixer to the detector (or the quenching mixer) is the reaction time. The length of the delay loop is used to control the reaction time.

the fact that aqueous solution is relatively incompressible. Compressible air bubbles in the system will severely compromise the apparatus's performance. Air bubbles are usually introduced to the system when loading solutions into the syringes. To prevent this from happening, the solutions should be degassed before they are loaded into the syringes, and the syringes should be oriented so that the solutions are pushed upward to aid in the removal of air trapped in the system (opposite to the syringe orientation showed in Fig. 3.4). The solutions are pushed to the mixers that mix the solution together. In the stopped-flow apparatus, an online detector is placed downstream of the mixer to monitor the reaction on the fly. In quenched-flow apparatus, the reaction that occurs in the mixed solutions is stopped when they flow to a second mixer where they mix with a quenching reagent. The quenched mixtures further flow downstream to a 6-port/3-way valve that directs the solution going to either waste or to a temporary storage loop. The solution in the temporary storage loop is then pushed into collection vials for further analysis offline. The speed of the push bar movement is monitored in real time electronically to calculate the reaction time. With stopped-flow apparatus, the reaction time is the time it takes for the solution moving from the mixer to detector. With quenched-flow apparatus, the reaction time is the time it takes for the solution moving from the first mixer to the second mixer. The limit on how fast of a reaction can be studied depends on the maximum pressure the system can withstand and the dead volume of the mixer. Typically, reaction speed the in millisecond scale can be studied. Because the time it takes to move the solution from mixer to the detector (or the quenching mixer) is the reaction time, the length of the delay loop between the two is used to control the reaction time. The stopped-flow method has the advantage of on-the-fly monitoring of the undisturbed reaction mixtures. The disadvantage of stopped-flow method is that only a limited number of systems can be analyzed on the fly (such as UV, fluorescence, circular dichroism). The advantage of the quenched-flow method is that it can analyze many different systems because of flexibility of analyzing samples offline. The disadvantages of this method are that the quenching process may disturb the system and the quenched reaction mixtures may change over time before the offline analysis.

Calculating Kinetic Parameters in Protein Binding Studies The binding process described in Scheme 3.3 is used here to illustrate how to obtain the kinetic parameters. The binding between protein and ligands involves two species and is a second-order reaction. To simplify the analysis, either ligand or protein should be made in large excess compared with their counterpart so that the reaction behaves as a pseudo-first-order reaction. In the majority of binding studies, the ligand is made in large excess and its concentration can be considered as constant. Thus, the reaction rate is

$$\frac{d[\mathrm{LE}]}{dt} = k_1[\mathrm{L}][\mathrm{E}] - k_{-1}[\mathrm{LE}] = k_1[\mathrm{L}]_0[\mathrm{E}]_0 - (k_1[\mathrm{L}] + k_{-1})[\mathrm{LE}] \tag{3.23}$$

Integrating Eq. (3.23) and applying the boundary conditions will lead to the following apparent first-order kinetic equation [compare with Eq. (3.18)]:

$$[\mathrm{LE}]_t = [\mathrm{LE}]_\infty(1 - e^{-k_{\mathrm{obs}}t}) \tag{3.24}$$

with

$$k_{\text{obs}} = k_1[\text{L}]_0 + k_{-1} \tag{3.25}$$

where $[\text{LE}]_t$ is the concentration of the bound complex at time t, $[\text{LE}]_\infty$ is the concentration of the bound complex at infinite time (when equilibrium is reached), k_{obs} is the apparent (or observed) first-order rate constant, and $[\text{L}]_0$ is the ligand concentration, which does not change during the reaction because it is in large excess. The half-life for the formation of the bound complex is $\ln 2/k_{\text{obs}}$. The k_{obs} can be obtained experimentally by measuring the reaction progressive curve and then fitting the data into Eq. (3.24).

Because of the symmetry between E and L in Scheme 3.3, the same equation [Eq. (3.24)] can be used in the situation when initial concentration of the protein is much higher than the ligand. However, k_{obs} will take the following form in this situation:

$$k_{\text{obs}} = k_1[\text{E}]_0 + k_{-1} \tag{3.26}$$

where $[\text{E}]_0$ is the protein concentration that does not change during the reaction because it is in large excess.

From Eqs. (3.24), (3.25), and (3.26), it is clear that the forward reaction kinetic parameter of the binding cannot be obtained by studying the association kinetics alone because k_{obs} is composed of both k_1 and k_{-1}. In practice, k_{-1} is obtained first by studying the dissociation of the bound complex that follows the first-order decay process. After k_{-1} is obtained, k_1 can be extracted from k_{obs} using Eq. (3.25) or (3.26). In addition to obtaining the k_{obs} to aid calculating kinetic parameters, Eq. (3.24) is very important in practice to determine the time for a system to reach equilibrium. Because true 100% equilibrium will never be reached in practice, deciding when the system has reached the practical equilibrium state is important. By convention, if the concentration of the bound complex is less than 3% of the expected concentration of the bound complex at true equilibrium, it is considered that equilibrium has been reached. For first-order reaction whose kinetics can be described by Eq. (3.24), practical equilibrium is considered reached if the incubation proceeds for 5 half-lives ($5 \times \ln 2/k_{\text{obs}} = 3.5/k_{\text{obs}}$).

In addition to using the kinetic studies alone to determine k_1 and k_{-1}, the kinetic parameters can also be determined from the measured k_{obs} value obtained in kinetic studies and the measured equilibrium constant (K_d) obtained in equilibrium studies. An example of this method is illustrated with a stopped-flow fluorescence polarization experiment in Chapter 5. This method of using both kinetic and equilibrium experiments to obtain a kinetic rate constant is very important in extreme situations when k_{-1} is too slow to prevent reliable measurement due to instability of the measurement devices over a long period of time. In this case, k_1 is obtained first. Because k_{-1} is too small, the k_{-1} term is eliminated from Eqs. (3.24) and (3.26); k_1 can be obtained from $k_{\text{obs}}/[\text{E}]_0$. The value of k_{-1} can then be obtained from k_1 and K_d. When k_1 is too high to exceed the diffusion limits, k_{obs} cannot be experimentally obtained. In this situation, k_{-1} can be obtained by studying the dissociation of the bound complex, which will follow the first-order dissociation process. The value of k_1 can then be obtained from k_{-1} and K_d.

3.4 ENZYME KINETICS

3.4.1 Enzymes and Enzyme Classification

Enzymes are proteins (except ribozymes) that catalyze a chemical reaction by lowering the activation energy of the transition state of the reaction intermediate. Enzymes are the most abundant proteins in an organism and carry many crucial functions. Inside the live cells, a tremendous amount of biochemical reactions are taking place constantly in the process of metabolism. All these chemical changes are made possible by enzymes. There are constant physical and chemical changes in live organisms as well, such as building of new tissue, conversion of food into energy, disposal of waste materials, and reproduction. Again, enzymes play an important role in these processes. The existence of enzymes has been known for over a century. Urease is the first enzyme obtained in pure form from jack bean in 1926 through isolation and crystallization processes. The use of enzymes in bioassays can be traced back to the 1940s when enzymes emerged as the essential tool in the diagnosis of diseases.

While some enzymes can catalyze the chemical reaction alone, other enzymes require the presence of nonprotein cofactors to be able to catalyze chemical reactions. The protein portion of the enzyme is called apoenzyme, and the entire functional enzyme with apoenzyme and cofactors is called holoenzyme. The cofactors can be divided into three types: a coenzyme, a prosthetic group, and metal ions. A coenzyme is an organic molecule loosely attached to the apoenzyme and is dialyzable. Nicotinamide adenine dinucleotide (NAD^+) and coenzyme A are good examples of coenzymes. A prosthetic group is an organic molecule firmly attached to the apoenzyme. Prosthetic groups are often vitamins or made from vitamins and are usually involved in the active site of enzymes. The common metal ions required for proper enzyme functions are Mg^{2+}, Mn^{2+}, Ca^{2+}, Zn^{2+}, Co^{2+}, Cu^{2+}, Mo^{3+}, Fe^{2+}, and Fe^{3+}.

Because of the large number of enzymes and the diverse functions they perform, it is important to classify all the enzymes into groups to facilitate their study. The most recognized enzyme classification and nomenclature scheme is the system proposed by the Enzyme Commission (EC) of the International Union of Biochemistry and Molecular Biology (IUBMB). This system recommends that enzyme names indicate both the substrate they act upon and the type of reaction they catalyze. For example, acetylcholinesterase is the enzyme with acetylcholine as the substrate, and the type of reaction it catalyzes is hydrolyzing the ester bond. The enzymes are classified according to the chemical reactions they catalyze in the EC system. Thus, all enzymes are classified into six top classes based on the following six types of the reaction: oxidation–reduction reactions, transfer a functional group, hydrolysis of various bonds, cleavage of various bonds by means other than hydrolysis and oxidation, isomerization, and joining two molecules covalently. The six top classes of enzyme names start with "EC" and are followed by numbers (see Table 3.2). The published information on all enzymes that have been classified in the EC scheme are collected in BRENDA

TABLE 3.2 Top-level Enzyme Classification in EC System

EC1	*Oxidoreductases*: catalyze oxidation/reduction reactions
EC2	*Transferases*: transfer a functional group
EC3	*Hydrolases*: catalyze the hydrolysis of various bonds
EC4	*Lyases*: cleave various bonds by means other than hydrolysis and oxidation
EC5	*Isomerases*: catalyze isomerization changes within a single molecule
EC6	*Ligases*: join two molecules with covalent bonds

database with the range of data that includes the catalyzed reaction, detailed description of the substrate, cofactor and inhibitor specificity, kinetic data, structure properties, information on purification and crystallization, properties of mutant enzymes, participation in diseases and amino acid sequences. The BRENDA database contains more than 1.3 million manually annotated data with on average 300 single entries per EC number. Enzymes from 7500 different organisms are covered with about 170,000 single data for human enzymes. In addition, the *Springer Handbook of Enzymes* series contains several dozen books dealing with most of the enzymes. Because of the large number of enzyme groups, it is impossible to discuss each of them in this book. Thus, we will only discuss assays for proteases that belong to EC3.4 subclass and assays for protein kinases that belong to EC2.7 (EC2.7.10 and EC2.7.11 to be specific) subclass to show the general assay approaches to enzymes.

3.4.2 Free Energy and Enzyme-Catalyzed Reactions

For any chemical transition from reactant(s) to product(s) to occur, it must follow the law of physics that the free energy always flows from a high state to a low state if there is no barrier to prevent the conversion. This is the case for the cleavage of a protein's peptide bond to form two peptides in the presence of a protease (Fig. 3.5*a*). However, proteins do not normally break down into pieces, though this is a thermally favored process. There is a transition barrier to prevent the automatic breakdown of proteins from happening. The presence of protease eliminates this transition barrier and enables the breakdown of peptides. When the reactant(s) has lower energy than the product(s), transition from the reactant(s) to the product(s) cannot proceed but the reverse will happen. This is the case that phosphoprotein cannot be made by simply mixing a protein and phosphates even in the presence of an enzyme. On the other hand, the reverse will happen that a phosphoprotein can be cleaved to make protein and phosphate in the presence of phosphatase (Fig. 3.5*b*). The addition of ATP to the protein gives enough energy to the reactant(s), which enables the conversion from protein to phosphoprotein in the presence of kinase. In the absence of the enzymes, none of the above conversion can happen because of the presence of a transition barrier. The enzymes simply eliminate the transition barrier for the conversion but do not change the energy states of the reactants nor the products. Because enzyme-catalyzed biochemical reactions are one-way reactions following the direction of the energy downhill, they are studied by the kinetic method. Though enzymes can also be treated as normal protein interaction with ligand

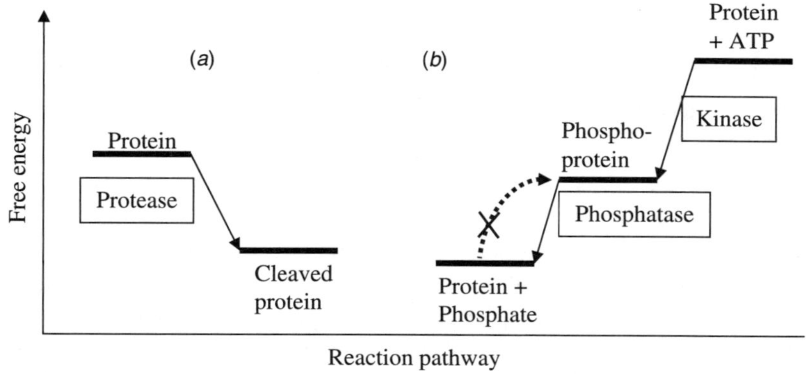

Figure 3.5 Enzyme-catalyzed reactions follow one-way downhill pathway to lower energy state. (*a*) Peptide bond cleavage is an energetically favored reaction. However the reaction can only proceed in the presence of protease due to transition barrier. (*b*) Protein and phosphate together cannot make phosphoprotein even in the presence of enzymes because it is an energetically unfavored reaction. However, the addition of energy input from ATP and in the presence of kinase to eliminate the transition barrier can make phosphoprotein. Breaking down phosphoprotein to protein and phosphate is energetically favored that can happen in the presence of phosphatase.

(commonly referred to as "substrate" in kinetic analysis) in equilibrium, enzyme kinetic study has more advantages and is the predominant method in studying enzymes. Since the first step in the interaction involves transiently formed bound complex between enzymes and substrate, some equations obtained in enzyme kinetics resembles the equation obtained in equilibrium studies (see the rapid equilibrium model of enzyme kinetics below).

3.4.3 Modeling Enzyme-Catalyzed Reactions

Experimental evidence suggested that enzyme (E) and substrate (S) first form a complex (ES) before the product (P) is formed in enzyme-catalyzed reactions. This can be described in following scheme:

$$\mathrm{S} + \mathrm{E} \underset{k_{-1}}{\overset{k_1}{\rightleftharpoons}} \mathrm{ES} \xrightarrow{k_2} \mathrm{E} + \mathrm{P}$$

3.4

The major objectives in enzyme kinetic studies are to obtain the kinetic rate constant (k_2) and the equilibrium rate constant between substrate and enzyme ($K_d = k_{-1}/k_1$), which is accomplished through the study of the rate of the reaction. As we learned from the chemical kinetics (see Section 3.3.1), the rate of the one-way product formation is proportional to the concentration of the enzyme–substrate complex:

$$v = k_2[\mathrm{ES}] \qquad (3.27)$$

The maximum velocity achievable is when all the available enzymes are bound to the substrate when the substrate concentration is very high compared with its association constant with the enzyme. Under this condition, the bound complex concentration is equal to the initial enzyme concentration ($[E]_0$) and the reaction will proceed at the maximum velocity:

$$v_{max} = k_2[E]_0 \tag{3.28}$$

Since $[E]_0$ is known, the first-order rate constant k_2 can be obtained by measuring v_{max} under this experimental condition. When the substrate concentration is not high enough to occupy all the binding sites on the enzyme, the bound complex concentration will be a fraction of the initial enzyme concentration. The concentration of the bound complex can be obtained based on two models: the rapid equilibrium model and the steady-state model. Both models are based on the assumption that substrate concentration is in large excess compared with the enzyme concentration. This assumption is met in commonly encountered enzyme-catalyzed reactions.

Rapid Equilibrium State Model This model assumes that the equilibrium between the enzyme and the substrate are rapidly established with the rate to form the product from the bound complex is much slower than the rate of dissociation ($k_2 \ll k_{-1}$). In the initial velocity study (substrate turnover is less than 10%), the substrate depletion is negligible and thus its concentration remains the same as its initial concentration ($[S]_0$) because of its large excess. Thus, we can ignore the substrate turnover part in Scheme 3.4 and only focus on the equilibrium part. The concentration of the bound complex [ES] is determined by the equilibrium between the substrate and the enzyme. From Eq. (3.12), which we discussed previously in Section 3.2.2, the concentration of the bound enzyme and substrate complex can be expressed as

$$[ES] = \frac{[E]_0[S]_0}{K_d + [S]_0} \tag{3.29}$$

By combining Eqs. (3.27), (3.28), and (3.29), we obtain

$$v = k_2[ES] = \frac{k_2[E]_0[S]_0}{K_d + [S]_0} = v_{max}\frac{[S]_0}{K_d + [S]_0} \tag{3.30}$$

Thus, a series of experiments can be designed in which the concentration of the enzyme is fixed but the starting concentration of the substrate is different. The initial velocity of the product formation is measured at different initial concentrations of the substrate. The value of the kinetic parameter (k_2) and the equilibrium constant K_d can be obtained by fitting the experimental data to Eq. (3.30) if the rapid equilibrium model is a true description of the system.

Steady-State Model The rapid equilibrium model requires the condition of $k_2 \ll k_{-1}$, which cannot always be met. Thus, the applications of the equilibrium model are limited. When performing enzyme kinetic studies in some special situation,

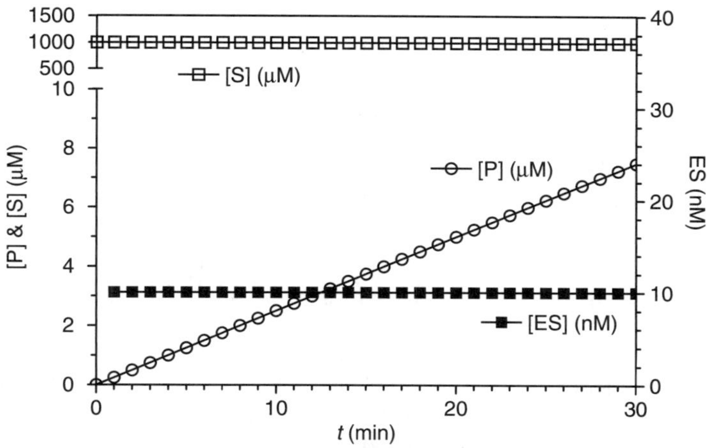

Figure 3.6 Illustration of steady-state model kinetics with a hypothetic experiment. Here the starting substrate concentration is 1 mM. The starting enzyme concentration is 10 nM. After mixing the enzyme and substrate, the steady state is already established at the first measurable point (1 min). The concentration of the product continued increasing linearly and at a rate of 0.25 μM/s and reaching 7.5 μM at 30 min. The starting concentration of substrate is 1 mM and remained relatively constant through the course of the initial velocity measurement because only 0.75% of the substrate is depleted at the end of the experiment. The bound enzyme–substrate complex remains 10 nM throughout the period.

the bound enzyme–substrate complexes can be observed. It was found that the concentration of the substrate–enzyme complex in many enzyme-catalyzed reactions remains constant throughout the initial velocity study period as illustrated in Figure 3.6. Based on these observations, a steady-state model was proposed. This model assumes that the concentration of the enzyme–substrate complex remains at constant concentration ($d[\mathrm{ES}]/dt = 0$) in enzyme-catalyzed reactions. Since the substrate concentration remains constant at $[\mathrm{S}]_0$ within the initial velocity studying period, we can obtain Eq. (3.31) based on the steady-state model:

$$\frac{d[\mathrm{ES}]}{dt} = k_1[\mathrm{S}]_0[\mathrm{E}] - (k_{-1}[\mathrm{ES}] + k_2[\mathrm{ES}]) = 0 \tag{3.31}$$

By rearranging Eq. (3.31), we obtain

$$[\mathrm{ES}] = \frac{k_1[\mathrm{E}][\mathrm{S}]_0}{k_2 + k_{-1}} \tag{3.32}$$

Substituting $[\mathrm{E}] = [\mathrm{E}]_0 - [\mathrm{ES}]$ into Eq. (3.32) leads to

$$[\mathrm{ES}] = \frac{[\mathrm{E}]_0[\mathrm{S}]_0}{[\mathrm{S}]_0 + \dfrac{k_2 + k_{-1}}{k_1}} = \frac{[\mathrm{E}]_0[\mathrm{S}]_0}{[\mathrm{S}]_0 + K_m} \tag{3.33}$$

where

$$K_m = \frac{k_2 + k_{-1}}{k_1} \tag{3.34}$$

and K_m is commonly referred to as the Michaelis constant. Thus, the initial velocity of an enzyme-catalyzed reaction based on the steady-state model is

$$v = k_2[\mathrm{ES}] = \frac{k_2[\mathrm{E}]_0[\mathrm{S}]_0}{K_m + [\mathrm{S}]_0} = v_{\max}\frac{[\mathrm{S}]_0}{K_m + [\mathrm{S}]_0} \tag{3.35}$$

This is the most important equation in studying enzyme kinetics. Comparison between the steady-state model [Eq. (3.35)] and the rapid equilibrium model [(Eq. 3.30)] reveals that the finial equations take the same form. If substituting the boundary condition in the rapid equilibrium model ($k_2 \ll k_{-1}$) to the steady-state model, K_m is reduced to K_d. Thus, the steady-state model is a more general model and the rapid equilibrium model is a special case of the steady-state model.

The above discussion was made based on a simplified model with only one binding site on the enzyme for the substrate (Scheme 3.4). The enzyme-catalyzed reaction can be much more complicated and may involve many steps to generate the final product after the initial binding. In addition, there may be more binding sites on the enzyme. Enzymologists extended the above model to describe more complicated enzyme-catalyzed reactions by dividing the process into two parts. The first part is the formation of the bound complex whose concentration can be described in a more generalized form as

$$[\mathrm{ES}] = \frac{[\mathrm{E}]_0[\mathrm{S}]_0^n}{K_m + [\mathrm{S}]_0^n} \tag{3.36}$$

where n is the Hill coefficient (or the number of binding sites). The second part is the transformation from bound complex to product. Realizing that many steps may happen during the process to form the final product, k_{cat} is used to substitute k_2 to account for the overall process. Thus, the general equation used to describe an enzyme-catalyzed reaction is

$$v = k_{\mathrm{cat}}[\mathrm{ES}] = \frac{k_{\mathrm{cat}}[\mathrm{E}]_0[\mathrm{S}]_0^n}{K_m + [\mathrm{S}]_0^n} = v_{\max}\frac{[\mathrm{S}]_0^n}{K_m + [\mathrm{S}]_0^n} \tag{3.37}$$

In enzyme kinetic studies, most enzyme-catalyzed reactions are described by this equation. When applying this equation, one must make sure that the measured velocity of reaction is the initial velocity because all the equations derived so far are based on this assumption. When more than one substrate is involved in the enzyme-catalyzed reaction, experiments can be designed to force the reaction into pseudo-first-order reactions by making all other substrates in large excess over one substrate. In this case, only one substrate is the focus of the studies and Eq. (3.37) can be applied. Further details on the treatment of enzymes with multiple substrates are beyond the scope of this book. Interested readers can study this subject with the reference books listed in the bibliography.

3.4.4 Utility of k_{cat} and K_m in Enzymology

As seen from the above discussion, k_{cat} is equal to $v_{max}/[E]_0$, that is, the rate of the reaction per unit of enzyme when all the substrate binding sites on the enzyme are occupied with substrate. Since v_{max} is usually expressed in the units of molar/minute and $[E]_0$ is in molar, this leads to k_{cat} with a unit of minute^{-1}, and it is referred to as the substrate turnover number per unit time. Thus, the k_{cat} value defines how efficient an enzyme can catalyze the conversion of a substrate into a product, which is not affected by the substrate's ability to bind to the enzyme. The value of K_m on the other hand is a measure of how well a substrate can bind to an enzyme. A substrate having a low K_m is a good binder to the enzyme, but it may not be a good substrate if the k_{cat} is too low. On the other hand, a substrate having a high k_{cat} value may not be a good enzyme substrate either if it is a poor binder (with a high K_m) for the enzyme. Therefore, the ratio of k_{cat}/K_m is often used to evaluate the substrate–enzyme pairs. The ratio is often used in optimizing the substrate for a given enzyme, in optimizing the recombinant enzyme in search for an enzyme to catalyze a specific reaction and in enzyme purification.

3.5 INHIBITION OF PROTEIN FUNCTION

In drug discovery, finding a molecule that interacts with a target protein is the first step leading to a drug. Ideally, scientists would prefer to have a universal technology that can directly measure the interaction between any molecule and a target protein. However, this is not achievable in practice with reasonable throughput because of technical limitations. Instead, scientists commonly rely on measuring the effect of the test molecules on either the binding between a target protein and a tracer known to bind to the target protein or the kinetics of enzyme-catalyzed substrate conversion.

In protein binding studies, the assay test system contains the test protein and a known labeled ligand (tracer) with measurable properties, such as radioactivity and fluorescence. The effect of the test molecules on the interaction between the target protein and the tracer is measured. This assay method is commonly referred to as competitive binding assays. Because the interaction between the test molecules and the target protein is only measurable when the test molecules bind at the same site as the tracer's binding site on the protein, competitive binding assays have serious limitation. Interactions between the test molecules and the target protein occurred outside of the tracer's binding sites are undetectable in competitive binding assays.

In enzymatic functional assays, the test system contains the target enzyme and its substrates. The baseline kinetic parameter in the absence of the test molecule is first established. The interaction between the test molecules and the enzyme can be measured by studying the changes in the kinetic parameters in the presence of the test molecules. Because enzymes are proteins too, competitive binding studies can also be employed to study the interaction between the test molecules and the enzyme.

However, the functional studies of enzyme inhibition offer many advantages in studying the test molecule's interaction with target protein.

3.5.1 Inhibition of Protein Function in Competitive Binding Studies

A potential protein inhibitor can be directly studied by accessing its ability to bind to the target protein. In direct binding studies between small-molecule inhibitor and target protein, the binding between the inhibitor and the target protein can be treated the same way as we discussed in ligand–protein binding section (see Section 3.2). The inhibitor's dissociation constant is defined as

$$K_i = \frac{[\mathrm{I}][\mathrm{E}]}{[\mathrm{EI}]} \tag{3.38}$$

where [I] is the concentration of the inhibitor and [EI] is the concentration of the enzyme–inhibitor complex. All the other equations derived in Section 3.2 are applicable here with the substitution of the ligand with the inhibitor.

Because the interaction between small-molecule inhibitors and target proteins rarely can be studied directly, a majority of the studies are done with competitive binding studies to determine the ability of the inhibitor to displace a tracer known to bind to the target protein. Competitive binding studies can be described with the following scheme:

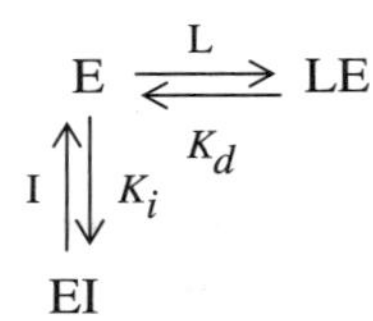

3.5

When performing a competitive binding experiment, a binding curve between the tracer and the target protein in the absence of the inhibitor is first obtained. From this curve, the $\mathrm{EC}_{50} = K_d$ value of the tracer is obtained. The concentration of the tracer is then fixed at $[\mathrm{L}] < K_d$ and the concentrations of the tracer-bound complex [LE] is measured at different concentrations of the inhibitor. The data is then plotted that will look like what is shown in Figure 3.7. This graph can be described with following equation:

$$f_b = \frac{\mathrm{IC}_{50}}{\mathrm{IC}_{50} + [\mathrm{I}]} \tag{3.39}$$

where f_b is the fraction of tracer-bound complex as compared with that in the absence of the inhibitor, and IC_{50} is the concentration of the inhibitor that caused 50% inhibition. In the simple case as outlined in Scheme 3.5:

$$\mathrm{IC}_{50} = K_i\left(1 + \frac{[\mathrm{L}]}{K_d}\right) \tag{3.40}$$

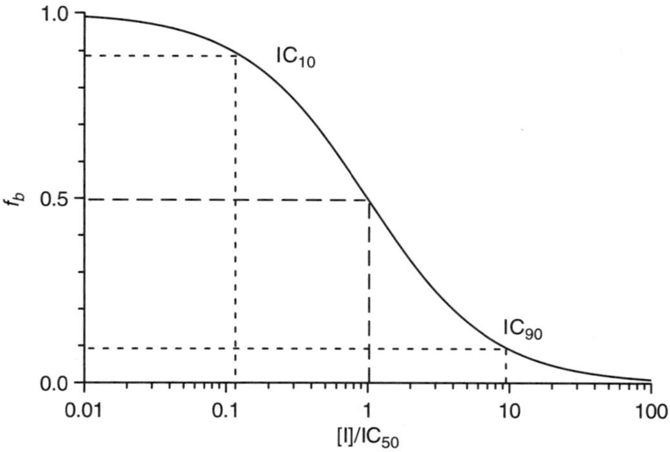

Figure 3.7 Illustration of one-site competition study. The concentration of a known labeled ligand (tracer) that binds to the target protein is fixed. The fraction of bound tracer/protein (compared with that in the absence of inhibitor) is measured at different concentration of inhibitor (in IC_{50} unit). The IC_{50} value is the inhibitor concentration that produced 50% inhibition. The IC_{10} value is the inhibitor concentration that produced 10% inhibition, and the IC_{90} value is the inhibitor concentration that produced 90% inhibition. The slope of this curve or the Hill coefficient is -1.

It is clear from Eq. (3.40) that the IC_{50} value increases with increasing concentrations of the tracer. Thus, higher tracer concentrations will cause the inhibition curve (see Fig. 3.7) to shift to the right. Such a shift will make weak inhibitors undetectable at reasonable testing concentrations. This is one of the reasons that the concentration of the tracer should be set as low as possible while making sure not to compromise the detection of the bound complex when designing competitive binding experiments. In practice, the competitive binding studies may be more complicated than what is shown in Scheme 3.5. A more general equation is usually used to describe the common competitive binding process:

$$f_b = \sum_i \frac{[\mathrm{I}]^{-n}}{\mathrm{IC}_{50}^i + [\mathrm{I}]^{-n}} \tag{3.41}$$

where n is the Hill coefficient that reflects the number of cooperative binding sites and i accounts for the number of independent binding sites on the protein. The negative value for the Hill coefficient reflects the fact that the fraction of tracer-bound complexes decreases with increasing concentrations of the inhibitor.

3.5.2 Inhibition of Enzyme Function

While the study of competitive binding relies on monitoring the tracer to detect the binding of the test molecules to the same binding site as the tracer, the study of the inhibition of enzyme function relies on monitoring the rate of the substrate conversion into final products. The study of inhibition of the enzyme function is of more advantage compared with competitive binding studies because functional assays can detect enzyme inhibition that happens at other sites on the enzyme in addition to the competitive binding site. Furthermore, the signals in the enzyme functional assays are amplified by the continuous substrate turnover to generate more products. The inhibition of the function of enzymes can be divided into two modes: reversible and irreversible inhibition. We will first discuss reversible inhibition and then discuss irreversible inhibition. Finally we will discuss how to experimentally determine reversibility.

Reversible Inhibition Reversible inhibition of the enzyme is further classified into three modes: competitive, noncompetitive, and uncompetitive inhibition (see Scheme 3.6).

A. Competitive

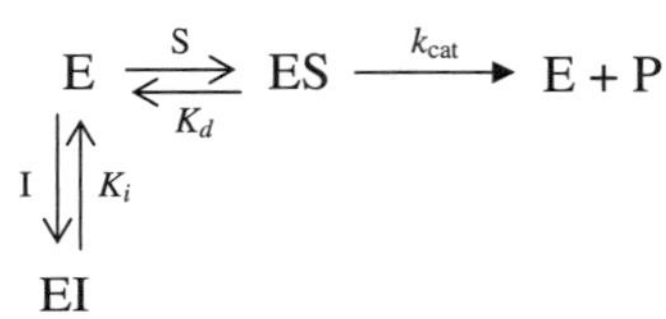

B. Uncompetitive

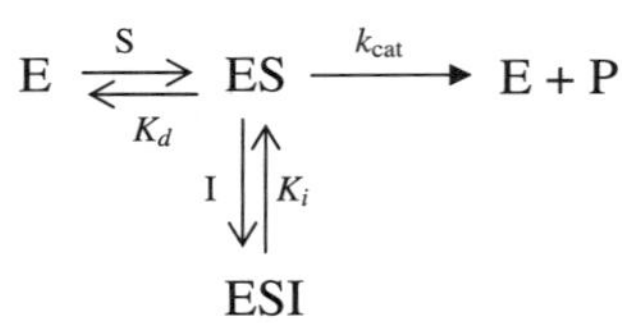

C. Noncompetitive

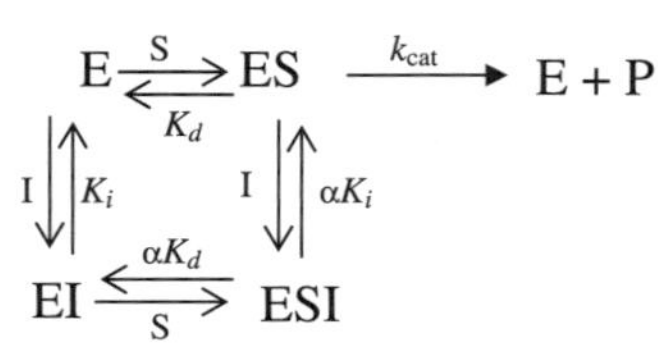

3.6

Competitive inhibitors bind to the free enzyme at the same site as the substrate binding site but do not bind to any enzyme complex (see Scheme 3.6A). The initial velocity in the competitive inhibition mode is described by the following equations:

$$v = v_{\max} \frac{[\mathrm{S}]_0}{K'_m + [\mathrm{S}]_0} = v_{\max} \frac{[\mathrm{S}]_0}{K_m\left(1 + \dfrac{[\mathrm{I}]}{K_i}\right) + [\mathrm{S}]_0} \tag{3.42}$$

$$K'_m = K_m\left(1 + \frac{[\mathrm{I}]}{K_i}\right) \tag{3.43}$$

where K'_m is the apparent Michaelis constant and K_m is the Michaelis constant in the absence of the inhibitor. From Eq. (3.42), it is clear that increasing the concentration of competitive inhibitors does not change the $v_{\max}$. This characteristic alone can distinguish competitive inhibition from the other two modes of inhibition because none of them has this property. From Eq. (3.43), it is clear that the apparent Michaelis constant increases with increasing concentrations of the competitive inhibitor. This is why increasing the concentration of competitive inhibitors will cause the rightward shift of the substrate concentration–response curve.

Uncompetitive inhibitors do not bind to free enzyme. Instead, they only bind to the enzyme and substrate complex (see Scheme 3.6B). The initial velocity in the uncompetitive inhibition mode is described by the following equations:

$$v = v'_{\max}\frac{[\mathrm{S}]_0}{K'_m + [\mathrm{S}]_0} = \frac{v_{\max}}{1 + \dfrac{[\mathrm{I}]}{K_i}} \times \frac{[\mathrm{S}]_0}{\dfrac{K_m}{1 + \dfrac{[\mathrm{I}]}{K_i}} + [\mathrm{S}]_0} \tag{3.44}$$

$$v'_{\max} = \frac{v_{\max}}{1 + \dfrac{[\mathrm{I}]}{K_i}} \tag{3.45}$$

$$K'_m = \frac{K_m}{1 + \dfrac{[\mathrm{I}]}{K_i}} \tag{3.46}$$

where $v'_{\max}$ is the apparent maximum velocity, K'_m is the apparent Michaelis constant, $v_{\max}$ is the maximum velocity in the absence of inhibitor, and K_m is the Michaelis constant in the absence of the inhibitor. From Eqs. (3.45) and (3.46), it is clear that increasing concentrations of uncompetitive inhibitors will decrease both the apparent maximum velocity and the apparent Michaelis constant by the same amplitude. Thus, the ratio between apparent maximum velocity and the apparent Michaelis constant will remain the same with increasing concentrations of uncompetitive inhibitor. This characteristic alone can distinguish uncompetitive inhibition from the other two modes of inhibition because none of the other two modes of inhibition have this property.

Noncompetitive inhibitors bind to all the available enzymes no matter if they are free or they are bound to substrates (see Scheme 3.6C). The factor α is a constant that accounts for the difference between the dissociation constant between the enzyme–substrate complex and enzyme–inhibitor–substrate complex. The value of α can be equal to 1 when there is no difference and can be larger or less than 1. The initial velocity for noncompetitive inhibition is described by the following equations:

$$v = v'_{\max} \frac{[\mathrm{S}]_0}{K'_m + [\mathrm{S}]_0} = \frac{v_{\max}}{1 + \dfrac{[\mathrm{I}]}{\alpha K_i}} \times \frac{[\mathrm{S}]_0}{K_m \left(\dfrac{1 + \dfrac{\mathrm{I}}{K_i}}{1 + \dfrac{[\mathrm{I}]}{\alpha K_i}} \right) + [\mathrm{S}]_0} \tag{3.47}$$

$$\frac{v'_{\max}}{K'_m} = \frac{v_{\max}}{K_m} \times \frac{1}{1 + \dfrac{\mathrm{I}}{K_i}} \tag{3.48}$$

Equation (3.47) shows that the apparent $v_{\max}$ will decrease with increasing inhibitor concentration. However, the apparent K_m can change in any direction depending on the value of α. The ratio of apparent $v_{\max}/K_m$ will decrease with increasing inhibitor concentration because of the relationship shown in Eq. (3.48).

The above analysis for reversible inhibition is summarized in Table 3.3. Experimentally, the substrate dose–response curve is obtained with different concentrations of inhibitors in the initial velocity measurement. The changes in apparent $v_{\max}$, K_m, and $v_{\max}/K_m$ at different concentrations of inhibitors are then compared with the expected changes for the three different modes of inhibition summarized in Table 3.3 to determine the mode of the inhibition. I would suggest not to transform the original data to fit to transformed linear equations as was done in the past. Instead, nonlinear fitting should be performed using a personal computer to extract the values of the above parameters, which is much more accurate. In addition, the nonlinear fit method will result in even more reliable assay results if a global fitting is performed with all the experimental data obtained at different concentration of inhibitors. The equations describing protein binding equilibrium and enzyme kinetics were derived long before the first personal computer went to the market. In those days, it was not practical to do a nonlinear fitting easily. Thus, many mathematical transformations of the original equations were derived to make a secondary variable that is a function of the primary measured variables (Lineweaver–Burk plots, Eadie–Hofstee plots, Hane–Wolff plots, etc.). Such a transformation allowed the presentation of the secondary variables in linear form that can be fit without a computer. However, transformations of original independent variables often distort the error associated with the measured data, leading to further distortion in the calculated results. The problem is especially severe in situations with limited data points. In comparison, it is not a problem to fit nontransformed data directly to kinetic

TABLE 3.3 Characteristics of Three Reversible Inhibition Modes with Increasing Inhibitor Concentration

	Competitive	Uncompetitive	Noncompetitive
$v_{\max}$	Constant	Decrease	Decrease
K_m	Increase	Decrease	Any
$v_{\max}/K_m$	Decrease	Constant	Decrease

equations to obtain more reliable data. With the arrival of powerful personal computers, those old transformations are obsolete. Many have suggested that scientists should stop using the old transformation methods to treat their data. This is one of the reasons that none of the famous transformations of the original kinetic equations are discussed in this book.

In the above rigorous kinetic analysis, the apparent $v_{\max}$ and K_m are measured experimentally by varying substrate concentrations with a set of fixed inhibitor concentrations. A convenient but less rigorous approach is often performed in practice in which the substrate is fixed at an arbitrary (with some boundary conditions) concentration while inhibitor concentrations are varied to obtain the IC_{50} values. In these IC_{50} studies, one arbitrarily picks a starting velocity (determined by substrate concentration) as the maximum and measures its reduction with increasing concentrations of inhibitor. This approach does not consider the $v_{\max}$ and hence does not provide sufficient information to deduce the inhibition mechanisms. Futhermore, the IC_{50} value is not constant because it is affected by many factors including substrate concentrations and the mode of inhibition. However, the IC_{50} value determination offers a quick way to evaluate the relative potencies of a large number of inhibitors under identical assay conditions. It is especially important in high-throughput screening mode that only one concentration of the inhibitor at one time point in the reaction progressive curve is measured with one fixed substrate concentration. Analysis of the relationship between the IC_{50} value and other experimental variables can help design more sensitive assays because an inhibitor at a fixed concentration can result in larger percentage of inhibition if the assay is designed with lower apparent IC_{50} value. Cheng and Prusoff in 1973 derived the following equations to calculate the IC_{50} values in different reversible inhibition modes.

For competitive inhibition:

$$IC_{50} = K_i\left(1 + \frac{[S]}{K_m}\right) \tag{3.49}$$

For uncompetitive inhibition:

$$IC_{50} = K_i\left(1 + \frac{K_m}{[S]}\right) \tag{3.50}$$

For noncompetitive inhibition:

$$IC_{50} = \frac{[S] + K_m}{\dfrac{K_m}{K_i} + \dfrac{[S]}{\alpha K_i}} \tag{3.51}$$

For noncompetitive inhibition, if $\alpha \approx 1$, then $IC_{50} \approx K_i$. In this case, IC_{50} is constant and is independent from substrate concentration. Equations (3.49) and (3.50) can be graphically represented in Figure 3.8. The plot is in log-log scale to show broader range. For competitive inhibitors, the ratio of IC_{50}/K_i starts at 1 and increases with increasing inhibitor concentrations. For uncompetitive inhibitors, the same ratio decreases reciprocally, and it reaches a minimum of 1 with increasing

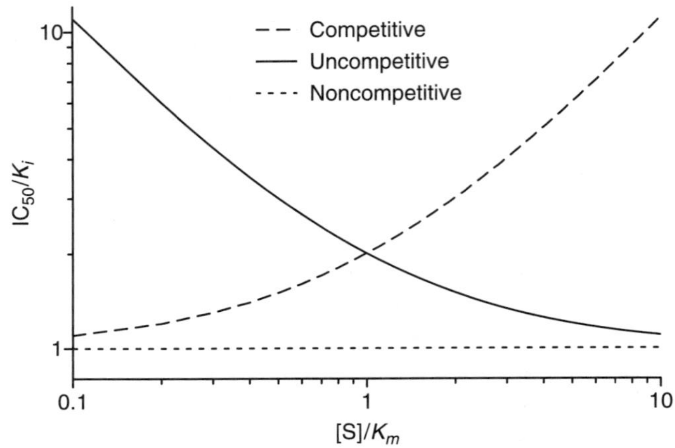

Figure 3.8 Relationship between IC_{50} and substrate concentration in three different reversible inhibition modes. The graph is in log-log scale to visualize a broad range.

inhibitor concentrations. Thus, if an assay is designed to screen a library of compounds for competitive inhibitors only, the substrate concentration should be kept as low as possible so that the assay has the lowest IC_{50} ($[S] \ll K_m$) value for a given inhibitor. On the other hand, the same rationale demands the substrate concentration be kept as high as possible ($[S] \gg K_m$) when screening for uncompetitive inhibitors. For noncompetitive inhibitors, the substrate concentration does not affect the IC_{50} values. When screening a large number of compounds, it is not practical to screen the library many times using different substrate concentration. Instead, a compromise is made that the substrate concentration is often chosen at equal to the K_m value in HTS screening because this is the point at which all inhibitors with different modes of inhibition have an equal chance.

Irreversible Inhibition Irreversible inhibition can be either from non-covalent tight binding (such as binding between streptavidin and biotin) or from covalent binding. Tight binding inhibition appears irreversible experimentally because the off-rate is too slow and the K_i is usually <1 nM. In theory, noncovalent tight binding inhibition is still reversible and follows the reversible inhibition kinetics if the experiment used to study the inhibition can meet the boundary condition when the inhibition equations were derived, that is, $K_i \gg [E]$. In practice, it is often impossible to perform kinetic experiments with such a low enzyme concentration ($[E] < K_i < 1$ nM), and scientists are forced to use experimental conditions with [E] closer or larger than the K_i. In this situation, the enzyme concentration will affect the IC_{50} value obtained experimentally. Stoichiometric balance must be accounted for by the following equation:

$$IC_{50} = K_i^{app} + \frac{[E]}{2} \tag{3.52}$$

Truly irreversible inhibition is when the inhibitor covalently binds to the enzyme, which results in enzyme inactivation. Covalent binding inhibition can be divided into random non-mechanism-based inactivation of proteins by reactive chemical species and mechanism-based inactivation of proteins by selective inhibitors. Mechanism-based inactivation occurs when the inhibitor is not reactive but only becomes reactive after binding to the active site of the enzymes where it was transformed by the enzyme, resulting in covalent modification of the enzyme at the active site. The mechanism-based inhibitors are sometimes referred to as "suicide substrates."

Determination of Inhibition Reversibility After an inhibitor of interest is obtained by HTS or other methods, the mechanism of inhibition is studied to help understand the molecule and to drive further SAR studies. The first study is to determine whether the inhibition is reversible or not. If it is reversible, further studies can be performed to determine the mode of inhibition as to competitiveness. If the inhibition is irreversible, further studies are needed to determine whether it is covalent binding or simple tight binding. Tight binding and mechanism-based covalent binding indicate that the leads are worthy of further studies. If the covalent inhibition is not mechanism based (reactive molecules randomly bind to many molecules), the compound is discarded. To determine reversibility, the first experiment is to determine the inhibitor concentration response curve in the same way as the determination of an inhibitor's IC_{50} value (see Fig. 3.7). A rapid dilution experiment is then designed based on this curve. When the Hill coefficient is -1, it takes an 81-fold concentration span to reverse from 90% inhibition to 10% inhibition. Thus, a 100-fold dilution from 90% inhibition condition would result in less than 10% inhibition. A dilution experiment is designed by mixing 100-fold of normal assay concentration of enzyme with inhibitor at a concentration that inhibits about 90% enzyme activity. This solution is instantly mixed with the assay solution containing normal concentration of enzyme substrate and other components at a ratio of 1 : 100. The product formation is then monitored with time and is compared with the control experiment that are performed identically except no inhibitor is present in the 100-fold concentrated enzymes. Three scenarios may happen in the rapid dilution study, which is shown in Figure 3.9. (a) *Rapid reversible inhibition*: Linear increase in product concentration with the initial velocity about 10% less than the initial velocity of the control, depending on the final inhibitor's concentration in the diluted solution when equilibrium is established. (b) *Irreversible inhibition*: Linear increase of product formation at a velocity that is about 90% less than the control, depending on the concentration of the inhibitor. The residue enzyme activity is from the enzymes that are not bound to the inhibitor. (c) *Slowly reversible inhibition*: Because the active enzyme concentration [E] is equal to kt where k is the enzyme–inhibitor complex dissociation rate, the concentration of the reaction product can be calculated by $[P] = [E]t = kt^2$. Thus, the initial increase in the product formation is proportional to t^2. This is followed by a linear increase in product formation when most of the inhibited enzymes are reversed to active form and equilibrium is established. The velocity in the final linear phase should be about 10% less than the control (if no substrate depletion yet in the reaction progressive curve).

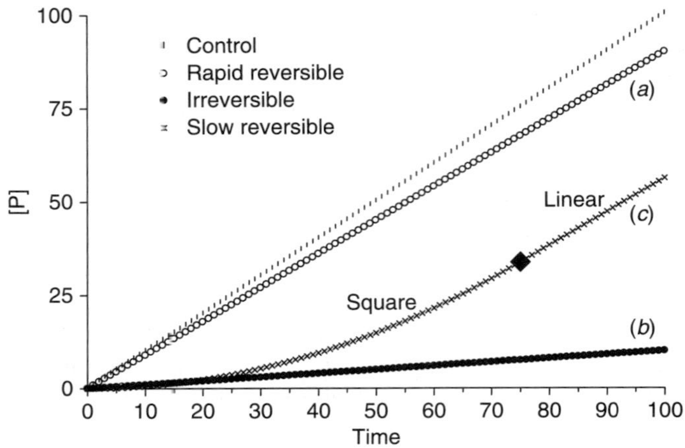

Figure 3.9 Graphical illustration of rapid dilution method to determine reversibility of inhibition. The time and product concentration are in arbitrary units. Three situations may be encountered in rapid dilution experiments. (*a*) *Rapid reversible inhibition*: Linear increase in product concentration with the initial velocity about 10% less than that of the control. (*b*) *Irreversible inhibition*: Linear increase of product formation at a velocity that is about 90% less than the control. (*c*) *Slowly reversible inhibition*: There are two phases with initial increase in the product concentration proportional to t^2. This is followed by the second phase with linear increase in product formation at velocity about 10% less than the control. The big diamond symbol in the graph separates the two phases.

3.6 ASSAY DEVELOPMENT WITH ISOLATED PROTEINS

The ultimate goal to study proteins is to understand their crucial role in live cells. Because the complexity of protein interactions in cells and the technical limitations, protein function studies were often done with isolated proteins (purified and partially purified) in the hope that the knowledge gained from isolated simple systems can be reassembled to lead to knowledge of the whole process in the cell. The advantages of studying isolated proteins include simplified system that makes the interpretation of results more reliable, no cell membrane barrier that exposes difficulties in studying intracellular proteins, and no need to deal with cell cultures. The downside of assays with isolated proteins is that the assay results may not correspond to what happens in cells because an isolated protein may lose critical functions when taken out from its native environment. In addition, isolated proteins may behave differently in the artificial assay conditions that are different from the physiological environment in cells where the protein normally functions. The assay development with isolated proteins is the process to evaluate all different parameters that may affect the assay to obtain an assay system that closely resembles what happens in cells with minimum perturbation and has adequate signal to noise to enable precise measurements. General factors are discussed below.

Protein Sources Target proteins used in bioassays can come from three major sources: (1) purified proteins from human tissues, animal tissues, or other biological sources, (2) purified recombinant human proteins expressed from mammalian cell lines, bacteria, yeast, and insect cells, and (3) partially purified proteins containing significant amounts of other cellular components (such as membrane preparations containing target receptors) from biological sources to maintain the protein's function. The assay target proteins should be as close to native protein as possible. For drug discovery, the desirable protein is the protein directly isolated from the human source. However, few human proteins except that derived from blood can be obtained due to limited availability of human tissues. Traditionally, most assays have to use proteins purified from animal tissues as a substitute for human proteins. Some proteins are abundantly present in native biological tissues that can be purified directly. For example, trypsin and trypsinogen can be isolated from the pancreas of bovine, rat, whale, and the like. The caveat of using purified proteins from animals in bioassays is that their pharmacological characteristics may not be the same as the human proteins even if they share relatively high sequence homology. Currently, human recombinant proteins are the most commonly used protein source in bioassays. Recombinant proteins can be mass produced reliably and cost effectively from cell lines. The protein purification is simpler because no need to deal with tissues. In addition, modification of the proteins can be made to introduce desirable functional groups. Several recombinant protein expression systems are available including mammalian cells, bacteria, yeast, and insect cells. Mammalian cells are the first choice to produce proteins for bioassays, especially for functional assays, because they contain the appropriate machinery for post-translational modification to produce properly functional proteins. However, it takes time to develope protein expression system in mammalian cells, and the system cannot match the performance of the expression system developed with bacteria or virus in terms of quantity and easy of operation. This is why many simple soluble proteins used in bioassays are commonly produced in large quantity in *Escherichia coli*. A caveat of protein expression in *E. coli* is the lack of adequate posttranslational modification that may result in loss of function of the expressed proteins. For example, many kinases expressed in *E. coli* are nonfunctional. Many functional kinases are expressed in sf9 insect cells that are transfected with baculovirus containing the recombinant kinase sequence. Partially purified proteins are used in some assays because the proteins are functional only when they are associated with other proteins in a complex or when they are in native membranes. Many membrane proteins, such as receptors and ion channels, are often assayed with crude membrane preparations.

Proteins are prepared at high concentration. They are then divided into small aliquots enough for single use and are stored in conditions that can maintain their function for a long time. Because all chemical activities are slowed down at lower temperatures, proteins are usually store at low temperature for long-term storage. Common storage temperatures for proteins are −80, −20 and 4°C depending on the stability of the protein in the particular storage media. Though storing proteins at −80°C can significantly slow down chemical activities, it is very inconvenient to retrieve the sample, and the picked sample has to go through a freeze–thaw cycle. Freeze–thaw cycles can cause proteins to denature and should be minimized.

In addition, freezers cooling to −80°C are very expensive and they occupy large laboratory space. When proteins are stored at 4°C, the sample is not frozen, making it convenient to access the sample. However, 4°C may not be cool enough to stabilize many proteins in aqueous solution. Many proteins can be stored at −20°C when they are in 50% glycerol, which enables the proteins to remain in the liquid phase and avoid the freeze–thaw cycle. However, not all proteins can tolerate 50% glycerol. In addition to temperature control, other additives may be added to the protein storage solution, such as molecules that may bind to the protein (cofactors, substrate, etc.) and large molecules (bovine serum albumin (BSA), polyethylene glycol (PEG), polylysine, collagen, and the like) that prevent protein from absorption to container surface.

Substrate in Enzyme Assays Using the enzyme's native substrate in an enzyme assay would be most desirable because of physiological relevance. However, using native substrate in enzyme assays may require extra assay development efforts. Native protein substrate can be difficult to obtain, resulting in increased assay development time and costs. If the protein substrate is not stable, the quality of the assay will suffer too. Many enzymes are not very selective and may have many native substrates. For example, trypsin will cleave many protein substrates if they have a positively charged arginine or lysine group. Thus, it may not be necessary to use native protein substrate in enzyme assays in situations when the enzyme is not selective and alternative substrate can be used. Synthetic peptides are often used as artificial substrates in enzyme assays. Small peptides can be easily synthesized and many modifications can be readily introduced to them during synthesis. The use of synthetic peptide substrates in enzyme assays enabled many assay technologies (chromogenic, fluorogenic, FRET, SPA, HTRF, etc.). Many enzymes only have a limited number of native substrates and are very selective to peptide substrates. In this case, substrate profiling from a peptide library has to be performed to find the best substrate. Many technologies, especially phage display technology, have greatly facilitated the process. The primary goal of an assay also determines whether native substrate or artificial substrate should be used. If the purpose of the assay is to find enzyme inhibitors, the substrate merely serves as a tracer of the assay to provide measurable signal and any substrate (native or artificial) can be used. If the assay is to find the best enzymes from a collection of recombinant enzymes to cleave a specific target protein, the native protein substrate should be used. Using artificial peptide substrate in this case may lead to wrong conclusions.

Small peptide substrate should be dissolved in high concentrations (>100-fold of the concentration to be used in assays). They can either be dissolved in an aqueous buffer (e.g., 10 mM HEPES, pH 7.2) or organic solvent (e.g., dimethyl sulfoxide (DMSO) and ethanol) if they are not soluble in aqueous solution at high concentrations. Usually no other additives are added to the solution. Aliquots of the solution are usually stored at −20°C and are kept in the dark to prevent photobleaching if a fluorophore is attached to the peptide. Peptide substrates should not be stored in aqueous solutions at low concentration for extended periods of time because they may be absorbed by the container. I have experienced a total loss of detectable fluorescence signal when a hydrophobic fluorescently labeled peptide was stored overnight at 10 nM concentration.

Other Factors Affecting Assays Many factors affect assay performance and they should be evaluated during assay development. These factors include assay buffer components, pH, ionic strength, and specific cations such as Ca^{2+} and Mg^{2+}. The compounds to be assayed are usually dissolved in a different media from the assay solution. The effect of the compound media at final assay concentration should be evaluated. For example, if the compound is originally dissolved in DMSO, after addition at 1 : 100 in volume ratio to the assay solution, there is still 1% DMSO (about 140 mM) in the final reaction solution. Substrate concentration can affect the IC_{50} value for reversible inhibitors and thus affects the sensitivity of the assay to detect inhibitors. Setting the substrate at concentrations lower than K_i will reduce the IC_{50} value for competitive inhibitors and thus increase the ability of the assay to detect weak inhibitors. However, the rate of enzymatic reaction also depends on the concentration of the substrate. An assay with too low a substrate concentration will generate too little product to be detected and may decrease the quality of the assay. The optimized substrate concentration should be established during assay development. In addition to the peptide (protein) substrate, the enzyme may need a second small substrate (or cofactor) for its proper function. This second substrate should be studied in the same way as the peptide substrate is studied.

Temperature is another significant factor in assay development because it can affect the rate of reaction and can also change the stability of the enzyme in assay solution, which may affect the quality of the assay. As a rule of thumb, the rate of a chemical reaction will double for every 10°C temperature increase. For an enzyme-catalyzed reaction, the temperature effect may be even bigger because the enzyme functions optimally in native temperature (e.g., 37°C is the optimal temperature for enzymes derived from humans). Some enzymes may not be very active at room temperature at which most assays were developed. Depending on a particular enzyme, the reaction rate at room temperature may be good enough. On the other hand, raising the temperature to 37°C may reduce the stability of the enzyme and cause enzyme inactivation during the assay. Typically, enzyme activity versus temperature show a bell-shaped curve as shown in Figure 3.10. Thus, temperature effect on the assay should be evaluated to find the optimal assay temperature. The pH at which the enzyme-catalyzed reactions take place also typically shows a bell-shaped curve similar to the curve shown in Figure 3.10 when plotting the enzyme activity versus the pH. Thus, the pH effect on the assay should be evaluated to find the optimal assay pH as well.

In addition to temperature and pH, there are other factors that may inactivate the enzyme and affect the assay. Enzyme inactivation should be tested during assay development to make sure that no enzyme inactivation occurs in the final assay conditions. Selwyn's test is one method to test enzyme inactivation. This test is based on the fact that the concentration of the product in enzyme-catalyzed reactions is a function of the product of time (t) and enzyme concentration [E], that is, the product concentration should be the same if $[E] \cdot t$ remains the same during the assay period. For example, the concentration of the product should be the same when the enzyme concentration is increased twofold while the reaction time is shortened by a factor of 2 and vice versa. Thus, Selwyn's test is performed to measure the product formation over time with a series of different enzyme concentrations

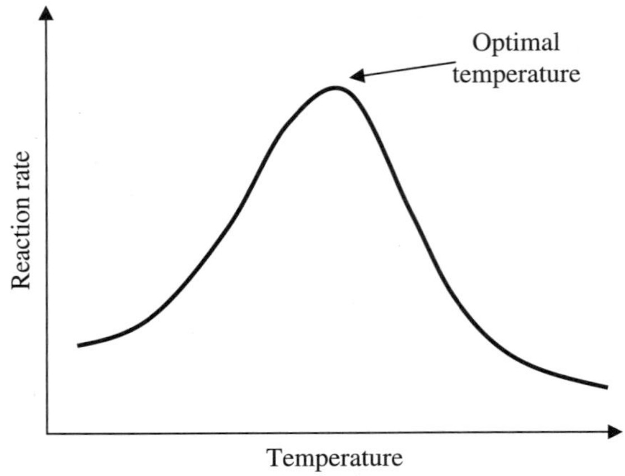

Figure 3.10 Typical effect of temperature on the rate of enzyme-catalyzed reactions. At lower temperature, the enzyme is not very active. At high temperature, the enzyme denatures and losses activity. There is an optimal temperature at which the rate of the enzyme-catalyzed reaction is the highest.

while the other components of the assay remain the same. If there is no enzyme inactivation, a plot of product concentration against $[E] \cdot t$ with all the data from different enzyme concentrations will result in one smooth curve as shown in Figure 3.11*a*. However, if enzyme inactivation occurs over time, the enzyme concentration will continue to decrease over time. The plot in Figure 3.11 will result in several nonoverlapping curves with data from each enzyme concentration forming its own curve (Fig. 3.11*b*). The experiment with the lowest starting enzyme concentration has the most enzyme inactivation because longer reaction time is required to obtain the same $[E] \cdot t$. In Figure 3.11*b*, the lowest curves is the one with the least starting enzyme concentrations.

From Primary "Hits" to High-Quality "Leads" Biochemical assay is the method of choice to advance molecules from primary "hits" to high-quality "lead" through SAR studies because there is no ambiguity on the hit molecule's target protein, and a good knowledge of the structure of the target protein is usually known or can be inferred from known structures. In addition to primary assays, many more assays are designed to obtain further information about the mechanism of inhibition from different angles. The first step in the process is to obtain the inhibitor's dose–response curve. From this curve, a study of the reversibility of the inhibition is performed. If the inhibition is reversible, further studies are carried out to determine whether the inhibition is competitive, uncompetitive, or noncompetitive. If the inhibition appears irreversible in the dilution experiment, more assays are performed to determine whether the interaction is covalent or not. Noncovalent irreversible inhibition means tight binding with a slow off-rate that is a good quality for potential drugs. If the interaction is covalent, additional assays are performed to determine whether the inhibition is mechanism based or not. Mechanism-based covalent inhibition may

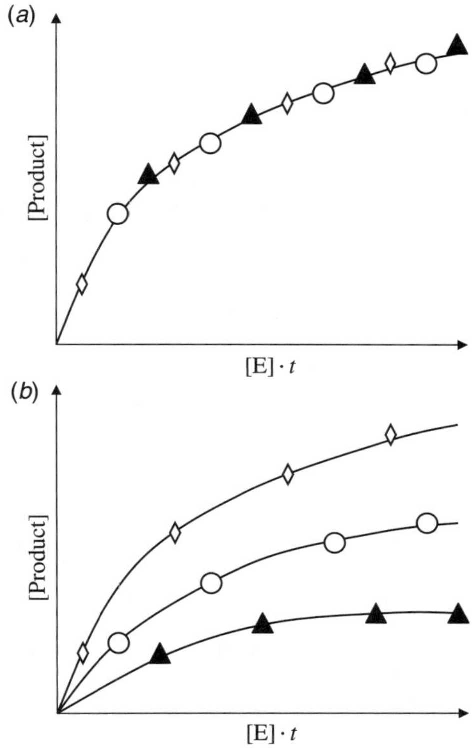

Figure 3.11 Testing enzyme inactivation during an assay with Selvyn's method. The product formation is measured at three different enzyme concentrations. The concentration of the product is then plotted against [E] · *t*. (*a*) When there is no enzyme inactivation, all the data from the three enzyme concentrations form a single curve. (*b*) When there is enzyme inactivation, the data from each enzyme concentrations form its own curve.

still be good drug candidates because many well-known drugs, such as penicillin, belong to this class. Nonselective covalent inhibition by reactive organic species or their precursors are of no use for further development. In addition to performing SAR to generate good mechanism-based inhibitors with lowest IC_{50}, the molecules are often evaluated for their drug-likeness with in vitro studies such as oil/water partition, ability to cross membrane, interaction with critical liver enzyme cytochrome P450 isoforms, and interaction with critical cardio-proteins such as hERG potassium channels.

Useful Websites

http://www.photophysics.com/

http://www.iubmb.org/

http://www.brenda-enzymes.org/

BIBLIOGRAPHY

Atkins, P. and de Paula, J. (2006) *Physical Chemistry*. Oxford University Press, Oxford.

Bisswanger, H. (2004) *Practical Enzymology*. Wiley-VCH, Weinheim.

Chang, A., Schomburg, D., and Schomburg, D. (eds.) (2002) *Class 3.4 Hydrolases II (Springer Handbook of Enzymes)*. Springer, New York.

Cheng, Y. and Prusoff, W. H. (1973) Relationship between the inhibition constant (K1) and the concentration of inhibitor which causes 50 per cent inhibition (I50) of an enzymatic reaction. *Biochem. Pharmacol.* **22**, 3099–3108.

Copeland, R. A. (2000) *Enzymes: A Practical Introduction to Structure, Mechanism, and Data Analysis*, 2nd ed. Wiley-VCH, New York.

Copeland, R. A. (2005) *Evaluation of Enzyme Inhibitors in Drug Discovery: A Guide for Medicinal Chemists and Pharmacologists*. Wiley, Hoboken, NJ.

Eisenthal, R. and Danson, M. (eds.) (2002) *Enzyme Assays: A Practical Approach*, 2nd ed. Oxford University Press, Oxford.

Goutelle, S., et al. (2008) The Hill equation: A review of its capabilities in pharmacological modelling. *Fund. Clin. Pharmacol.* **22**, 633–648.

Greis, K. D. (2007) Mass spectrometry for enzyme assays and inhibitor screening: An emerging application in pharmaceutical research. *Mass Spectrom. Rev.* **26**, 324–339.

Jelesarov, I. and Bosshard, H. R. (1999) Isothermal titration calorimetry and differential scanning calorimetry as complementary tools to investigate the energetics of biomolecular recognition. *J. Mol. Recog.* **12**, 3–18.

Kersey, P. J., et al. (2004) The International Protein Index: An integrated database for proteomics experiments. *Proteomics* **4**, 1985–1988.

Motulsky, H. and Mahan, L. (1984) The kinetics of competitive radioligand binding predicted by the law of mass action. *Mol. Pharmacol.* **25**, 1–9.

Price, N. C. and Stevens, L. (1999) *Fundamentals of Enzymology: The Cell and Molecular Biology of Catalytic Proteins*, 3rd ed. Oxford University Press, Oxford.

Rahman, A.-U., Choudhary, M.I., and Thompson, W. J. (2001) *Bioassay Techniques for Drug Development*. Harwood Academic, Amsterdam.

Reymond, J.-L. (ed.) (2006) *Enzyme Assays: High-Throughput Screening, Genetic Selection and Fingerprinting*. Wiley-VCH, Weinheim.

Segel, I. H. (1993) *Enzyme Kinetics: Behavior and Analysis of Rapid Equilibrium and Steady-State Enzyme Systems*. Wiley, New York.

Selwyn, M. J. (1965) A simple test for inactivation of an enzyme during assay. *Biochim. Biophys. Acta* **105**, 193–195.

Sumner, J. B. (1926) The isolation and crystallization of the enzyme urease. *J. Biol. Chem.* **69**, 435–441.

Tinoco, I., Sauer, K., Wang, J. C., and Puglisi, J. D. (2001) *Physical Chemistry: Principles and Applications in Biological Sciences*, 4th ed. Prentice Hall, Upper Saddle River, NJ.

CHAPTER 4

SEPARATION-BASED TECHNIQUES IN BIOASSAYS

ALL BIOASSAYS can be divided into heterogeneous and homogeneous assays. Heterogeneous assays involve separation steps that separate the component(s) to be measured from the rest of the assay components in an assay system. Because the removal of other components that interfere with the final reading in the assay mixtures, heterogeneous assays can achieve a high signal-to-background ratio. On the other hand, the heterogeneous assays are usually more complicated, and the assay variations can be high due to the separation step in the assay. In contrast, homogeneous assays do not involve the separation step. In homogeneous assays, the measurements are based on the distinct physical/chemical properties of the analyte or the unique interactions between the analyte and its surrounding environment. The operation of homogeneous assays is very simple because they only involve "mix and read." Since the measurements were made in the presence of all the other assay components, homogeneous assays suffer from high background caused by interference to the measurement from other components in the assay mixture. Typical homogeneous assays have a signal-to-background ratio of less than 10. Heterogeneous assays are usually performed because either there is no applicable homogeneous assay or the homogeneous assay does not meet the signal-to-background ratio requirement for a particular situation. Because heterogeneous assays are an important part of bioassays and they are based on separation, we will discuss the commonly used separation techniques here.

4.1 WASHING SOLID SUPPORTS TO REMOVE IMPURITIES

A common method to separate the analyte from the other components in the assay system is to first tightly bind (covalent or noncovalent) the analyte in an assay system to a solid support and then wash away all other components in the assay system. The most well-known heterogeneous assay based on this method is the enzyme-linked

Assay Development: Fundamentals and Practices. By Ge Wu

immunosorbent assay (ELISA). In ELISA, the analyte is selectively bound to the surface of a microplate. The binding between the analyte and the microplate is made possible through a capturing reagent that must be immobilized on the surface of the microplate first. Streptavidin and antibodies are commonly used as the capturing reagents. The immobilization of the capturing reagent to the surface of the microplate is achieved either by noncovalent absorption of the capturing reagent to a high binding surface in the microplate or by covalent binding of the capturing reagent with chemically activated microplate surfaces through the reactive groups (e.g., amino group, sulfhydryl group, and carboxyl group) in the capturing reagent. In the covalent binding scheme, the capturing reagent is first mixed with the activated microplate to allow covalent binding to the surface of the microplate. The microplate is then deactivated with a reagent that destroys all the active sites on the surface of the microplate. The microplate is then subjected to several cycles of washing (usually more than three cycles) to remove unbound capturing reagents. In the noncovalent binding scheme, the capturing reagent is simply added to a high binding microplate and incubated to allow association to the microplate. After incubation, the unbound capturing reagent is removed by several cycles of washing. The subsequent treatments of the microplate are the same for both types of immobilization of the capturing reagent. A block reagent is then added to the microplate to block any nonspecific binding sites on the microplate surface. The excess block reagent in the microplate is then removed by several cycles of washing (usually more than three cycles). After these treatments, the microplate is ready to capture analyte that either contains the epitope for the antibody or is labeled with biotin.

To alleviate the inconvenience in the tedious washing cycles, several dedicated automatic microplate washers were developed to facilitate the washing process. These instruments usually contain separate aspiration and dispensing fluidic paths. The washing buffer is stored in a reservoir that is connected to the dispensing path. An empty tank is connected to the aspiration path to collect the waste. Preset programs that can be stored in the instrument's memory control the wash operation. There are two types of microplate washers: strip washer and 96/384-well washer. A strip washer may have 8, 12, or 16 aspiration and dispensing channels arranged in a line while a 96/384-well washer usually has 96 aspiration and dispensing channels arranged in a matrix that fits standard microplate format. For operation with 384-well microplates, the microplate shifts its position sequentially in each of the four quadrants so that the 96-channel manifold can access all the wells in the microplate. A good automatic washer should be able to control dispensing volumes, dispensing flow rate, aspirating rates, and sometimes dispensing angle to handle different washing operations from gentle cell washing to vigorous washing in biochemical binding assays. Overflow of the buffer is a common problem with some automatic washers. The overrun buffer may get into the instrument and cause damage. A good automatic washer should have a good liquid sensing module to prevent buffer overflow. Another common problem with most automatic washers is the clogging of the dispensing or aspiration channels. To prevent this problem, the washer must be thoroughly washed with water and all the channels are soaked in water after each use. A good example of microplate washer is the Elx405 microplate washer from Biotek that is widely used in HTS operations.

Immobilization of the analyte on the surface of magnetic beads offers an alternative method to immobilize the analyte. This method is referred to as biomagnetic separation. It was first invented in the 1980s and marketed by Dynal (now part of Invitrogen). Dynal's technology employs uniform spherical polymer beads with imbedded magnetic material. In addition to Dynal, several other companies also market magnetic beads (but may not have uniformed size) for biomagnetic separation-based assays. With the biomagnetic separation method, the surface of the magnetic bead is functionalized with capture reagents, such as streptavidin and antibodies, to selectively capture the analyte in the assay mixture. With this technology, the biological reactions are carried in solution with suspended beads instead of on the surface of a microplate. The wash operation is performed when the magnetic beads are attracted to a magnet. After the wash step, the beads are released to the solution as suspension by removing the magnet. Biomagnetic bead-based separation has been applied to separating DNA, RNA (ribonucleic acid), protein, and cells. The key advantages of biomagnetic separation are that the reaction process happens in three-dimensional solution instead of in two-dimensional microplate surface, and the larger capturing capacity in unit volume (larger total surface area of the total suspended beads than limited microplate surface). In addition, this method is gentle to the cells as compared with centrifugation, making it a superior technology in separating cells.

There are several disadvantages with assays that employ washing as a method for separation. Uneven washing among the wells in a microplate is the major source of large variations in the assay results. The problem is more serious when the binding between the binding partners is not strong, and uneven washout of the analyte may occur with uneven washing among the wells in a microplate, leading to large variations in the assay results among different wells in the microplate. In addition, many cycles of washing create a large volume of waste buffer. Disposal of a large volume of biological waste solution is harmful to the environment and is costly, especially if radioactivity is present in the waste buffer.

4.2 ORGANIC SOLVENT EXTRACTION OF HYDROPHOBIC MOLECULES

Extraction is a technique used to separate substances based on their different solubility between two immiscible solvents. In format, one of the solvents is the aqueous assay buffer that contains many salts, and the other solvent is the water-immiscible organic solvent (such as chloroform, hexane, methylene chloride, and ethyl acetate). Organic solvent extraction is used in assays when the analyte is hydrophobic with higher solubility in organic solvent than that in water. When thoroughly mixed in the two extraction solvents, the hydrophobic analyte partitions to the organic solvent while the other hydrophilic substances and salts in the assay buffer remain in the aqueous phase. Since the two extraction solvents do not mix, they are separated in two layers after settling. Whether the aqueous layer is on the top or at the bottom depends on the density of the organic solvent. Chloroform and methylene chloride are heavier than water and will stay at the bottom. Hexane and ethyl acetate are lighter than water and will stay at the top.

In addition to assay of a neutral hydrophobic molecule, organic extraction is also widely used to assay an organic acid or base that can be changed to a neutral molecule by adjusting the pH in aqueous solution. For example, the analysis of fatty acid as a product in an enzymatic reaction was commonly performed using an organic extraction method. The disadvantage of an organic extraction method is its low throughput. The assays employing organic extraction are usually performed in microtubes. The mixing of the two immiscible solvents is particularly challenging when the assay is performed in a microplate. In case the assay is performed in a microplate, a lighter organic solvent, such as hexane, is commonly used to extract the analyte for ease of operation. Because the lighter organic solvents stay at the top in the wells in a microplate, the analyte in the top organic layer can be taken out for further analysis with less contamination from interfering substances present in the bottom aqueous layer.

4.3 CENTRIFUGATION TO REMOVE DENSE PARTICLES

Centrifugation is one of the most important and widely applied research techniques in biochemistry, cellular and molecular biology, and in medicine. Centrifugation is a process that uses the centripetal force to separate a mixture containing particles suspended in a liquid medium. The particles can be cells, subcellular organelles, viruses, biomembranes, and large molecules (such as proteins and nucleic acids). Centrifugation separates substances on the basis of the particle size and the density difference between the liquid and solid phases. Sedimentation of spherical material in a centrifugal field is described by

$$\nu = \frac{d^2(\rho_s - \rho_l)\omega^2 r F_s}{18\eta} \tag{4.1}$$

where ν is the rate of sedimentation, d is the diameter of the particle, ρ_s is the density of the particle, ρ_l is the density of the solution, ω is the angular velocity in radians per second, r is the radius of rotation, η is the viscosity of suspension, and F_s is a correction factor for particle interaction during hindered settling. F_s depends on the volume fraction of the solids present that is approximately equal to 1, 0.5, 0.1 and 0.05 for 1, 3, 12, and 20% solids volume fraction, respectively. The force of centrifugation applied to the sample ($=\omega^2 r$) is determined by the speed of rotation and the distance of the sample from the axis of rotation. The centrifugal force is usually compared to the force of gravity and is reported as the relative centrifugal force (RCF) in g units. The following formula relates the RCF to the rotor speed (N: rpm) and radius of rotation (r: cm):

$$\text{RCF} = 1.117 \times 10^{-5} \times N^2 \times r \tag{4.2}$$

One important component in centrifugation technology is the rotor that holds the sample tubes. Pelleting efficiency of a rotor is called the K factor, which is defined as

$$K = \frac{2.53 \times 10^{11} \ln(R_{\max}/R_{\min})}{N^2} \tag{4.3}$$

where $R_{\max}$ is the maximum radius from the axis of rotation, $R_{\min}$ is the minimum radius from the axis of rotation, and N is the rotor speed in rpm. The smaller the K factor is, the higher pelleting efficiency. There are two types of rotors commonly used in laboratories engaged in assay development: fixed-angle rotor and swing-bucket rotor. In a fixed-angle rotor, the materials are forced against the side of the centrifuge tube and then slide down the wall of the tube. This leads to faster separation but also leads to abrasion of the particles along the wall of the centrifuge tube. In the swing-bucket rotor, the materials must travel through the entire length of the centrifuge tube that is filled with the media down to the bottom of the tube. Since the media is usually a viscous substance, the settling rate is slower. The swing-bucket rotor is very useful to separate molecules or organelles on the basis of their movements through viscous field. With the same centrifugation tubes, a fixed-angle rotor has smaller $R_{\max}/R_{\min}$ than a swing-bucket rotor. Thus, the fixed-angle rotor is more efficient for pelleting.

For applications in bioassays, throughput is very important. This requires that centrifugation must be able to handle microplates. A swing-bucket rotor that holds microplate is commonly used in bioassays. Swing-bucket rotors can enable all the wells in the microplate to experience the same centrifugal force. For example, cells used in cell-based assays are commonly separated from other assay components by centrifugation with a swing-bucket rotor holding microplate at low rotation speed (3000 to 5000 rpm) for a few minutes. Assays based on immunoprecipitation also use centrifugation to separate the immunoprecipitated components from the rest of the assay components. For example, when performing kinase assays in live cells, the kinase of interest must be separated from the other kinases in the cell lysate after stimulation of the cells. This is achieved by immunoprecipitating the target kinase, which is followed by centrifugation. The isolated kinase is then tested for its activity to phosphorylate its substrate. Assays involving centrifugation step are not widely used because of the low throughput.

4.4 MEMBRANE FILTRATION

Membrane filtration is the separation process based on the use of thin, selective, semipermeable barrier. In membrane filtration, the solution to be filtered passes through a filter membrane either under positive pressure from the feed or under vacuum from the filtrate end. The separation occurs at the surface of the filter. The membrane selectively retains some components in the assay mixture based on the particle size, charge, hydrophobicity, or specific interactions (such as streptavidin/biotin and antibody/epitope). For application in bioassays performed with a microplate, membranes are placed at the bottom of a 96-well microplate with porous support to create a filter plate. Filter plates are used for high-throughput screening applications for many types of assays. One example of filtration-based assay is the radioactivity filtration assay for kinases (see Chapter 7). In this assay, the filter membrane at the bottom of the filtration plates is negatively charged phosphocellulose. The negatively charged phosphate group on the filtration membrane binds to the positively charged residues on the protein/peptide kinase substrates and retains them on the membrane. The negatively charged radioactive ATP passes through the membrane. After washing the membrane for a few cycles, the amount of the phosphorylated substrate retained on the membrane is measured by counting the radioactivity on the membrane.

A common problem encountered with filtration assays in microplate format is the uneven filtration among different wells. Though the same pressure or vacuum is applied to all the wells in a microplate, different wells may have different resistance to the liquid flow because of uneven blockage of the pores on the membrane and other factors. After filtration, it is often observed that some wells are wetter than others. Attempts to completely dry up the few remaining wet wells by increasing pressure or vacuum often lead to breakage of the membrane.

4.5 LIQUID CHROMATOGRAPHY

4.5.1 Introduction to Chromatography

Chromatography is a separation process whereby a mixture of molecules, dissolved in a solvent, is separated from one another by the differential distribution of the molecules between the mobile phase and the stationary phase inside a column or in a thin layer of flat solid material. When the stationary phase is a thin layer of flat solid material, it is called thin layer chromatography (TLC). When the stationary phase is packed inside a column, there exists gas chromatography (GC) if the mobile phase is gas, liquid chromatography (LC) if the mobile phase is liquid, and supercritical-fluidic chromatography. When high pressure is applied to drive the liquid phase in LC, it is called high-performance (or pressure) liquid chromatography (HPLC). In HPLC column packing, the stationary phase is supported by particles with typical diameters of 3 to 10 μm. This small-particle packing increases the separation efficiency but restricts the flow of the mobile phase. This is the reason high pressure (up to 8000 psi) is used to drive the flow of the mobile phase. Early HPLC columns are packed with irregular-shaped particles made of silica or alumina. To push the speed of separation even higher, ultra-performance liquid chromatography (UPLC) emerged recently, which uses packaging particles with diameters less than 2 μm and are driven by pressures up to 15,000 psi. HPLC and GC are commonly used in a quality control function to characterize the purity and the identity of drug molecules. TLC and GC are rarely used in bioassays. Small LC columns have been fit to 96-well format, which can generate acceptable throughput for some applications. A high-throughput HPLC system with 24-parallel columns has been applied to bioassays with some successes. Unfortunately, the company that pioneered this application (Nanostream) ceased operation in 2008. The following discussions will focus on LC and HPLC.

4.5.2 Basic Theories of Chromatography

To help understand what happens inside the column when performing chromatography separation, we assume that we can continuously monitor two colored analytes (red and green) along the column as shown in Figure 4.1. We further assume that the two compounds have equal absorbance and they are loaded in the column at same concentration. Two processes occur inside the column. One process is the partition of the analytes between the mobile phase and the stationary phase. This interaction will affect the time it takes for the analyte to reach the end of the column (referred to as retention time). Different compounds have different retention times in liquid chromatography, which is how the two compounds are separated chromatographically. The other process is the broadening of the zone that the analyte occupies inside the column. This is caused by diffusion and unevenness of the column packaging media. Even if there is no flow, the initial loading zone will become broader with time because of diffusion. At the loading zone (position 1), the two compounds overlap in the same narrow zone. After moving down the column to position 2, the two compounds are separated incompletely with some overlapping between the two zones occupied by the two compounds. In position 2, the zones become broader and the peaks are shorter for both compounds. The shortening of the height of the two peaks is a result of the broadening of the two peaks because the total amounts of the analytes remain the same for each compound. In other words, the area under the peak, which is proportional to the total amount of analyte introduced in the column, should remain the same throughout the separation process. At position 3, the two peaks are completely separated. However, the zones are even broader and the height of the peaks is shorter for both of them. At the end of the column (position 4),

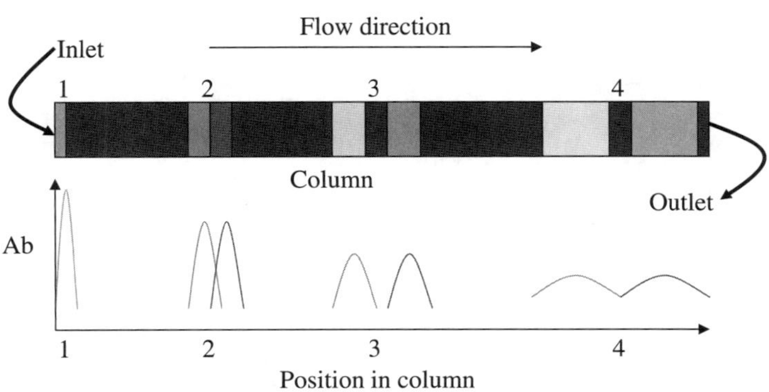

Figure 4.1 Separation of two compounds inside a column in chromatography. Two compounds with equal absorbance and at the same concentration are loaded in a column. At the loading zone (position 1), the two compounds are overlapping in the same narrow zone. After moving down the column to position 2, the two compounds are separated but with some overlapping. The zones occupied by the compounds become broader and the peaks become shorter. At position 3, the zones are even broader and the peaks are shorter. At the end of the column (position 4), the zones are very broad and the peaks are very low. (See color insert.)

there is further separation between the two peaks. However, the zones are very broad and the height of the peaks is very short, which may affect the accurate quantitation. The ideal separation condition should be based on the balance between the peak separation (favorable with a longer column or a slower flow rate) and the zone broadening (favorable with a shorter column or a higher flow rate). In this example, the ideal chromatographic condition should be between position 2 and position 3.

Below we will introduce some useful terms in chromatography. The column efficiency for a given compound can be quantitatively described by plate height (H) and plate number (N), which are related by Eq. (4.4):

$$N = \frac{L}{H} \tag{4.4}$$

where L is the length of the column packing. In ideal chromatography with symmetric peaks (Gaussian shape, see Fig. 4.2), the plate height can be calculated using Eq. (4.5):

$$H = \frac{LW^2}{16t_R^2} \tag{4.5}$$

where W is the length at the base of the analyte peak in time domain and t_R is the retention time of the analyte. The plate number is

$$N = 16\left(\frac{t_R}{W}\right)^2 \tag{4.6}$$

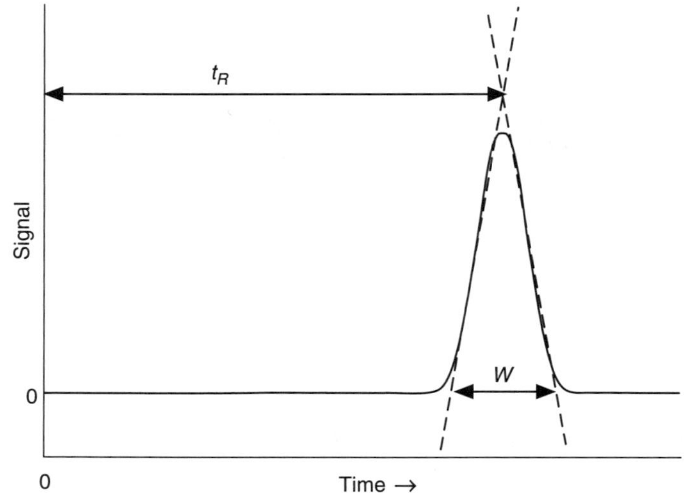

Figure 4.2 Determination of plate height and plate number with an analyte that display Gaussian-shaped chromatography. W is the base width of the triangle in the analyte elution zone and t_R is the retention time of the analyte.

The plate height and plate number are dependent on the identity of the analyte used to measure these quantities. Thus, to compare the efficiency of the two columns, the same analyte must be used to measure these quantities.

4.5.3 Types of Liquid Chromatography

There are four basic types of liquid chromatography: adsorption chromatography, partition chromatography, ion-exchange chromatography, and size-exclusion chromatography. The principle of separation for most types of chromatography is based on the chemical interaction among the analytes, the stationary phase, and the mobile phase, except for size-exclusion chromatography where separation is based on the physical sieving process.

Adsorption chromatography (also referred to as liquid–solid chromatography) is the earliest type of chromatography where bare silica or alumina particles are used as stationary phase. The separation of analytes is based on their adsorption to the highly polar silica or alumina surfaces in the stationary phase. The elution solvents are usually a mixture of organic solvents (such as hexane, toluene, and methylene chloride). More polar compounds have stronger interaction with the stationary phase and thus have longer retention times. This trend is the same as the normal phase partitioning chromatography that will be discussed next. Adsorption chromatography has a unique strength in separating isomeric compounds. This unique property of adsorption chromatography is not shared by the other chromatographic methods. In addition, adsorption chromatography is particularly powerful in separating different types of phospholipids (e.g., phosphatidylcholine, phosphatidylserine, and phosphatidylethanolamine). In contrast, it is very difficult for the other chromatographic methods to separate these phospholipids. An example of using adsorption chromatography to separate phosphatidylserine from other phospholipids is shown in Figure 4.3.

Partition chromatography has a layer of bound phase on the surface of the solid support. The analyte partitions between the bound phase and the mobile phase as it moves along the column. Different analytes have different partition coefficients

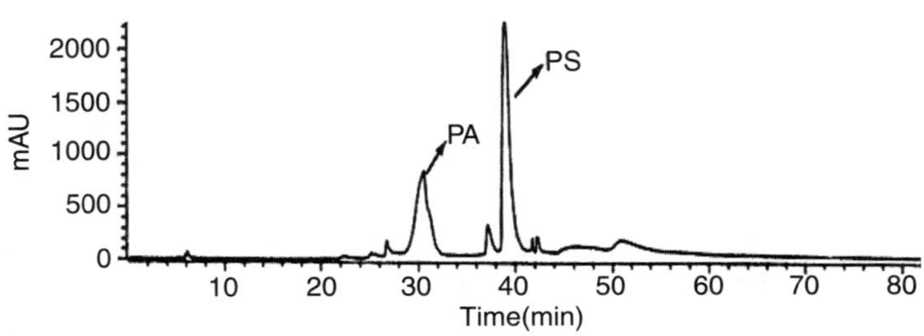

Figure 4.3 Separation of phosphatidylserine (PS) from other phospholipids with adsorption chromatography. Silica column MicroPak Si-10 with dimensions of 300 × 8 mm is used. The mobile phase is a linear gradient of 100% solvent A (hexane : isopropanol : sulfuric acid at 97 : 3 : 0.02) to 100% solvent B (hexane : isopropanol : water : sulfuric acid at 40 : 50 : 10 : 0.02). UV detection at 214 nm is used. Phosphatidic acid (PA) is one of the impurities observed here.

between the bound phase and the mobile phase and thus have different retention times. When polar compounds are bound on the stationary surface, less polar solvents (such as hexane and toluene) are usually used as the mobile phase. Polar compounds have stronger partition with the stationary phase and thus have longer retention time. This type of chromatography is called *normal-phase chromatography* because the analytes are eluted in the same order as the absorption chromatography. When non-polar compounds are bound to the solid support (such as hydrocarbons with different chain length, C4, C8, C18, etc.), more polar solutions are used as the mobile phase (water, methanol, acetonitrile). In this case, polar compounds have less partition with the stationary phase and thus have less retention time. This type of chromatography is referred to as *reverse-phase chromatography* because the order of the retention times for different analytes is the reverse of the normal-phase chromatography. The stationary phase in reverse-phase chromatography has very high surface density that causes strong interactions with hydrophobic analytes, and thus organic solvents are required to desorb the analytes. The pore size of the stationary phase in reverse-phase chromatography should be larger than 300 Å for protein separation and about 100 Å for peptide separation. The column length and binding condition is not very critical in reverse-phase chromatography as long as enough capacity exists for the analyte. The analyte is adsorbed in the column through hydrophobic interactions when starting with aqueous buffer. The analyte is eluted out of the column in organic gradient in a very narrow range. This is similar to the "on-and-off" mode. Reverse-phase chromatography is the most useful chromatography for biological applications and accounts to about three/fourths of all the HPLC operations. For example, most proteins and peptides can be analyzed with a reverse-phase column using a gradient of water [containing 0.1% trichloroacetic acid (TCA)] and acetonitrile (containing 0.1% TCA) as the mobile phase.

Ion-exchange chromatography uses ion-exchange resin as the stationary phase. The resin can be functionalized with anions (such as strong acidic sulfonic acid —$SO_3^-H^+$ and weak acidic carboxylic acid —COO^-H^+) to make a cation-exchange column or be functionalized with cations (such as a strong basic tertiary amine group or a weak basic primary amine group) to make an anion-exchange column. The separation basis of ion-exchange chromatography is the Coulombic interaction between the analytes and the stationary phase. The mobile phase in ion-exchange chromatography commonly contains two buffers. One is the adsorption buffer with a low-salt concentration at a pH that makes the analytes charged. The other is the desorption buffer with increasing salt concentration gradient that can elute the analytes from the column. One critical factor in ion-exchange chromatography is the selection of pH in the adsorption buffer. For small molecules, the pH is adjusted to make the analyte charged. The idea is to make the analytes have larger charge difference from the contaminant molecules. The anion or cation-exchange column is then chosen based on the charge of the analytes. For proteins, the situation is complicated because a protein can be positively charged when the pH is below its isoelectric point, negatively charged when the pH is above its isoelectric point, and neutral when the pH is equal to its isoelectric point. For analysis of a protein with a known isoelectric point, test at several pH values in the adsorption phase below or above its isoelectric point should be performed with corresponding cation- or anion-exchange columns to determine the optimal pH. Another important factor in-ion exchange chromatography is the selection of the pore size in the stationary phase. The analyte must be allowed to access the interior surface of the porous matrix, which is much larger than the exterior surface. Generally, the pore size should be at four times the size of the analyte. For a protein with 50 kDa molecular weight, its diameter is about 30 Å. The stationary phase with a pore size of 120 Å should be used. For larger proteins, stationary phase with a pore size of 300 Å may be used.

Ion-exchange chromatography was originally developed to separate inorganic ions. However, it has seen increased use for separation of organic ionic compounds in biological applications (such as pharmaceutical products, sugars, vitamins, and body fluid) and for separation of proteins. Ion-exchange columns have a very large capacity compared with other types of chromatography. This enables the use of very short columns to efficiently handle a large quantity of samples with high speed. For example, a miniaturized, anion-exchange chromatography method was developed to measure inositol phosphate accumulation in cells using 96-well-formatted anion-exchange minicolumns. Another application using an ion-exchange column to remove a tracer ion (Rb^+) from the bulk media in functional ion channel assay will be discussed in Chapter 9.

Size-exclusion chromatography separates analytes based on their size when they flow through a stationary phase made of beads with defined porosity. The mobile phase has access to both the volume inside the pore of the beads and the volume external to the beads. The analytes, depending on their size, may or may not be able to access the volume inside the pore of the beads. This provides the basis for separating molecules based on their size. The preferred packing stationary media are hydrophilic, nonionic, and porous uniform particles that do not interact with the analytes. They should be chemically stable in an elution solution at an operating pH and be rigid enough to withstand operating pressure. Similar to partition chromatography, the flow rate is very critical for successful separation of analytes. Too high a flow rate will lead to incomplete separation. For low-pressure applications, dextran, agarose, and acrylamide are commonly used to form the matrix with different properties (such as Sephadex, Superose, Superdex, and Trisacryl GF). The linear flow rate is usually in the range of 5 to 15 cm/h. For HPLC applications, the packing beads must withstand high pressure and the silica base matrix is commonly used in this situation. The linear flow rate is usually in the range of 300 to 400 cm/h in HPLC applications. Most common columns for size-exclusion chromatography are designed to separate proteins in the range between 30 and 300 kDa. Some columns, such as polyhydroxyethyl A from PolyLC, together with chaotropic mobile phase can resolve peptides with molecular weight from a few thousands down to less than one hundred. These columns are very useful to analyze hormones, metabolites, and amino acids.

Size-exclusion columns are often used for rapid desalting or removal of excessive small labels that did not react with the protein after labeling a protein. Many commercial small disposable premade cartridges filled with various size-exclusion media are available. Use of these columns can save a lot of time compared with the traditional dialysis method in the same applications (a couple of hours versus more than a day). An example is shown here to separate excessive small activated biotin molecules from biotin-labeled IgG antibody using a dextran-based desalting column with a 5-mL bed-volume and molecular exclusion limit of 5 kDa (Pierce Cat#43230). The column was first washed with 25 mL (5 × bed volume) of Dulbecco's phosphate-buffered saline (DPBS) at pH 7.4. After loading 340 μL biotinylation reaction mixtures to the column, the column was eluted with a DPBS buffer and fractions of eluent were collected at ~0.5 mL interval. A total of 8 fractions were collected. The absorbance of each collected fraction was measured with a Beckman Coulter UV-Vis spectrometer at 280 nm (Fig. 4.4). Fractions 4, 5, and 6 were pooled together that contain purified biotin-labeled IgG. This chromatography process resulted a dilution of the sample from the original volume of 340 μL to the final volume of 1500 μL.

In addition to the above four basic types of chromatography, *affinity chromatography* is often encountered in biological applications. In affinity chromatography, the stationary phase in the column is derived with molecules or functional groups that bind selectively to the analyte. After loading the sample containing the analyte into the column, the analyte will attach to the solid support in the column. The unbound species in the sample is washed away from the column with loading buffer. The bound analyte is then eluted out of the column with a specific eluent that can interrupt the binding between the analyte and the stationary phase. Though columns are used, affinity chromatography is different from the normal chromatography in the sense that there is no continuous interaction between the stationary phase and the analyte when the mobile phase flows through the column. The interaction between the analyte and the stationary phase is in the on-and-off mode. The basic theory of chromatography does not apply here. An important characteristic of the affinity column is the adsorbent capacity that can be estimated by frontal analysis which is a test of a small volume of adsorbent (1 mL) with a continuous flow of real test samples

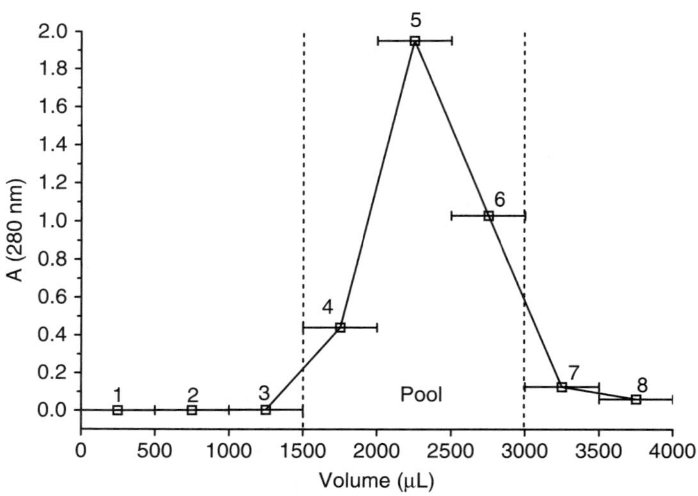

Figure 4.4 Separation of IgG antibody from small labeling reagents: 340 μL reaction mixture was loaded into the column. The column is eluted with DPBS buffer at pH 7.4. Fractions were collected at 0.5-mL intervals. The absorbance at 280 nm of collected fractions is used to monitor the elution of IgG. Fractions 4, 5, and 6 are pooled together to obtain the final labeled protein.

containing the analyte that has affinity to the adsorbent. The active molecules that are linked to the stationary phase usually bind to the analyte noncovalently. In this case, the immobilized molecule can be a ligand for a protein, an epitope for an antibody, protein A, an antibody, and so forth. Affinity chromatography is widely used to purify a specific protein, such as the purification of IgG by the protein A-derived column that specifically binds to the F_c subunits of IgG and the purification of GST fusion proteins by glutathione-derived columns. The molecules that are linked to the stationary phase can also covalently bind to the analyte of interest. For example, the affinity column with Ni^{2+} chelates linked to stationary phase through nitrilotriacetic acid (NTA) can selectively bind to fusion proteins with polyhistidine tags. This column is commonly used to purify recombinant proteins with poly-His tags.

4.6 ELECTROPHORESIS

4.6.1 Introduction to Electrophoresis

Electrophoresis is a separation method based on the difference in the migration rate of charged species in the presence of the electric field. The migration velocity (cm/s) of a charged species in an electric field is expressed as

$$v = \mu_e E \tag{4.7}$$

where E is the field strength (V/cm^{-1}) and μ_e is the electrophoretic mobility (cm^2/sV). The migration rate is proportional to the electric field strength, which in turn is proportional to the voltage applied between the two electrodes (V) and is inversely proportional to the length of the conducting fluidic path between the two electrodes. The electrophoretic mobility is proportional to the ionic charge carried by the analyte and is inversely proportional to friction retarding factors that include the size and shape of the analyte and the viscosity of the medium. The plate number (N) in electrophoresis is given by

$$N = \frac{\mu_e V}{2D} \tag{4.8}$$

where D is the diffusion coefficient of the analyte. The higher the voltage can be applied, the higher the plate number and hence higher resolution of separation. Electrophoresis can be performed in three different formats: slab electrophoresis (two dimensional), capillary electrophoresis (one dimensional), and interconnect microfluidic channels (two or more dimensional).

4.6.2 Electrophoresis on Slab Gels

Slab gel electrophoresis is carried out in a thin layer of semisolid porous gel containing aqueous electric conducting solution within the pores. The gel serves both as the physical support for the slab and as the sieving media to help separating closely related molecules. A slab gel usually has many wells that enable simultaneously separation of many samples. Slab gel electrophoresis offers very high resolution for charged macromolecules and is widely used in biological applications for proteins and nucleic acids. Because slab gel electrophoresis was the only technology that had the resolution power to separate DNAs that differ only by one nucleotide, early DNA sequencing was all based on slab gel separation. For protein analysis, the proteins are dissolved in a buffer containing sodium dodecyl sulfate (SDS) that denatures protein together with dithiothreitol (DTT) that reduces all the disulfide bonds. The negatively charged SDS binds to proteins at a ratio of about one SDS for every two amino acid residues. This leads to about the same ratio of charge to mass for most protein–SDS complexes. Thus, the separation of different proteins in slab gel is primarily based on the sieving function of the gel matrix. Slab gel made by polyacrylamide is commonly used to separate negatively charged SDS–protein complexes based on their differential migration when the electric field is applied to both ends of the slab gel. It was found that the migration of the SDS–protein complexes through the porous gel matrix is proportional to the logarithm of the mass of the protein. This method of separating denatured proteins based on their mass is commonly referred to as SDS–PAGE, which stands for sodium dodecyl sulfate–polyacrylamide gel electrophoresis. Another common method to separate proteins with slab gel is isoelectric focusing, which is based on the differential isoelectric point (pI) of different proteins. In this method, the proteins to be separated are placed in a slab gel with pH gradient. The proteins will first migrate in the electric field and then stop migration when they reach the positions at which their pIs are equal to the pH. The combination of SDS-PAGE and isoelectric focusing in two-dimensional electrophoresis is the most powerful method to separate samples with more than thousands of proteins and is widely used in proteomic studies. Slab gel electrophoresis has several disadvantages. It involves many manual steps that are difficult to automate. It is labor intensive that involves pouring the gel (though there are precast gels for some applications), loading samples, applying voltage, dissembling the gel assembly, staining the gel, destaining the gel, and imaging the gel. In addition, Joule heating caused by the electric current passing through the gel is a significant problem for slab gel electrophoresis. To reduce Joule heating, the current flowing through the gel must be reduced. The relationship between current (I) and electric field strength (E) can be expressed as

$$I \sim Er^2\kappa \tag{4.9}$$

where r is the thickness of the slab gel (or the radius of capillary) and κ is the conductivity of the solution. Thus, the solution conductivity and the thickness of the slab gel should be lowered to reduce Joule heating. However, too low a thickness will reduce the mechanical strength of the slab gel. With normal slab gel at 1 to 2 mm thickness, only 5 to 10 V/cm can be applied before excessive heating. This limits the voltage applied across the gel to less than ~500 V.

4.6.3 Capillary Electrophoresis

Capillary electrophoresis offers automated electrophoresis operation by mimicking HPLC. Here the column is a capillary with a diameter of 10 to 100 μm and a length of 40 to 100 cm. Voltage is applied between the two ends of the capillary to cause the differential migration of ionic species based on their polarity and charge-to-mass ratio. An online detector (UV, fluorescence, or conductivity) is present to allow instant data acquisition. Capillary electrophoresis can be performed with either an open tube or a polymer-filled tube. In ideal open-tube capillary zone electrophoresis, the analytes are separated solely based on the ratios of their charges to friction retarding factors (or mass with approximation). Coating of the wall of the capillary is usually required to prevent nonspecific interactions between the wall and the analytes. Capillary filled with gel has extra sieving power and is widely used in DNA sequencing and in protein sizing.

Because the current passing through the capillary is small (5 to 30 μA) and the capillary is very thin (10 to 100 μm), Joule heating is not significant at normal operating electric field strength of between 100 and 1000 V/cm in capillary electrophoresis. The high electric field strength results in a very high plate number in capillary electrophoresis [see Eq. (4.8)]. Capillary electrophoresis can have a plate number in the range of 100,000 to 200,000 as compared to 5000 to 20,000 plates for a typical HPLC chromatography. Another property of capillary electrophoresis is the small sample volume (<10 nL) introduced into the capillary. A negligible amount of sample is consumed in capillary electrophoresis. In comparison, slab gel and HPLC both require a sample in the microliter range. However, this low consumption of sample for each analysis in capillary electrophoresis can hardly be translated into sample savings because a large volume of samples in the sample reservoir is required to interface with the capillary. More than 10 μL samples are usually required in the reservoir because the issue of positioning of the capillary to contact the sample and to maintain the constant concentration and conductivity of the sample that are affected by evaporation. The same amount of sample in the reservoir is needed whether the sample is analyzed once or a thousand times (if the experiment can be done within a short period of time with negligible evaporation of the sample).

Electroosmotic flow (EOF) is a phenomenon that the bulk solution flow occurs when an electric field is applied between two ends of the capillary that is filled with buffer and the capillary wall is charged. The cause of the bulk flow is the electric double layer that develops at the column–solution interface. With a fixed voltage polarity, the direction of EOF depends on the charge on the surface of the inner column. Capillary columns made with silica are the most commonly used columns in electrophoresis. Silica is negatively charged due to the ionization of the surface silanol group. The cations in the buffer congregate in the electrical double layer adjacent to the negative surface of the silica. When a high voltage is applied, these cations are attracted to the cathode (the negative end of the applied voltage). Because the cations are solvated, they drag the bulk solvent along with them flowing toward the cathode. The velocity (v) of EOF is given by Eq. (4.10):

$$v = \mu_{eo} E \tag{4.10}$$

where μ_{eo} is the electroosmotic mobility that takes the same form as the electrophoretic mobility discussed before and E is the electrical field. In the presence of electroosmosis, the velocity of an ion is the sum of the electrophoretic velocity and electroosmotic velocity as shown in Eq. (4.11):

$$v = (\mu_e + \mu_{eo})E \tag{4.11}$$

The direction of EOF may be different from the electrophoretic flow of a particular ion. For example, the direction of EOF for an anion in silica capillary is the opposite of its electrophoretic flow direction. If the velocity of EOF is greater than the electrophoretic flow, the anion that is expected to elute out at one end of the capillary (based on electrophoretic flow) may not appear at all because it flows in the opposite direction. EOF usually is not desirable in electrophoretic separation experiments because it complicates the situation and may deteriorate the performance of the electrophoresis when the EOF is in the same direction as the ions' electrophoretic flow. Coating the capillary surface with a neutral reagent can eliminate EOF. For example, trimethylchlorosilane can react with the silanol group to eliminate the negative charge on the capillary surface and thus eliminate EOF. Sometimes the covalent modification of the capillary surface may not be strong enough and the modification reagents may be eluted out of the surface in some buffers after long periods of use and the EOF will reemerge. This is a common problem when using the same capillary in high-throughput screening applications for a long time.

Electroosmotic flow is desirable in some situations because of its superior flow profile over the flow profile of pressure-driven flow (see Fig. 4.5). In pressure-driven flow, a shear flow can increase the effective diffusivity of a species. Essentially, the shear acts to smear out the concentration distribution in the direction of the flow, enhancing the rate at which it spreads in that direction. This effect is called Taylor dispersion. Thus, pressure-driven flow has a broad parabolic flow profile. In comparison, the EOF-driven flow has a flat pluglike profile. The flat profile in EOF flow is the reason for the superior separation power of capillary electrochromatography (CEC). CEC is similar to HPLC but uses a packed capillary instead of a packed column. Both porous and nonporous particles (1 to 2 μm) can be used to pack the capillary. In CEC, the electric field is employed to inject the sample and to drive the bulk flow through the column. Since the flow profile is flat, the analytes in the column are in a pluglike narrow band versus the broad parabolic band in pressure-driven HPLC. Thus, the column efficiency in CEC is at least an order of magnitude higher than that in HPLC.

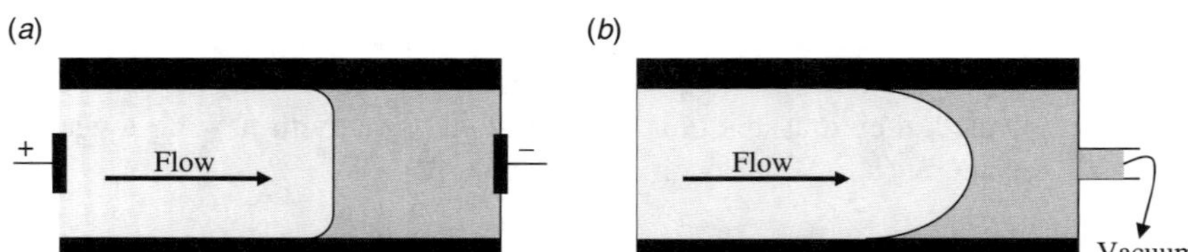

Figure 4.5 Comparison of flow profiles driven under (*a*) electroosmotic flow and (*b*) pressure. The EOF driven flow has a flat pluglike profile with a narrow band for an analyte while the pressure-driven flow has a parabolic profile resulting in significant peak broadening for an analyte.

4.6.4 Electrophoresis in Microfluidic Systems

Microfluidic systems are capillary networks fabricated in silicon, glass, or poly... substrates. The channels on the substrate are usually in the range between 20 and 100 μm, similar to the diameter of the capillary used in capillary electrophoresis. However, the lengths of the channels are much shorter (less than 10 cm) than a typical capillary in electrophoresis (longer than 40 cm). While samples in capillary electrophoresis flow in a single one-dimensional capillary, they flow in two or three dimensions in microfluidic systems. Flow of liquid in microfluidic systems can be controlled by the combination of electroosmotic flow and pressure-driven flow. Because the electroosmotic flow rates depend on the charge densities on the walls of capillaries, they are influenced by many factors including substrate material, fabrication processes, surface pretreatment procedures, and buffer additives. The separation of analytes in microfluidic systems is based on the same principle of electrophoresis. Proteins, peptides, and nucleic acids can be separated because of their different electrophoretic mobility. The detection is usually achieved with fluorescence detectors to detect labeled molecules. There are systems that use mass spectrum as the detection method to eliminate the need to label the analytes with fluorescent tags. Separation of analytes in microfluidic systems has been demonstrated in the analysis of nucleic acids, enzymes, and immunoassays. The microfluidic systems are also referred to as "lab-on-a-chip" or micro total analysis systems (μTAS). We will discuss this system further in Chapter 14.

When applying microfluidic technology in bioassays, it is important to understand the liquid flow pattern and the mixing process of liquids coming from different channels. Normally, liquid flow in a pipe can adopt one of the two different patterns of flow: turbulent flow or laminar flow. In fluid dynamics, turbulent flow is a fluid regime characterized by chaotic, stochastic property changes. This includes low-momentum diffusion, high-momentum convection, and rapid variation of pressure and velocity in space and time. The turbulent flow has the characteristic of mixing the fluidic inside the pipe. The opposite is laminar flow, also known as streamline flow. Laminar flow occurs when a fluid flows in parallel layers, with no disruption between the layers. In fluid dynamics, laminar flow is a flow regime characterized by high-momentum diffusion, low-momentum convection, and pressure and velocity independent from time. The laminar flow has the characteristic of no active mixing of fluid inside the pipe. The flow pattern of fluid in a pipe is determined by the Reynolds number (Re), which is a dimensionless number that gives a measure of the ratio of inertial forces to viscous forces. Reynolds number is calculated with Eq. (4.12):

$$\mathrm{Re} = \frac{\rho V D}{\eta} \tag{4.12}$$

where ρ is the density of the fluid, V is the mean fluid velocity, D is the diameter of the pipe, and η is the dynamic viscosity of the fluid. For fluid flowing in a pipe, Reynolds numbers of more than 2300 are generally considered to be of a turbulent type and Reynolds numbers of less than 500 are generally considered to be of a laminar type. The properties of the buffer used in bioassays can be estimated by using the properties of water. Thus, we have $\rho = 1\ \mathrm{g/mL}$ and $\eta = 0.01\ \mathrm{g/(cm \cdot s)}$ in the typical bioassay buffer. If we use the microfluidic device with the diameter of the channel at 20 μm and with a flow rate of 1 mm/s, Re = 0.02 is obtained. When the Reynolds number is much less than 1, creeping motion or Stokes flow occurs. This is an extreme case of laminar flow where viscous (friction) effects are much greater than inertial forces. Because of Stokes flow, there is no active mixing in the microfluidic channel at all. The interaction among different reactants (such as substrate and enzyme) must rely on the passive diffusion process. To enable interaction between them within a few seconds, at least one of the reactants should be relatively small so that it can access the larger reactant. Thus, a small peptide substrate is usually used to study enzymatic reactions in microfluidic assay format.

Useful Websites

http://www.biotek.com/

http://www.thermo.com/com/cda/landingpage/0,10255,128,00.html

http://beckmancoulter.com/products/Discipline/Life_Science_Research/pr_disc_gen_centrif.asp?bhcp=1

http://www.millipore.com/catalogue/module/C7781

http://www.pall.com/34696_40537.asp

http://www4.gelifesciences.com/aptrix/upp01077.nsf/Content/Products?OpenDocument&ParentId=549

BIBLIOGRAPHY

Benjamin, E. R. et al. (2004) A miniaturized column chromatography method for measuring receptor-mediated inositol phosphate accumulation. *J. Biomol. Screen.* **9**, 343–353.

Bilitewski, U., Genrich, M., Kadow, S., and Mersal, G. (2003) Biochemical analysis with microfluidic systems. *Anal. Bioanal. Chem.* **377**, 556–569.

Bruin, G. J. M. (2000) Recent developments in electrokinetically driven analysis on microfabricated devices. *Electrophoresis* **21**, 3931–3951.

Dang, F. et al. (2006) Hybrid dynamic coating with *n*-dodecyl beta-D-maltoside and methyl cellulose for high-performance carbohydrate analysis on poly(methyl methacrylate) Chips. *Anal. Chem.* **78**, 1452–1458.

Everley, R. and Croley, T. (2008) Ultra-performance liquid chromatography/mass spectrometry of intact proteins. *J. Chromatogr. A* **1192**, 239–247.

Gomis, D. B., Junco, S., Expósito, Y., and Gutiérrez, D. (2003) Size-based separations of proteins by capillary electrophoresis using linear polyacrylamide as a sieving medium: Model studies and analysis of cider proteins. *Electrophoresis* **24**, 1391–1396.

Graham, J. (2001) *Biological Centrifugation (The Basics)*. BIOS Scientific Publishers, Oxford.

Gritti, F. and Guiochon, G. (2008) Ultra high pressure liquid chromatography. Column permeability and changes of the eluent properties. *J. Chromatogr. A* **1187**, 165–179.

Huang, B., Kim, S., Wu, H., and Zare, R. N. (2007) Use of a mixture of *n*-dodecyl-beta-D-maltoside and sodium dodecyl sulfate in poly(dimethylsiloxane) microchips to suppress adhesion and promote separation of proteins. *Anal. Chem.* **79**, 9145–9149.

Huikko, K., Kostiainen, R., and Kotiaho, T. (2003) Introduction to micro-analytical systems: Bioanalytical and pharmaceutical applications. *Eur. J. Pharm. Sci.* **20**, 149–171.

Kakehi, K. (ed.) (1998) *High Performance Capillary Electrophoresis*. Wiley, New York.

Karger, B. L. and Hancock, W. (eds.) (1996) *High Resolution Separation and Analysis of Biological Macromolecules: Fundamentals*, Vol. 270, Part A. Academic, San Diego.

Kay, R. G., Gregory, B., Grace, P. B., and Pleasance, S. (2007) The application of ultra-performance liquid chromatography/tandem mass spectrometry to the detection and quantitation of apolipoproteins in human serum. *Rapid Commun. Mass Spectrom.* **21**, 2585–2593.

Kostal, V., Katzenmeyer, J., and Arriaga, E. A. (2008) Capillary electrophoresis in bioanalysis. *Anal. Chem.* **80**, 4533–4550.

Kraly, J. et al. (2006) Bioanalytical applications of capillary electrophoresis. *Anal. Chem.* **78**, 4097–4110.

Nagata, H., Tabuchi, M., Hirano, K., and Baba, Y. (2005) High-speed separation of proteins by microchip electrophoresis using a polyethylene glycol-coated plastic chip with a sodium dodecyl sulfate-linear polyacrylamide solution. *Electrophoresis* **26**, 2687–2691.

Nichol, L. W. (1974) Evaluation of equilibrium constants by affinity chromatography. *Biochem. J.* **143**, 435–443.

Roddy, E. S., Xu, H., and Ewing, A. G. (2004) Sample introduction techniques for microfabricated separation devices. *Electrophoresis* **25**, 229–242.

Tabeling, P. (2005) *Introduction to Microfluidics*. Oxford University Press, Oxford.

Vandaveer IV, W. R., Pasas-Farmer, S. A., Fischer, D. J., Frankenfeld, C. N., and Lunte, S. M. (2004) Recent developments in electrochemical detection for microchip capillary electrophoresis. *Electrophoresis* **25**, 3528–3549.

Wu, S. et al. (2007) Staining method for protein analysis by capillary gel electrophoresis. *Anal. Chem.* **79**, 7727–7733.

Yu, K., Little, D., Plumb, R., and Smith, B. (2006) High-throughput quantification for a drug mixture in rat plasma—A comparison of ultra performance liquid chromatography/tandem mass spectrometry with high-performance liquid chromatography/tandem mass spectrometry. *Rapid Commun. Mass Spectrom.* **20**, 544–552.

CHAPTER 5

GENERAL PROTEIN BINDING ASSAY FORMATS

VIRTUALLY EVERY cellular process involves the binding of a protein to another protein, a DNA sequence, a RNA sequence, a peptide, or an endogenous small molecule. Studying the binding between these binding partners plays a important role in elucidating the cellular processes, such as transcription, translation, metabolism, or signal transduction pathways. In addition, the study of binding between synthetic small molecules to target proteins is a critical step in small-molecule drug discovery. Many methods have been developed to study the protein binding processes. Some of the methods are only applicable to the interaction between a large protein and a smaller binding partner (e.g., protein polarization assays and equilibrium dialysis) while other methods do not have this limitation (e.g., surface plasmon resonance and ELISA). The most desirable method should be able to study the binding partners with minimum perturbation of the system and generate robust signals over a large range of protein concentrations and with adequate temporal resolution. Because of the need to quantify the components in the binding system, many binding assays either perturbed the equilibrium of binding by separating the binding components or chemically modified some of the binding partners. Few methods, such as the well-established equilibrium dialysis method, have negligible perturbation to the binding system. However, the throughput in an equilibrium dialysis study is too slow to be useful for a large number of test compounds. In this chapter, we will discuss the commonly used biochemical technologies to study binding processes with focus placed on the underlying principles, practices, and limitations. Cell-based assays to study protein–protein interactions in vivo [such as the yeast two-hybrid systems and protein fragment complementation assay (PCA) method] are beyond the scope of this chapter and will not be discussed. The basic principles and general equations describing binding process are discussed in Chapter 3.

Assay Development: Fundamentals and Practices. By Ge Wu

5.1 EQUILIBRIUM DIALYSIS

Equilibrium dialysis is a well-established method to study the binding between a small molecule and a large protein. The principle of this assay is based on a semipermeable membrane that separates two chambers with equal volume (Fig. 5.1). Old equilibrium dialysis chambers are large, and two sampling ports are present to allow sampling of solutions. Newer disposable dialysis apparatuses are physically very small and some of them can be adapted into 96-well format. The two chambers are separable and no sampling port is present. After equilibrium is reached in the two chambers, the two chambers are instantly separated, allowing the access to the samples in each chamber. The volume required to perform equilibrium dialysis with these disposable apparatuses can be as low as 25 μL, which offers significant savings for the valuable test samples. These small disposable dialysis apparatuses for both single use, and in 96-well format are commercially available from several vendors, such as Harvard Apparatus and Thermo Scientific/Pierce Biotechnology.

In equilibrium dialysis experiments, the protein and the small ligand are initially placed in each chamber, respectively. The membrane's permeability is based on the size of the two molecules that bind to each other. Thus, a membrane is chosen so that only the small ligand can freely pass through the membrane while the much larger protein is confined in the original chamber where it is originally placed.

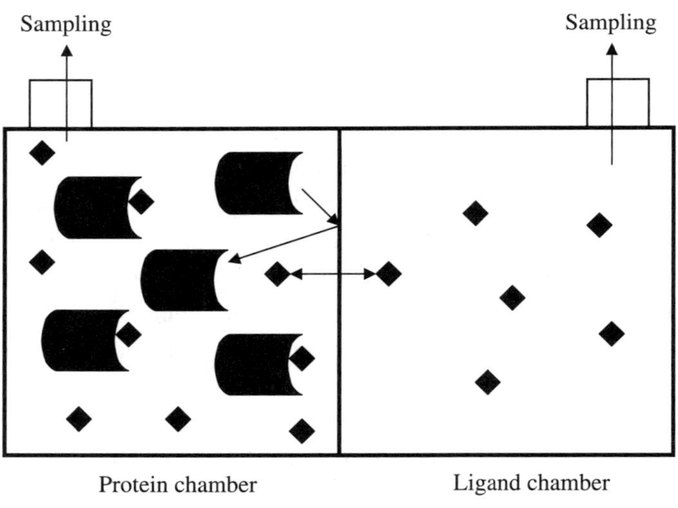

Figure 5.1 Equilibrium dialysis scheme. The protein chamber and the ligand chamber are separated by a semipermeable membrane that is permeable to the small ligand but not to the bigger protein. Proteins and ligands are loaded to their respective chambers initially. While proteins stay in the same chamber, small ligands diffuse through the membrane and reach equilibrium after shaking for a time period that can be experimentally determined. A small volume of samples is taken out from each chamber for offline analysis. Newer disposable dialysis apparatus are very small (holding ~200 μL). After equilibrium, the two chambers are separated and no sampling port is required.

Usually, low binding cellulose membranes with molecular weight cut-off (MWCO) between 5000 and 10000 Da are used in equilibrium dialysis apparatuses. The small ligand from the ligand chamber will diffuse through the semipermeable membrane to the protein chamber and bound to the protein. At equilibrium, the free ligand concentration in both chambers should be the same. However, the Gibbs–Donnan effect may result in accumulation of charged ligand near a semipermeable membrane. This phenomenon sometimes leads to the charged ligand unevenly distributed across the two sides of the membrane. To suppress this effect, 100 mM NaCl should be added to the binding buffer when the ligand is charged. Common equilibrium dialysis buffer may contain 20 mM HEPES, pH 7.4, 100 mM NaCl, and 0.1% nonionic detergent. After equilibrium is established, the two chambers are separated and a small amount of the samples from each chamber is removed from the chamber for offline analysis. The concentration of the ligand from each chamber can be analyzed by mass spectrum if it cannot be measured by molecular spectra such as fluorescence or absorbance. The equilibrium concentration of the free ligand [L] is measured from the ligand chamber. The measurement of the total ligand concentration in the protein chamber depends on the analytical methods. If the ligand is radiolabeled, directly counting the sample can be made. If HPLC is used to analyze the sample, the protein-bound ligand should be released from the protein first. This can be done by subjecting the sample to protein denaturing conditions, such as heat, organic solvent, ionic detergent, and the like. The bound ligand concentration [LE] is obtained by subtracting the free ligand concentration from the total ligand concentration measured in the protein chamber. The free protein concentration [E] can be obtained by subtracting the bound ligand concentration (if there is only one binding site on the protein) either from the initial protein concentration placed in the chamber or from the measured protein concentration in the chamber, which is more accurate because it accounts for nonspecific absorption of the protein to the chamber wall and membranes. The equilibrium constant can then be calculated according to Eq. (3.7), using the free ligand concentration, the free protein concentration, and the bound protein concentration. Alternatively, the equilibrium dialysis experiment can be performed by mixing the protein and the ligand together first, and the resulting mixture is then placed in the protein chamber. The free ligand is allowed to diffuse to the other chamber and reach equilibrium. The subsequent procedures and the analysis are the same as the method discussed above. The time for the binding partners to reach equilibrium depends on the design of the apparatus (volume-to-membrane surface ratio), the properties of the ligand, and how the whole apparatus is shaken. Generally, 4 hours are the minimum time allowed for the binding/dissociation processes to reach equilibrium between the two chambers.

The advantage of equilibrium dialysis is that the binding equilibrium is always maintained throughout the experiment. With older apparatus with large chambers, the removal of a small volume in each chamber for further analysis has negligible effect on the equilibrium. With newer small disposable chambers, the disruption of equilibrium for offline measurement of the samples does not affect the equilibrium analysis because the two chambers are instantly separated. In addition, because the samples are analyzed offline, many techniques, such as mass spectrum, can be used to analyze the sample without the need to chemically modify the ligands. Thus, equilibrium dialysis has minimum perturbation to the system and offers most reliable results for the equilibrium study. Equilibrium dialysis is inexpensive and easy to perform, and yet it proves to be the most accurate method available. It is widely used in the characterization of candidate drugs in serum binding assays and in the detailed study of antigen–antibody interactions. Equilibrium dialysis also offers the ability to study low-affinity interactions that are undetectable using other methods. The disadvantage of equilibrium dialysis is the low throughput. It is not applicable to a large number of samples. Its application is limited to studying binding processes with a small and a large molecule that can be separated by a membrane. Equilibrium dialysis is not applicable to study protein–protein interactions when the two proteins are of similar sizes. Because equilibrium dialysis is an end point assay at equilibrium, no kinetic parameters about the binding process can be obtained with this method.

5.2 COMPETITIVE BINDING ASSAYS WITH RADIOACTIVE OR OTHER LABELED LIGANDS

Competitive binding assays are based on the displacement of a known labeled ligand from the protein it binds to by test molecules. The known labeled ligand serves as a "tracer" to monitor the binding progress. A saturation binding experiment is usually performed first to establish the binding constant of the target protein and the known labeled ligand. In this experiment, the concentration of the labeled ligand is continuously increased to achieve saturation of all the binding sites on the target protein. The quantity of the label (radioactivity count, fluorescence, etc.) associated with the target protein is measured at different concentrations of the labeled ligand and the results are plotted to obtain a Langmuir isotherm graph. The concentration of the target protein and the labeled ligand is then fixed, usually with both concentrations much less than the determined K_d value from the isotherm graph. This system is then used to assay potential molecules that may displace the labeled ligand from the target protein in competitive binding studies. The mathematical model for competitive binding and related equations was discussed in Chapter 3.

Radioactive isotope-labeled ligands are the most used ligands in early competitive binding studies. One example is radioimmunoassay (RIA), which was originally developed to quantify the antigen by measuring the displacement of radioactive or "hot" antigen from its antibody–antigen complex by nonradioactive "cold" antigen. Figure 5.2 shows the experimental scheme to perform RIA. The hot antigen at a fixed concentration and the cold antigen at various concentrations are mixed with the antibody. After equilibrium is established, the large free- and ligand-bound antibody are separated from the small hot and cold antigen based on their size difference by either membrane filtration under vacuum (Fig. 5.2*a*) or gel filtration chromatography (Fig. 5.2*b*). These two separation technologies are discussed in Chapter 4. After separation, the radioactivity associated with the antibody is measured. By varying the concentration of the cold antigen, a dose–response curve can be plotted and the IC_{50} value can be obtained. This strategy can be used to measure the binding between other proteins and their binding partners. For example, the binding between receptors in biomembranes isolated from cells and their ligand can be studied in this way. When

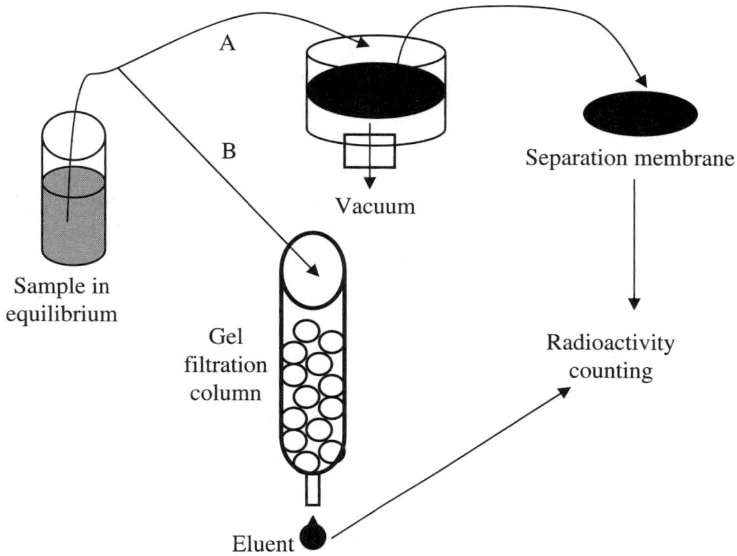

Figure 5.2 Illustration of radioimmunoassay (RIA). The radioactive ligand, nonradioactive ligand, and protein are mixed and allowed to reach equilibrium. The mixture is then applied to membrane separation (A) or gel filtration chromatography (B) to separate the protein-bound radioligand from the free radioligand. The protein-bound radioligand is then measured with liquid scintillation counting. The techniques of gel filtration and membrane filtration are discussed in Chapter 4.

studying interaction between purified proteins and radioligand, equations obtained from Chapter 3 can be directly applied to analyze the data. However, special treatment must be performed if the study involves membrane preparation that will be discussed below.

Receptors are cell membrane proteins that signal many cellular changes when they bind to small ligands. Cell surface receptors include nuclear receptors, G protein-coupled receptors (GPCR), neurotransmitter receptors, receptors tyrosine kinases (RTKs), and so forth. Because receptors are membrane proteins, they cannot be purified the same way as normal soluble proteins. They are usually studied with crude membrane preparations that contain the receptors and many other proteins in the membrane. The binding between membrane-associated receptor and ligand has been studied extensively with a radioligand as a tracer. Unlike with purified proteins, binding assays performed with receptors in biomembrane are complicated by the large amount of nonspecific binding of the radiolabeled tracer to the membrane. It was observed that the magnitude of nonspecific binding increases with increasing concentration of the tracer with no apparent saturation, and the more hydrophobic the tracer is the more nonspecific the binding. These evidences indicate that the membrane may be the primary source for nonspecific binding because the membrane favors absorption of hydrophobic molecules, and the membrane provides a huge reservoir to absorb molecules with no apparent saturation. The nonspecific binding significantly reduced the detection window between signal and background of the assay, which is the major caveat of using crude membranes in the assay. To account for the contribution from nonspecific binding to the total measured radioactive signal, the magnitude of nonspecific binding at different concentrations of the tracer has to be measured and be subtracted from the total signal. One way to obtain the nonspecific binding data is to oversaturate the binding sites on the receptor with cold ligand and then measure the radioactivity associated with the membrane at different concentrations of the radioactive tracer. Because the nonspecific binding is not saturatable, the high concentration of the cold ligand has a minimum effect on the nonspecific binding of the radioactive tracer. A typical radioactive ligand binding isothermal graph is shown in Figure 5.3.

When performing competitive binding assays with radioligands, the ligand saturation experiment is performed first to obtain the K_d value. If directly determining the binding between the protein and hot ligand by scanning through all concentrations of hot ligand, too much radioactivity would be used, especially when high concentrations of radioligand are used to oversaturate the protein binding site to obtain a complete curve. Alternatively to the saturation experiment, the K_d value can be obtained by performing a competition assay when the binding between a fixed concentration of the hot ligand and the protein is measured with different concentrations of the cold ligand. From Eq. (3.40) and taking the fact that $K_d = K_i$ since the hot and the cold ligands are the same molecule with different isotope substitution, we obtain

$$\mathrm{IC}_{50} = K_i\left(1 + \frac{[\mathrm{L}]}{K_d}\right) = K_d + [\mathrm{L}] \tag{5.1}$$

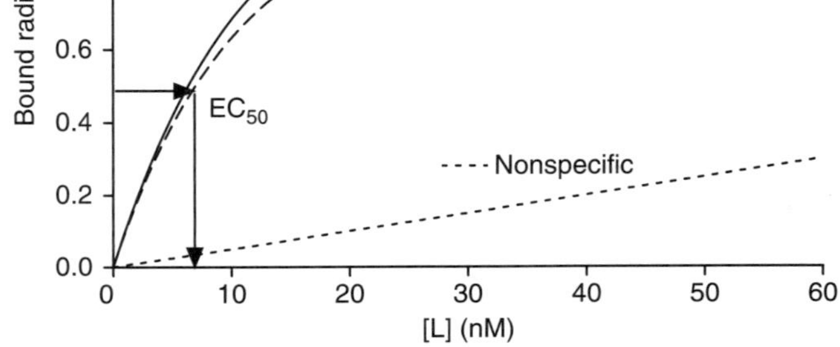

Figure 5.3 Typical receptor binding isotherm graph. The measured total membrane-bound radioactivity comes from both the radioligand bound to the receptor and the radioactivity due to nonspecific binding to membranes. The total bound radioactivity and the nonspecific radioactivity can be obtained experimentally at different concentrations of the ligand. The true receptor-bound radioactivity can be obtained by subtracting the nonspecific radioactivity from the total radioactivity. The EC_{50} or the K_d value is then obtained.

where K_d is the dissociation constant for the binding between the tracer and the protein and [L] is the concentration of the hot ligand. It is obvious that if the hot ligand concentration is much higher than the K_d, the IC_{50} value will remain relatively constant and K_d cannot be determined with certainty. In practice, a reasonable concentration of hot ligand is picked first based on previous experience. The IC_{50} value is than determined. If the obtained IC_{50} value is close to [L], the experiment is repeated with lower hot ligand concentration, and the IC_{50} value is redetermined until it is more than twice the hot ligand concentration. The K_d value can then be calculated using Eq. (5.1).

Because of environmental concerns, many efforts have been made to substitute the radioligand with fluorescence or time-resolved fluorescence ligand. Two major obstacles need to be overcome to replace the radioisotope with other kind of tracers. First, the new tracer should be detectable at low concentration because high concentration of the ligand will significantly reduce the assays sensitivity by shifting the IC_{50} value in competitive binding assay as shown is Eq. (5.1). Up to now, very few technologies have the lower detection limit that is comparable with radioactivity detection (e.g., DELFIA technologies from PerkinElmer, which employ time-resolved fluorescence detection of lanthanide). Second, all other types of label will have to introduce a new detectable group into the original ligand, which may change the binding character of the ligand. Experiments have to be performed to qualify the newly modified ligand.

The competitive binding assay discussed above relies on the separation of proteins and their binding partners based on their difference in size. Thus, this method is only applicable to studying the binding between a large protein and a small molecule. In addition, the separation step significantly perturbed the equilibrium by disrupting the equilibrium between the small molecule and the protein before the separation is completed. In this assay scheme, the off-rate for the dissociation of the bound complex must be low enough so that the tracers are not being continuously washed away in a membrane separation method or continuously dissociate in the gel filtration column. In addition, the number of wash cycles and the duration of each wash cycle in membrane separation method and the elution volume in the gel filtration chromatography have to be controlled to be consistent between experiments. Except radiolabeled tracers, the other tracers rely on chemical modification of the original ligand structure and cause further perturbation to the original equilibrium system.

5.3 APPLICATION OF SPA AND FLASHPLATE IN BINDING STUDIES

The radioactive competitive binding assay can also be performed in SPA and FlashPlate formats (both were discussed in Chapter 2). When performing competitive binding assay in SPA or FlashPlate formats, the target protein is first attached to the surface of the beads or the surface of the FlashPlate. The binding of the target protein to the surface can be direct covalent binding, direct noncovalent absorption, or through a capturing reagent that was attached to the solid surface. For example, SPA beads and FlashPlate with attached streptavidin are commercially available. The target protein can be labeled with biotin first and is then attached to the surface of SPA beads or

FlashPlate through the binding between biotin moiety and streptavidin. After the target protein is attached to the surface of the SPA beads or FlashPlate, radiolabeled ligands and potential competitive inhibitors are mixed with the SPA beads or are placed in FlashPlate. After equilibrium is reached, the signal can be measured with scintillation counting.

The advantage of performing RIA in these two assay formats over traditional competitive binding studies is that no separation of the binding partners is required and thus there is no disruption of equilibrium. Chromatographic separation and solid surface washing techniques are known to cause large variations in measurement. The elimination of the separation step not only simplified the assay but also reduced the overall errors associated with the assay. The disadvantage of RIA assays in these two assay formats is that the attachment of the target protein may alter the properties of the protein. In addition, the target protein in these two assay formats is attached to a two-dimensional surface that may change the kinetic and equilibrium of binding to its ligand.

5.4 APPLICATION OF FLUORESCENCE POLARIZATION (FP) IN BINDING STUDIES

We discussed the principle of FP and its application in binding studies in Section 2.8. An example is presented here to illustrate how FP is applied in the study of binding process between a peptide and a protein.

5.4.1 Background of the Assay

The *p53* gene is a well-known tumor suppressor gene. It is located on chromosome 17p 13.1 and codes for a 393-amino-acid nuclear phosphoprotein. Inactivation of *p53* is a common event in the development of most types of human cancers. About half of the cancer cases analyzed thus far involve mutation of one *p53* allele combined with the deletion of the second allele, and many of the remaining cases involve a functional inactivation of p53 protein through nonmutational mechanisms. The p53 protein is a critical participant in a signal transduction pathway that mediates a G1 cell cycle arrest and apoptotic cell death in mammalian cells after stress. Inactivation of the p53 tumor suppressor protein has been established as a major mechanism of cancer progression.

The protein coded by the murine double minute 2 gene (*mdm2*) is a cellular phosphoprotein with an apparent molecular mass of 90 kDa (p90) that forms a complex with both mutant and wild-type *p53*. The *mdm2* gene enhances the tumorigenic potential of cells when it is overexpressed. The p53 protein can be inactivated in human cancers through binding *hdm2* (the human version of *mdm2*). The *hdm2* gene is frequently amplified and overexpressed in sarcomas and gliomas. It was hypothesized that the interruption of the binding between p53 and hdm2 may lead to cancer therapy. The mdm2 protein was found to bind with great specificity to short synthetic peptides derived from the N-terminus of p53. Several synthetic peptide libraries derived from p53 sequences and an alanine substitution series at the optimal binding site of p53 have been used to establish the mdm2 binding site to the sequence TFSGLW (aa 18–23) in mouse and TFSDLW in humans (aa 18–23). Free synthetic peptides derived from the phage-selected sequences proved to be up to 100 times stronger inhibitors of the p53–mdm2 interaction than the p53-derived wild-type peptide in ELISA assays. Because ELISA assays are tedious and have a large error in the assay due to several wash steps, an FP assay for the binding between the p53-derived peptide and hdm2 is developed for screening small molecule libraries to find molecules that may interrupt the interaction between hdm2 and the p53-derived peptide.

5.4.2 FP Assay Design

A small peptide derived from the phage display study that contains the consensus binding sequences in the amino terminus of p53 and has the highest affinity for mdm2 is chosen as the labeled binding partner for hdm2. The labeled ligand (Pept1L) has an FITC moiety attached at a lysine group in the carboxyl terminus. The amino terminus is capped by acetylation and the carboxyl terminus is modified to amide. It has the following sequence: Ac-M-P-R-F-M-D-Y-W-E-G-L-N-**K***-NH2. The peptide is dissolved in DMSO. The binding protein employed in this study is a GST fusion protein with residues 1–188 of hdm2 (GST-hdm2). Based on the principle of FP, the labeled peptide should have a low anisotropy value when it is free in solution and should have a large anisotropy value when it is bound to hdm2. Inhibitors that compete with the labeled peptide for the binding site on the GST-hdm2 will release the labeled peptide to the solution and result in reduction in the overall measured anisotropy value. Because of the unique properties of the FP assays, the labeled ligand should be fixed at a low concentration and the protein concentration is varied to obtain the EC_{50} value. Both equilibrium and kinetic studies are demonstrated here. In the equilibrium studies, the saturation binding is studied first, followed by the competitive binding studies with known peptides as the controls. Kinetic studies are performed with a stopped-flow device coupled with the fluorescence polarization detection.

5.4.3 Saturation Binding Experiment

A 120-μL assay buffer (50 mM NaCl, 10 mM HEPES, pH 7.6, 1 mM ethylenediaminetetraacetic acid (EDTA), 0.1% NP-40, 2 mM DTT, 4% DMSO) is added to each well in a 96-well black polypropylene plate. The 15-μL 100 nM Pept1L in the assay buffer was transferred to each well. The 15-μL GST-hdm2 at different concentrations in assay buffer was then added to each well. The final assay volume was 150 μL/well with 4% DMSO and 10 nM Pept1L. The mixture was incubated for 30 min at room temperature. The amplitudes of fluorescence in parallel and vertical planes were measured with an excitation wavelength at 485 nm and an emission wavelength at 530 nm. In a parallel control experiment, Pept1L was substituted by fluorescein and the identical experiment was performed to verify that the binding to GST-hdm2 was from the peptide sequence but not from the fluorescein. The data are shown in Figure 5.4. The experimental data are fitted to the Hill equation using nonlinear regression with a floating Hill coefficient. The following parameters are

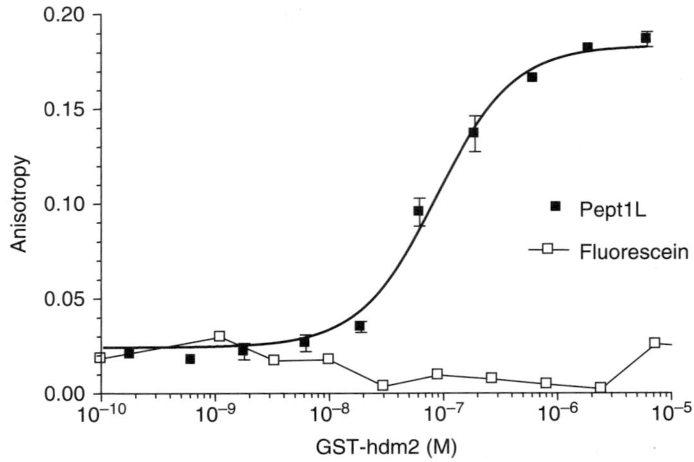

Figure 5.4 Saturation binding between Pept1L and GST-hdm2. 10 nM of Pept1L or fluorescein was mixed with different concentrations of GST-hdm2. The anisotropy was then measured. The data for Pept1L is nonlinear fitted to the Hill equation. The Hill coefficient is 1.2 ± 0.1 and the EC50 (K_d) is 88 ± 10 nM. There was no binding between fluorescein and GST-hdm2.

obtained: Hill coefficient is 1.2 ± 0.1 and the EC_{50} (K_d) is 88 ± 10 nM. There is no binding between fluorescein and GST-hdm2. Another similar control experiment was also performed to demonstrate that Pept1L did not bind to GST. In this experiment, the fusion protein GST-hdm2 was substituted by GST and the same FP measurement was performed with Pept1L. Thus, the binding between GST-hdm2 and Pept1L is specific.

5.4.4 Stopped-Flow FP Measurement of the Binding Kinetics Between Pept1L and GST-hdm2

In previous fluorescence polarization measurements, Pept1L and GST-hdm2 were mixed and incubated for about half an hour to reach equilibrium. We assumed that the equilibrium between the two binding partners was reached in that time window and the binding constant, K_d, is obtained. In order to understand the binding kinetics between Pept1L and GST-hdm2 and hence to determine the minimum incubation time required for the binding, continuous measurement of the changes in anisotropy immediately after mixing was performed. Because many protein–protein binding processes reach equilibrium in a subsecond time scale, stopped-flow technique with the same temporal resolution is ideal to study the binding kinetics. In FP measurements, the total light output, from both parallel plane and perpendicular plane, is collected and thus the total fluorescence intensity changes are measured simultaneously with the FP measurements. Because fluorescence intensity often changes when the fluorophore changes from an aqueous to a hydrophobic environment in the protein binding process, total fluorescence intensity measurement offers another way to monitor the binding process.

The stopped-flow experiment was performed by first placing equal volumes of 30 nM Pept1L and 620 nM GST-hdm2 in two separate syringes. The two solutions were then rapidly mixed in the millisecond scale, and the mixed solution was excited with radiation at 485 nm and with 4.65 nm bandwidth that is selected by two tandem monochromators. The emission light was collected with a 520 nm longpass cut-on filter. The light emission from parallel and perpendicular planes was collected by a photodiode. Figure 5.5*a* shows the anisotropy changes immediately after Pept1L

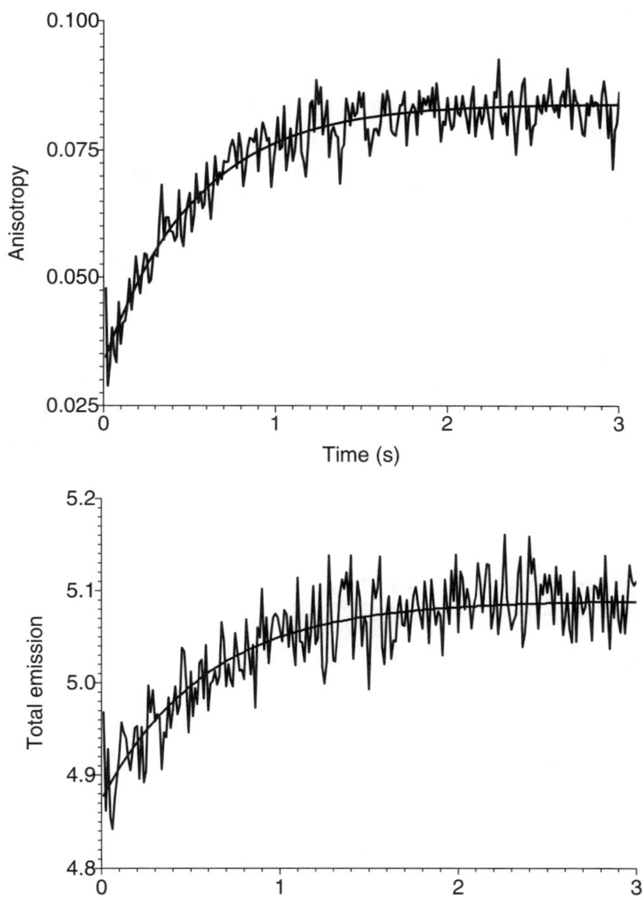

Figure 5.5 Stopped-flow kinetic studies of the binding between Pept1L and GST-hdm2. Equal volumes of 30 nM Pept1L and 620 nM GST-hdm2 in assay buffer were rapidly mixed and the binding was measured with FP. The sample was excited at 485 nm with polarized light, and the emission was detected with a 520-nm longpass cut-on filter on parallel and perpendicular planes. The emission was collected within 1 ms after mixing the two samples. The anisotropy and the total emission was plotted against mixing time. The k_{obs} from anisotropy measurement and from the total emission measurement were $1.9 \pm 0.1\ s^{-1}$ and $1.7 \pm 0.1\ s^{-1}$, respectively.

was mixed with GST-hdm2. The increase in anisotropy is expected since the molecular motion of Pept1L decreases after it binds to GST-hdm2. The data fit well to first-order kinetics with one binding site indicating that only one binding site on GST-hdm2 exists to bind to Pept1L. The binding has an observed $k_{obs} = 1.9 \pm 0.1\ s^{-1}$. Figure 5.5*b* shows the total fluorescence intensity changes in the same experiment. A small increase ($\sim$2%) in total fluorescence intensity is observed. This may be due to the change of the fluorophore from the aqueous environment to a hydrophobic environment after binding to a protein. The observed $k_{obs} = 1.7 \pm 0.1\ s^{-1}$ is the same as obtained with the anisotropy measurement. This is expected because the two measurements were monitoring the same binding process with the same sample. Because the change in total fluorescence intensity is very small after Pept1L bound to GST-hdm2, no attempt was made to modify the molar fraction (it is approximately the same as the emission intensity fraction in the anisotropy measurement). If the change is significant, the molar fraction of the bound peptide obtained above must be calculated differently to factor in the changes in fluorescence emission (see Chapter 2). From Eq. (3.26), we obtain $k_{obs} = k_1 * [\text{GST-hdm2}] + k_{-1} = 1.9\ s^{-1}$. In the stopped-flow experiment, the concentration of GST-hdm2 was diluted by a factor of 2 when it was mixed with an equal volume of the ligand. Thus, [GST- hdm2] = 320 nM. From the saturation binding experiment, we calculated that $K_d = k_{-1}/k_1 = 88$ nM. Substituting these values into the above equation leads to $k_1 = 4.6 \times 10^6\ M^{-1}\ s^{-1}$ and $k_{-1} = 0.41\ s^{-1}$. These values are within the range of typical binding kinetics between a peptide and a protein. Though the fluorescence total intensity change was also a true measurement of the process, the signal window was much smaller and the signal-to-noise ratio was much smaller as well compared with the anisotropy measurement. Thus, the anisotropy measurement is the preferred choice in this situation.

5.4.5 Competition of Other Peptides for the Pept1L Binding Site on GST-hdm2

A few peptides from the phage display library were used to evaluate the competitive binding in the fluorescence polarization assay. The assay condition for competitive binding was chosen at 10 nM Pept1L and 500 nM GST-hdm2 based on the binding curve shown in Figure 5.4. With this condition, a little over 50% of the Pept1L labels bound to GST-hdm2 could be displaced. If the chosen concentration of GST-hdm2 is too high, it would take a higher concentration of competing peptide to displace the Pept1L and the assay would be less sensitive. If the chosen concentration of GST-hdm2 is too low, there would not be many Pep1L-bound GST-hdm2 at the start. Thus, the assay window would be too small. After setting up the assay conditions, concentrations of the competing peptides over a wide range were used to compete for the binding site on GST-hdm2. Figure 5.6 shows the competition curve of four peptides having different affinity for GST-hdm2. The IC_{50} values for each of the peptides were determined by nonlinear regression using Eq. (3.41). The most potent competing peptide is 12/1, which gave an IC_{50} value of $2.4 \pm 0.4\ \mu M$.

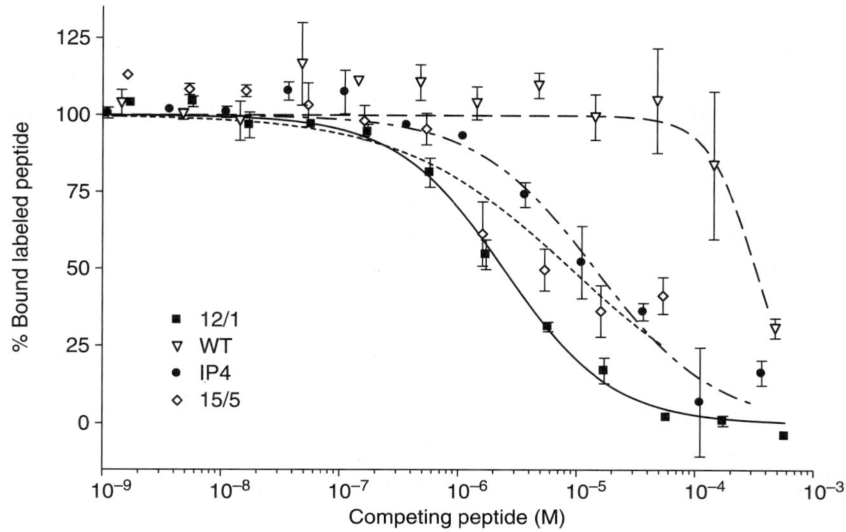

Figure 5.6 Competitive binding studies. Four peptides were tested for their competition for the labeled Pept1L binding to GST-hdm2. The concentration of Pept1L was set at 10 nM and the concentration of GST-hdm2 was set at 500 nM. The concentration of the competing peptides was varied up to 500 μM.

5.5 APPLICATION OF FRET ASSAYS IN BINDING STUDIES

Fluorescence resonance energy transfer can only happen between two small organic fluorophores when the distance between the two is within 30 Å (see Chapter 3). This is not very useful for many applications. On the other hand, FRET between a lanthanide chelate and other fluorophores can happen when the pair is separated by 100 Å or more. The long distance makes the FRET technology very useful for bioassays. In addition to FRET, lanthanide chelates as the donor also enable TR-FRET because of their long half-life. The most commonly used TR-FRET assays are HTRF (CysBio) and Lance (PerkinElmer). The two assay formats are very similar. In addition to FRET technology, AlphaScreen from PerkinElmer, which will be discussed in Chapter 7, can detect protein binding using similar strategy as FRET. We introduced the strategy to study protein–protein binding in HTRF format using *p53* and *hdm2* as an example in Chapter 2 (Fig. 2.14). More details of this particular assay are discussed here to illustrate the general strategy to apply HTRF in ligand–protein binding studies.

5.5.1 HTRF Assay Design

In FP assays discussed previously, the highest affinity peptide derived from p53 bound to hdm2 with a K_d of 88 nM. Using this peptide as the tracer in the FP assay format, the GST-hdm2 concentration had to be raised to about 500 nM in order to perform the competition study. Such a high concentration of protein results in a very insensitive assay as evidenced by the experimentally obtained high IC_{50} values in the competitive binding assay. In comparison, the concentration of GST-hdm2 is not required to be much higher than its binding partner in an HTRF assay. However, the proper ratio between concentration of the biotin-labeled peptide and the concentration of GST-hdm2 must be established. To obtain the proper ratio between the two, the initial concentration of GST-hdm2 is set at 5 nM and different concentrations of biotin-p53 up to 20 nM is scanned. Because the binding stoichiometry between biotin and XL665-derived streptavidin is approximately 2 (modified streptavidin sometimes only has 2 of the original 4 binding sites left after derivatization), SA-XL665 concentration should be higher than half of the biotin-p53 concentration. The concentration of anti-GST-Eu was recommended by the manufacturer to give 10,000 to 20,000 B counts (emission at 620 nM), which is proportional to total europium concentration (usually between 0.5 and 2 nM). From the results obtained in a kinetic FP study, the binding between the highest affinity peptide and GST-hdm2 has fast on and off rates in the time scale of seconds. It is reasoned that the binding between these peptide and the GST-hdm2 should reach equilibrium in the second scale as well. Thus, the four components in the HTRF assay can be mixed together. When enough time was given to allow both sample mixing and interaction (e.g., 30 min), the mixture would have enough time to reach equilibrium. When the mixture in equilibrium was excitation at 340 nm, the emission signal was acquired at 665 nm (A channel) and 620 nm (B channel) and their ratio A/B was plotted as shown in Figure 5.7. The purpose of the ratio treatment is to normalize the FRET signal with excited europium to account for any differences in the concentration of europium between samples in the microplate. From results obtained in this experiment, it was determined that ~7 nM

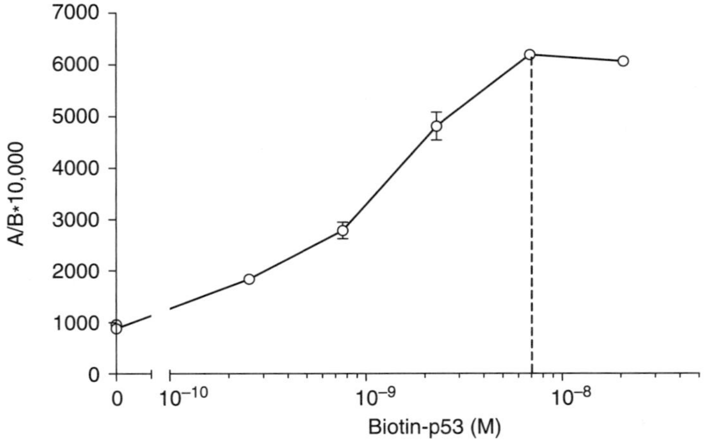

Figure 5.7 Testing assay conditions in HTRF assay for the binding between a p53-derived peptide and GST-hdm2. Concentration of anti-GST-Eu is fixed so that it gave B count of 15,000. The concentration of GST-hdm2 was 5 nM and the concentration of SA-XL665 was 25 nM.

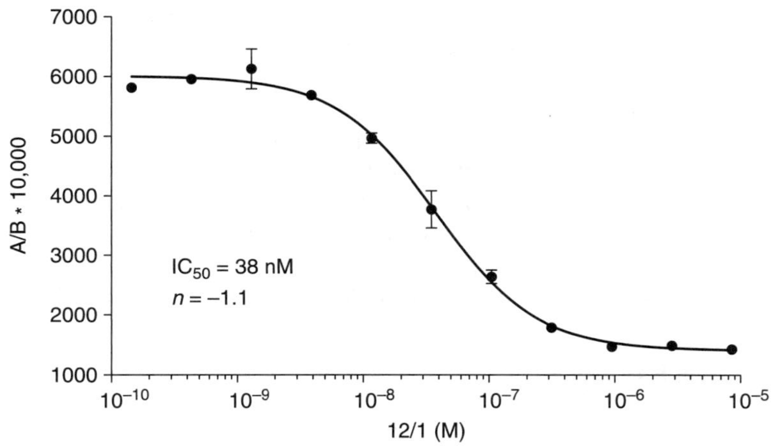

Figure 5.8 Competition assay in HTRF format. The experiment was performed at room temperature by incubating a p53-derived peptide and the competing 12/1 peptide with the master assay solution for 30 min. The 12/1 peptide blocked the binding between the biotin-p53 peptide and GST-hdm2 and reduced the HTRF reading. The IC_{50} of 12/1 peptide is 38 nM in this study.

biotin-p53 should be used in the assay. The concentration of the SA-XL665 could be reduced accordingly to match the concentration of biotin-p53.

5.5.2 Competitive Binding Assay

The final concentrations of each component in the HTRF assay obtained from the above experiment were: 7 nM biotin-p53, 5 nM GST-hdm2, 10 nM SA-XL665, and anti-GST-Eu (giving 15,000 B count). These reagents were mixed together to form a master assay solution. To measure the activity of a competing peptide with this assay, different concentrations of the competing peptide (e.g., peptide 12/1) were mixed with the master assay solution and the mixture was incubated at room temperature for 30 min. The FRET signal was then measured. Figure 5.8 shows the data obtained when peptide 12/1 was tested with the HTRF assay. Nonlinear regression was performed to fit the data to the Hill equation and the following parameters were obtained: $IC_{50} = 38 \pm 8$ nM and $n = -1.1$. In comparison, the IC_{50} value for the same peptide was 2.4 μM in FP assays. Thus, the HTRF assay is about 63 times more sensitive than the FP assay in this application.

5.6 APPLICATION OF ELISA IN BINDING STUDIES

The enzyme-linked immunosorbent assay (ELISA) was developed in the 1960s based on the principle of immunoassay with an enzyme rather than radioactivity as the reporter label. The use of an enzyme allowed amplification of the signal by enzymatic reactions that convert a large amount of substrate into product. Commonly used enzymes in ELISA are horseradish peroxidase (HRP) and alkaline phosphatase (AP). HRP is usually preferred to AP because the size of HRP is smaller relative to AP and it is less prone to nonspecific binding to solid surfaces. The small size of HRP also makes it less likely to interfere with immunological recognition of the epitope by enzyme–antibody conjugates. The substrates of the enzymes used in ELISA are molecules that generate optically detectable colorimetric, chemiluminescent, or fluorescent products. ELISA has been widely used to quantify proteins, peptides, hormones, and other small molecules. There are several ways to configure ELISA. We will discuss four commonly employed ELISA formats here as shown in Figure 5.9.

5.6.1 Direct Assay

In this ELISA format, the enzyme is directly linked to the primary antibody. The antigen can be either directly linked to the surface of a microplate by covalent bond or by absorption to a high binding surface. After blocking remaining nonspecific

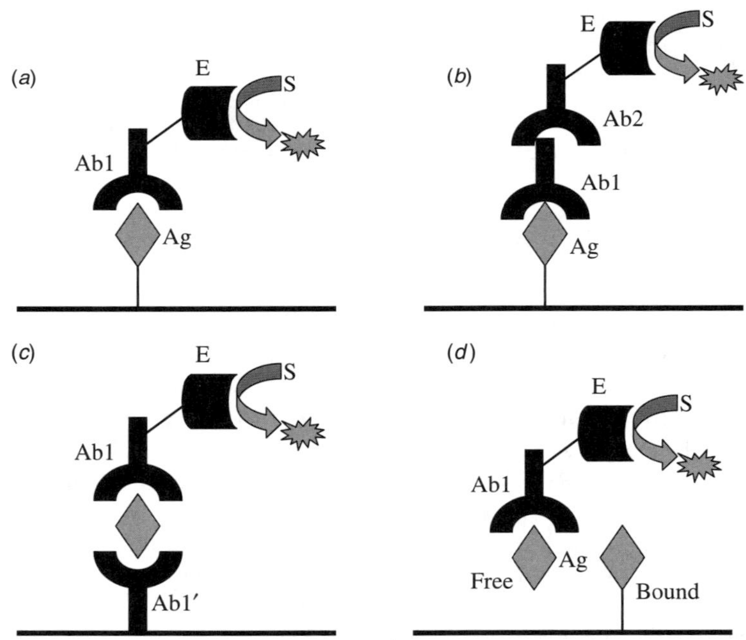

Figure 5.9 Illustration of four common ELISA formats. (*a*) Direct assay: The enzyme is directly linked to the primary antibody. (*b*) Indirect assay: The enzyme is linked to a secondary antibody that recognizes the primary antibody. (*c*) Sandwich assay: Two primary antibodies for different epitopes in the same antigen are used. One of the antibody serves as capturing reagents to fix the antigen onto the microplate surface and the other antibody is used for detection. (*d*) Competitive assay: Free antigen competes with the surface-bound antigen for the binding to primary antibody resulting in reduced the signal. Ag: antigen; Ab1 and Ab1′: primary antibody; Ab2: secondary antibody; E: Enzyme; S: substrate.

binding sites on the surface and several wash cycles, the enzyme-linked primary antibody is added to the microplate and allowed to reach equilibrium. After several wash cycles, the substrate for the enzyme is added to the microplate and the mixture is incubated until sufficient signal is generated for detection. The microplate is then measured. The advantages of direct assay are the short procedure and the elimination of secondary antibody that may cause potential problems of cross-reactivity with components in the antigen sample. However, direct labeling of the primary antibody requires extra assay development efforts and may reduce the antibody's immunoreactivity by the labeling process. There is less flexibility in this assay format because each primary antibody is labeled for a given assay.

5.6.2 Indirect Assay

In this ELISA format, the early procedures are similar to the direct assay. However, the primary antibody is not linked to the enzyme. A secondary antibody, usually from a different species that recognize the Fc subunits of the primary antibody (e.g., rabbit to goat), is linked to the enzyme. This procedure added one more step to the assay. The advantage of indirect assay is no need to covalently link the enzyme to the primary antibody. This assay offers the flexibility to quickly develop many ELISA assays because a wide variety of enzyme-liked secondary antibodies are readily available commercially. The disadvantage of this assay is the potential problems of cross-reactivity of the secondary antibody with components in the antigen samples.

5.6.3 Sandwich Assay

In this ELISA format, a primary antibody against the antigen is attached to the microplate surface instead of directly attaching the antigen to the surface. The antigen is then captured to the microplate by the immobilized primary antibody. Another primary antibody recognizing different epitopes on the antigen is added to the microplate. The remaining procedures are the same by following either direct or indirect assays. Though more assay steps are added to the procedure, the sandwich assay provides more specificity for recognizing the antigen because of double antibody selection. Thus, the signal-to-noise ratio is much higher in this assay.

5.6.4 Competitive Assay

In this ELISA format, the enzyme is attached to either the primary antibody or the antigen. The competition between the nonlabeled primary antibody or antigen and the labeled counterpart of the binding partner is then measured. In the example shown in Figure 5.9, the primary antibody is labeled with the enzyme. The analyte that can bind to the antibody competes with the surface-bound antigen resulting in a reduced signal.

These ELISA formats can be adapted to study protein–ligand binding. For example, the binding between a biotin-labeled peptide derived from p53 (biotin-p53) and GST-hdm2 can be adapted in the ELISA format as shown in Figure 5.10. The biotin-labeled p53-derived peptide (biotin-p53) can be captured

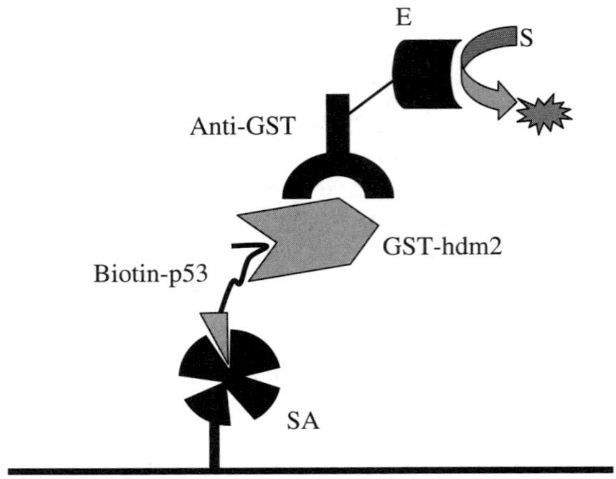

Figure 5.10 ELISA assay for the binding between p53-derived peptide and GST-hdm2. Streptavidin (SA)-labeled microplate was used to capture biotin labeled p53-derived peptide (biotin-p53). Hdm2 protein co-expressed with GST (GST-hdm2) bound to biotin-p53. Horseradish peroxidase (HRP) linked anti-GST antibody then bound to the GST moiety of GST-hdm2. Addition of a fluorogenic substrate of HRP generated a fluorescent signal. The potential inhibitors that could disrupt the binding between biotin-p53 and GST-hdm2 would reduce the fluorescent signal.

with a streptavidin (SA)-labeled microplate, which is commercially available. After washing the microplate to remove the unbound biotin-p53 peptide, GST-hdm2 is added to the microplate to allow its binding to biotin-p53. After washing away the unbound GST-hdm2, anti-GST antibody linked with HRP (anti-GST-HRP) is added to the microplate to allow binding to the GST moiety in GST-hdm2. After washing away unbound anti-GST-HRP, a substrate for HRP is added to the microplate and the signal is measured.

Different substrate for HRP can be used to generate the absorption, fluorescence, or chemiluminescence signal. However, the signal strength and the noise produced in these three substrates are different as shown in Figure 5.11. Because absorption measurements are less sensitive, more than tens of micromolar of light-absorbing product must be generated by HRP to produce a detectable signal. This in turn requires longer incubation time between HRP and the substrate. In comparison, a fluorescent signal is detectable with only a few nanomolar fluorescent products. This in turn requires only a short incubation time for the accumulation of enough fluorescent products for detection. If the reaction kinetic parameters are the same, fluorescence-generating substrate can produce thousand-folds more detectable signals than the chromatic substrate within the same incubation time. HRP substrates that generate chemiluminescence in the presence of HRP are commonly used as well. Unlike fluorescence and absorbance measurement that benefit from the accumulation of signal over time because of the accumulation of the light absorbing or fluorescent product, lights produced from chemiluminescent reaction does not accumulate. Thus, the signal does not increase with prolonged incubation in either flash luminescence or stable luminescence. With luminescence substrate, the signal should be measured as

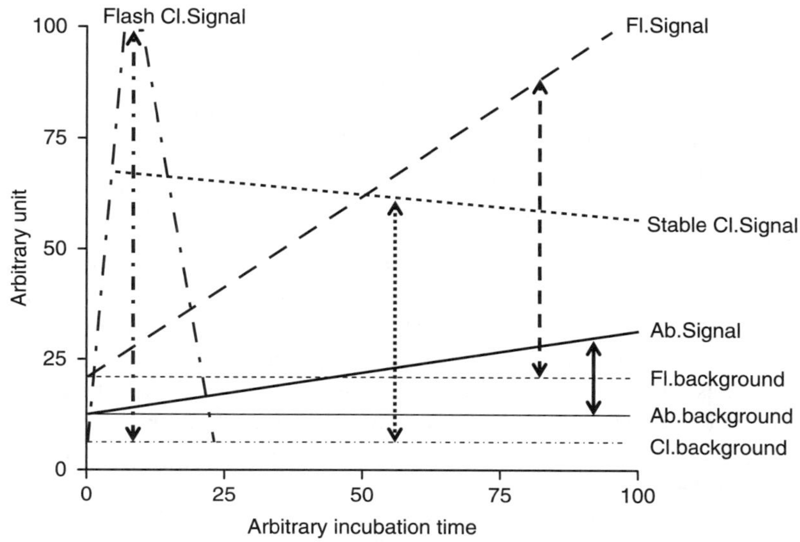

Figure 5.11 Comparison of the absorption, fluorescence, and chemiluminescence signals generated from different substrates of HRP. The background of fluorescence (Fl.background) is represented by the thin dash line; the background of absorption (Ab.background) is represented by the thin solid line; and the background of chemiluminescence (Cl.background for both flash and stable chemiluminescence) is represented by thin dash/dotted line. The signal of flash chemiluminescence is represented by thick dash/dotted line; the signal of stable chemiluminescence is represented by thick dotted line; the signal of fluorescence is represented by thick dash line, and the signal of absorption is represented by thick solid line.

soon as the peak light output is produced. However, chemiluminescence detection offers the lowest noise among the three detection methods.

5.7 SURFACE PLASMON RESONANCE (SPR) TECHNOLOGY AND ITS APPLICATION IN BINDING STUDIES

5.7.1 Principles of SPR

When a beam of light passes through material with a high refractive index (e.g., glass) into material with a low refractive index (e.g., water), some light is reflected and some light is refracted from the interface. However, when the incident light is above a certain *critical angle*, the light is completely reflected at the interface with no light being refracted. This phenomenon is referred to as "total internal reflection." While incident light is totally reflected internally, the electromagnetic field component penetrates a short distance into a medium of a lower refractive index creating an evanescent wave. If the surface of the glass facing the water side is coated with a thin film of a metal (e.g., gold), and the incident light is monochromatic and polarized, the intensity of the reflected light is reduced when the angle of incident light falls into a narrow range due to the resonance energy transfer between evanescent wave and surface plasmons. This phenomenon is called surface plasmon resonance. Surface plasmons are oscillating plasma (mobile electrons) waves at the surface of the metal film. When the evanescent wave vector matches the wavelength of the surface plasmons, the electrons resonate and produce SPR. When plotting the reflected light intensity versus the angle of incident light, a dip is observed. The surface plasmon resonance angle (θ_{spr}) is the angle of incident light at which the largest reflected light intensity loss occurs.

The evanescent (decaying) electrical field associated with the plasma wave can only travel a short distance (~300 nm) into the medium from the metallic film. Thus, SPR only depends on the material absorbed onto the metal film within this depth. This is an ideal situation for bioassays because it enables selective observation of interactions within a predefined distance without the need for separation (similar strategy with SPA, HTRF, AlphaScreen, etc.). In addition, a linear relationship is found between resonance energy and the mass of biological molecules, such as proteins, sugars, and DNA, attached to the surface. This technology enables the detection of binding process by monitoring the changes in mass and no label is required for detection. Thus, SPR is also a label-free detection technology.

5.7.2 Biacore Technology

Biacore (which became part of GE Healthcare in 2006) carried the most successful implementation of SPR technology in bioassay and had revolutionized the field of protein binding and kinetics studies. Biacore combined SPR and microfluidic technology to create a very sensitive and stable platform. The Biacore chip is mounted onto flow cells inside the Biacore instruments. Sample injection is automated and switching assay solution is rapid. It is especially useful for rapid kinetic studies. In a typical Biacore chip, the gold surface is coated with one layer of dextran polymer. The surface is functionalized so that a molecule of interest can be attached to the surface. Biacore offers chips with surfaces containing several commonly used capturing reagents, such as streptavidin. Biacore instruments use response units (RUs) as the measure for the changes in θ_{spr} with 1 RU equivalent to a shift of 10^{-4} degrees. Empirical measurements have shown that the binding of 1 ng/mm^2 of protein to the sensor surface leads to a response of ~1000 RUs. Since the dextran matrix is ~100 nm thick, this represents a protein concentration of 10 mg/mL within the matrix. When designing Biacore experiments for binding studies, it is preferable to immobilize the smaller binding partner on the chip surface and allow the larger binding partner to flow through the surface because of the balance of mass in binding. For example, when the binding between an analyte with a molecular weight of 200 Da and a protein with a molecular weight of 40,000 Da (with only one binding site for the analyte) is studied, the mass ratio between the two components in the pair is 200. This means to obtain 50 RUs of signal if the small molecule is used in the flow phase, 10,000 RUs of proteins should be immobilized on the sensor chip. Achieving this level of immobilization is very difficult because of the limited space on the dextran matrix. If lowering the protein concentration on the sensor chip, the detectable signal will be lower than 50 RUs, which may be problematic with older Biacore instruments (such as Biacore 2000). In contrast, only 5 RUs of small 200-Da molecules needs to be immobilized to obtain 1000 RUs of binding signal from the large 40,000-Da protein when the small molecule is bound to the chip surface. Though newer Biacore instruments, such as T100 and S51, can detect a few RUs reliably, immobilizing smaller binding partners on the chip surface still offers advantages in terms of assay quality. Overall, biological needs will dictate the way the assay is designed. In the applications section, an example of immobilizing a large FGFR fusion receptor on the surface to evaluate the association of a series of ligands is presented.

Biacore experiments are performed by first mounting the sensor chip to the instrument's flow cell interface. The sensor chip usually contains two to four parallel channels, and solution flow can be controlled so that it can sequentially go through all the channels or it can go through any one of the channels individually. This flexibility allows each channel surface to be modified differently in the immobilization step, and allows the same solution containing the binding partners to flow sequentially through all the channels. The experiment results obtained from different experiments performed on the different surfaces of the channel are more consistent when the same test solution is used. Usually the first channel that the solution flows through does not contain the immobilized binding partner to serve as a control. After mounting the chip, the solution already in the flow cell is exchanged by continuously injection of running buffer (HBS: 10 mM HEPES, 150 mM NaCl, 3 mM EDTA, 0.005% P-20). The RU is always measured continuously in Biacore experiments. After the RU is stabilized, a small volume (usually 5 to 100 μL) of solution that contains the binding partner to be immobilized on the chip is injected in each channel individually and is followed by injection of the running buffer. When the RU is stabilized, the new RU is compared with the RU before immobilization to estimate the amount of the immobilized substance in each channel. After a satisfactory level of immobilization of one of the binding partners on the chip is achieved, the analyte (the other binding partner) is injected into the flow cell. The association of the analyte with its binding partner on the chip is monitored in real time by following the change in RU. After a predefined time, the solution is switched to running buffer and the dissociation of the analyte from its binding partner is measured (refer to Fig. 5.12). Many of the binding pairs do not dissociate or dissociate slowly, and a chip regeneration solution (usually with low pH) is required in most cases to break the bond to regenerate the chip for the next experiment. The Biacore chip is not disposable because of the cost and the effort spent to generate a good working chip. A good chip usually can last a long time and can be used to perform many experiments. In comparison, many of the other non-SPR-based technologies to study binding process employ disposable detection materials (such as the special microplates used in Epic and Bund instruments and the special tips used in Octet instruments, which will be discussed in next section).

The detection limit for the binding kinetics with Biacore technology covers a great range. For association measurements, Biacore can detect k_a at about $5 \times 10^6\ M^{-1}\ s^{-1}$, which is only about an order lower than the diffusion limit. The detection limit for dissociation measurements is at k_d about $10^{-4}\ s^{-1}$. This is limited by the instruments stability over a long period of time that may cause the RU shifted by the same amplitude as the decrease of RU caused by dissociation in the same time period.

Because the SPR measurement is critically dependent on the detection of reflected light, the refractive index of the sample solution should match that of the running buffer when performing experiments with Biacore instruments. Small refractive index differences can be corrected by subtracting the signal from the control channel. However, large differences in the refractive index are problematic. DMSO, which is commonly used as a solvent for many small organic molecules, and glycerol, which is commonly used in protein stock solutions, have refractive indexes significantly different from water and should be avoided in Biacore assays. The analyte samples should be dissolved and diluted, if necessary, in the running buffer to match the refractive index between the sample solutions and the running buffer solution.

5.7.3 Applications of Biacore Technology

Two examples are shown here to illustrate the application of Biacore technology in binding studies. The first example shows how Biacore is applied in studying the binding process between p53 and hdm2, which has been employed as an example to demonstrate different technologies throughout this chapter. Because there was no biological requirement constraint, the assay was configured by attaching the smaller peptide to the chip surface to take advantage of the mass balance as discussed earlier. These experiments were performed with the old Biacore 2000 instrument. The second example discusses the application of Biacore to evaluate the affinity of potential natural ligands [all the known fibroblast growth factors (FGFs)] for a particular FGF receptor. Newer Biacore T100 and Biacore X instruments were used in this study. These instruments are very sensitive and can detect changes of 1 RU (about 1 pg/mm^2 changes of mass on the sensor surface) reliably if the change is over a short period (faster than the instrument drifting). Because of the high sensitivity of the instruments and the biology requirement, the large FGF receptor was immobilized on the chip surface and different FGFs were included in the mobile phase. The basic theory and assay procedure in Biacore technology remains the same, but the sensitivity of the instruments has increased dramatically from Biacore 2000 to T100.

Competition Assay for the Binding Between a p53-Derived Peptide and GST-hdm2 A Biacore chip with streptavidin immobilized on the surface was used in this experiment. After mounting the chip to the liquid path in the Biacore 2000 instrument, the solution in the chip was changed with the running buffer (HBS buffer and other components) by injecting a large volume of running buffer. A biotin-labeled p53-derived peptide (biotin-p53) was then injected into the chip and the level of biotin-p53 immobilized onto the chip surface was monitored through the change in RU. A small amount of biotin-p53 was injected initially, and it was gradually increased to achieve an appropriate level of immobilization as judged by the change in RU. The first channel on the chip was not subjected to biotin-p53 and it served as the control. The remaining three channels were immobilized with different amounts of biotin-p53 so that the binding of GST-hdm2 can be studied at three different levels of biotin-p53. A large volume of running buffer was then injected to achieve a stable baseline. Different concentrations of GST-hdm2 were injected and the association process between the GST-hdm2 with the surface bound biotin-p53 was monitored. The assay results obtained from one channel with appropriate level of biotin-p53 was shown here (Fig. 5.12). After switching to the running buffer, the

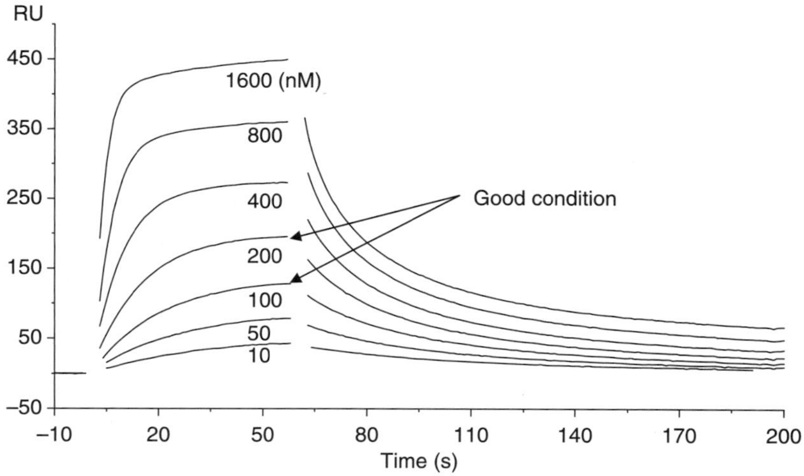

Figure 5.12 Association and dissociation between different concentrations of GST-hdm2 and the biotin-p53 peptide attached to the surface on the Biacore chip. With the present level of biotin-p53 immobilized on the chip, GST-hdm2 in the concentration range between 100 and 200 nM could be used to screen for competitive inhibitors.

dissociation process was monitored. Based on the results in this experiment, the assay conditions to study competitive binding by other molecules to GST-hdm2 were determined. It was found that the GST-hdm2 concentrations between 100 and 200 nM give a sufficient level of RU (120 to 180) changes that could be reliably measured by a Biacore 2000 instrument with this particular level of biotin-p53 immobilized on the chip surface. Higher concentrations of GST-hdm2 resulted in a large amount of nonspecific binding (they were quickly washed away when switching to running buffer and complicated the analysis of the data). Potential inhibitors were screened with this condition by premixing them with GST-hdm2, and the mixtures were allowed to flow through the chip surface. Figure 5.13 shows the inhibition of the binding of 200 nM GST-hdm2 to the chip by one of the p53-derived peptides known to bind to hdm2.

Binding Between an Fc-FGFR Fusion Receptor (180 kDa) and FGFs (∼20 kDa) There are four FGF receptors and about 20 FGFs discovered so far. The purpose of this project is to use Biacore to screen all available FGFs against a particular human Fc-FGFR fusion protein to rank the pair according to the association constant. Because the Fc-FGFR contains Fc, the strategy is to immobilize the Fc-FGFR to the chip surface through protein A, which is covalently attached to the chip surface. Each FGFs are then allowed to flow through the surface to measure its association to and dissociation from the bound Fc-FGFR. This strategy allows the proper orientation of the Fc-FGFR so that the binding sites are available for FGFs. The chip can be regenerated by well-established protocol to break the binding between protein A and the Fc portion of the fusion protein with a buffer at pH 2.5. There are

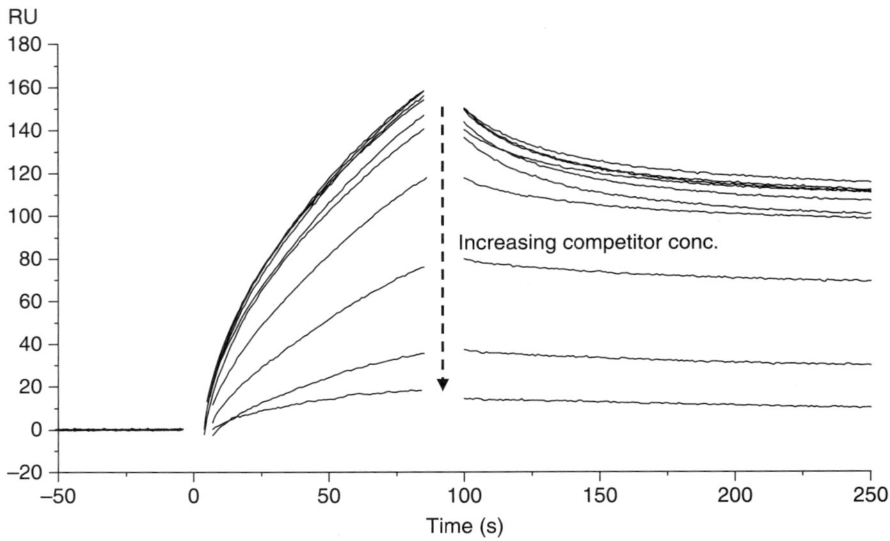

Figure 5.13 Competition for the binding between GST-hdm2 and a p53-derived peptide immobilized on the Biacore chip. The 200-nM GST-hdm2 was mixed with different concentrations of an inhibitor peptide. The mixture was then allowed to flow through the surface of the Biacore chip with a immobilized p53-derived peptide.

many advantages to immobilize the receptor on the surface for this experiment. Because many FGFs will be used to scan their interaction with the receptor, one chip with immobilized receptor is required for the whole experiment. In contrast, the same number of chips as the number of FGFs under study ($\sim$20) will be required to perform the whole experiment if FGFs are immobilized. Generating so many chips not only is expensive and time consuming, it is very difficult to compare the assay results as well because the FGFs immobilized on each chip will have different immobilization levels. In addition, there is no proper way to orient the FGFs on the chip surface and a random activation method must be used. The FGF binding site may be destroyed during the modification process or it may not be accessible to their binding partner after immobilization due to stereo inference. A typical binding and dissociation curve between FGFs and FGFR is shown in Figure 5.14. The detailed procedure of performing the experiment with Biacore is shown below, which is similar to most general Biacore-based binding studies.

1. *Activation of CM5 Chip and Covalent Attachment of Protein A (Fig. 5.15)* A solution containing 200 mM 1-ethyl-3-(3-dimethylaminopropyl)-carbodiimide (EDC) and 50 mM N-hydroxysuccinimide (NHS) in 10 mM NaAc at pH 4.5 was freshly made and injected immediately to the system at 10 μL/min for 7 min over the CM5 sensor chip. After washing with the running buffer, 205-RU shift was obtained. Protein A at a concentration of 50 μg/mL in 10 mM NaAc at pH 4.5 was then injected at 20 μL/min for 5 min. After washing with running buffer, 2950 RUs of protein A were immobilized on the chip surface.

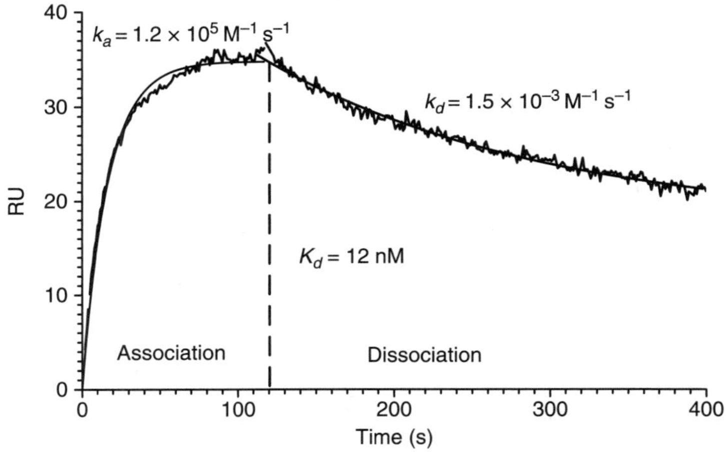

Figure 5.14 Measurement of binding kinetics between an Fc-FGFR fusion protein and FGF19 with Biacore technology. The Fc-FGFR was immobilized to the chip surface through its binding to protein A that was covalently attached to the chip surface. FGF19 was then injected to the chip and the association process was measured. After about 120 s, the flow fluid was switched to an HBS running buffer and the dissociation was measured.

2. *Blocking Remaining Reactive EDC on Surface of Chip* (*Fig. 5.16*) A solution of 1 M $EtNH_2$ at pH 8.5 was injected at flow rate of 10 μL/min for 7 min, which destroyed all remaining reactive EDC on the chip surface. The chip regeneration solution containing 10 mM glycine at pH 2 was then injected at 10 μL/min for 1 min. This procedure is repeated once.

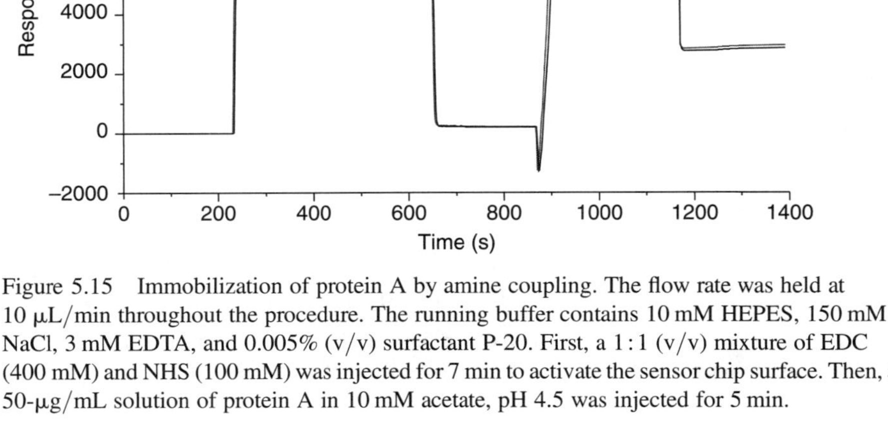

Figure 5.15 Immobilization of protein A by amine coupling. The flow rate was held at 10 μL/min throughout the procedure. The running buffer contains 10 mM HEPES, 150 mM NaCl, 3 mM EDTA, and 0.005% (v/v) surfactant P-20. First, a 1 : 1 (v/v) mixture of EDC (400 mM) and NHS (100 mM) was injected for 7 min to activate the sensor chip surface. Then, a 50-μg/mL solution of protein A in 10 mM acetate, pH 4.5 was injected for 5 min.

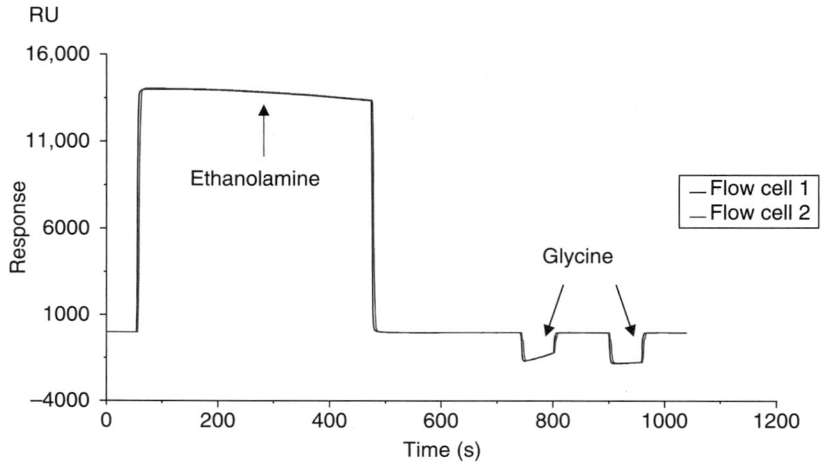

Figure 5.16 Blocking and regenerating chip surface. 1 M ethanolamine, pH 8.5 was injected to block any remaining activity on the surface. This is followed by two injections of the 10-mM glycine pH 2.0 regeneration solution. Each regeneration injection was performed at 10 μL/min for 1 min. The procedure resulted in each flow cell having approximately 2900 RU of protein A immobilized on each flow cell.

3. *Immobilization of Fc-FGFR onto Chip Surface* (*Fig. 5.17*) A solution of 2 μL Fc-FGFR at 10 μg/mL in HBS buffer containing 0.1% glycerol was injected at 10 μL/min to channel 2, leaving channel 1 as control. The total Fc-FGFR attached to the surface is not high enough as judged by the RU. Another 3 μL were injected. Finally 628 RUs were obtained.

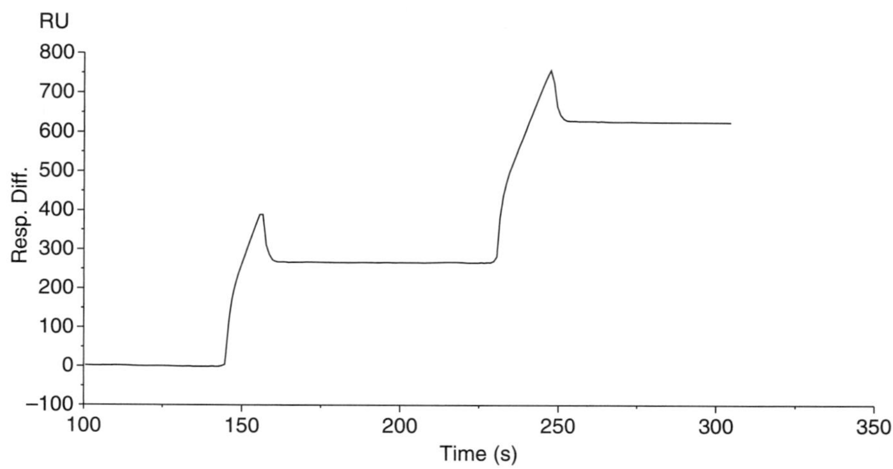

Figure 5.17 Capture of Fc-FGFR onto the chip surface with protein A attached. 2 μL and 3 μL of 10 μg/mL Fc-FGFR were injected sequentially to obtain final 628 RU.

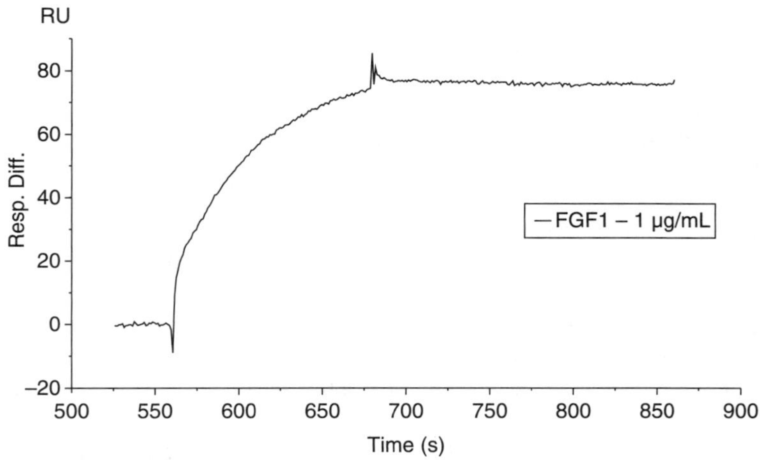

Figure 5.18 Association and dissociation of FGF1 binding to Fc-FGFR. FGF1 in HBS buffer containing 1.25 mg/mL heparin was injected at 30 μL/min for 2 min, followed by 3 min of dissociation by washing with running buffer. This sensorgram is reference subtracted.

4. *Test Binding Between Surface-Bound Fc-FGFR and FGF1 (Fig. 5.18)* A 60-μL solution of FGF1 at 1 μg/mL was injected at 30 μL/min for 2 min to obtain the association curve. The running buffer is then injected to obtain the dissociation curve. The surface was then regenerated with 10 mM glycine pH 2.0 for next use. The association and dissociation constant is obtained by the analysis software provided by Biacore. In this case, the dissociation is too slow and the dissociation constant cannot be obtained reliably.

5.8 APPLICATION OF LABEL-FREE TECHNOLOGIES IN BINDING STUDIES

Though Biacore has dominated the field of protein binding kinetic studies for over a decade, many companies have tried to introduce competing technologies in similar applications. Corning has developed an instrument (EPIC) to measure the binding of molecules as small as ~300 Da to its binding partner that is attached to a sensor surface. The principle of the technology is based on the refractive index changes within 200 nm of the sensor surface. The sensor surface incorporates nanostructured optical grating. The grating of sensor reflects only a single wavelength resonant light that is dependent on the refractive index of the sensor surface. When a biomolecule or cell binds to the biosensor surface, the refractive index will change and the reflected wavelength will increase. Real-time binding can be observed by monitoring the shift in the wavelength of the reflected light over time. The detection limits is 5 pg/mm^2 changes of mass on the sensor surface according to the company. Both biochemical and cell-based assays can be performed with this assay format. This technology is compatible with microplates because there is no fluid path and is implemented in 96- or 384-well microplates. Thus, this technology has found increasing use in high-throughput screening applications. In comparison, the Biacore technology offers a very slow throughput. In addition to Corning's EPIC systems, SRU Biosystems has developed similar technology (BUND) based on the same principles.

ForteBio offers a label-free system to detect the binding process of a biomolecule to a biosensor surface. The instrument is called Octet and utilizes proprietary single-use biosensors that look like pipette tips. The biosensors are made of optical fibers that can transmit light. There is an optical coating layer at the tip of the sensor. The optical surface is coated with a biocompatible matrix that interacts with the solution to be tested in a microplate well. The Octet instrument shines white light from the top of the tip down to the biosensor surface and collects the light reflected back. The reflected light is originated from two surfaces. One component of the reflected light is from the interface with the optical layer and the other reflected light is from the surface of the biocompatible layer. The two light waves, propagating back from the two reflecting surfaces, interact with each other, resulting in constructive or destructive interference depending on the phase and wavelength. While the wavelength and the phase of the reflected light from the optical surface remain constant, the phase of the reflected light from the surface of the biocompatible layer will change depending on the thickness of the layer that in turn is dependent on the mass of the binding molecules. This interference is captured by a spectrometer as a pattern of intensity variation by wavelength with a characteristic profile of peaks and troughs. Any change in the number of molecules bound to the biosensor tip changes the optical path, causing a shift in the interference pattern that can be measured (as $\Delta\lambda$) in real time. Because the diameter of the Octet biosensor tips is very small, only a small volume of the test solution that is close to the end of the tip is sampled. Shaking of the sample in the solution is required to allow homogenous measurement of all the components in the sample. On the other hand, this property enables the measurement of a variety of samples that may contain particulates by letting the sample settle and only the solution phase is measured. In contrast to other label-free technologies mentioned above, the measurement with Octet is not dependent on the refractive index. The technology is still new, and its sensitivity is still not high enough to detect small molecules at the current stage. However, it does offer many advantages over the well-established Biacore technology in terms of ease of use, not dependent on media refractive index, and cost effectiveness, especially when studying the association of large biomolecules (e.g., antibodies) to the biosensor surface.

Useful Websites

http://biacore.com/lifesciences/index.html

http://www.corning.com/lifesciences/us-canada/en/epicsystem/

http://www.srubiosystems.com/

http://www.fortebio.com/

http://www.harvardapparatus.com/

http://www.piercenet.com/

www.nestgrp.com

BIBLIOGRAPHY

Allen, M., Reeves, J., and Mellor, G. (2000) High throughput fluorescence polarization: A homogeneous alternative to radioligand binding for cell surface receptors. *J. Biomol. Screen.* **5**, 63–69.

Banks, P., Gosselin, M., and Prystay, L. (2000) Impact of a red-shifted dye label for high throughput fluorescence polarization assays of G protein-coupled receptors. *J. Biomol. Screen.* **5**, 329–334.

Banks, P., Gosselin, M., and Prystay, L. (2000) Fluorescence polarization assays for high throughput screening of G protein-coupled receptors. *J. Biomol. Screen.* **5**, 159–167.

Böhm, H.-J. and Schneider, G. (eds.) (2003) *Protein–Ligand Interactions: From Molecular Recognition to Drug Design*. Wiley, New York.

Cabantous, S. and Waldo, G. S. (2006) In vivo and in vitro protein solubility assays using split GFP. *Nat. Methods* **3**, 845–854.

Chen, X., et al. (2007) High-throughput screen for small molecule inhibitors of Mint1-PDZ domains. *Assay Drug Dev. Technol.* **5**, 769–784.

Colquhoun, D. (2006) The quantitative analysis of drug-receptor interactions: A short history. *Trans Pharmacol. Sci.* **27**, 149–157.

Connors, K. A. (1987) *Binding Constants: The Measurement of Molecular Complex Stability*. Wiley, New York.

Cooper, M. A. (2006a) Non-optical screening platforms: The next wave in label-free screening? *Drug Disc. Today* **11**, 1068–1074.

Cooper, M. A. (2006b) Optical biosensors: Where next and how soon? *Drug Disc. Today* **11**, 1061–1067.

Davenport, A. P. (ed.) (2005) *Receptor Binding Techniques*, 2nd ed. Humana, Totowa, NJ.

Fang, Y. (2006) Label-free cell-based assays with optical biosensors in drug discovery. *Assay Drug Dev. Technol.* **4**, 585–595.

Fu, H. (ed.) (2004) *Protein–Protein Interactions: Methods and Applications*. Humana, Totowa, NJ.

Goodrich, J. A. and Kugel, J. F. (2007) *Binding and Kinetics for Molecular Biologists*. Cold Spring Harbor Laboratory Press, Cold Spring Harbor, NY.

Goutelle, S., et al. (2008) The Hill equation: A review of its capabilities in pharmacological modelling. *Fund. Clin. Pharmacol.* **22**, 633–648.

Harding, S. E. and Chowdhry, B. Z. (eds.) (2001) *Protein–Ligand Interactions: Hydrodynamics and Calorimetry*. Oxford University Press, Oxford.

Karlsson, R. (2004) SPR for molecular interaction analysis: A review of emerging application areas. *J. Mol. Recognit.* **17**, 151–161.

Klussmann, E. and Scott, J. (eds.) (2008) *Protein–Protein Interactions as New Drug Targets*. Springer, Berlin.

Lane, P. M. F. a. D. P. (2004) Small-molecule inhibitors of the p53 suppressor HDM2: Have protein–protein interactions come of age as drug targets? *Trends Pharmacol. Sci.* **25**, 343–346.

Lequin, R. M. (2005) Enzyme immunoassay (EIA)/enzyme-linked immunosorbent assay (ELISA). *Clin. Chem.* **51**, 2415–2418.

Newton, P., Harrison, P., and Clulow, S. (2008) A novel method for determination of the affinity of protein: Protein interactions in homogeneous assays. *J. Biomol. Screen.* **13**, 674–682.

Park, S. H. and Raines, R. T. (2004) Fluorescence polarization assay to quantify protein–protein interactions. *Methods Mol. Biol.* **261**, 161–166.

Phillips, K. and Cheng, Q. (2007) Recent advances in surface plasmon resonance based techniques for bioanalysis. *Anal. Bioanal. Chem.* **387**, 1831–1840.

Porstmann, T. and Kiessig, S. T. (1992) Enzyme immunoassay techniques. An overview. *J. Immunol. Methods* **150**, 5–21.

Sadana, A. (2006) *Binding and Dissociation Kinetics for Different Biosensor Applications Using Fractals*. Elsevier, Amsterdam.

Stefan, E., et al. (2007) Quantification of dynamic protein complexes using Renilla luciferase fragment complementation applied to protein kinase A activities in vivo. *Proc. Nat. Acad. Sci.* **104**, 16916–16921.

Strange, P. G. (2008) Agonist binding, agonist affinity and agonist efficacy at G protein-coupled receptors. *Br. J. Pharmacol.* **153**, 1353–1363.

Wells, J. A. and McClendon, C. L. (2007) Reaching for high-hanging fruit in drug discovery at protein–protein interfaces. *Nature* **450**, 1001–1009.

Winzor, D. J. and Sawyer, W. H. (1995) *Quantitative Characterization of Ligand Binding*. Wiley-Liss, New York.

Wong, S. and Witte, O. N. (2004) The BCR-ABL story: Bench to bedside and back. *Annu. Rev. Immunol.* **22**, 247–306.

Wyllie, D. J. A. and Chen, P. E. (2007) Taking the time to study competitive antagonism. *Br. J. Pharmacol.* **150**, 541–551.

CHAPTER 6

FUNCTIONAL ASSAYS WITH ISOLATED PROTEASES

6.1 INTRODUCTION TO PROTEASES AND THEIR SUBSTRATES

Proteases, also known as peptidases, proteinases, and proteolytic enzymes, are enzymes that catalyze the hydrolysis of peptide bonds in proteins. Proteases occur naturally in all organisms and encoded by about 2% of genes in all kinds of organisms. They are necessary for the survival of all living creatures. Proteases are one of the largest classes of enzymes with more than 500 protease genes encoded in human genomes. Proteases are involved in a multitude of physiological reactions from simple digestion of food proteins (metabolic process) to highly regulated cascades (e.g., the blood-clotting cascade, the complement system, and apoptosis pathways). The activities of proteases are regulated by hundreds of protein inhibitors in an organism. Proteases are an exceptionally important group of enzymes in biology, medical research, and biotechnology.

Proteases have a binding pocket for their substrate. There are many interactions between the amino acid residues in the protease binding pocket and the amino acid residues around the peptide bond to be hydrolyzed, the *scissile bond*. The terminology to describe the interaction between a protease and its substrate in the binding pocket was originally created by Schechter and Berger to describe the specificity of papain in 1967. This system is widely adopted in protease research nowadays. In this system, the substrate binding site on the protease is envisioned as a series of subsites, and each subsite interacts with one amino acid of the substrate. This concept is supported by crystallographic structures of proteases showing that the active site is commonly located in a groove on the surface of the molecule between adjacent structural domains, and the substrate is arranged along the groove on one or both sides of the catalytic site that is responsible for hydrolysis of the scissile bond. The active site on the protease can accommodate between 5 and 7 amino acid residues of the substrate. By convention, the subsites on proteases are numbered from the catalytic site, $S_1, S_2, \ldots, S_n$ toward the N-terminus of the substrate, and $S'_1, S'_2, \ldots, S'_n$ toward the C-terminus. The corresponding amino acid residues of the substrate that subsites accommodate are numbered $P_1, P_2, \ldots, P_n$ from the scissile bond toward the N-terminus, and $P'_1, P'_2, \ldots, P'_n$ toward the C-terminus (Fig. 6.1). Proteases can

Assay Development: Fundamentals and Practices. By Ge Wu

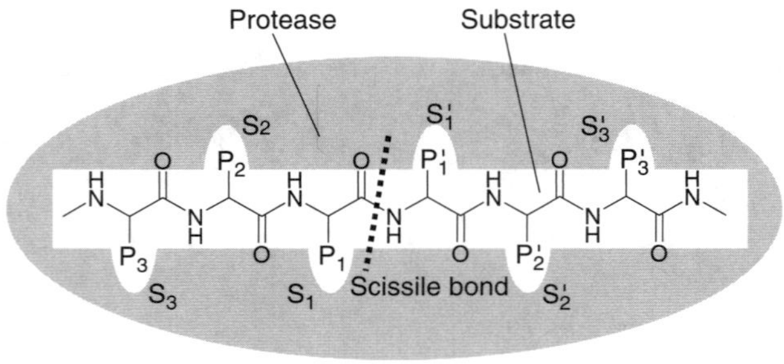

Figure 6.1 Conventional nomenclature for binding of a peptide substrate to a protease. The protease is reppresented as the shaded area. P and P′ are the amino acid side chains on the substrate. S and S′ are the corresponding subsites on the protease.

either selectively hydrolyze specific peptide bonds in proteins (referred to as limited proteolysis) or nonselectively hydrolyze proteins into amino acids (referred to as unlimited proteolysis).

Though all proteases catalyze the same reaction of hydrolyze peptide bonds, studies revealed that some proteases share similar properties, making them distinct from other proteases. Based on the their distinct properties, proteases can be classified into different groups. There are three methods of nomenclature and classification for proteases: (a) by the chemical mechanism of catalysis, (b) by the relative position the protease substrate that was cleaved (EC system), and (c) by molecular structure and sequence homology (the MEROPS system).

6.1.1 Protease Classified by the Chemical Mechanism of Catalysis

Hartley initiated the concept of grouping protease by their catalytic mechanisms. With this system, proteases are commonly classified into four major groups according to their catalytic active sites: serine proteases, cysteine proteases, aspartic proteases, and metalloproteases. In serine proteases the catalytic mechanism depends on the triad of a serine, a histadine, and an aspartic acid in the catalytic site. The hydroxyl group of the serine residue acts as the nucleophile that attacks the peptide bond in the protease substrate. A covalent acyl intermediate is formed between the substrate and the enzyme, which is subsequently hydrolyzed to release the cleaved peptide substrate. Trypsin, chymotrypsin, and proteases involved in blood clotting cascade are examples of serine proteases. Cysteine proteases have a cysteine residue in place of the serine residue in the triad of the catalytic active site. The catalytic mechanism is the same as the serine proteases except that the sulfhydyl group of the cysteine acts as the nucleophile. Caspases and cathepsin B are examples of cysteine proteases. Aspartic proteases have two aspartic residues in the catalytic active site to coordinate the hydrolysis of the peptide bond of the substrate. The aspartic acid residues do not directly form bonds with the substrate protein. Instead, the carboxyl groups of the aspartic residues assist an activated water molecule that acts as a nucleophile to attack the peptide bond. Renin, cathepsin D, pepsin, and HIV (human immunodeficiency virus) proteases are examples of aspartic proteases. Metalloproteases have a divalent metal ion in the catalytic active site. Zinc is the most common ions in metalloproteases. Sometimes iron, manganese, or cobalt ion can be found in the active site too. Many metalloproteases contain a HEXXH motif in the catalytic active site that binds to a tetrahedrally coordinated zinc atom. The metal ion in metalloproteases does not directly participate in the breaking of the substrate peptide bond. Instead, it functions to activate a water molecule that attacks the peptide bond of the substrate. Matrix metalloproteases and the ADAM family of proteases are examples of metalloproteases. In addition to the four major groups, there are a few small groups of proteases. Some proteases have the serine residues in the catalytic site substituted by threonine residues. These proteases are sometimes classified as a subgroup of serine protease but sometimes are classified as a stand-alone threonine proteases group. Similarly, some proteases have the aspartic residues in the catalytic site substituted by glutamic residues. These proteases are sometimes classified as a subgroup of aspartic protease but sometimes are classified as a stand-alone glutamic proteases group. In addition, there are a few proteases of unknown catalytic type. Of all the proteases in the human genome, 36% are metalloporteases, 32% are serine proteases, 23% are cysteine proteases, and 3% are aspartic proteases.

6.1.2 Protease Classified by Substrate Cleavage Position

Though all proteases catalyze the same reaction by hydrolyzing a peptide bond, none of them will hydrolyze all peptide bonds because of their selectivity for the scissile bond. The selectivity can be from particular amino acids in nearby positions of the scissile bond or a particular position of the scissile bond in the polypeptide chain of the substrate molecule. On the basis of scissile bond position on the substrate, proteases can be classified into endopeptidases and exopeptidases (Fig. 6.2). The EC

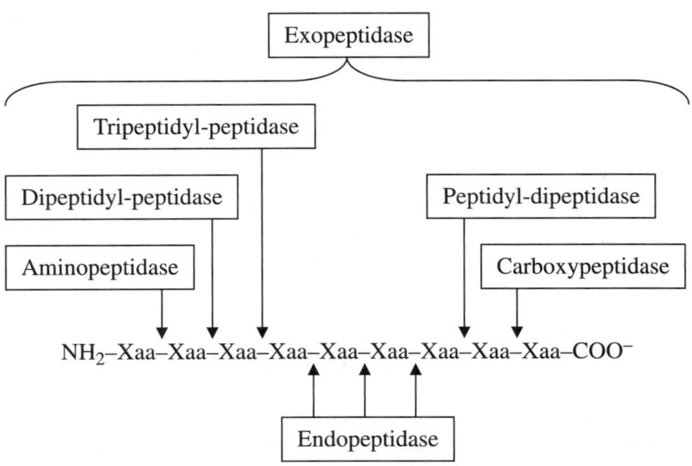

Figure 6.2 Classification of proteases according to the relative cleavage position on the substrate based on EC method of classification.

TABLE 6.1 Proteases Classified into 14 Subclasses in EC Classification

EC 3.4	Peptidases
EC 3.4.11	Aminopeptidases
EC 3.4.13	Dipeptidases
EC 3.4.14	Dipeptidyl-peptidases and tripeptidyl-peptidases
EC 3.4.15	Peptidyl-dipeptidases
EC 3.4.16	Serine-type carboxypeptidases
EC 3.4.17	Metallocarboxypeptidases
EC 3.4.18	Cysteine-type carboxypeptidases
EC 3.4.19	Omega peptidases
EC 3.4.21	Serine endopeptidases
EC 3.4.22	Cysteine endopeptidases
EC 3.4.23	Aspartic endopeptidases
EC 3.4.24	Metalloendopeptidases
EC 3.4.25	Threonine endopeptidases
EC 3.4.99	Endopeptidases of unknown catalytic mechanism

system (by Nomenclature Committee of IUBMB) is a classification based on substrate cleavage position and the reaction catalyzed (Table 6.1). This classification system is very helpful in assay development to choose a protease substrate for a specific protease of interest because the information of the protease substrate cleavage position is built into the name of the proteases.

An endopeptidase hydrolyses internal peptide bonds more than three amino acids away from the N-terminus or C-terminus in a polypeptide substrate. Endopeptidases are further divided into serine endopeptidases, cysteine endopeptidases, aspartic endopeptidases, metalloendopeptidases, and threonine endopeptidases. Well-known examples of endopeptidases are chymotrypsin, trypsin, pepsin, and papain. In addition, there are a few endopeptidases that act at a fixed distance from one terminus of the substrate, an example being mitochondrial intermediate peptidase. Some endopeptidases, termed oligopeptidases, act only on small peptide substrates, for example, thimet oligopeptidase.

The exopeptidases require the scissile bond to be within three residues from either one of the terminus in the substrate, plus a free N-terminal amino group, a free C-terminal carboxyl group, or both. The exopeptidases are further divided into aminopeptidases (remove a single amino acid from free N-terminus), carboxypeptidases (remove a single amino acid from free C-terminus), dipeptidyl-peptidases (remove a dipeptide from free N-terminus), tripeptidyl-peptidases (remove a tripeptide from free N-terminus), peptidyl-dipeptidases (remove a dipeptide from free C-terminus), and dipeptidases (cleave only dipeptides with free terminuses on both ends).

6.1.3 Proteases Classified by Molecular Structure and Sequence Homology

Both of the systems of classifying proteases discussed above have great strengths, but they also have limitations. For example, the serine proteases include many proteases with very different molecular structures, and they are by no means all homologs of each other. Rawlings and Barrett developed a protease classification by grouping the proteases according to their structural features and evolutionary relationship. This classification system formed the bases for the MEROPS database, which is the most comprehensive protease database. Over 2000 individual proteases and nearly 400 inhibitors were collected in the MEROPS database. In MEROPS, all proteases are grouped at the top level by catalytic type. Within each catalytic type, the proteases are grouped into families based on their sequence homology. The families are assembled into clans according to their common ancestry, which is based on their similarity in tertiary structures. The MEROPS system is arranged in four levels from top to bottom: catalytic type, clan, family, and individual proteases with unique ID.

6.2 FUNCTION OF PROTEASES AND THEIR ROLE IN DRUG DISCOVERY

Proteases were originally known for their function in nonselectively breaking down proteins in food into amino acids (unlimited proteolysis). In food processing, endopeptidases randomly cut large peptides into smaller pieces and generate many free amino- and carboxyl-termini. Exopeptidases further process these smaller peptides into amino acids. In addition to food digestion, unlimited proteolysis also happens with intracellular proteins. The function of proteases here is to recycle old, malfunctioned, or misfolded proteins into amino acids. In some cases the destruction of proteins is not a result of malfunctioning but is a result of control of cellular regulatory switches. Proteins to be degraded are first tagged with polyubiquitins that are conjugated to an internal lysine residue or less commonly to the free amino terminus of the substrate. The tagged proteins are digested by proteosomes to generate small peptide fragments. A collection of exopeptidases degrades these small peptide fragments into amino acids that can be reused in the synthesis of new proteins. Extracellular proteins that are endocytosed do not enter the cytoplasm but are instead directed to lysosomal compartments. In the lysosomes, the cathepsins are responsible for degrading the internalized protein into pieces. These fragments are subsequently degraded to amino acids by lysosomal exopeptidases. In addition to unlimited proteolysis to reduce proteins into amino acids, proteases also perform limited proteolysis that selectively cut a protein at a few defined locations. For example, trypsin, chymotrypsin, and elastase are synthesized in the pancreas and excreted into the intestine for food digestion. It was later realized that these enzymes are made as pro-enzymes (also referred to as zymogens) that are inactive enzyme precursors. These enzymes are activated in the intestine by limited proteolytic degradation. It was discovered that precise cleavage of proteins by proteases leads to the regulation in many key physiological processes, such as cell-cycle progression, cell proliferation and apoptosis, DNA replication, tissue remodeling, blood coagulation, wound healing, and the immune response. Because protein cleavage is an irreversible process, protease signaling differs from most other signaling pathways (such as protein phosphorylation, which will be discussed in Chapter 7) that are reversible. Because proteases are involved in many key physiological processes in humans, modulation of human protease activities is an attractive area for drug discovery. In addition, many viral proteases are critical for activation of their surface proteins and membrane fusion. Consequently, inhibitors targeting viral protease play a significant role in controlling viral infection.

The most successful protease inhibitors as drugs are the angiotensin-converting enzyme (ACE, also referred to as peptidyl dipeptidase A) inhibitors with annual sales exceeding US$6 billion. The drug targeting ACE, Captopril from Bristol-Myers Squibb, was approved several decades ago. Since then, there are more than a dozen ACE inhibitors currently on the market (Table 6.2). ACE inhibitors are used to treat cardiovascular conditions including hypertension, heart failure, and heart attack. ACE is a zinc metalloproteinase and is one of the central proteases in the renin–angiotensin system. It catalyzes the conversion of angiotensin I into angiotensin II by removing a dipeptide from the carboxyl terminus of agiotensin I. This step is required for angiotensin receptor activation. Protease inhibitors targeting proteases in the blood coagulation cascade (thrombin, Factor II, VII, IX, and Xa) are another successful class of drugs to treat thrombosis. All the proteases in the blood coagulation pathways are serine proteases. Several protease inhibitors targeting thrombin (Argatroban from Mitsubishi Pharma, Lepirudin from Aventis, and Desirudin from Novartis) and Factor X (Fondaparinux sodium from Sanofi Synthelabo) are marketed drugs.

The HIV protease inhibitors as drugs are success stories for targeting nonhuman proteases. There are eight approved drugs targeting HIV proteases (Table 6.3) with annual sales of these drugs exceeding billions of dollars. The HIV protease inhibitors represent one of the best examples of the application of rational drug design to obtained marketable drugs. Soon after the discovery of HIV, the genetic code of the virus was revealed and the structural and regulatory gene products were identified. Codons for HIV protease, reverse transcriptase, and integrase, all of which are essential to the maturation of the virus, are identified within the *pol* region of the virion's genetic structure. The HIV protease is an aspartic protease enzyme that cleaves

TABLE 6.2 Marketed Drugs Targeting ACE

Drug	Company
Captopril	Bristol-Myers Squibb
Enalapril	Merck
Lisinopril	AstraZeneca
Trandolapril	Abbott
Zofenopril	Menarini
Ramipril	Aventis
Moexipril	Boehringer Mannheim
Imidapril	Trinity
Perindopril	Daiichi
Qinapril	Pfizer
Fosinopril	Bristol-Myers Squibb
Benazepril	Novartis
Cilazapril	Roche

TABLE 6.3 Marketed Drugs Targeting HIV Proteases

Drug	Company
Ritonavir	Abbott
Amprenavir	Vertex Pharmaceuticals, GlaxoSmithKline
Fosamprenavir	GlaxoSmithKline
Atazanavir	Bristol-Myers Squibb
Lopinavir	Abbott
Indinavir	Merck
Saquinavir	Hoffmann-La Roche
Nelfinavir mesylate	Pfizer

polyproteins of the virus into essential functional protein products during the maturation process of the virion. This critical process occurs as each new virion buds forth from the membrane of an HIV-infected cell and continues after the immature virus is released from the cell. If the HIV protease is inhibited, the virus will not mature and is incapable of infecting a new cell. To develop HIV inhibitors, the HIV protease gene product was expressed in large quantity and crystallized. The three-dimensional structure of the protease was then resolved. Inhibitors for HIV protease were then computationally designed that specifically targeted the active site of the HIV protease. Because HIV protease activity is unique for HIV proteins, and there is virtually no cross-reactivity in the host between the HIV protease and normal human aspartic protease, the designed HIV protease inhibitors can specifically inhibit HIV proteases without inhibiting the function of normal host proteases.

Though many proteases are potentially drug targets, approved drugs in the past only target a small number of proteases. It is estimated that 10 to 15% of drug targets that pharmaceutical companies are working on are proteases. Some of them are novel proteases targets with no approved drugs yet. This includes targeting metallomatrix proteins (MMPs) for cancer, β-amyloid precursor proteins cleaving enzyme (BACE) for Alzheimer's disease, cathepsin K for osteoporosis, cathepthin S and TNFα-converting enzyme (TACE) for inflammation and autoimmune diseases, tryptase and chymase for allergy and asthma, and methionine aminopeptidase-2 (MetAP2) for cancer. Some of the efforts spent on novel protease targets have paid off. Valcade (bortezomib) targeting proteosome was approved in 2003 for treating multiple myeloma. Januvia (sitagliptin) targeting dipeptidyl peptidase IV (DPP IV) was approved in 2006 for treating type II diabetes.

6.3 PROTEASE ASSAYS

6.3.1 Protease Assays with Protein as Substrate

Proteases assays can be carried out either with proteins as substrates or with synthetic peptides as substrates. Many old protease assays were developed using proteins as substrates. Using proteins as protease substrates is complicated by the fact that protein substrates possess many peptide bonds and many of them may potentially be hydrolyzed by the protease under study. Each potential scissile bond behaves like an individual substrate, and the observed reaction is a collection of many proteolytic reactions with many substrates. The observed overall kinetics is very difficult to interpret because each of the cleavage reaction has its own kinetics. To complicate the situation further, as the proteolysis occurs, new protease substrates are generated that have their own kinetics in the next round proteolysis. This makes the reaction progressive curve not useful to extract the initial velocity because each individual reaction is not known and the contribution from each individual reaction to the overall generation of final detectable signal is not known. Because of these complications, the enzyme kinetic treatment described in Chapter 3 may not be useful in this situation. Though most modern protease assays are performed with synthetic peptide substrates with ideally only one scissile bond designed specifically for the protease under study, protease assays with protein substrate is still useful in special situations, such as assaying proteases with unknown specificity (detecting protease activity with uncharacterized crude samples), assaying proteases for which there are no other known substrates, or when the protein substrate has only one scissile bond.

Historically, a few well-characterized unmodified proteins with high purity that are readily available in large quantities have been used as protease substrates. These protease substrates include native hemoglobin, casein, and gelatin. Because there may exist several scissile bonds in the substrate for the protease to be assayed, many fragments of the protein substrate may be generated by the protease. There is no simple method to distinguish the products from the substrates. One way to detect the protease products is by separation of the products from the substrate based on their difference in size. This can be done with HPLC, CE, or SDS–PAGE. The advantage of this method is that all the protease products can be potentially monitored (depending on the separation conditions). Thus, a full picture of the proteolytic process is obtained. This method is particularly useful to assay endopeptidases where several large peptide products may be produced. However, this method is time-consuming and with lower throughput. Alternatively, the protease products can be detected with antibodies that can bind specifically to the newly generated amino acid terminus and the nearby residues with ELISA or other methods (see example in Section 6.7). This method can be used for both endopeptidases and exopeptidases. The disadvantages of this method are that it is dependent on the availability of the antibody and it can only monitor the hydrolysis of one scissile bond at a time. A third method to assay the protease activity is to measure the free amino acids generated by the action of exopeptidases. There are many well-established methods to measure free amino acids, such as HPLC or amino acid coupled reactions that generate fluorescent signals. The applicability of this assay is limited to exopeptidases. For example, carboxypeptidase U was assayed by measuring the free lysine (see example in Section 6.7)

In addition to the unmodified native protein substrate, many protease assays employ the protein substrates that are modified with chromophores or fluorophores. One way to assay protease is to label the protein substrates with the azo group. For example, azocasein, which is commercially available from many vendors, is used as a nonspecific protease substrate. Hydrolysis of the azocasein by protease releases small amino acids or peptide products with the attached azo group. The large unhydrolyzed azocasein and the large fragments of hydrolyzed products are separated from the small amino acids or peptide products containing the azo group after precipitating with reagents, such as 2 to 8% tricholoroacetic acid, which is followed by membrane filtration or centrifugation. The azo group-containing small amino acids and peptides in solution is then detected by absorbance at 440 nm. With fluorescence-based assays using the same strategy, the azo group is substituted with fluorophores, preferably pH insensitive fluorophores such as BODIPY and Oregen Green. Protease assays using fluorescently labeled protein substrate is much more sensitive than using the corresponding chromophore-labeled protein substrates. Whether chromophores or fluorophores are used, the major disadvantage of this method is the involvement of protein precipitation and separation steps. Molecular Probes (now part of Invitrogen) offers two methods to assay proteases based on fluorescently labeled protein substrates without the need to physically separate the proteolytic products from the substrate. The first method is based on the fluorescence self-quenching phenomenon. The protein substrates are heavily labeled with fluorophore, and thus the fluorescence is self-quenched to less than 3% of the fluorescence of the corresponding free fluorophore. The proteolytic cleavage of the protein substrate releases the fluorophores from the substrate into solutions that emit strong fluorescence. Many protein substrate assay kits based on this method are available, such as DQ BSA, DQ collagen, DQ ovalbumin, and DQ gelatin. The second method is based on florescence polarization (see Chapter 2). The protein substrate is labeled with an adequate amount of fluorophore per protein without internal self-quenching. The fluorophore attached to the protein substrate has a large anisotropy value due to the long rotational correlation time of the large protein. After the protein substrate is hydrolyzed and the small peptide or amino acids with attached fluorophore is released, the anisotropy is reduced due to the fast rotational correlation time of the smaller peptides or amino acids. The reduction of anisotropy can be monitored in real time to follow the protease-catalyzed reaction.

All the protease assays using chromophore- or fluorophore-labeled protein substrate are based on the assumption that small peptides or amino acids are generated by multiple cleavages on the protein substrate. They are very powerful tools for unlimited hydrolysis of the protein substrates. However, they may not be useful with limited hydrolysis where one or a few scissile bond cleavages may not result in small peptides or amino acids that carry a chomophore or fluorophore. In those situations, chromatography or SDS–PAGE may be used to resolve the substrate and the products.

6.3.2 Protease Assays with Synthetic Substrate

Fluorogenic and Chromogenic Assays One popular strategy is to use a chromophore or a fluorophore whose absorption or excitation wavelengths are shifted dramatically when a primary amine group in the molecule is modified to form an amide bond. The hydrolysis of this particular amide bond regenerates the original chromophore or fluorophore (Fig. 6.3*a*). This method is commonly referred to as chromogentic or fluorogenic assays. 4-Nitroaniline (410 nm) and 2-naphthylamine (340 nm) are commonly used as a chromophore in chromogenic assays. 7-Amino-4-methylcoumarin (AMC) and a related molecule, 7-amino-4-trifluoromethylcoumarin

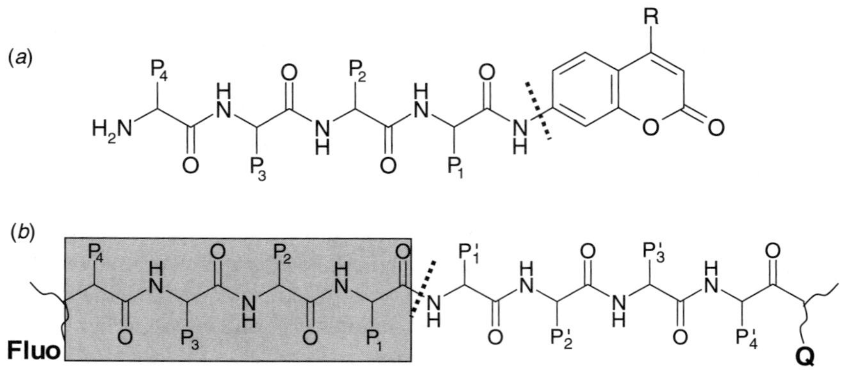

Figure 6.3 Protease assays with fluorescently labeled peptide substrates. (*a*) Fluorogenic assay. A peptide is label at the carboxyl terminus with a 7-aminocoumarin through an amide bond. Protease cleavage of this bond will release 7-aminocoumarin, which can be measured with a fluorimeter. This substrate can be used in PSSCLs to study substrate specificity by scanning the amino acid residues on position P_1 to P_4 (*b*). Quenched FRET assay. A fluorescent probe and a quencher are placed at opposite ends of the scissile bond. Cleavage of any peptide bonds, not just the scissile bond, between the fluorophore and the quencher will generate fluorescent signal. When the amino acid residues on the amino side of the scissile bond are fixed (shaded area), the amino acids on prime side from P_1' to P_4' can be studied in PSSCLs by scanning.

(AFC), are commonly used as a fluorophore in fluorogenic assays. Both compounds have only one primary amine that is attached to the C-terminus of a peptide through an amide bond to form the protease substrate. When this peptide bond is hydrolyzed, the original fluorophore is regenerated. AFC has longer excitation and emission wavelengths (ex 400 nm/em 500 nm) than AMC (ex 380 nm/em 460 nm), which may be of advantage in some applications. Rhodamine 110, a visible light-excitable dye with stronger absorbance than AMC, has been used as a fluorophore in fluorogenic assays as well. However, rhodamine-110-based substrates comprise two amide bonds and both of them have to be hydrolyzed to generate the originally rhodamine 110. This may cause a problem in interpreting the kinetics.

The amide bond formation between the dye and the carboxyl group of the peptide does not make the fluorophore nonfluorescent as the name "fluorogenic" may suggest. The amide bond formation simply changed the excitation and emission wavelength of the fluorophore (see example in Section 6.6). The excitation and emission spectra of the substrate and the released fluorophore should be studied carefully to define the best excitation wavelength for the released fluorophore while minimizing the interference from the fluorescence emitted by the labeled substrate. This is particularly important at low substrate conversion when the molar ratio between the substrate and the released fluorophore can exceed 99 to 1 in the sample to be measured. One special property of fluorogenic or chromogenic assays is that only the hydrolysis of the amide bond where the dye is attached gives a signal. Any hydrolysis of other peptide bonds does not give a signal. However, the silent hydrolysis will still affect the reaction kinetics because they will behave like competitive inhibitors for the signal producing hydrolysis. This is different from the quenched FRET assay (discussed below) where the hydrolysis of any peptide bonds between the fluorophore and the quencher will generate a fluorescent signal.

Because the fluorogenic and chromogenic reagents depend on their amino group to form the scissile bond, these peptide substrates have to terminate with the dye at the carboxyl terminus of the scissile bond. This requirement restricted these protease substrates with only the amino acid residues at the amino terminus of the scissile bond. Thus, the application of fluorogenic and chromogenic assays are restricted to endopeptidases with substrate specificity at the N-terminus of the scissile bond and aminopeptidases with the substrate specificity at the P_1 position. Because of the required modification of the carboxyl group at the C-terminus with the dye, fluorogenic and chromogenic assays cannot be used to study carboxypeptidases that require a free carboxyl terminus. When assaying endopeptidases, the amino terminus of the substrate peptides is usually capped with the acetyl group. A caspase-3 assay is discussed in Section 6.6 to illustrate fluorogenic assays with endopeptidases. When studying aminopeptidases with fluorogenic or chromogenic assays, only an amino acid with an amide bond linked to a dye can be used as a substrate because of the requirement of free amino ends for aminopeptidases. Using a modified amino acid as a substrate may not reflect the true biology of the proteases. For example, MetAP-2 (EC 3.4.11.18) has been assayed by monitoring the hydrolysis of methionine-*p*-nitroanilide (MetpNA) or methionine-7-amino-4-methylcoumarin (MetAMC). The assay mixture consisted of 50 mM Tris-HCl (pH 7.5), 0.25 mM methionine-*p*-nitroanilide and methionine aminopeptidase. The reaction mixture was incubated at 37°C for 30 min and left on ice for 15 min. The proteolytic product, *p*-nitroaniline, was measured by its absorption at 405 nm. Though this assay can detect protease activity, the amino-acid-derived reagent may not be a good protease substrate. Yang and co-workers (2001) found that MetAP-2 prefers a tripeptide or larger peptide substrates. The catalytic efficiency of dipeptide substrates was found to be at least 250-fold lower than those of the tripeptides. The amino-acid-based substrates MetAMC and MetpNA were found to further diminish the protease's catalytic efficiency substantially.

Quenched FRET Assays Another popular strategy to assay protease activity is by the FRET-based quenching of a fluorescent donor by a nonfluorescent acceptor placed on opposite side of the scissile bond (Fig. 6.3*b*). The fluorescence is quenched due to the close proximity between the fluorophore and the quencher. After the substrate cleavage by the protease, the fluorophore and the quencher are separated and strong fluorescence is generated. The quencher and fluorophore pair should be selective so that the absorption spectrum of the quencher matches the emission spectrum of the fluorophore, such as the EDANS/DABCYL pair. The reason for selecting a nonfluorescence acceptor instead of a fluorescence acceptor is that measuring product generation is usually preferred to measuring substrate decreasing in assay development. By using a nonfluorescence acceptor, the substrate turnover at any level can be measured as long as sufficient amount of fluorescence is generated. On the other hand, if the FRET signal is measured with a fluorescent acceptor, a significant amount ($>10\%$) of substrate turnover is required to detect the decrease in the FRET signal from a large background. One caveat of quenched FRET assays is that the hydrolysis of any peptide bonds, not just the intended scissile bond, between the fluorophore and the quencher will generate a fluorescent signal. The enzyme kinetics will be difficult to interpret when there are more scissile bonds between the fluorophore and the quencher. One example of the application of the quenched FRET method is the HIV protease assay. The native HIV protease substrate gag-pol polyprotein that contains Asn-Tyr + Pro-Ile-Val-Gln sequence can be cleaved at the scissile bond between Tyr and Pro. A synthetic substrate containing EDANS and DABCYL at each end of the peptide can be used to assay HIV proteases. For example, Arg-Glu(EDANS)-Ser-Gln-Asn-Tyr-Pro-Ile-Val-Gln-Lys(DABCYL)-Arg was used in several commercial HIV protease assay kits, such as those from Invitrogen, Sigma Aldrich, and AnaSpec. Many commercial renin assay kits are also designed using quenched FRET methods.

Fluorogenic and quenched FRET are widely used in protease assays. Each technology has its own advantages and disadvantages. The substrates in quenched FRET assay do not emit fluorescence when excited at any wavelength, whereas the fluorogenic substrate simply shifted its excitation and emission wavelength and will emit fluorescence when excited at the proper wavelength. The quenched FRET assays require two labels on the peptide substrates that can be placed at many positions, whereas fluorogenic assays have only one fluorophore in the substrate and it can only be placed at the scissile bond. Because the quenched FRET assay can use substrates with amino acid residues on both ends of the scissile bond, it can be used for many proteases that may require specific amino acid residues on both sides of the scissile bond. In comparison, the fluorogenic assays can only be used when the substrate specificity is at the amino terminus of the scissile bond.

Assays for carboxypeptidases are challenging. Though fluorogenic and the quenched FRET methods are powerful in application to endopeptidases and aminopeptidases, they cannot be used to assay carboxypeptidases because of the requirement for the free carboxyl end in the substrate. To assay carboxypeptidases, the protease-catalyzed reaction products, namely the released amino acid or the peptide product that is one amino acid shorter than the substrate, are monitored. There are many established methods to detect the released amino acid. One of the methods is to use ninhydrin to react with the released amino acid that yields a blue-colored product. This method may be interfered with if the substrate contains primary amines, such as lysine and free N-terminal amine. A more selective method to measure the released amino acid is to measure the amount of hydrogen peroxide generated when the amino acid to be assayed is oxidized in the presence of the amino-acid-specific oxidase. In addition to assaying the released amino acid, antibodies specifically designed to recognize the cleaved product that is one amino acid shorter than the substrate can be used to measure the peptide product. Both of these assays for carboxypeptidases are illustrated in the assays for carboxypeptidase U in Section 6.7.

6.4 PROTEASE SUBSTRATE PROFILING

The ability of proteases to initiate and regulate biological pathways is due to the substrate specificity of proteases. The substrate specificity is primarily determined by the substrate binding pocket in the active site of the proteases. Consequently, the amino acid residues near the scissile bond in a peptide determine the substrate specificity for a particular protease. For example, trypsin prefers the positively charged lysine and arginine on the amino side of the scissile bond while chymotrypsin prefers aromatic or bulky side chains at the same position. In addition to the composition of the substrate, the temporal and spatial expression of the proteases, substrates, cofactors, activators, and endogenous inhibitors can also determine the substrate specificity in vivo. Except for a few proteases, such as renin that has only one known substrate, most proteases readily cleave small peptides with broad substrate specificity in vitro. Though some preferences to specific peptides exist, they usually are not very stringent. For example, the preference of chymotrypsin A (EC3.4.21.1) for the amino acid at the amino side of the scissile bond includes phenylalanine, tyrosine, leucine, tryptophan, methionine, asparagine, and alanine, which accounts for about one third of the 20 available amino acids. This broad tolerance to different substrate made it easier to develop in vitro assays with isolated proteases. For the purpose of developing assays to screen for enzyme inhibitors, it is not necessary to use an optimal peptide substrate. Any substrate can be used as long as enough detectable signal can be generated within normal assay conditions. However, increased incubation time or increased enzyme and substrate concentration may be required to obtain an adequate amount of signal in the assay if the substrate is not optimal for the protease. There is a limitation in incubation time and protease concentration. In general, the incubation time should be no more than over night and the enzyme concentration should be no more than 0.1 μM. Obtaining optimal peptide substrate for a protease is of advantage for assay development because of lower substrate concentration requirement and higher turnover rate. In addition, the identification of the optimal substrate peptide for a protease can help understand the biology of the protease and aid the design of inhibitors for the protease.

Traditionally, the substrate specificity of a protease is determined by the analysis of the digestion pattern of standard proteins, such as hemoglobin, insulin B-chain, gelatin, or casein. The limitation of this approach is the limited number of sequences whose standard proteins possess and potential important cleavage pattern may be missed. To overcome this limitation, evaluation of the digestion of the synthetic peptide library containing systematically substituted amino acids in each position near the scissile bond is widely used. In these studies, cysteine is usually not studied and methionine is usually substituted with norleucine due to technical reasons. This requires 19 amino acids to be randomly placed in each position near the scissile bond. A limitation to this approach is that only a limited (usually less than 5) number of positions of the substrate can be studied due to the large number of peptides that is required to achieve true randomization. For example, to randomize 19 amino acids at 4 positions will require $19^4 = 130{,}321$ peptides and at 5 positions will require $19^5 = 2{,}476{,}009$ peptides. In practice, substrate specificity is often studied with positional scanning synthetic combinatorial libraries (PSSCLs). With this method, each position from the scissile bond is fixed with one of the 19 amino acids in sequence while the other positions (up to 3) is substituted with all the 19 amino acids. This will create many pools with each pool containing many peptides. If the number of peptides in a pool is too large, the concentration of each peptide in a pool will be lower because the total concentration of peptides in a pool is limited in the peptide synthesis. If the concentration of each peptide is too low, the assay results may not be meaningful because the final read is contributed from all the peptides in the pool. Fixing more than one position at a time will create larger numbers of pools and reduce the number of peptides in each pool. The lower number of peptides in a pool will increase the concentration of individual peptides and result in more reliable assay data. Another advantage for fixing two positions at the same time is that the synergy, if there is any, between the two sites can be detected. However, the larger number of pools means more assays need to be performed.

The assays to study protease substrate specificity with PSSCLs are shown in Figure 6.3. When scanning the amino acid residues on the amino side of the scissile bond, fluorogenic assays with 7-aminocoumarin or its derivatives linked to the carboxyl side through the 7-amino group can be used (see Fig. 6.3*a*). With this method, the amino acids on positions P_1 to P_4 are varied. After the scissile bond is hydrolyzed, 7-aminocoumarin is released and its fluorescence can be measured. Because 7-aminocoumarin terminated the peptide at the scissile bond, the substrate specificity at the prime side of the substrate cannot be studied with this method. To study prime side or both sides of the amino acid residues, quenched FRET assays can be used (see Fig. 6.3*b*). A fluorescence donor and a nonfluorescence acceptor are placed at positions beyond the fourth position from the scissile bond. As discussed earlier, no more than 5 positions can be studied with PSSCLs due to the large number of peptide combination, this strategy cannot be practically used to scan all the 8 positions on both sides of the scissile bond. Usually the residues on the amino side are studied first using the peptide library with covalently linked coumarin derivatives. These positions are then fixed with optimal residues obtained from this study (shaded area in Fig. 6.3*b*). The amino acids on the prime side from P'_1 to P'_4 are then studied with PSSCLs to obtain the optimal substrate.

6.5 PROTEASE INHIBITORS

In enzymatic assays, obtaining an inhibitor to serve as a control for the assay is a critical step in addition to obtaining enzyme substrates. Because proteases are among the most well-studied enzymes, inhibitors are available for most proteases. The protease inhibitors can be classified into protein-based natural inhibitors and synthetic small-molecule inhibitors. They can also be classified into broad nonspecific inhibitors, class-level specific inhibitors, and specific inhibitors that only inhibit one or a few closely related proteases. Endogenous protease inhibitors occur naturally to regulate proteases' functions in vivo by inhibiting protease activity. For example, serine protease inhibitors (serpins) and Kunitz-type protease inhibitors (such as aprotinin) inhibit some of the serine proteases in blood clotting, the tissue inhibitors of metalloproteases (TIMPs) inhibit matrix metalloproteases. Many endogenous inhibitors of proteases are known and the information can be found in the MEROPS database. Endogenous protease inhibitors can be very selective to inhibit particular protease. For example, TIMPs can form a tightly bond complex with some members of the MMPs and only inhibit these MMPs but not metalloproteases belonging to other families.

However, these protein-based inhibitors can be very costly to be used as a control in bioassays. There are many selective synthetic inhibits to specific proteases, especially those proteases that are therapeutic targets, such as HIV proteases and ACE. For the purpose of serving as a control in assays with isolated proteases, nonselective inhibitors that inhibit many classes of proteases and class-specific inhibitors have many advantages such as reducing assay development time and cost-effectiveness.

6.5.1 Nonselective Cross-Class Protease Inhibitors

The inhibitor EDTA binds nonselectively to many divalent ions including Ca^{2+}, Mg^{2+}, Mn^{2+}, Co^{2+}, and Zn^{2+}. This property makes EDTA a universal inhibitor for all metalloproteases. In addition, proteases in other classes that require divalent ions, such as Ca^{2+}, for their function will be inhibited too. α-Macroglobulins are endogenous universal inhibitors found in the circulation of many species (vertebrate and invertebrate) that inhibit endopeptidase in every protease class. The mechanism of the inhibition is different from active site inhibition. The protease will bind to α-macroglobulin and then cut-off the "bait region" in α-macroglobulin (~20 amino acid residues). The resulting α-globulin tightly binds to proteases and blocks the proteases from accessing large protease substrates. However, the protease activity still remains and it can hydrolyze accessible small peptide substrates. Small peptide aldehydes (2 to 4 residues) are reversible inhibitors that inhibit serine, threonine, and cysteine proteases. The aldehyde group in aqueous solution reacts with water to form a tetrahydral intermediate that resembles the transition-state of the reaction catalyzed by serine, threonine, and cysteine proteases and thus inhibits the protease activity. Molecules containing chloromethyl ketones can covalently bind to the hydroxyl and sulfhydryl groups in the active site of serine, threonine, and cysteine proteases and inhibit their function. Selectivity can be obtained by placing different groups to the other side of the ketone. A good example is tosylphenoalanine chloromethyl ketone (TPCK), which has selectivity to chymotrypsin.

6.5.2 Class-Specific Protease Inhibitors

Metalloproteases can be inhibited by molecules containing metal (more specifically, Zn^{2+}) chelating groups, such as carboxylates, thiols, phosphorous, and hydroxamates. A range of inhibitors, such as 1,10-phenanthroline, bestatin, and phosphoramidon, can be chosen that offers different selectivity. Different from EDTA that nonselectively binds to many divalent ions, 1,10-phenantroline has high affinity for Zn^{2+} but has little affinity for Ca^{2+}. This property makes it specifically inhibit metalloprotease. Betastatin selectively inhibits metalloaminopeptidases and phosphoramidon selectively inhibits metalloendopeptidases.

There are many specific serine protease inhibitors. Diisopropylphosphofluoridate (DFP) is a well-known irreversible serine protease inhibitor and has been widely used in the past. However, it has a high toxicity and a low reaction rate. Phenylmethylsulfonyl fluoride (PMSF) is another well-known irreversible serine protease inhibitor. A more water-soluble version of PMSF, 4-[2-aminoethyl]benzenesulfonyl fluoride (AEBSF), is more widely used. Both molecules are not stable in aqueous solution and can lose activity over time. PMSF and its derivatives may also react reversibly with some cysteine proteases. Isocoumarins are mechanism-based irreversible serine inhibitors that inhibit most serine proteases except complement factor B and C2. Peptide boronic acids, similar to peptide aldehydes, resemble the transition-state of the reaction catalyzed by serine, threonine, and cysteine proteases. However, peptide boronic acids are much more selective for serine proteases and they are very potent. In addition to the small-molecule inhibitors, there are many protein inhibitors of serine proteases, such as serpins (a family of natural proteins), aprotinin, avian ovomucoids, and soybean trypsin inhibitors.

Cysteine proteases have a similar catalytic structure with serine proteases except that the serine at the active site in serine proteases is substituted by a cysteine. Because the sulfhydryl group in cysteine is a stronger nucleophilic group than the hydroxyl group in serine, cysteine protease-specific inhibitors exploit this property to achieve selectivity over serine proteases. E64 is a natural product containing an epoxide group. It is isolated from the extract of *Aspergillus japonicus*. The sulfhydryl group in the active site of cysteine proteases attaches to the electrophilic carbon of the epoxide group in E64 and its synthetic analogs to form a covalent bond. Most cysteine proteases in family C1 are inhibited by E64. Another type of inhibitor targeting cysteine protease contains a methyl ketone motif with a leaving group X attached to the electrophilic methyl group [$-C(O)CH_2-X$]. The leaving group $-X$ can be fluoro ($-F$), iodo ($-I$), or acyloxy [$-O-C(O)R$]. Peptides with carboxyl-terminus substituted with diazomethane ($-CH{=}N_2$ or $-CH^{-}-N_2^{+}$) are found to be cysteine protease inhibitors as well. The difference of this class of inhibitors from the inhibitors with methyl ketone motif is that the carbon atom on the methane group is nucleophilic instead of electrophilic. The detailed reaction mechanism is not fully understood. In addition to the small-molecule inhibitors, cystatins (a family of natural proteins) can selectively inhibit the papain family of cysteine proteases.

Aspartic proteases can be inhibited by pepstatin A, a hexa-peptide containing the unusual amino acid statine [Sta, (3S,4S)-4-amino-3-hydroxy-6-methylheptanoic acid]. It has the sequence of Iva-Val-Val-Sta-Ala-Sta. It was originally isolated from cultures of various species of *Actinomyces*. It can inhibit nearly all aspartic proteases with high potency. Pepstatin A is insoluble at any concentration in methanol or DMSO and is only sparingly soluble in water. However, it can be dissolved at 1 mg/mL in 1 : 10 mixtures of acetic acid and methanol. The interaction between aspartic proteases and pepstatin A has been studied extensively and many synthetic aspartic protease inhibitors, especially HIV protease inhibitors, were derived from the active motif in pepstatin A.

6.6 ASSAY DEVELOPMENT FOR CASPASES WITH A FLUOROGENIC SUBSTRATE

Caspases are cysteine proteases and are key components of the cellular apoptotic pathway. Several disease states, including cancer, inflammation, and autoimmune disorders, involve apoptosis. Because alteration of the apoptotic process might offer treatment for these diseases, caspases have become important drug discovery targets. Active caspases are tetramers consisting of two ~20-kDa subunits and two ~10-kDa subunits that are formed after the zymogen polypeptide has been activated by proteolysis. Each active caspase contains two independent active sites, each defined by aspects of both 20-kDa and 10-kDa subunits and containing a critical free-thiol Cys residue. Caspases are endopeptidases and their substrate specificity universally requires an aspartic acid residue at P_1 position. Due to the broad interest in caspases, many caspases assays are commercially available in microplate format. Here a fluorogenic assay for caspase-3 is discussed using the fluorogenic substrate, Ac-DEVD-AMC, from Calbiochem.

As discussed before, fluorogenic substrate only shifted the excitation and emission wavelength from the original fluorophore. The first experiment should be to study the fluorescence spectra of the fluorogenic substrate and the released fluorescent product AMC. Figure 6.4 shows the absorption spectra of the AcDEVD-AMC and AMC. In this experiment, the AcDEVD-AMC solution was placed in two wells in a microplate. Caspase-3 was added to one well and an equal volume of buffer was added to another well. The final AcDEVD-AMC concentration was 2 μM. After all the substrate in the first well was converted into AMC, the absorbance spectra of the two samples were taken. This ensured that the two samples contained equal molar concentration of substrate and product for easy comparison. These spectra showed no changes in the AMC absorbance coefficient when it forms an amid bond, but the absorption maximum changed from 345 to 325 nm. The spectra showed that AMC could be selectively excited at 365 nm or longer, and its fluorescence could be measured with minimal interference from the substrate. The fluorescence spectra of AMC and AcDEVD-AMC are shown in Figure 6.5 when excited at 365 nm. It was clear that AcDEVD-AMC would not interfere with AMC fluorescence over a wide emission wavelength. Alternatively, AcDEVD-AMC could be selectively monitored when the samples were excited at 310 nm with some degree of interference from AMC. The fluorescence spectra of AMC and AcDEVD-AMC are shown in Figure 6.6 when excited at 310 nm. Though AMC was excited at 310 nm as well, its emission peaks at 460 nm are far away from the

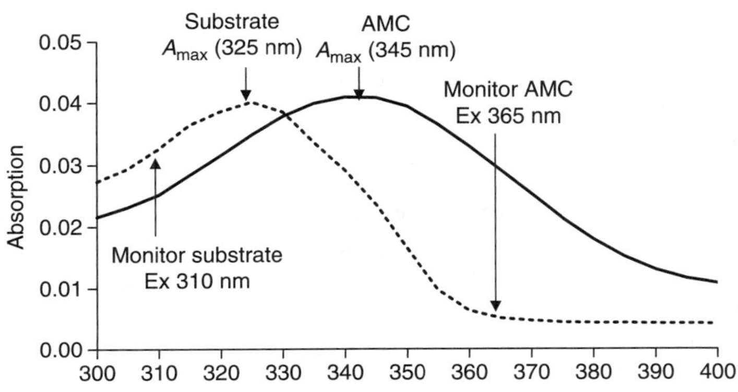

Figure 6.4 Absorption spectra of caspase-3 substrate (AcDEVD-AMC) and product (AMC) in the buffer containing 100 mM HEPES, pH 7.5, 0.1% Triton X-100, 1 mM EDTA, 10 mM DTT. The spectra were collected separately with identical condition with the same concentrations of the substrate and AMC.

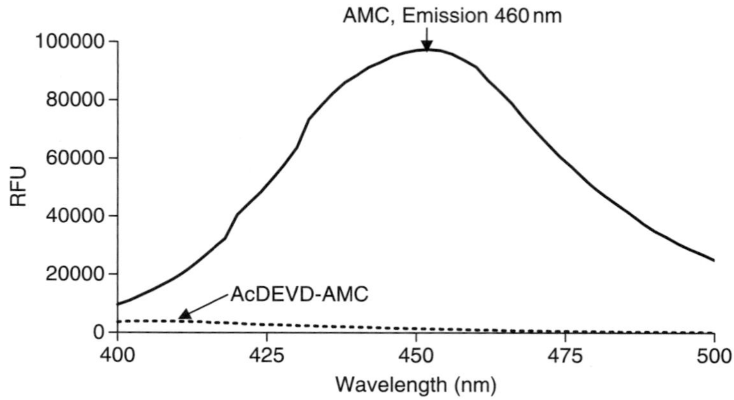

Figure 6.5 Emission spectra of AMC and AcDEVD-AMC when they were excited at 365 nm. Equal volumes of the samples from the absorption measurement were diluted by a factor of 10 to 200 nM in the same buffer.

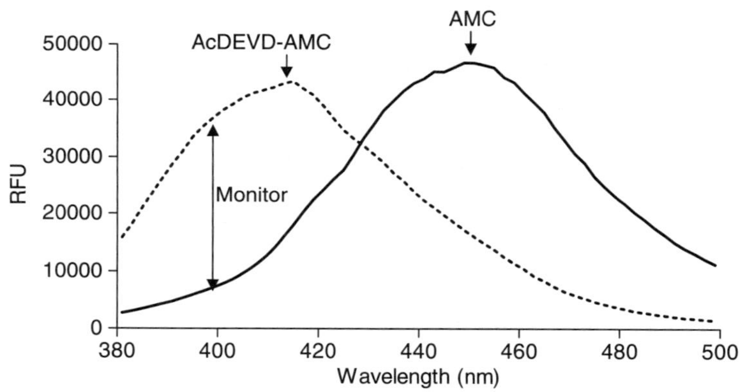

Figure 6.6 Emission spectra of AMC and AcDEVD-AMC when excited at 310 nm. Equal volumes of the samples from the absorption measurement were diluted by a factor of 10 to 200 nM in the same buffer. The substrate could be selectively monitored when emission at 400 nm was measured.

emission peak of AcDEVD-AMC. Thus, AcDEVD-AMC could be selectively monitored at 400 nm when the sample was excited at 310 nm with negligible interference from AMC. Based on the above fluorescence spectra studies, the caspase-3-catalyzed hydrolysis of AcDEVD-AMC could be studied by monitoring the product formation and substrate depletion simultaneously with dual wavelength excitation and emission. Figure 6.7 shows the time course of substrate depletion and product formation in the caspase-3-catalyzed reaction. The reaction mixture contained 2 μM AcDEVD-AMC and 3 nM caspase-3 in a microplate. The microplate was immediately measured after mixing with a Flexstation reader (Molecular Devices) after a lag time of about 30 s. The reaction was allowed to proceed at room temperature. Dual wavelengths were used to measure the substrate (ex 310 nm and em 400 nm) and product (ex 365 nm and em 460 nm).

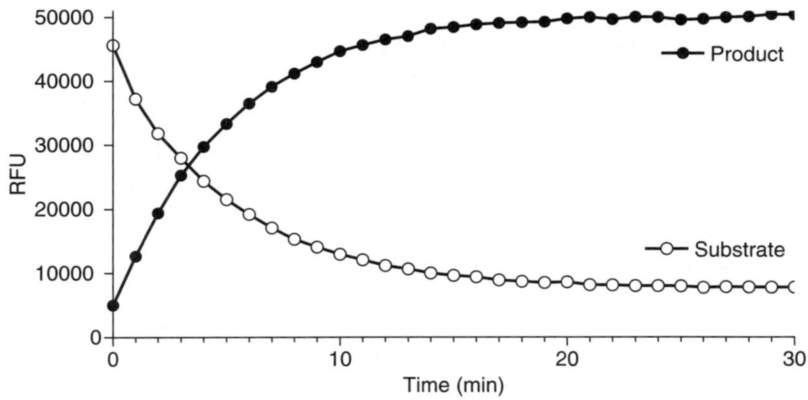

Figure 6.7 Kinetic study of caspase-3-catalyzed peptide hydrolysis. Substrate (AcDEVD-AMC) depletion and product (AMC) formation were simultaneously measured with dual wavelength (ex 310 nm/em 400 nm and ex 365 nm/em 460 nm).

6.7 ASSAY DEVELOPMENT FOR CARBOXYPEPTIDASE U (EC 3.4.17.20)

Carboxypeptide U, also referred to as thrombin-activable fibrinolysis inhibitor (TAFI), arginine carboxypeptidase, and plasma carboxypeptidase B, is a single-chain 60-kDa metallocarboxypeptidases secreted by the liver and found in human plasma as a zymogen. TAFI is *N*-glycosylated, and the attached glycans account for ~20% of the overall size of the protein. Proteolytic cleavage of TAFI by thrombin/thrombomodulin generates a 35-kDa active enzyme (TAFIa), which downregulates fibrinolysis by removing C-terminal lysine residues from partially degraded fibrin. TAFIa has also been implicated in the regulation of the inflammatory response by inactivating complement-derived inflammatory peptides. Thus, inhibitors of TAFIa may have therapeutic value.

TAFIa belongs to metallopeptidases and has Zn^{2+} at its active site. TAFIa has substrate specificity for a peptide with lysine or arginine at the P_1' position and with free carboxyl C-terminus. Because TAFIa activity depends on the free carboxyl group in the substrate, fluorogenic assay and quenched FRET assay cannot be used. The assay design can focus on the quantitation of the proteolytic products. Potato tuber carboxypeptidase inhibitor (PTCI), ε-aminocaproic acid, and 2-guanidinoethylmercaptosuccinic acid (GEMSA) can be used as a positive control in the assay. TAFIa is thermally unstable with a half-life of about 2 h at room temperature after activation. The presence of 150 μM GEMSA can significantly extend its half-life at room temperature. TAFIa should be generated just before performing assays. Because of its instability at room temperature, a fast assay that can be finished within 1 h at room temperature is desired.

One strategy to assay the carboxypeptidase activity of TAFIa is to measure the released amino acid from the carboxyl terminus of the substrate. ActiScreen (marketed by American Diagnostica Inc.) is a functional assay for TAFIa carboxyeptidase activity using hippuryl-L-lysine as the substrate. The proteolytic product, lysine, is detected by the specific activity of lysine oxidase that selectively oxidize lysine to 6-amino-2-oxo-hexanoic acid and hydrogen peroxide is produced in the process. The generated hydrogen peroxide can be measured with horseradish peroxidase in the presence of luminnescent, fluorescent, or colorimetric substrate. Willemse and co-workers (2005) developed another assay that detects the arginine formed as the hydrolysis product from a TAFIa peptide substrate containing C-terminal arginine. This method uses arginine kinase, pyruvate kinase, and lactate dehydrogenase as auxiliary enzymes to detect free arginine.

Another strategy to assay the carboxypeptidase activity of TAFIa is to measure the peptide product generated from the peptide substrate after the arginine or lysine at the carboxyl terminus is removed. A peptide substrate can be designed so that an antibody can recognize the resulting peptide product. For example, biotin-RRGLMVGGVVR was used as a TAFI substrate. The peptide product, biotin-RRGLMVGGVV could be assayed by using a monoclonal antibody (G2-10) that can recognize the epitope GLMVGGVV-OH but cannot bind to the peptide substrate. With this antibody, the peptide product could be assayed with technologies such as ELISA or HTRF. When the biotin group is substituted with a Cy5 dye, the peptide product could be monitored by fluorescence polarization technology.

An example of the assay development of TAFIa assay in HTRF format is shown here. Figure 6.8 shows the scheme to detect the peptide product with HTRF format. The biotin moiety binds to the streptavidin moiety in SA-XL665 and the europium

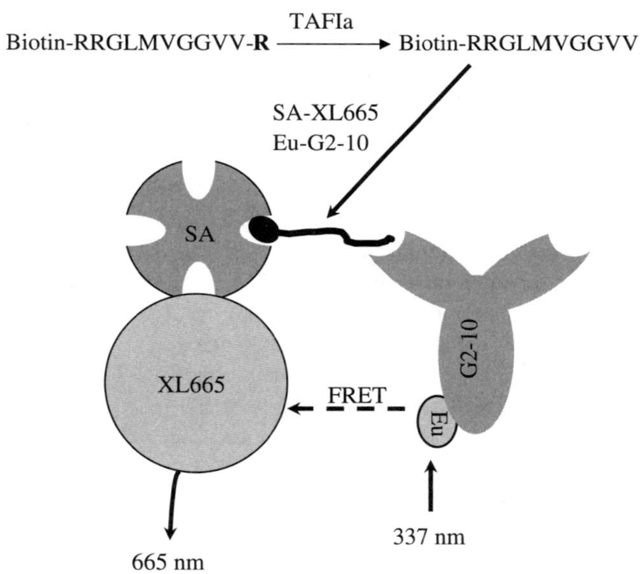

Figure 6.8 TAFIa assay with HTRF technology. The biotin-labeled substrate peptide biotin-RRGLMVGGVV-**R** was cleaved by TAFIa. The resulting peptide bound to a europium-labeled monoclonal antibody G2-10. The biotin group on the peptide bound to XL665-labeled streptavidin. This brought the europium and XL665 together, and the FRET signal was measured that was proportional to the concentration of the peptide product.

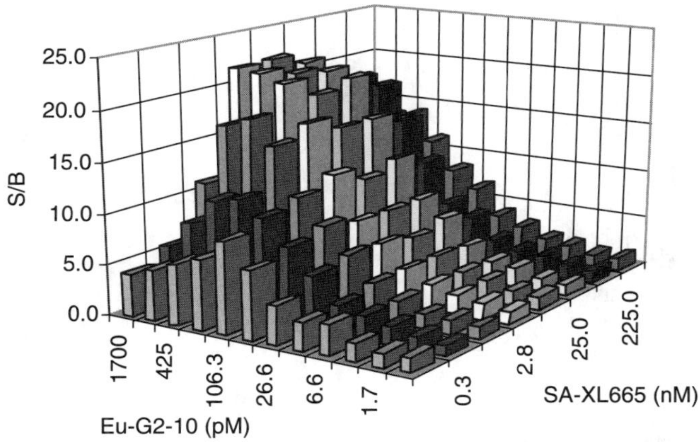

Figure 6.9 Evaluating assay conditions to detect 10 nM biotin-RRGLMVGGVV. The concentration of europium-labeled G2-10 and SA-XL665 were varied. The assay signal to background (S/B) was plotted at different conditions. Any conditions with S/B at >15 could be chosen as the final assay condition. (See color insert.)

labeled mAb G2-10 (Eu-G2-10) binds to the GLMVGGVV sequence. This brings the europium and XL665 together and the FRET signal can be measured that is proportional to the concentration of the peptide product. To investigate the optimal detection conditions, the pure synthetic peptide product biotin-RRGLMVGGVV at fixed concentrations was subjected to different combinations of SA-XL665 and Eu-G2-10. The HTFR signal was then collected and the signal-to-background ratio (S/B) was used as a criteria to select the optimal detection reagent combinations. The result of a typical experiment is shown in Figure 6.9. There are many reagent

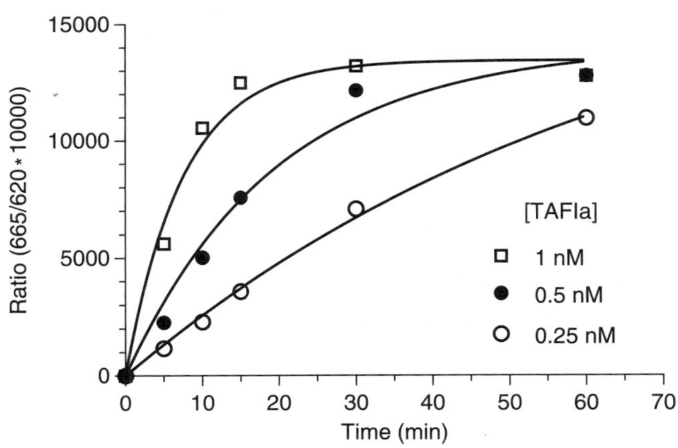

Figure 6.10 Time course of substrate hydrolysis at different concentrations of TAFIa. 60 nM substrate biotin-RRGLMVGGVV**R** was used. After quenching with 200 nM PTCI at different time, 80 nM SA-XL665 and 1.7 nM Eu-G2-10 were added to the solution and incubated for 30 min. The FRET signal at 620 and 650 nm were then measured.

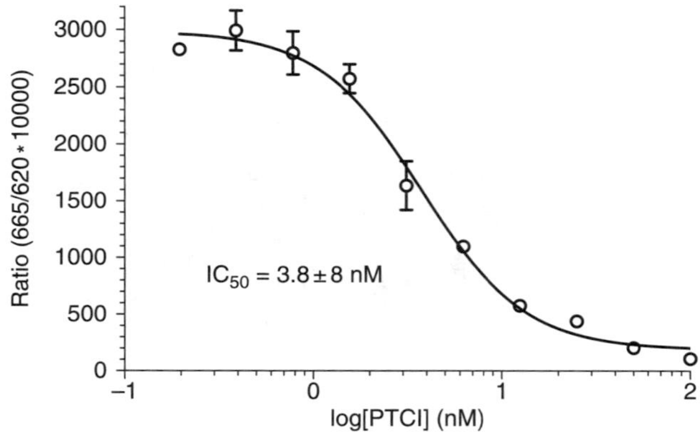

Figure 6.11 Inhibition of TAFIa-catalyzed hydrolysis of substrate biotin-RRGLMVGGVVR hydrolysis by PTCI. The concentrations of TAFIa and the substrate were 0.25 and 50 nM, respectively. After termination of the reaction, 62 nM SA-XL665 and 1.7 nM Eu-G2-10 were used to detect the proteolytic product.

combinations that can lead to acceptable assays. In general, assay conditions with S/B larger than 15 were used in subsequent studies. Because of the thermal instability, an assay that could generate enough detectable product within one hour is desirable. The reaction time course at different concentrations of TAFIa was tested with substrate concentrations at 60 nM (Fig. 6.10). If the reaction proceeded too fast, it would exert many operation constraints and the assay would be difficult to carry out. Assay time between 15 to 30 min is ideal for this assay. A concentration of 0.25 nM TAFIa was picked that gives sufficient signal and the initial velocity could be easily obtained because of the linear kinetics in this region. With the developed assay conditions, the IC_{50} of the known inhibitor PTCI was measured. The result is shown in Figure 6.11. The obtained IC_{50} value for PTCI was 3.8 nM, which matched the value in the literature very well.

Useful Websites for Proteases

http://www.chem.qmul.ac.uk/iubmb/

http://merops.sanger.ac.uk/

http://www.worthington-biochem.com/

http://www.brenda-enzymes.org/

https://www.sigmaaldrich.com/catalog/search/ProductDetail/SIGMA/A2765

http://www.invitrogen.com/site/us/en/home.html

http://www.emdbiosciences.com/html/CBC/home.html

http://www.anaspec.com/

http://www.worthington-biochem.com/

BIBLIOGRAPHY

Abbenante, G. and Fairlie, D. P. (2005) Protease inhibitors in the clinic. *Med. Chem.* **1**, 71–104.

Ascenzi, P., et al. (2003) The bovine basic pancreatic trypsin inhibitor (Kunitz inhibitor): A milestone protein. *Curr. Protein Pept. Sci.* **4**, 231–251.

Augustyns, K., Veken, P. V. d., Senten, K., and Haemers, A. (2005) The therapeutic potential of inhibitors of dipeptidyl peptidase IV (DPP IV) and related proline-specific dipeptidyl aminopeptidases. *Curr. Med. Chem.* **12**, 971–998.

Bajzar, L., Manuel, R., and Nesheim, M. E. (1995) Purification and characterization of TAFI, a thrombin-activable fibrinolysis inhibitor. *J. Biol. Chem.* **270**, 14477–14484.

Berg, J. M., Tymoczko, J. L., and Stryer, L. (2006) *Biochemistry*, 6th ed. W. H. Freeman, New York.

Bernier, S. G., et al. (2004) A methionine aminopeptidase-2 inhibitor, PPI-2458, for the treatment of rheumatoid arthritis. *Proc. Nal. Acad. Sci.* **101**, 10768–10773.

Bernier, S. G., Taghizadeh, N., Thompson, C. D., Westlin, W. F., and Hannig, G. (2005) Methionine aminopeptidases type I and type II are essential to control cell proliferation. *J. Cell. Biochem.* **95**, 1191–1203.

Beynon, R. and Bond, J. S. (2001) *Proteolytic Enzymes: A Practical Approach*, 2nd ed. Oxford University Press, Oxford.

Black, R. A. (1997) A metalloproteinase disintegrin that releases tumour-necrosis factor-[alpha] from cells. *Nature* **385**, 729–733.

Blum, G., et al. (2005) Dynamic imaging of protease activity with fluorescently quenched activity-based probes. *Nat. Chem. Biol.* **1**, 203–209.

Bode, W., Gomis-Ruth, F. X., Huber, R., Zwilling, R., and Stocker, W. (1992) Structure of astacin and implications for activation of astacins and zinc-ligation of collagenases. *Nature* **358**, 164–166.

Bode, W. and Huber, R. (1992) Natural protein proteinase inhibitors and their interaction with proteinases. *Eur. J. Biochem.* **204**, 433–451.

Boffa, M. B. and Koschinsky, M. L. (2007) Curiouser and curiouser: Recent advances in measurement of thrombin-activatable fibrinolysis inhibitor (TAFI) and in understanding its molecular genetics, gene regulation, and biological roles. *Clin. Biochem.* **40**, 431–442.

Boulware, K. T. and Daugherty, P. S. (2006) Protease specificity determination by using cellular libraries of peptide substrates (CLiPS). *Proc. Natl. Acad. Sci.* **103**, 7583–7588.

Bredemeyer, A. J., et al. (2004) A proteomic approach for the discovery of protease substrates. *Proc. Nal. Acad. Sci.* **101**, 11785–11790.

Celerier, J., et al. (2000) Characterization of a human angiotensinogen cleaved in its reactive center loop by a proteolytic activity from Chinese hamster ovary cells. *J. Biol. Chem.* **275**, 10648–10654.

Cooper, A. C., et al. (2006) A novel methionine aminopeptidase-2 inhibitor, PPI-2458, inhibits non-Hodgkin's lymphoma cell proliferation in vitro and in vivo. *Clin. Cancer Res.* **12**, 2583–2590.

Coussens, L. M., Fingleton, B., and Matrisian, L. M. (2002) Matrix metalloproteinase inhibitors and cancer—trials and tribulations. *Science* **295**, 2387–2392.

Coussens, L. M. and Werb, Z. (1996) Matrix metalloproteinases and the development of cancer. *Chem. Biol.* **3**, 895–904.

Deaton, D. and Kumar, S. (2004) Cathepsin K inhibitors: their potential as anti-osteoporosis agents. *Prog. Med. Chem.* **42**, 245–375.

Ding, L., et al. (1995) Origins of the specificity of tissue-type plasminogen activator. *Proc. Natl. Acad. Sci.* **92**, 7627–7631.

Docherty, A. J. P. (1985) Sequence of human tissue inhibitor of metalloproteinases and its identity to erythroid-potentiating activity. *Nature* **318**, 66–69

Ferrer, M., et al. (2003) Miniaturizable homogenous time-resolved fluorescence assay for carboxypeptidase B activity. *Anal. Biochem.* **317**, 94–98.

Foley, J. H., Kim, P., and Nesheim, M. E. (2008) Thrombin-activable fibrinolysis inhibitor zymogen does not play a significant role in the attenuation of fibrinolysis. *J. Biol. Chem.* **283**, 8863–8867.

Glickman, M. H. and Ciechanover, A. (2002) The ubiquitin-proteasome proteolytic pathway: destruction for the sake of construction. *Physiol. Rev.* **82**, 373–428.

Gosalia, D. N., Salisbury, C. M., Maly, D. J., Ellman, J. A., and Diamond, S. L. (2005) Profiling serine protease substrate specificity with solution phase fluorogenic peptide microarrays. *Proteomics* **5**, 1292–1298.

Hardy, J. A., Lam, J., Nguyen, J. T., O'Brien, T., and Wells, J. A. (2004) Discovery of an allosteric site in the caspases. *Proc. Natl. Acad. Sci.* **101**, 12461–12466.

Harris, J. L., Peterson, E. P., Hudig, D., Thornberry, N. A., and Craik, C. S. (1998) Definition and redesign of the extended substrate specificity of granzyme B. *J. Biol. Chem.* **273**, 27364–27373.

Hartley, B. S. (1960) Proteolytic enzymes. *Annu. Rev. Biochem.* **29**, 45–72.

Hellen, C. (1994) Assay methods for retroviral proteases. *Methods Enzymol.* **241**, 46–58.

Hsu, H.-J., et al. (2008) Factor Xa active site substrate specificity with substrate phage display and computational molecular modeling. *J. Biol. Chem.* **283**, 12343–12353.

Hu, J., Van den Steen, P. E., Sang, Q.-X. A., and Opdenakker, G. (2007) Matrix metalloproteinase inhibitors as therapy for inflammatory and vascular diseases. *Nat. Rev. Drug Discov.* **6**, 480–498.

Jensen, C., Herold, P., and Brunner, H. R. (2008) Aliskiren: The first renin inhibitor for clinical treatment. *Nat. Rev. Drug Discov.* **7**, 399–410.

Kaufmann, S. H., Lee, S. H., Meng, X. W., Loegering, D. A., Kottke, T. J., Henzing, A. J., Ruchaud, S., Samejima, K., and Earnshaw, W. C. (2008) Apoptosis-associated caspase activation assays. *Methods* **44**, 262–272.

Ke, S.-H., et al. (1997) Distinguishing the specificities of closely related proteases. *J. Biol. Chem.* **272**, 16603–16609.

Lah, T. T., Duran Alonso, M. B., and Van Noorden, C. J. (2006) Antiprotease therapy in cancer: Hot or not? *Expert Opin. Biolog. Ther.* **6**, 257–279.

Lopez-Otin, C. and Matrisian, L. M. (2007) Emerging roles of proteases in tumour suppression. *Nat. Rev. Cancer* **7**, 800–808.

Lopez-Otin, C. and Overall, C. M. (2002) Protease degradomics: A new challenge for proteomics. *Nat. Rev. Mol. Cell Biol.* **3**, 509–519.

Matthews, D. and Wells, J. (1993) Substrate phage: Selection of protease substrates by monovalent phage display. *Science* **260**, 1113–1117.

Nakagawa, T., Akaki, J., Satou, R., Takaya, M., Iwata, H., Katsurada, A., Nishiuchi, K., Ohmura, Y., Suzuki, F., and Nakamura, Y. (2007) The His-Pro-Phe motif of angiotensinogen is a crucial determinant of the substrate specificity of renin. *Biol. Chem.* **388**, 237–246.

Nazif, T. and Bogyo, M. (2001) Global analysis of proteasomal substrate specificity using positional-scanning libraries of covalent inhibitors. *Proc. Natl. Acad. Sci.* **98**, 2967–2972.

Neefjes, J. and Dantuma, N. P. (2004) Fluorescent probes for proteolysis: Tools for drug discovery. *Nat. Rev. Drug Discov.* **3**, 58–69.

Niles, A. L., Moravec, R. A., and Riss, T. L. (2008) Caspase activity assays. *Methods Mol. Biol.* **414**, 137–150.

Obermajer, N., Doljak, B., and Kos, J. (2006) Cysteine cathepsins: Regulators of antitumour immune response. *Expert Opin. Biol. Ther.* **6**, 1295–1309.

Okun, I., et al. (2006) Screening for caspase-3 inhibitors: effect of a reducing agent on identified hit chemotypes. *J. Biomol. Screen.* **11**, 694–703.

Overall, C. M. and Blobel, C. P. (2007) In search of partners: Linking extracellular proteases to substrates. *Nat. Rev. Mol. Cell Biol.* **8**, 245–257.

Pinilla, C., Appel, J. R., Blanc, P., and Houghton, R. A. (1992) Rapid identification of high affinity peptide ligands using positional scanning synthetic peptide combinatorial libraries. *Bio. Techniques* **13**, 901–905.

Puente, X. S., Sanchez, L. M., Overall, C. M., and Lopez-Otin, C. (2003) Human and mouse proteases: A comparative genomic approach. *Nat. Rev. Genet.* **4**, 544–558.

Rawlings, N. D., Morton, F. R., Kok, C. Y., Kong, J., and Barrett, A. J. (2008) MEROPS: The peptidase database. *Nucl. Acids Res.* **36**, D320–325.

Reed, J. C. (2002) Apoptosis-based therapies. *Nat. Rev. Drug Discov.* **1**, 111–121.

Robinson, S., et al. (2008) A mass spectrometry-based strategy for detecting and characterizing endogenous proteinase activities in complex biological samples. *Proteomics* **8**, 435–445.

Rowan, A. D., Litherland, G. J., Hui, W., and Milner, J. M. (2008) Metalloproteases as potential therapeutic targets in arthritis treatment. *Expert Opin. Ther. Targets* **12**, 1–18.

Schechter, I. and Berger, A. (1967) On the size of the active site in proteases. I. Papain. *Biochem. Biophys. Res. Commun.* **27**, 157–162.

Selvakumar, P., et al. (2004) High expression of methionine aminopeptidase 2 in human colorectal adenocarcinomas. *Clin. Cancer Res.* **10**, 2771–2775.

Sieber, S. A., Niessen, S., Hoover, H. S., and Cravatt, B. F. (2006) Proteomic profiling of metalloprotease activities with cocktails of active-site probes. *Nat. Chem. Biol.* **2**, 274–281.

Sottrup-Jensen, L. (1989) Alpha-macroglobulins: Structure, shape, and mechanism of proteinase complex formation. *J. Biol. Chem.* **264**, 11539–11542.

Southan, C. (2001) A genomic perspective on human proteases. *FEBS Lett.* **498**, 214–218.

Sun, H., Panicker, R. C., and Yao, S. Q. (2007) Activity based fingerprinting of proteases using FRET peptides. *Peptide Sci.* **88**, 141–149.

Thornberry, N. A. and Lazebnik, Y. (1998) Caspases: Enemies within. *Science* **281**, 1312–1316.

Thornberry, N. A., et al. (1997) A combinatorial approach defines specificities of members of the caspase family and granzyme B. *J. Biol. Chem.* **272**, 17907–17911.

Tomkinson, B. (1999) Tripeptidyl peptidases: Enzymes that count. *Trends Biochem. Sci.* **24**, 355–359.

Tung, C.-H. (2004) Fluorescent peptide probes for in vivo diagnostic imaging. *Pept. Sci.* **76**, 391–403.

Turk, B. (2006) Targeting proteases: successes, failures and future prospects. *Nat. Rev. Drug Discov.* **5**, 785–799.

Turk, B. E., Huang, L. L., Piro, E. T., and Cantley, L. C. (2001) Determination of protease cleavage site motifs using mixture-based oriented peptide libraries. *Nat. Biotech.* **19**, 661–667.

Valnickova, Z., Thogersen, I. B., Potempa, J., and Enghild, J. J. (2007) Thrombin-activable fibrinolysis inhibitor (TAFI) zymogen is an active carboxypeptidase. *J. Biol. Chem.* **282**, 3066–3076.

Wang, J., et al. (2008) Correlation of tumor growth suppression and methionine aminopetidase-2 activity blockade using an orally active inhibitor. *Proc. Natl. Acad. Sci.* **105**, 1838–1843.

Willemse, J. L. and Hendriks, D. F. (2006) Measurement of procarboxypeptidase U (TAFI) in human plasma: A laboratory challenge. *Clin. Chem.* **52**, 30–36.

Willemse, J. L., Leurs, J., Verkerk, R., and Hendriks, D. (2005) Development of a fast kinetic method for the determination of carboxypeptidase U (TAFIa) using C-terminal arginine containing peptides as substrate. *Anal. Biochem.* **240**, 106–112.

Wolf, B. B. and Green, D. R. (1999) Suicidal tendencies: Apoptotic cell death by caspase family proteinases. *J. Biol. Chem.* **274**, 20049–20052.

Xu, X., et al. (1998) Detection of programmed cell death using fluorescence energy transfer. *Nucl. Acids Res.* **26**, 2034–2035.

Yang, G., et al. (2001) Steady-state kinetic characterization of substrates and metal-ion specificities of the full-length and N-terminally truncated recombinant human methionine aminopeptidases (type 2). *Biochemistry* **40**, 10645–10654.

Zaman, M. A., Oparil, S., and Calhoun, D. A. (2002) Drugs targeting the renin-angiotensin-aldosterone system. *Nat. Rev. Drug Discov.* **1**, 621–636.

CHAPTER 7

FUNCTIONAL ASSAYS FOR PROTEIN KINASES

7.1 INTRODUCTION TO PROTEIN KINASES

One of the most important methods in cellular communication is the addition or removal of phosphate groups in proteins or lipids along the signal transduction pathways in eukaryotic cells. Phosphorylation of proteins can control their enzymatic activity, their interaction with other proteins and molecules, their location in the cell, and their propensity for degradation by proteases. Dysregulation of many kinases has been linked to disease development, especially in cancer development.

Kinases catalyze the transfer of γ-phosphate groups from ATP to a hydroxyl group in proteins or lipids. The hydrolysis of ATP generates adenosine 5′-diphosphate (ADP) in this process as a by-product. Opposite to the phosphorylation, the removal of the phosphate group from phosphorylated proteins is catalyzed by phosphatases. The γ-phosphate group in ATP can be transferred to either serine/threonine or tyrosine residues in a substrate protein in the presence of protein kinases. Based on this, the Enzyme Committee classifies protein kinases into two large families: Ser/Thr kinases (EC2.7.11) if the phosphate group is transferred to serine or threonine residues and Tyr kinases (EC2.7.10) if the phosphate group is transferred to tyrosine residues.

Protein genome-wide searches revealed that there are a total 518 putative kinases genes in the human genome. Of these, 478 genes encode protein kinases, and 40 genes encode atypical protein kinases that are proteins having biochemical kinase activities but lacking sequence similarity to protein kinase domain. Fifty of the 478 protein kinases lack at least one of the conserved catalytic residues (Lys^{30}, Asp^{125}, and Asp^{143}) and are likely to be catalytically inactive. Thus, only about 468 catalytically active protein kinases can phosphorylate cellular proteins. Based on the classification method originally developed by Steven K. Hanks (1988), human protein kinases can be classified into seven groups based on sequence similarity and catalytic domains (AGC, CaMK, CK1, CMGC, STE, TK, and TKL). The AGC group includes the PKA, PKC, and PKG families of kinases. The CaMK group comprises the kinases that are regulated by calmodulin, which are important in the process of neural transmission. The CK1 group comprises the casein kinase 1 family kinases. The CMGC group contains the cyclin-dependent kinases (CDKs) and mitogen-activated protein kinases (MAPKs). These kinases are key components of a number of signal

Assay Development: Fundamentals and Practices. By Ge Wu

transduction pathways including the responses of cells to external stimuli. The STE group contains homologs of yeast Sterile 7, Sterile 11, and Sterile 20 kinases. The TK group comprises both membrane-bound and soluble protein tyrosine kinases (EGFRs, FGFRs, PDGFRs, InsR, etc.). The TKL group contains tyrosine kinase-like kinases.

After identifying all the protein kinases, the major challenge in establishing the function of the kinases inside cells is to relate all protein kinases to their natural protein substrates. This turned out to be a very complicated and difficult task. Exposure of cells with a stimulation factor may result in phosphorylation of thousands of intracellular proteins. For example, 6600 phosphorylation sites on 2244 proteins were detected with epidermal growth factor (EGF)-stimulated HeLa cells (Olsen, 2006). Fourteen percent of the phosphorylation sites are modulated at least twofold by EGF. As of July 2007, 4026 phosphorylated substrate proteins from different species covering 2079 tyrosine, 11,993 serine, and 2356 threonine sites have been reported in the Phospho.ELM database. A larger number of 7117 nonredundant phosphrylated proteins are claimed in the PhosphositePlus database. In consideration of the total number of active protein kinases of 468, one kinase on average must phosphorylate many proteins and even more phosphorylation sequences. For example, more than 100 phosphopeptide sequences collected from mass spectrum studies are assigned to be phosphorylated by PKA.

Because all signal transduction pathways are regulated at some level by phosphorylation, kinases are attractive drug targets in cancer and other proliferative diseases, inflammatory diseases, metabolic disorders, and neurological diseases. A large number of kinase inhibitors are therefore currently under investigation in preclinical and clinical trials. This makes the kinase family of proteins the most studied drug targets after the GPCR family of proteins. All the currently approved drugs targeting kinases act on the ATP binding pocket of kinases. Initially, it was thought that the ATP binding pocket in different kinases is structurally similar, and the kinase inhibitors targeting ATP binding pocket will be too toxic because of their inability to distinguish the target kinase from the other physiological important normal kinases. It turned out that the ATP binding pocket in many kinases, especially the kinases that are not in the same group or family, is structurally different enough for small-molecule inhibitors to selectively inhibit a few closely related kinases. In addition, it was discovered that less selectivity of some inhibitors actually has better therapeutic effects. Since the approval of the first kinase inhibitor, imatinib mesylate (Gleevac), several kinase inhibitors have been approved or are in Phase III clinical trials. As of this writing, seven kinase inhibitors have been approved as anticancer drugs so far. This further validates kinases as drug targets.

7.2 SUBSTRATES FOR IN VITRO KINASE ASSAYS

Identification of molecules that can be phosphorylated by the isolated kinase is the most important part of kinase assay development. Two categories of substrates in protein kinase assays exist: intact proteins and synthetic peptides with typically 10 to 15 residues. We will discuss both separately below.

7.2.1 Intact Proteins as Kinase Substrate

Isolated native proteins were the original substrates used in in vitro kinase assays. The proteins were picked as kinase substrates because they were phosphorylated in vivo and were readily available in large quantity. Commonly used protein substrates include histones, myelin basic proteins (MBP), and casein. In addition, each of these proteins is known to be substrates of not just one but several kinases in vivo. For in vitro assays, they can be phosphorylated by even a greater number of kinases because isolated kinases usually display a high degree of promiscuity with regard to substrate. This promiscuity may be caused by the artificial in vitro assay conditions and the loss of regulatory mechanism to the kinase when only the catalytic subunits are used in assays. The kinases having multiple substrates in vivo usually recognize a short primary sequence motif. These kinases are more promiscuous with substrate and they can phosphorylate a small peptide derived from the active site sequence. The fact that a kinase has on average more than 10 substrates in vivo hints that the majority of the kinases must have multiple substrates. This is why the majority of the kinases can use peptides as a substrate. PKA, AKT, and MAP kinase are good examples that they phosphorylate many substrates on the serine or threonine residues in vivo, and they can phosphorylate corresponding synthetic peptide substrates in vitro. There are a small number of kinases that recognize their substrates based on the three-dimensional structure. In addition to the primary sequence on the phosphorylation site, they also interact with amino acid residues with primary sequences far away from the phosphorylation site. MEK, Raf, Csk, and JNK1 are good examples of this type of kinases. The peptide substrates derived from their phosphorylation sites are poor substrates for these kinases. In this situation, only intact protein substrate can be used in assays before the identification of novel synthetic peptide substrates.

The major challenges in using intact proteins as substrates are the limited number of commercially available proteins that can serve as substrates (e.g., histones, MBP, and casein), the identification of a substrate protein for a particular kinase, the affordability, and the limited choice of assay technologies that can be applied to protein substrates. There are several databases that match kinases with known native substrates (e.g., NetworKin and Kinasource). As of May, 2008, 134 kinases and 686 substrates are collected in the Kinasource database with 1357 distinct kinase/substrate pairs. However, if the identified substrate protein is not commercially available, it will take some effort to make the substrate protein in large quantity. In addition, a majority of the modern kinase assay technologies are not applicable to assays with protein as substrates. This limits the kinase assays with intact proteins as substrates to a few technologies, mainly the radioactivity filtration assay (will be discussed in Section 7.4). Because of all these limitations, only a relatively small number of kinase assays can be performed with commercially available intact proteins as the substrate.

7.2.2 Synthetic Peptides as Kinase Substrate

Using peptides as kinase substrates solves many of the problems associated with intact protein substrates. Peptides with well-defined chemical compositions can be readily obtained in large quantity economically with any desired sequences and modification. This flexibility made it possible for the development of many kinase assay technologies based on peptide substrates. Currently, the majority of in vitro functional kinase assays are performed with synthetic peptides with typically 10 to 15 residues as the substrates. These peptide substrates commonly contain a phosphorylation site that is occupied by a serine (serine is always preferably phosphorylated compared with threonine) or tyrosine as the acceptor for the phosphate group to transfer to. In addition, they contain the sequences that are required for the kinase's sequence specificity (also referred to as consensus sequences). A consensus sequence for a particular kinase may contain one or more determinant amino acids at specific positions relative to the phosphorylation site. By convention, the positions on the consensus sequences are denoted as $-1, -2, -3, \ldots$ by counting from the phosphorylation site to the amino terminal direction and $+1, +2, +3, \ldots$ by counting to the carboxyl terminal direction. For example, the consensus sequence for PKA is R-(R/K)-X-S*-B, meaning there is a preference for Arg at -3, Arg or Lys at -2, and hydrophobic residues at $+1$. There is no preference for any amino acids at the -1 position. There are two strategies to obtain peptide substrates for kinases. The first strategy is based on the studies of the amino acid sequences of the kinase's natural substrates. This strategy depends on the existing knowledge of the kinase and the substrate protein. The second strategy is based on screening a library of synthetic peptides without the requirement of preexisting knowledge of the peptide sequences. In general, a good peptide substrate should have a K_m value of less than 100 μM and V_{max} comparable to its natural protein substrates.

Peptide Substrates Derived from Natural Proteins Peptide substrates for a given kinase were originally obtained by studying the kinase's natural protein substrates. If multiple substrates exist for a particular kinase, the conserved residues around the phosphorylation site may help to derive the consensus sequence. Many early known peptide substrates for kinases were successfully obtained this way. A good example of this approach is the well-known kemptide (LRRAS*LG), which is a substrate for PKA. Kemptide was derived from the phosphorylation site on the porcine liver pyruvate kinase. It is an excellent substrate for PKA with kinetic parameters comparable to many intact natural protein substrates. Sometimes, a more potent peptide substrate than the original peptide derived from the natural sequences may be obtained by mutation studies on selected residues. For example, SAMS (HMRSAMS*GLHLYKRR) peptide was derived from rat acetyl CoA carboxylase, which is a substrate of AMPK. Mutation of this substrate by substituting most of the noncritical amino acid residues to alanine leads to a more potent peptide substrate AMARA (AMARAAS*AAALARRR). In addition to the peptide derived from the sequences surrounding the phosphorylated sites, peptides derived from sequences that are known to interact with the active site of kinases can also be good substrates. For example, the pseudosubstrate on the regulatory subunits of PKC can become a substrate when a serine is introduced to sequence to substitute the nonphosphorylatable alaline.

Peptide Substrates from Screening Synthetic Peptide Library When the kinase's natural substrates are not known or if the synthesized peptide substrate derived from its native phosphorylation sequence does not work, screening of a peptide library offers an alternative way to find a peptide substrate. It has been demonstrated that protein kinase substrate motifs can be successfully obtained by screening a random library of penta- and hepta-peptides on solid beads. However, it is believed that peptides with at least 9 to 12 amino acid residues may be required to contact the active site cleft of a kinase. It is not practical to synthesize and screen all the possible peptides with more than 10 amino acids because 20^{10} peptides are required. A method with oriented degenerated peptide library was developed that significantly reduced the size of the library. This method can find a favorable amino acid in specific position relative to the phosphorylation site based on the average of many sequences in the pool. After favorable amino acids in each position are identified, a peptide substrate can be built. However, this peptide substrate may not be the optimal substrate because the method does not take account to interaction between each residue in the peptide. With this method, the contribution from individual sequence is small. Even if there exists a particular optimal sequence in the pool, its contribution to the total signal will not be significant. Nevertheless, this method has been successfully shown to obtain good peptide substrates for many kinases.

Other Methods to Obtain Peptide Substrates Because of the huge interest in assay kinases to obtain kinase inhibitors that may lead to drugs, many peptide substrates for many kinases have been demonstrated experimentally in in vitro assays. These sequences can be obtained by searching the literature or databases. If no known substrate has been demonstrated experimentally in the past for a particular kinase under study, a substrate may be obtained by computational prediction or by screening a small, diverse, and well-curated library of peptides known to be phosphorylated. Recent developments in studying the phosphoraylation of native protein by mass spectrometry have provided a large number of the native phosphopeptide sequences. This information is assembled in several databases, such as PhosphositePlus and Phospho.ELM database. These known phosphopeptide sequences can help identify peptide substrates for kinases. One challenge is to match the more than 400 functional kinases with the more than 10,000 known native peptide phosphorylation sites. There are over 10 databases collecting information on kinases and their substrates. For example, Prodikin is a searchable database that predicts the −3 to +3 residue specificity for a kinase. Based on kinase sequences, it searches for substrate-determining residues (SDR). SDRs are conserved amino acid residues, located in the catalytic domain of a serine/threonine protein kinase, which determine whether a protein is a likely substrate for the kinase. After the algorithm determines the SDRs in a kinase, it searches a database to finds kinases with similar SDRs and then retrieves substrates for those kinases and builds a scoring matrix. In addition to computation methods, many companies use screening methods to find peptide substrates. With screening methods, a libary of a few hundred to a thousand peptides is assembled with proprietary methods. The kinase under study is then assayed with this small library of peptides and the peptide substrates are obtained if they are phosphorylated. Libraries of peptide substrate candidates are commercially available from companies such as JPT Peptide Technologies. For in vitro kinase assay development applications, it is possible to find a reasonably good, though not optimal, peptide substrate for a given kinase by screening a peptide library.

Many of the tyrosine kinases used in drug discovery are the recombinant catalytic domains of the tyrosine kinases. Tyrosine kinases, in general, lack stringent substrate specificity as compared with serine/threonine kinases and are more promiscuous with substrate. It was shown that synthetic random polymers containing tyrosine and negatively charged glutamic acid residues are excellent substrates for many tyrosine kinases. For tyrosine kinase assay development, a negatively charged peptide Poly(Glu_4Tyr)$_n$ is often used as a potential substrate in initial tyrosine kinase activity testing. Despite the promiscuity of the tyrosine kinases, they can display dramatic different kinetics with different peptide substrates. Thus, it is recommended to always test a few peptide substrates in tyrosine kinase assay development.

7.3 KINASE ASSAY DEVELOPMENT STRATEGIES

Kinases can be studied either as isolated proteins or in intact cells. Functional in vitro kinase assays with isolated kinases are convenient to perform and the interpretation of the assay results is straightforward. However, isolated kinases lock the native regulated environment and the assay results may not reflect the true biology in vivo. In addition, many kinases employed in bioassay are expressed in artificial expression systems that may not faithfully reproduce the covalent modification of the kinase. Further, many kinases, especially receptor tyrosine kinases (RTK), are studied in vitro with only the expressed catalytic domain that greatly perturbed the native system. The results obtained from these studies must be carefully interpreted. In cell-based kinase assays, it is difficult to unambiguously demonstrate that the phosphorylation of a protein substrate in the cell is due to the enzymatic activity of a particular kinase because of the large number of protein substrates for a given kinase. Thus, the detection of the modification of the kinase itself as a result of its activation is sometimes used to gauge the activity of the kinase. This is usually done by the detection of the phosphorylation level of the target kinase because of correlation of phosphorylation and activation. Alternatively, the kinase of interest can be isolated from the cells after the cell-based experiment is done. The kinase activity is then measured by measuring its phosphorylation of a known substrate. In this chapter, we will focus on functional kinase assays with isolated kinases only.

There are many ways to configure a biochemical assay for a kinase of interest. A binding assay can be configured by the methods discussed in Chapter 5. However, because of the availability of a large collection of functional kinase assays, binding assays are rarely performed with kinases. Functional kinase assays are predominant assay formats for kinases that can be configured in many ways. They are classified here into five categories based on the kinase-catalyzed chemical reaction scheme: (1) measurement of the generation of phosphorylated product, (2) measurement of the generation of ADP, (3) measurement of ATP consumption, (4) measurement of the peptide or protein substrate consumption, and (5) measurement of both peptide substrate consumption and phosphorylated product generation simultaneously. In theory, monitoring the generation of product from zero to some value is preferred to monitoring the depletion of ATP or protein–peptide substrate. Because it is usually not practical to measure reliably a 10% decrease in the ATP or the protein–peptide substrate from its original value in end-point assays due to the error associated with the assays, these assays often require the substrate turnover of more than 50%, which may cause some concerns. In discussing all these kinase assays, many specialized technologies are introduced as well. These technologies may be used beyond the context of kinase assay development.

Tyrosine kinases have some unique properties that are useful for bioassay development. Tyrosine kinases in general are more promiscuous in substrate phosphorylation. Thus, the negatively changed peptide poly(Glu_4Tyr)$_n$ can often be a reasonably good substrate for many tyrosine kinases. In addition, there are many anti-Tyr-PO_4^{2-} antibodies that are widely available. These antibodies can be used universally to detect any phosphotyrosine without much interference from surrounding amino acid sequences though they may display different affinity to different sequences. In comparison, there is no universal antibody for phosphorylated serine/threonine. Most anti-Ser/Thr-PO_4^{2-} antibodies recognize both the phosphorylated Ser/Thr residues and several adjacent amino acid residues. This requires the development of an antibody for every Ser/Thr substrate sequences.

Kinases have two substrate binding sites. One site binds to the native protein substrate (this site is also assumed to be the artificial peptide substrate binding site) and the other site binds to ATP. For enzyme kinetic studies, it is preferable to set the experimental conditions so that the kinetics will follow pseudo-first-order kinetics instead of second-order kinetics. This is done by setting the concentration of one substrate significantly higher than the concentration of other substrates (at least 10-fold in excess).

Isolated kinases obtained from cells expressing recombinant kinases may not always be active after isolation. Some kinases' activity is controlled by phosphorylation/dephosphorylation, such as AKT (also referred to as PKB). AKT is only active when it is phosphorylated on Thr308 and Ser473 residues. Some kinases' activity is regulated by regulatory subunits that inhibit the kinase activity. PKA is a good example with regulatory subunits. Removal of the regulatory subunits makes the catalytic subunit of PKA constitutively active. Other cofactors may be needed for the kinases' activity, for example, PKC requires the presence of Ca^{2+} and lipid activator. Most kinases require divalent ions, Mg^{2+} or Mn^{2+}, for proper function.

A proper control is important in kinase assay development. There are many kinase inhibitors commercially available, such as H9 and H89. Some of these inhibitors have broader inhibition ability across many kinases and some are specific for a few kinases. Because all the kinases require divalent ions for proper function, EDTA is a universal inhibitor for all kinases, and it is the most widely used inhibitor as a control in developing kinase assays.

7.4 KINASE ASSAY BASED ON DETECTION OF PHOSPHORYLATED PRODUCT

This is the most commonly employed strategy to assay kinase activity because it is based on the direct detection of the presence of the transferred phosphate group on the protein–peptide substrates. Three general strategies have been employed to detect the presence of the phosphate group in a peptide–protein sequence: (1) the detection of the radioactive phosphate group that is transferred from radiolabeled [γ^{32}P or γ^{33}P]ATP; (2) the detection of the phosphate group on the kinase substrate by an antibody; and (3) the detection of the phosphate group on the kinase substrate based on the affinity of the phosphate group in the phosphorylated substrate for trivalent ion chelates, such as Ga^{3+} or Fe^{3+} chelates. Monitoring the transfer of radiolabeled ATP to a substrate is the most straightforward method. Assays based on radiolabeled ATP have fewer limitations except the limitation on ATP concentration. Detection of the phosphate group in a phosphorylated product by antiphospho antibody works well with tyrosine kinases but has limited use for Ser/Thr kinases because of the availability of antibodies against the many substrate sequences. Affinity of trivalent ion chelates to the phosphate group in phosphorylated peptide–protein is universal and is applicable to both phospho-Ser/Thr and phospho-Tyr. Because of the universal applicability, this method has gained widespread use recently. One drawback of the method is that the affinity is not very specific and may be interfered with by a phosphate group containing molecules in the assay system, such as high concentration of ATP.

7.4.1 Radioactivity Filtration Assay for Kinase Activity

This assay uses radioactive [γ^{32}P or γ^{33}P]ATP to monitor the kinase-catalyzed transfer of the γ-phosphate group from ATP to the peptide–protein substrate. The assay is performed by incubating a mixture of kinase, peptide–protein substrate, and radiolabeled ATP. The reaction mixture is then filtered through a membrane under vacuum. The membrane is chosen so that the radiolabeled substrate is retained on the membrane. After several washes, the membrane is removed and the radioactivity in the membrane is counted. The measured radioactivity is proportional to the amount of the phosphorylated kinase product. In this assay, the phosphate-accepting substrates can be large native proteins or small peptides. The key requirement is that the peptide–protein substrates have to be able to attach to the filtration membrane for this method to work. The filtration membrane can be chosen based on the charge, size, hydrophobicity, and other properties (such as biotin on the substrate) of the peptide–protein substrates. The most common method uses phosphocellulose membrane that contains many negatively charged phosphate groups that can bind to positively charged residues on the protein–peptide substrates. This assay requires the protein–peptide substrate contain two or more basic residues. Most protein substrates satisfy this requirement. For peptide substrates, a patch of basic residues can be added to either end of the terminus that does not affect the phosphorylation site. Both unlabeled- and labeled-substrate retains on the membrane while the highly negatively charged radioactive ATP passes through the membrane. However, this method does not always quantitatively retain the substrate on the membrane, especially when the substrates have less positively charged residues. Confirming the quantitative binding of the substrate to the membrane is important when developing kinase assay with this method. Another filtration method is based on the fact that most proteins precipitate in 25% trichloroacetic acid (TCA). The peptide–protein substrate can be precipitated with appropriate concentration of TCA after phosphorylation. The reaction mixture is then filtered with a filtration paper that retains the peptide–protein substrate on the paper while allowing the ATP to pass through. Promega offers the SignaTECH kinase assay system that uses SAM biotin capture membrane to selectively retain the biotin-labeled peptide substrate on the membrane. Kinase assays using native proteins as substrates, such as histones and myelin basic protein, are commonly performed by filtration methods. Radioactive filtration assay is a classic kinase assay and is regarded as a gold standard assay. Newly developed kinase assays are usually validated by comparing the results with what is obtained from this assay. The advantages of this assay are the broad applicability to any kinases and the assay development is straightforward. The disadvantages of this assay are the radioactivity involved, the filtration step is not high throughput, and the very narrow window of ATP concentration can be used. Because of the narrow ATP concentrations in this assay, the K_m value for ATP sometimes cannot be obtained in this assay.

7.4.2 SPA and FlashPlate Assay for Kinase Activity

SPA technology (see Chapter 2) can be applied to kinase assays. The same strategy as the radioactivity filtration assay discussed above is used in SPA format without the tedious membrane filtration and washing steps. Instead of using membrane filtration to separate the radioactivity in the phosphorylated substrate from that in ATP, SPA beads are used to bind to the substrate and selectively detect the radioactivity from the phosphorylated product because of proximity. Radioactive ATP in solution will not produce the SPA signal. This assay requires the kinase substrate be able to bind to the SPA beads. The most common application of SPA in kinase assays have the kinase substrate labeled with biotin, which can bind to streptavidin-derived SPA beads. SPA-based kinase assays share many of the properties of the radioactivity filtration kinase assays but have some unique properties. The SPA format is a homogeneous assay because of the elimination of the tedious separation and washing step. [γ^{33}P]ATP has to be used because of lower energy compared with [γ^{32}P]ATP. Even with [γ^{33}P]ATP, the distance the radioactivity can travel is still long so that a large background signal is produced. One way to reduce the background is centrifugation, which separates the beads from the bulk solution containing [γ^{33}P]ATP.

FlashPlate (see Chapter 2) can also be applied in kinase assays the same way as the SPA. In application to kinase assays, FlashPlate has an advantage over SPA because the scintillant and the capture group are at the bottom of the plate. The radiolabeled phosphorylated kinase product will bind to the bottom of the plate to produce signal for detection as compared to the floating SPA beads in the solution. No centrifugation step is required in FlashPlate-based kinase assays. Further, the signal-to-background ratio will get a boost if the FlashPlate is washed.

7.4.3 Kinase Assays Based on Fluorescence Polarization (FP) Method

Fluorescence Polarization kinase assays are based on the binding of a large molecule to a small fluorescently labeled phosphorylated peptide substrate. The large molecule does not bind to nonphosphorylated substrate. Traditionally, anti-phospho antibodies against the phosphorylated kinase product serve as the large molecule in competitive FP assays. In this method, the phosphorylated products generated from kinase-catalyzed reactions compete with a phosphorylated peptide tracer for the

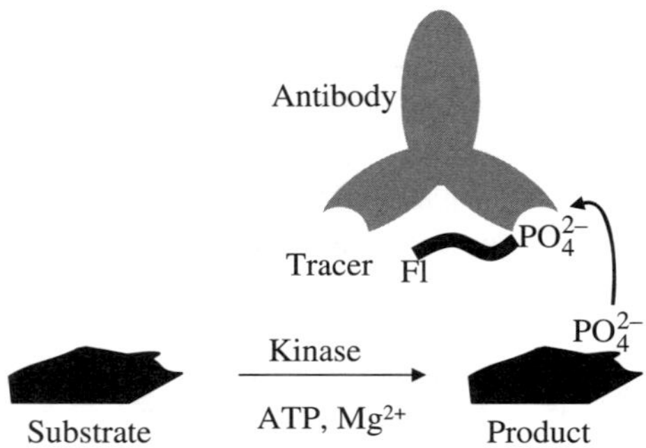

Figure 7.1 Schematic illustration of competitive fluorescence polarization (FP) assay for kinases. A fluorescently labeled phosphorylated peptide is used as the tracer, which can be recognized by antibody against the phosphorylated kinase product. In the absence of kinase product, all the tracers are bound to the antibody and give a large FP value. When the kinase products are generated, they compete with the antibody's binding site. The unbound tracer in solution will give a small FP value.

binding site in the antibody (Figure 7.1). Panvera's (now part of Invitrogen) PolarScreen assay kits and Upstate's (now part of Millipore) KinEASE are based on this method. In addition to the competitive FP assays based on anti-phospho antibody, a direct FP assay (IMAP), which measures the binding between small fluorescently labeled phosphorylated peptides and large beads coated with chelate of trivalent ions (Ga^{3+}) in the presence of fluorescently labeled substrate peptide, is marketed by MDS (formerly Molecular Devices). There are some theoretical concerns about this assay because of the intrinsic properties of FP that will be discussed below. Nonetheless, the product was quite successful in the marketplace in high-throughput applications, partly may due to less stringent assay requirements in the high-throughput screening mode.

Competitive FP Kinase Assays Competitive FP assays use a small fluorescently labeled phosphorylated peptide as a tracer. The phosphorylated peptide tracer has a low molecular weight and thus a low fluorescence polarization value when it is free in solution. The phosphorylated peptide tracer is allowed to bind to a phospho-specific antibody to form a high-molecular-weight complex and results in a high polarization value. In a kinase reaction, a peptide–protein substrate (nonfluorescently labeled) is phosphorylated by the kinase in the presence of ATP and magnesium ions. The unlabeled phosphorylated peptide–protein product competes with the tracer for binding to the phospho-specific antibody. When increasing amounts of phosphorylated product are formed from the kinase reaction, decreases in the fluorescence polarization value are observed because of the reduction of the bound tracer–antibody complex and increased free tracer in solution. The key requirement for this assay is to find a good combination of a tracer and its binding antibody. The binding must be very tight so that most of the tracers are bound to the antibody at low tracer concentration. This will give a large FP value to start with (a good pair can have an FP value of 0.4). Otherwise, the starting FP value will be low because of the contribution from the free tracer in solution, which will result in a lower signal-to-background ratio for the assay. The fluorescently labeled peptide in solution has an FP value of about 0.06. Thus, the maximum assay window in competitive FP assay is from 0.06 to 0.4 with a good pair of tracer and antibody. If the kinase product binds to the antibody equally as well as the tracer, equal amounts of product to the tracer is required to compete off half of the tracer. If the tracer concentration is low, few kinase products are needed to compete off all the tracers and less substrate turnover is required.

Direct FP Kinase Assay IMAP is the only commercial kinase assay product that directly labels a peptide substrate with fluorescence. In a kinase-catalyzed reaction, a phosphate group is added to the fluorescently labeled peptide substrate. The resulting fluorescently labeled phosphorylated peptide product is selectively bound to large functionalized beads by the affinity between the phosphate group on the peptide and the trivalent ions on the beads. The binding here is not based on antiphosphate antibody. Instead, the binding is based on the affinity between Ga^{3+} chelates and the phosphate group on the peptide product (same technology as in TruLight and IQ assays, which will be discussed later). As discussed in Chapter 2, an FP measurement is not directly proportional to the product concentration, and it must be converted to anisotropy (A). A fluorescently labeled peptide in solution typically has $A_f = 0.05$ and the bound peptide has $A_b = 0.3$. When the peptide substrate is labeled fluorescently, the final anisotropy is contributed from both the substrate and the bound product, which is governed by following equation [see Eq. (2.13), assuming that the fluorescence intensities of the ligand do not change after binding to the beads]:

$$A = f_f A_f + f_b A_b$$

When substrate turnover is 10%, which traditionally is the maximum an enzymologist will comfortably allow because of substrate depletion, the fraction of the fluorescence signal contributed from the kinase product (f_b) is 10%, and the fraction of the fluorescence signal contributed from the substrate (f_f) is 90%. If all 10% of the fluorescently labeled product is bound to the beads, we have the final $A = 0.9 \times 0.05 + 0.1 \times 0.3 = 0.075$ [Eq. (2.13)]. Thus, at a substrate turnover of 10%, the maximum achievable signal is 0.075 in direct FP assay, which is not much different from the starting background signal of 0.05. With experimental error, it is impossible to detect such small signal changes with end-point measurement. If we accept 50% substrate turnover, the IMAP assay will have a maximum assay signal of $A = 0.5 \times 0.05 + 0.5 \times 0.3 = 0.175$. Because the assay background is 0.05 and the error associated with the assay, there is still not enough assay windows for an end-point assay. This is why all FP assays were done with competitive FP assays before IMAP.

In order to obtain a measurable signal in direct FP assay, the substrate turnover has to be more than 50% because it is an end-point measurement. The IMAP application brochures describing the assay typically have $A = 0.26$ as the assay signal and $A = 0.05$ as the background. Substituting these values into Eq. (2.13), the substrate turnover would be about $(0.26 - 0.05)/(0.3 - 0.05) = 84\%$ at the conditions described in the brochures. Thus, to achieve high enough signal-to-background window in this assay, the substrate turnover was at around 80%. Currently, there is no consensus in assay community to support or disapprove such assay conditions.

When developing Caliper's off-chip kinase assay, I was facing a similar problem with substrate turnover limitation of 10% because the need of 40 to 50% substrate turnover to get a good signal-to-background ratio in my proposed off-chip kinase assay. I derived a theory to demonstrate that 50% substrate turnover is acceptable in HTS applications. Higher than 50% substrate turnover may be a concern depending on the substrate's concentration in relation to the enzymes K_m value. This will be discussed in Chapter 14.

7.4.4 Kinase Assays Based on TR-FRET

The principle of TR-FRET was discussed in Chapter 2. When applied to kinase assays, the aim is to quantify the phosphorylated product generated from the kinase-catalyzed reaction. This can be accomplished by attaching a fluorescent donor and a fluorescent acceptor molecule to the kinase-generated product. The strategy of using TR-FRET in kinase assays resembles sandwiched ELISA. The differences are that no component of the detection system is attached to a surface and the assay is homogeneous in TR-FRET format. The assay can be carried either in HTRF format or in Lance format. Since there is not much difference between the two formats, only the HTRF format is discussed here (Fig. 7.2). In contrast to the FP assays, the substrate here can be small peptides or large proteins. The only requirement is that the fluorescent donor molecule and the fluorescent acceptor molecules are within 100 Å. The selected substrate is then biotinylated (biotin-S) at one end that is far away from the phosphorylation site to avoid interference with kinase. An antiphospho antibody is then

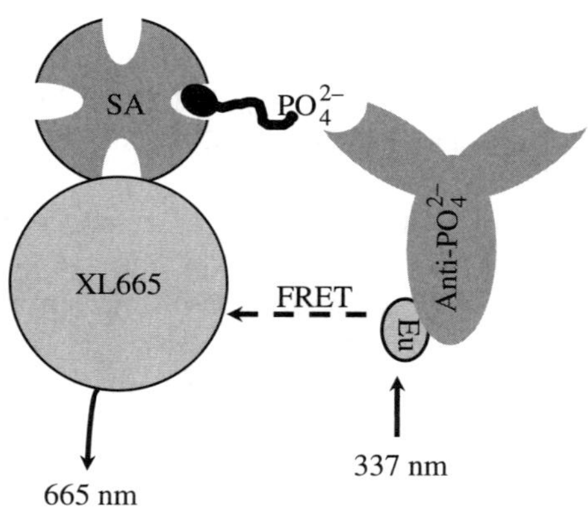

Figure 7.2 Functional kinase assay with HTRF format. The assay employs a biotin-labeled kinase substrate. The phosphorylated and biotinylated product is detected. A europium-labeled antiphosphate antibody is used to bind to the phosphate group, and XL665-labeled streptavidin is used to bind to the biotin group. This complex brings the europium chelates and the XL665 together to enable FRET. The concentration of the product is proportional to the emission at 665 nm when the europium is excited at 337 nm.

obtained and labeled with europium chelates (Eu-Ab). The streptavidin-labeled XL665 (SA-XL665) is commercially available. The europium-labeled antiphospho antibody only binds to the phosphate group in the phosphorylated product (biotin-P). The SA-XL665 binds to the biotin group on both the substrate and the product. Only the complex formed around the product brings the europium chelates and the XL665 together to enable FRET. Thus, the concentration of the product is proportional to the emission at 665 nm when the europium is excited at 337 nm if the assay is properly developed.

The successful application of kinase assays in HTRF format depends on the availability of a good antibody against the phosphate group on the phosphorylated product. For tyrosine kinase assays, many antibodies against the phosphotyrosine group are commercially available. These antibodies usually are not influenced by the amino acid sequence around the phosphorylation site. Nevertheless, all the antiphosphotyrosine antibodies do not performed equally in HTRF assays, and it may be necessary to test a few antibodies to decide the best one. In addition, some antibodies may be more prone to lose activity in the process of covalently attaching the europium chelate. For Ser/Thr kinase assays, it is difficult to find an antibody because there is no universal antibody against the phospho-Ser/Thr. All the anti-phospho-Ser/Thr antibodies are sequence dependent. Thus, for a given substrate for Ser/Thr kinase, an antibody has to be developed. Because developing an antibody takes time and resources, application of HTRF format in Ser/Thr kinases is limited.

The boundary conditions of the HTRF assay are set by the concentration of SA-XL665 and Eu-Ab. If the starting concentration of biotin-S is oversaturating all the available binding sites on SA-XL665 or the generated product (biotin-P) is saturating the binding on Eu-Ab, a decreased signal will be observed with increasing concentration of the phosphorylated product above the saturated concentration. On the other hand, the assays background increases with increasing SA-XL665 concentration. Thus, in the final developed HTRF assay, the concentration of SA-XL665 (in terms of available biotin binding sties) should be about the same as the concentration of biotin-S. The attachment of XL665 may block one or two of the four binding sites on streptavidin. The number of binding sites for biotin in XL665-labeled streptavidin may be determined experimentally. The recommended concentration of the Eu-Ab used in HTRF assay is at about 10,000 to 20,000 B counts with Packard's (now part of PerkinElmer) Discovery instrument. This corresponds to 0.5 to 5 nM Eu-Ab depending on the labeling efficiency. Usually, the HTRF experiment is done with a few nanomolar Eu-Ab, ~100 nM SA-XL665, and ~300 nM biotin-S. Thus, HTRF usually has a narrow dynamic range with regard to the concentration of the kinase product. Because of the many components involved in HTRF assays, the assay development is not trivial. An example of developing a kinase assay in HTFR format is shown in Section 7.7.

7.4.5 Kinase Assays Based on AlphaScreen

AlphaScreen is a technology marketed by PerkinElmer. The acronym ALPHA stands for Amplified Luminescent Proximity Homogeneous Assay. The principle of this assay is illustrated in Figure 7.3. The technology is based on hydrogel-coated latex

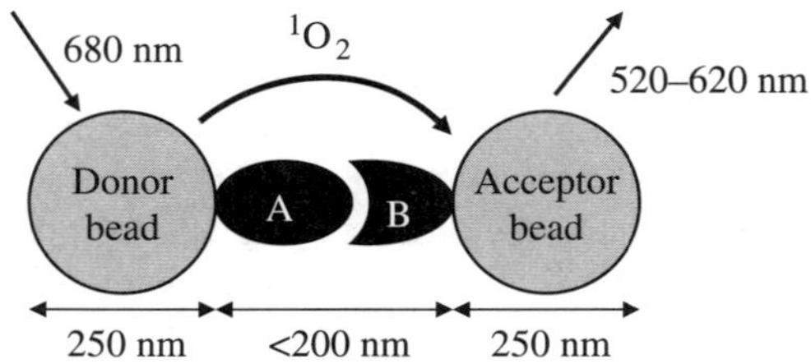

Figure 7.3 Principle of AlphaScreen. The technology is based on modified latex beads with diameter of 250 nm. The donor bead is coated with a molecule that generates reactive singlet oxygen species upon excitation at 680 nm. The unstable reactive singlet oxygen has a half-life that allows it to diffuse no longer than 200 nm in aqueous media. The acceptor bead is coated with another molecule that emits lights with wavelengths between 520 and 620 nm when it reacts with singlet oxygen. In bioassays, donor bead is attached to molecule A and acceptor bead is attached to molecule B. If A and B are binding partners, donor bead and acceptor bead can be brought together within 200 nm. When excited with 680 nm light, the light between 520 and 620 nm can be measured that is proportional to the quantity of the bound A/B complex.

beads with a diameter of 250 nm. The donor bead is derived with a molecule that generates reactive singlet oxygen species upon excitation at 680 nm. The unstable reactive singlet oxygen has a half-life of 4 μs that allows it to diffuse about 200 nm in aqueous media. The acceptor bead is coated with molecules that emit luminescent lights with wavelengths between 520 and 620 nm through a chain reaction upon encountering with singlet oxygen. In bioassays, donor bead is attached to molecule A and acceptor bead is attached to molecule B. If A and B are binding partners, donor bead and acceptor bead can be brought together within 200 nm. When excited at 680 nm, the light emission between 520 to 620 nm can be measured that is proportional to the quantity of the bound A/B complex.

Since the scheme of AlphaScreen is very similar to TR-FRET, most bioassays in TR-FRET can be adapted in AlphaScreen and vice versa. The difference is that the donor–acceptor beads in HTRF are used to substitute donor–acceptor fluorescent molecules. An example of adapting AlphaScreen to detect the biotinylated phosphorylated product catalyzed by epidermal growth factor receptor (EGFR) kinase domain is shown in Figure 7.4. The donor bead is labeled with streptavidin and the acceptor bead is labeled with antiphosphotyrosine antibody (P-Tyr-100). The kinase substrate used here for EGFR is biotin-labeled poly(Glu_4Tyr)$_n$. After kinase-catalyzed reaction, the reaction mixture containing the biotinylated and phosphorylated product is mixed with the labeled donor and acceptor beads. This will bring the donor beads and the acceptor beads together. When excited at 680 nm, the emitted light between 520 and 620 nm can be measured that is proportional to the quantity of the kinase product. The suggested concentration of beads used in the final detection solution is about 1 to 2 μg/mL for kinase assays. In experiment with EGFR, a serial diluted EGFR solution is placed in a 384-well microplate. The kinase solutions were incubated with 10 nM poly(Glu_4Tyr) in a buffer containing 25 mM HEPES, pH 7.4, 100 mM NaCl, 5 mM $MgCl_2$, 0.5 mM ATP, 0.5 mM DTT for 1 h at room temperature. This reaction is stopped by addition of a quenching solution containing 50 mM EDTA and the

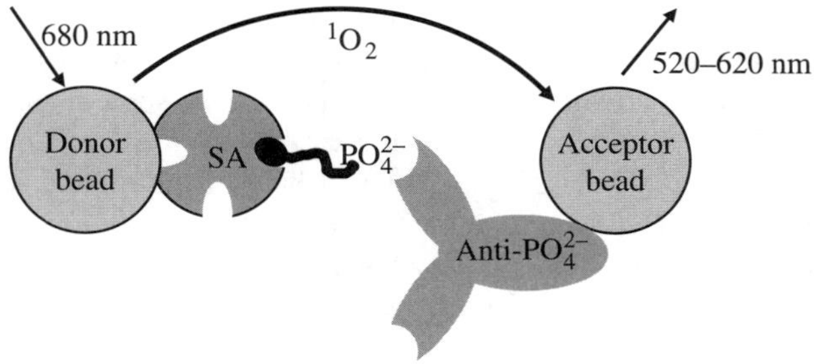

Figure 7.4 Adapting AlphaScreen to kinase assay. The donor bead is labeled with streptavidin and the acceptor bead is labeled with antiphosphate antibody. The kinase substrate is labeled with biotin. After kinase reaction, the biotinylated and phosphorylated product is mixed with the derived donor and acceptor beads. This will bring the two beads together. When excited at 680 nm, the emission between 520 and 620 nm can be measured that is proportional to the quantity of the kinase product.

donor–acceptor beads to the reaction mixture. The solution is incubated for 1 h. The plate is read with a Fusion reader (a multimode reader marketed by PerkinElmer). The data are shown in Figure 7.5. There is a linear increase in the detected signal with increasing EGFR concentration. This signal corresponds to the linear increase to the phosphorylated product formation. This indicates that there is no substrate depletion and the generated products do not oversaturate the reagents in the detection system under these assay conditions.

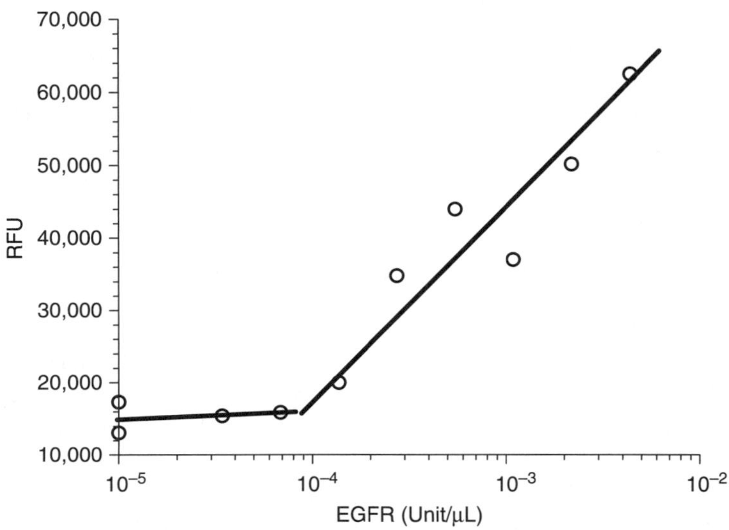

Figure 7.5 Detection of the EGFR-catalyzed reaction product. The substrate is biotin-poly(Glu_4Tyr)n. The phosphorylated product is detected by the labeled AlphaScreen beads. There is a linear increase in the detected signal with increasing EGFR concentration.

AlphaScreen has some unique advantages over TR-FRET. TR-FRET happens only within a 10-nm distance between the donor and acceptor fluorescent molecules. This distance is quite small and is close to the diameter of some large biomolecules (e.g., antibody and hemoglobin). In comparison, AlphaScreen can detect the signal when the binding pair is within 200 nm distance. Thus, the AlphaScreen can pickup more signal than TR-FRET when the binding pairs can adopt different geometries. In addition, the excitation light in AlphaScreen has longer wavelengths than the detection wavelength. This should give lower background in AlphaScreen compared with TR-FRET. Because AlphaScreen does not use expensive reagents, such as the europium, the cost of AlphaScreen can be lower. AlphaScreen beads are coated with a layer of hydrogel that minimizes nonspecific binding and self-aggregation. Because the beads used in AlphaScreen are much smaller (250 nm) than SPA beads (2 to 10 μm), they can stay in biological buffers for a long time during the assay with no precipitation. The bead suspensions look like homogeneous solution to the naked eye. This allows easy dispensing of the suspension with automated liquid handling devices. One disadvantage of AlphaScreen is that the beads are sensitive to red light. Dispensing of beads and plate handling should be done in rooms with green filter or simply dim light.

7.4.6 Kinase Assays Based on Enzyme Fragment Complementation (EFC)

EFC (marketed by DiscoverX) is based on the complementation of two genetically engineered ß-galactosidase fragments to gain enzyme activity when the two inactive fragments are associated with each other. The large inactive protein fragment is referred to as EA and the small inactive peptide fragment is referred to as ED. In solution, EA and ED rapidly recombine to form an active β-galactosidase enzyme that hydrolyzes substrate to produce a detectable chemiluminescent or fluorescent signal depending on the β-galactosidase substrate in the assay system. ED is a small peptide (4 to 11 kDa) with no tertiary structure that can be covalently attached to a variety of small molecules or proteins without changing their properties and functions. ED can also be recombinantly expressed in cells as a fusion protein for measuring translocation and protein expression. The large EA fragment is usually not modified in EFC assays.

The biochemical in vitro assays based on EFC are usually formatted as competitive immunoassays. The application of EFC in kinase assay is illustrated in Figure 7.6. ED is conjugated to a phosphopeptide as a tracer that binds to antiphospho antibody. The large antibody associated with ED will block its binding to EA and thus no complement complex is formed and no enzyme activity is observed. When phosphorylated product is generated in kinase-catalyzed reactions, the phosphorylated product will compete for the binding sites on the antiphospho antibody and release the ED tracer. The free ED tracer in solution will complement with EA and regain the β-galactosidase activity. Depending on the project, the binding protein can be an antibody or other proteins such as a receptor. The EFC competition assay shares some properties with the FP competition assay discussed before. High-affinity binding between the antibody and the tracer is required. Otherwise, a free tracer in solution

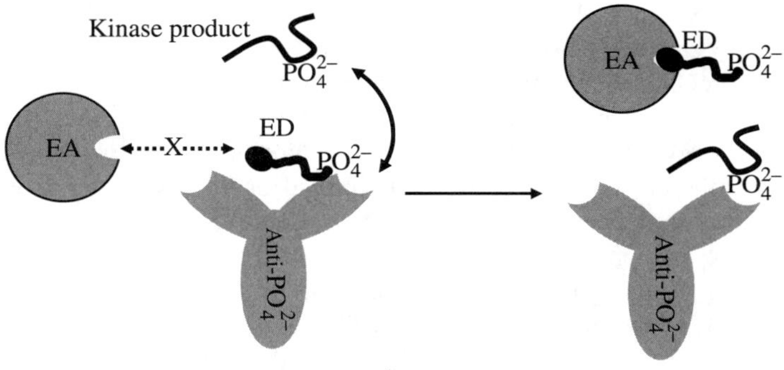

Figure 7.6 Illustration of kinase assay with EFC technology. A phosphorylated ED peptide is used as a tracer that binds to a large antiphosphate antibody. The large antibody prevents ED from binding to EA. The phosphorylated product generated from kinase-catalyzed reaction competes for the binding to the antibody. The displaced ED tracer in solution binds to EA that restore the β-galactosidase activity.

will give a large background signal. The signal window can be larger in EFC assay than in competitive FP assays because the maximum window in FP assays is limited by the maximum anisotropy value of about 0.3 with tracer–protein complex.

7.4.7 Kinase Assays Based on Fluorescence Quenching

The principle of fluorescence quenching was discussed in Chapter 2. Two kinase assays based on fluorescence quenching to monitor the kinase product are discussed here. The Antibody Beacon technology (marketed by Invitrogen) is a tyrosine kinase assay kit. The key to this tyrosine kinase assay is an Oregon Green 488-labeled tracer ligand, whose green fluorescence is quenched when the ligand is bound to an antiphosphotyrosine antibody. When phosphotyrosine product is generated in a kinase-catalyzed reaction, it competes off the bound Oregon Green 488-labeled tracer from the antibody. The tracer in solution is not quenched and gives off strong fluorescence when excited. Antibody Beacon provides a very simple and robust solution-based competitive assay to measure tyrosine kinase activity.

Trulight technology (marketed by EMD) is based on the affinity between trivalent ions (Ga^{3+}) and the phosphate group on the phosphorylated kinase product, similar to IMAP and IQ, which will be discussed later. Microspheres coated with fluorescent polymers and Ga^{3+} chelates are used to detect phosphorylated kinase product. TruLight does not label the kinase substrate with fluorescence and avoids the problems associated with substrate depletion assays, such as IQ. Instead, the kinase substrates are conjugated with a fluorescence quencher. When the kinase converts the quencher-containing substrate into a phosphorylated product, the generated kinase product will bind to the Ga^{3+} chelates on the microsphere and bring the quencher on the phosphorylated product close to the fluorescent molecules on the microsphere, resulting in fluorescence quenching and a decrease in the fluorescent signal.

The fluorescent molecules on the polymer microsphere emit fluorescence at 490 nm upon excitation. No information on the quencher is available. However, the Trulight brochure indicates that there is a transfer of energy from the polymer to the quencher, and this produces an FRET signal at 600 nm. The ratio of the intensity of the two wavelengths is used to monitor the kinase product formation, which helps minimizing interference from colored or fluorescent compounds. The fluorescently labeled microspheres carry many fluorescence molecules and Ga^{3+} chelates. Thus, one microsphere can bind to many quenchers. This results in many folds of signal amplifications. The ATP interference with the assay is always an issue with assays depending on Ga^{3+}/Fe^{3+} chemistry. Very high concentrations of ATP are prone to interfere with the binding of the microsphere sensor to the phosphorylated substrate. This interference is dependent on the structure of the peptide substrate and the composition of the assay buffer. Most of the assays exhibit optimal performance in the range of 50 to 100 μM ATP and can tolerate up to 1 mM ATP.

7.4.8 Multiplexed Kinase Assays

Multiplexed assays are methods that analyze multiple analytes simultaneously. When assays share similar properties, such as the same buffer and similar assay procedure, multiplexed assay can save time and cost. Two technologies, ECL and Beadlyte have been adopted in multiplexed kinase assays and they will be discussed here.

ECL was discussed in Chapter 2. The dominant player in applying ECL multiplexed assays is Meso Scale Discovery (MSD). In this format, the bottom of the wells in ECL plates is loaded with many reagent spots with each of them individually addressable. Each well having many spots allows simultaneously exposing the spots in the well to many reagents. For example, 100 spots/well in 24-well microplates can assay 2400 samples in one plate and 4 spots/well in 384-well microplate can assay 1536 samples in one plate. For kinase assays, the substrates can be captured at the bottom of the ECL plate by direct absorption or through biotin/streptavidin binding. Different substrates for different kinases can be placed as spots in the same well. A mixture of different kinases to be assayed in the assay solution is then added to the well. After a predetermined time, the reaction is terminated and the detection antibodies carrying ruthenium chelates and other required reagents are added to the plate. The luminescence signal is then detected after applying electrical current.

Beadlyte (marketed by Millipore) is also a multiplexed assay. Here special individually addressable beads (marketed by Luminex) are used to capture different kinase substrates. The beads are then mixed together with different kinases in kinase assay buffers containing the necessary components for kinase reaction to generate the kinase product. After the reaction, the beads are mixed with different fluorescently labeled antibodies against each kinase product. The bead mixtures are then read with Luminex100 instruments that can distinguish individual beads in a solution containing up to 100 different beads.

7.5 KINASE ASSAYS BY MEASURING THE GENERATION OF ADP

Adenosine 5′-diphosphate is the co-product generated in kinase-catalyzed phosphorylation of a peptide–protein substrate. There is a one-to-one ratio between generated phosphorylated product and ADP. Thus, ADP is an attractive universal tracer for the phosphorylated product. ADP measurement has the advantage that it is not a depletion measurement. Traditional ADP measurement was based on its involvement in the conversion of phosphoenolpyruvate (PEP) into pyruvate. The pyruvate can be coupled to other reactions, such as its conversion into lactate with the consumption of NADH, to generate a measurable signal (Fig. 7.7). PEP is used as a substrate that reacts with ADP to make pyruvate and ATP in the presence of pyruvate kinase. For every ADP molecule consumed in the reaction, one molecule of pyruvate is made. In the coupled reaction, the pyruvate reacts with NADH to make lactate and NAD^+ in the presence of lactate dehydrogenase. The absorption of NADH at 340 nm can be measured. The decrease in NADH concentration is proportional to the concentration of ADP. The disadvantages of this assay are that the measurement of NADH at 340 nm is prone to interference from many compounds and the absorption measurement is not sensitive as compared with other detection techniques. ADPHunter assay (marketed by DiscoverX) is a newer ADP assay based on similar strategy. In this assay, ADP is allowed to react with PEP and other cofactors in the presence of both pyruvate kinase and pyruvate oxidase to generate hydrogen peroxide. The generated hydrogen peroxide can be detected by measuring the fluorescence generated when Amplex Red peroxidase substrate is used in the presence of peroxidase.

The Transcreener (developed by BellBrook Labs) kinase assay is a competitive ADP assay based on TR-FRET. The key components of the assay are a terbium-chelate-labeled monoclonal antibody against ADP and a fluorescein-labeled ADP tracer. The turbium-labeled antibody binds to the fluorescein-labeled ADP tracer and enables TR-FRET. As ADP is produced, it displaces the fluorescein-labeled ADP tracer from the Tb-labeled ADP antibody and results in the decreased TR-FRET signal (Fig. 7.8). Other pairs of TR-FRET reagents can be used too, such as

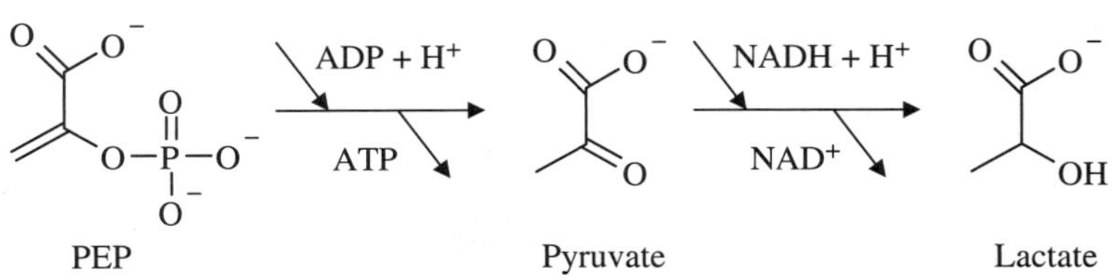

Figure 7.7 Coupled reactions to detect ADP. Phosphoenolpyruvate (PEP) is used as the substrate that reacts with ADP to make pyruvate and ATP in the presence of pyruvate kinase. For every one ADP molecule consumed in the reaction, one molecule of pyruvate is made. In the coupled reaction, the pyruvate reacts with NADH to make lactate and NAD+ in the presence of lactate dehydrogenase. The absorption of NADH at 340 nm can be measured. The decrease in NADH concentration is proportional to the concentration of ADP.

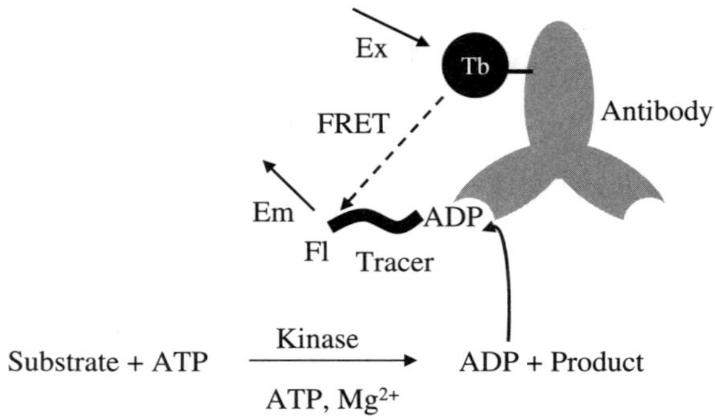

Figure 7.8 Schematic illustration of Transcreener assay for kinases. A fluorescently labeled ADP is used as the tracer that can be recognized by antibody against the ADP. In the absence of kinase product, all the tracers are bound to the antibody and give a large TR-FRET signal. When the kinase products are generated, they compete with the antibody's binding site. The displaced tracer in solution will not give TR-FRET signal because it is more than 10 nm away from the terbium.

europium-labeled anti-ADP antibody and Alexa Fluor 647-labeled ADP tracer (marketed by Invitrogen as Adapta).

Chiku and colleagues (2006) developed an ADP assay based on ATPβS (sulfur on the β-phosphorous). It is known that most kinases accept ATPβS as a substrate in place of ATP. After the transfer of the γ-phosphate group to the substrate, ADPßS is generated as a product. ADPβS can selectively react with sulfhydryl group detection reagents, such as Ellman's reagents while the ATPβS is not reactive. The major advantage of this assay is that the generated ADPβS can be continuously measured. In contrast, most of the kinase assays are end-point assays. One potential concern for this assay is how much perturbations will be caused to the native system by the substitution of the oxygen atom with a sulfur atom on the β position of ATP.

Adenosine 5′-diphosphate assays have many advantages. The assay is universal to any kinases and any kinase substrate can be used (small peptide, large protein, and even lipids). Because ADP generation is concurrent with the phosphorylated product generation, the ADP measurement-based kinase assays are not depletion assays. A potential concern with ADP-based kinase assay is that the ADP may be generated by a process other than a kinase-catalyzed reaction, such as contaminant ATPase activity in the assay system.

7.6 KINASE ASSAYS BY MEASURING THE DEPLETION OF ATP

Adenosine 5′-triphosphate is a reactant in the kinase-catalyzed reaction. For each phosphorylated product generated, one ATP is consumed. Kinase assays based on ATP depletion are attractive because the availability of many very sensitive and easy to perform luciferase-based ATP assay kits. The principle of ATP measurement is shown in Figure 7.9. Beetle luciferin is oxidized to oxyluciferin by oxygen in the

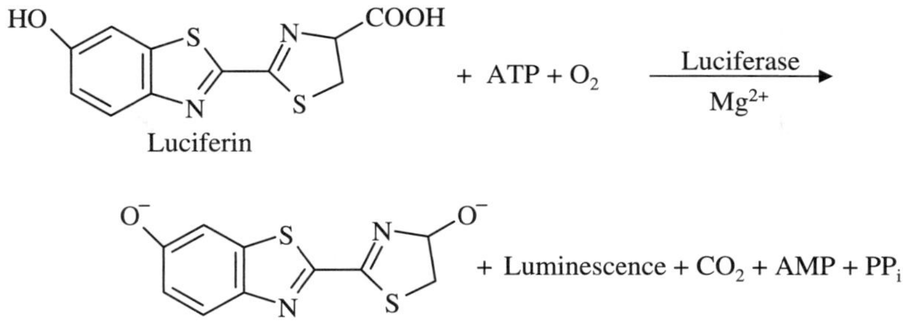

Oxyluciferin

Figure 7.9 Principle of ATP measurement with luciferase. Beetle luciferin is oxidized to oxyluciferin by oxygen in the presence of ATP and the magnesium ion. The reaction is catalyzed by luciferase. The reaction consumes ATP and generates luminescence. When the assay is properly configured, the light output is proportional to ATP concentration.

presence of ATP and magnesium ion. The reaction is catalyzed by luciferase. The reaction consumes ATP and generates luminescence. When the assay is properly configured, the light output is proportional to ATP concentration. In the past, the reaction can only be measured with a luminometer capable of measuring flash luminescence because the luminescence only lasted for a few seconds. This requires the luminometer to have integrated liquid-handling systems. Through genetic engineering, the reaction is significantly slowed down and the luminescence is steadily generated. New technology allows the luminescence to last several hours without losing the signal significantly. With steadily generated luminescence, the assay is less sensitive to detect ATP compared with the flash luminescence version of ATP detection. However, the sensitivity in steady luminescence is still far more sensitive than most assays would require.

Adenosine 5′-triphosphate detection kits from different manufacturers have different qualities. An evaluation of different kits with identical conditions, such as the same ATP standard solution and the same reader, should be performed before using the kit in an assay. The evaluation of two ATP detection kits from two different vendors is shown in Figure 7.10. They are tested at the condition recommended by the vendors. The same ATP solution is tested with the two kits at the same time in different rows in a microplate. The ATP concentration is made from ATP powder with no further calibration. The absolute value of ATP is not certain but unlikely to be off by more than a factor of 2. The plate is then read with the Fusion multimode reader. The data is plotted in log-log scale to cover a wide range. KinaseGlo (marketed by Promega) has almost four orders of linear dynamic range and the lower detection limit is lower. For kinase assays, there is no need to detect ATP concentration at low nanomolar range. Both kits will work fine in this regard. However, there may be a need to start with 10 μM or higher ATP concentration in some assay conditions as will be discussed below. The ATP detection kit from vendor 2 will not be useful in this situation.

There are some theoretical concerns with assays that measure substrate depletion. The treatments here are applicable to both ATP depletion and peptide

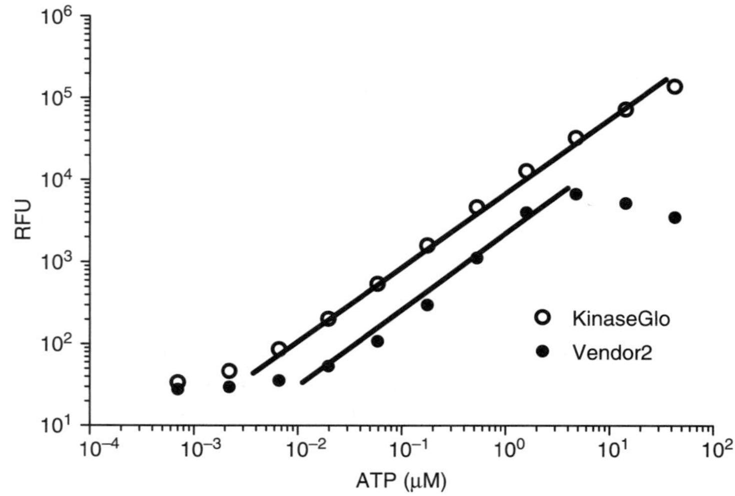

Figure 7.10 Evaluation of two commercial ATP measurement kits. Two commercial kits are tested at the condition recommended by the vendors. The same ATP solution is tested with the two kits at the same time in different rows in a microplate. The plate is then read with a luminometer. KinaseGlo has large dynamic range and lower detection limits. The ATP concentration is made from ATP powder with no further calibration. The absolute value of ATP is not certain but unlikely to be off by more than a factor of 2.

substrate depletion, which will be discussed in the next section. Substrate depletion measurement requires the detection of signal changes from a high baseline value. If substrate turnover is 10%, the detected signal will be a 10% reduction of the baseline. Only continuous measurement of the same sample may be able to distinguish reliably the 10% signal change from the baseline value. Considering the error associated with a given assay, end-point measurement can hardly distinguish reliably this small reduction in signal because the measurement has to be made up of two different samples with one measurement as the baseline and the other as the signal. In other words, the assay window is too small. To obtain an adequate assay window, the substrate turnover may need to be close to 50% or higher. The consequence of high substrate turnover depends on the K_m value of the substrate and the starting concentration of the substrate. A hypothetical kinase with $K_m = 7\ \mu M$ for ATP is used here to illustrate the potential concerns in two situations: zone A and zone B (Fig. 7.11). Let us consider zone A first. If the starting ATP concentration is at 60 μM, 50% turnover will reduce ATP concentration to 30 μM. The change in reaction velocity is negligible in this zone and a linear decrease in ATP concentration should be expected in reaction progression curve. The reaction velocity can be measured at any point in the reaction period as long as the decrease in ATP concentration can be measured reliably (>10% ATP depletion). Though the kinetic parameters can be obtained reliably, the caveat in this zone is that the ATP concentration is about 9 times of the K_m. From the discussion in Chapter 3, the IC_{50} for the ATP competitive inhibitor will right-shift to 10 times the K_i, which makes the assay insensitive to ATP

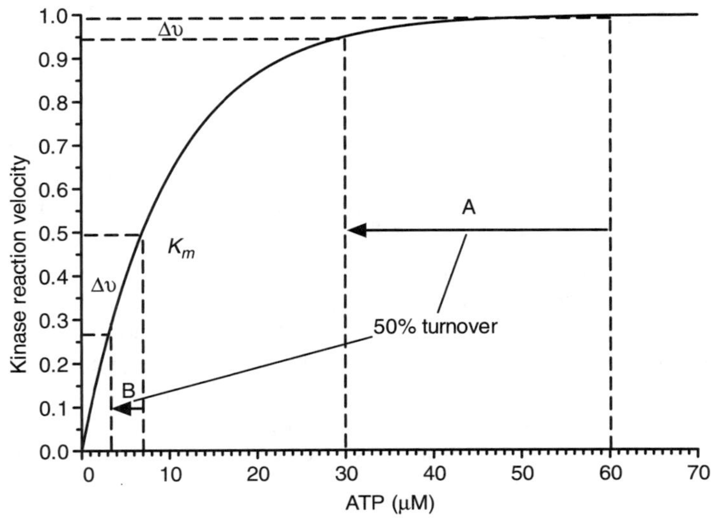

Figure 7.11 Illustration of kinase reaction velocity reduction in ATP depletion assay. There are two separate situations that are treated as zone A and zone B. The hypothetical kinase has a $K_m = 7$ μM for ATP. Zone A: If the starting ATP concentration is at 60 μM, 50% turnover will reduce ATP concentration to 30 μM. The change in reaction velocity is negligible and a linear decrease in ATP concentration should be expected in reaction progression curve. Zone B: If starting ATP concentration is at K_m, 50% ATP turnover will cause the reduction of reaction velocity by half. It is expected that the reaction progression curve will not be linear.

competitive inhibitors. Because most clinically interesting kinase inhibitors are ATP competitive, this condition is not useful in screening for ATP competitive inhibitors. In order to screen for ATP competitive inhibitors, the ATP concentration has to start at lower concentration, that is, in zone B. If the starting ATP concentration is at K_m, 50% ATP turnover will result in the reduction of the reaction velocity by half. In the reaction period, the reaction velocity will continuously decrease with decreasing ATP concentration. Thus, the reaction progression curve will not be linear but curved in the reaction period. The caveat for the assay in zone B is that the reaction rate obtained from the end-point measurement will be a variable depending on when the measurement is made.

Though there are several theoretical concerns with ATP depletion assays, this assay is still used in screening for kinase inhibitors with the assay condition set in zone B. When doing this assay, one must make sure that the concentration of the peptide–protein substrate is much higher than the concentration of ATP. Otherwise, the reaction will behave as a second-order reaction and depletion of both kinase substrates will cause even greater reaction rate reduction over time. ATP depletion assay has been adapted in HTS because it is universal to any kinases and only a kinase substrate is needed for a given kinase without the need of other reagents (e.g., antibody, beads, labels). One caveat in the ATP depletion assay is the nonspecific hydrolysis of ATP catalyzed by either ATPase activity or when the assay media turned to basic or acidic.

7.7 KINASE ASSAYS BY MEASURING THE DEPLETION OF PEPTIDE SUBSTRATE

There is only one commercial kinase assay that is based on measurement of the depletion of the peptide substrate. IQ technology (marketed by Pierce) uses fluorescently labeled peptides as kinase substrates. After a period of kinase-catalyzed reaction, some of the fluorescently labeled substrate peptide is converted into fluorescently labeled phosphorylated product. The fluorescence on the phosphorylated product is selectively quenched, leaving only the unconverted substrate to emit fluorescence upon excitation. The net effect is the measurement of depletion of the fluorescently labeled peptide substrate. The selective quenching of the fluorescence associated with the phosphorylated product is based on the affinity of the phosphate group in the product to Fe^{3+} chelates that carry a fluorescence-quenching moiety.

Because IQ is a substrate depletion assay, it possesses many similar advantages and concerns with the ATP depletion assay discussed above. However, it has its own unique properties because the reaction may be run in the zone A condition. If the goal of the assay is not to find competitive inhibitors for the peptide binding site, the peptide substrate concentration can be set at high starting concentration so that the reaction rate can be obtained reliably at any time because of linear kinetics. To do this, the ATP concentration should be set at a concentration that is at least 10 times higher than the substrate to ensure no depletion of ATP and thus linear kinetics. In order not to affect the sensitivity to detect ATP competitive inhibitors, the K_m for ATP must be much higher than the K_m for the peptide so that the high ATP concentration is still less than its K_m. If the above condition cannot be met, the assay has to be run at zone B where the same concerns with ATP depletion exist.

While ATP detection is straightforward and well established, quenching of fluorescently labeled phosphorylated product is not. The success of this technology depends on the fluorescence quenching efficiency so that there is little or no residual fluorescence signal from the product that may interfere with the assay. For example, if the substrate turnover is 10%, the resulting reaction mixture will contain 90% fluorescent substrate and 10% fluorescent product. If the quenching efficiency is close to 100%, a 10% reduction in the fluorescence should be observed. The assay thus has to be able to detect 10% signal reduction from a high fluorescence background. In reality, the quenching efficiency may not even be close to 90%. From the data shown in IQ handbooks from the vendor, the quenching efficiency is about 85% with the IQ technology. This will lead to 8.5% signal reduction with 10% substrate turnover. Considering the noise associated with the assay, it is impossible to detect the reaction at 10% substrate turnover. If the substrate turnover is at 50%, the detected reduction in the fluorescent signal will be 42.5%. Whether the assay has sufficient assay window at this condition will depend on the noise associated with the assay.

7.8 KINASE ASSAYS BY SIMULTANEOUS MEASUREMENT OF BOTH PRODUCT AND SUBSTRATE

Kinases catalyze the addition of a phosphate group to a substrate. The resulting product will have a net charge of -2 changes from the substrate. This charge difference between the substrate and the product will lead to different retention times for the product and substrate in capillary electrophoresis. If fully resolved, the product peak and the substrate peak can be observed. The two peaks with varying amplitudes at different reaction times correspond to the degree of substrate turnover. At the start, only the substrate peak is observed. As the reaction progresses, the product peak appears and gradually increases over time. This is the basis of kinase assay to measure the product and substrate simultaneously. The diameter of the capillary in a capillary electrophoresis instrument usually is between 50 and 100 μm. This narrow pathlength makes it difficult to measure UV absorption by peptide or protein. Thus, the substrate is usually labeled with fluorescence that gives a large signal. The −2 charge difference may not be significant to cause large separation between the product and the substrate if the substrate is a large protein because the separation in capillary electrophoresis is based on mass/charge ratio. Thus, a fluorescently labeled peptide substrate is preferred in kinase assay with capillary electrophoresis. Because of the instrument limitations, kinase assay with capillary electrophoresis does not offer enough throughput for HTS applications.

Kinase assays based on capillary electrophoresis in micro fluidic format can significantly increase throughput and still maintain the properties of detecting both kinase substrate and product simultaneously. The development of Caliper Technologies' off-chip kinase product will be discussed in Chapter 14 as a case study.

7.9 EXAMPLE OF A KINASE ASSAY DEVELOPMENT IN HTRF FORMAT

7.9.1 Background

Angiogenesis represents an important therapeutic target for cancer. The principal growth factors driving angiogenesis are the vascular endothelial growth factor (VEGF), bFGF, and the hepatocyte growth factor/scatter factor. Although stimulators of angiogenesis were considered for the treatment of cardiovascular diseases in the past, inhibitors of angiogenesis received a lot attention lately due to their ability to slow tumor progression. Vascular endothelial growth factor receptor 2 (VEGFR2), also referred to as KDR and FLK1, is a major player in angiogenesis. VEGFR2 is exclusively expressed in endothelial cells and appears to play a pivotal role in endothelial cell differentiation and vasculogenesis. VEGFR2 is a member of the receptor tyrosine kinase family. It is hypothesized that inhibiting the kinase activity of VEGFR2 may lead to cancer therapy. The goal here is to develop an assay for screening inhibitors that can inhibit the kinase activity of the recombinant VEGFR2 kinase domain.

7.9.2 Assay Design and Testing

The assay scheme is shown in Figure 7.12. There are two steps in the assay. First, an appropriate amount of VEGFR2 and biotin-S are mixed to generate enough biotin-P in the desired time window (about 1 h) for detection. Second, the reaction is stopped and the detection solution containing an appropriate amount of SA-XL665

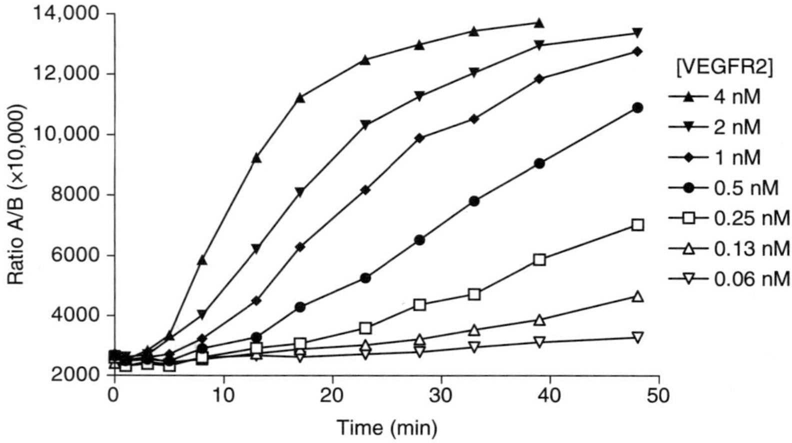

Figure 7.12 Reaction progression curves at different concentrations of VEGFR2. There is an initial lag phase that is followed by linear kinetics. The apparent saturation of the reaction is caused by the limited amount of detection reagents.

and Eu-Ab is added to the reaction mixture for a final read. In the first step, the assay buffer should be developed. From existing knowledge, the kinase assay solution may contain 20 mM HEPES pH 7.4, 4 mM Mg^{2+}, 1 μM to 4 mM ATP depending on what mode of inhibitor is sought, 1 mM DDT, 0.05% BSA. The reaction can be stopped by 30 mM EDTA that depletes Mg^{2+}, which is crucial for the kinase reaction. This solution can be mixed with the reagents used for detection (SA-XL665, Eu-Ab, 0.2 M KF, 0.1% Triton X-100). HTRF assay development can be started with the manufacturer's recommended detection conditions as outlined in Section 4.4 of this chapter (500 nM SA-XL665 and Eu-Ab giving about 10,000 to 20,000 B counts). Nonoptimal initial reaction progression curves can be obtained with this system to detect the kinase product at different concentrations of VEGFR2 and biotin-S. After a few rounds of fine-tuning of the system, assay conditions are obtained that are used to further study the reaction kinetics. The optional VEGFR2 concentration determination is shown in Figure 7.12. The reaction has an initial lag that is followed by linear product generation. The top value at about 12,000 counts may be limited by the limited amount of Eu-Ab in the detection mixture. The concentrations between 0.5 and 1 nM of VEGFR2 give sufficient signal that will be used in future experiments.

The reaction progression curve (Fig. 7.13) at different concentrations of ATP is studied with the final developed assay system. Again the reaction has initial lag. The reaction progressed quickly and reached a plateau, which is due to the limited amount of the detection reagent Eu-Ab. Though the concentration of the peptide substrate biotin-S is only at 0.5 μM, which is much lower than the ATP concentration, the concentration of Eu-Ab (a few nanomolar) is even lower. This dictates that the substrate turnover within the detection window is less than 10%. Thus, the initial linear phase corresponds to the initial velocity of the reaction. When the ATP concentration is tested at 4 mM, the results are abnormal. Thus, only the ATP concentrations up to 2.1 mM are analyzed. The initial velocity at different concentrations of

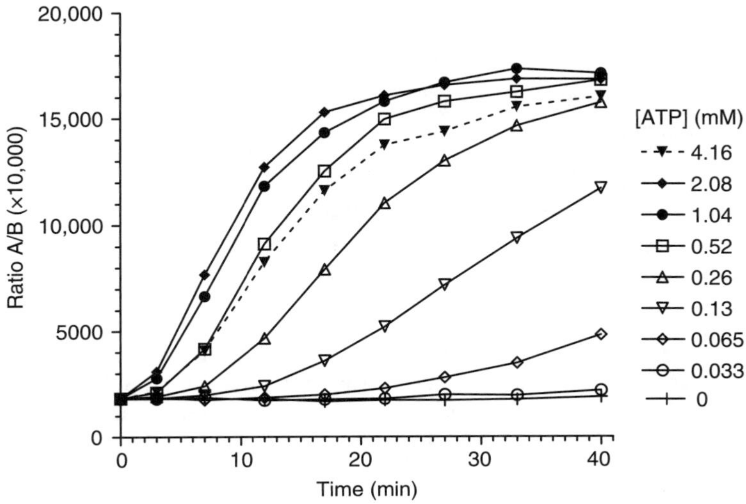

Figure 7.13 Reaction progression curve at different concentrations of ATP. The procedure from mixing to finish reading limits the first time point at 3 min. When ATP concentration is higher than 2 mM, the apparent rate unexpectedly slowed down that may be due to the tolerance of the system to high ATP concentration.

ATP is obtained by fitting the linear region of the reaction progression curve. The results are shown in Figure 7.14. Nonlinear fitting of the data in Figure 7.14 into Eq. (3.37) leads to $K_m = 0.39$ mM for ATP. Because the HTRF assay requires comparable number of binding sites on SA-XL665 for biotin-S, the SA-XL665 concentration has to be adjusted according to the biotin-S concentration if the

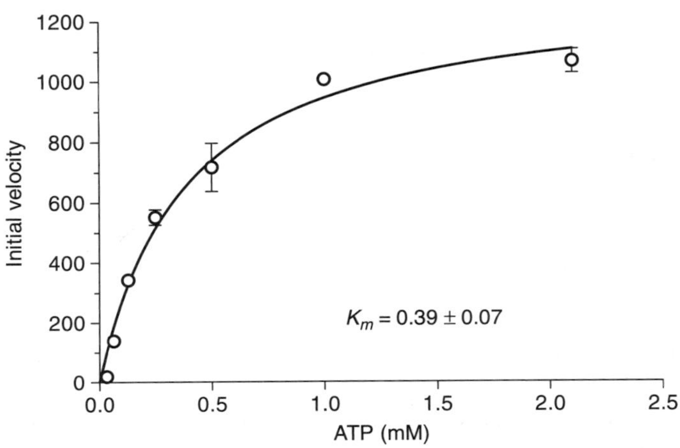

Figure 7.14 Determination of K_m for ATP. The initial velocity at different concentrations of ATP is obtained by using the data in the linear region in Figure 7.9. The data obtained from ATP concentrations higher than 2.1 mM is not used for this analysis due to abnormal behavior.

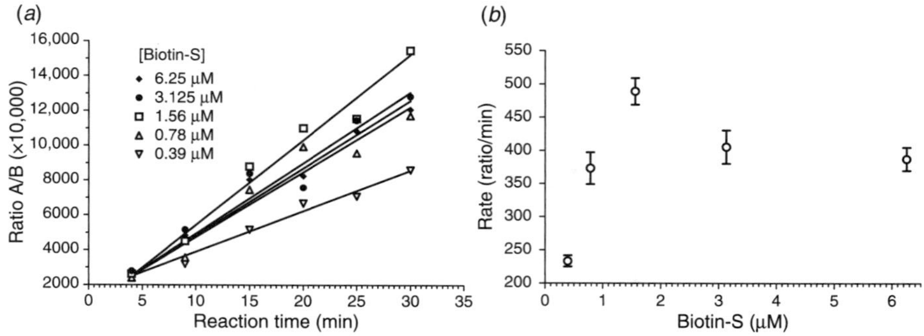

Figure 7.15 Relationship of initial velocity and the biotin-S concentration. (*a*) Initial velocity measurement at different concentrations of the substrate biotin-S. (*b*) Initial velocity vs. [Biotin-S]. At 1.56 μM biotin-S, the detected velocity is the highest and then starts decreasing with higher biotin-S. This may due to the fact that at the binding sites on SA-XL665 is limited.

concentration of biotin-S is varied. Thus, K_m determination for biotin-S may not be meaningful in this assay. Figure 7.15 shows the relationship between the biotin-S concentration and the reaction rate. The bell-shaped curve observed here is common in many HTRF-based assays. After testing several other parameters, such as stability of the detection mixture (room temperature and at 4°C) and DMSO effects on the detection, the assay is ready to be used in HTS. After the inhibitors are obtained in initial screening, the inhibition mechanism can be tested with this assay as well. The profiles of two inhibitors with different inhibition modes are shown in Figure 7.16. Both inhibitors were tested at 0.2 mM ATP (below K_m of ATP, which is 0.39 mM) and 2 mM (above K_m of ATP), respectively. Inhibitor 1 has the IC_{50} of 9 ± 2 nM at 0.2 mM ATP and 42 ± 7 nM at 2 mM ATP concentration (Fig. 8.12a). The Hill coefficient is forced at 2 in both ATP concentrations. Inhibitor 2 has IC_{50} value of 8 ± 5 nM and the Hill coefficient of -0.8 ± 0.3 at 2 mM ATP. The IC_{50} is 10 ± 7 nM and the Hill coefficient is -0.7 ± 0.2 at 2 mM ATP.

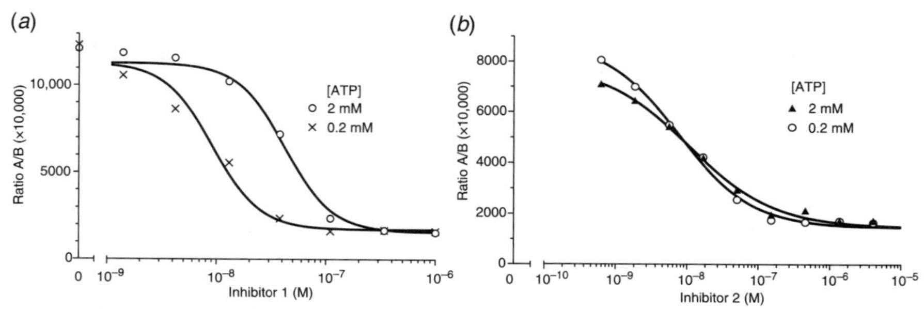

Figure 7.16 Profile of two inhibitors with different inhibition modes. (*a*) Inhibitor 1: The IC_{50} is 9 ± 2 nM at 0.2 mM ATP and 42 ± 7 nM at 2 mM ATP concentration. The Hill coefficient is forced at 2 for both cases. (*b*) Inhibitor 2: The IC_{50} is 8 ± 5 nM and the Hill coefficient is -0.8 ± 0.3 at 2 mM ATP. The IC_{50} is 10 ± 7 nM and the Hill coefficient is -0.7 ± 0.2 at 2 mM ATP.

Substituting these values into Eq. (3.49) ($IC_{50} = (1 + [S]/K_m)K_i$), the apparent IC_{50} value can be expressed in terms of K_i. This leads to the theoretical value of $IC_{50} = 1.5K_i$ when [ATP] = 200 μM and $IC_{50} = 6.3K_i$ when [ATP] = 2 mM. Thus, theoretically, the IC_{50} value is expected to change by a factor of $6.3/1.5 = 4.2$ when the concentration of ATP is increased from 0.2 to 2 mM with ATP competitive inhibitors. Experimentally, the measured IC_{50} values changed from 9 to 42 nM for inhibitor 1, which is a change of 4.6-fold. This is very close to the magnitude of the theoretically predicted IC_{50} shift. Thus, inhibitor 1 is an ATP competitive inhibitor. Inhibitor 2 does not cause detectable changes in the IC_{50} values when ATP concentration is changed from 0.2 to 2 mM. Thus, inhibitor 2 is a noncompetitive inhibitor with regard to ATP. Uncompetitive inhibitors, which should have decreased IC_{50} values with increasing ATP concentrations, are not observed in this assay.

Useful Websites

http://networkin.info/search.php

http://www.phosida.com/

http://phospho.elm.eu.org/

http://kinase.com/

http://www.phosphosite.org/homeAction.do;jsessionid=E7C36B6D0752C189BF8E290EC2967CA5

http://www.kinasource.co.uk/Database/welcomePage.php

http://predikin.biosci.uq.edu.au

http://www.cellsignal.com/

http://www.sigmaaldrich.com/Area_of_Interest/The_Americas/United/States.html

http://www.millipore.com/pathways/pw3/pathwayshome

http://www.emdbiosciences.com/html/CBC/home.html

http://www.carnabio.com/english/

http://www.invitrogen.com/site/us/en/home.html

http://www.jpt.com/products/enzyme_profiling/kinase_profiling/kinase_profiling.htm

BIBLIOGRAPHY

Abdel-Ghany, M., El-Gendy, K., Zhang, S., and Racker, E. (1990) Control of src kinase activity by activators, inhibitors, and substrate chaperones. *Proc. Natl. Acad. Sci.* **87**, 7061–7065.

Baselga, J. (2006) Targeting tyrosine kinases in cancer: The second wave. *Science* **312**, 1175–1178.

Blethrow, J. D., Glavy, J. S., Morgan, D. O., and Shokat, K. M. (2008) Covalent capture of kinase-specific phosphopeptides reveals Cdk1-cyclin B substrates. *Proc. Natl. Acad. Sci.* **105**, 1442–1447.

Bogoyevitch, M. A. and Fairlie, D. P. (2007) A new paradigm for protein kinase inhibition: Blocking phosphorylation without directly targeting ATP binding. *Drug Discov. Today* **12**, 622–633.

Capdeville, R., Buchdunger, E., Zimmermann, J., and Matter, A. (2002) Glivec (STI571, imatinib), a rationally developed, targeted anticancer drug. *Nat. Rev. Drug Discov.* **1**, 493–502.

Chen, J., Zhang, X., and Fernandez, A. (2007) Molecular basis for specificity in the druggable kinome: Sequence-based analysis. *Bioinformatics* **23**, 563–572.

Chiku, T., Pullela, P. K., and Sem, D. S. (2006) A dithio-coupled kinase and ATPase assay. *J. Biomol. Screen.* **11**, 844–853.

Cohen, P. (2002) Protein kinases—The major drug targets of the twenty-first century? *Nat. Rev. Drug Discov.* **1**, 309–315.

Costa, L. J. and Drabkin, H. A. (2007) Renal cell carcinoma: New developments in molecular biology and potential for targeted therapies. *Oncologist* **12**, 1404–1415.

Crews, C., Alessandrini, A., and Erikson, R. (1992) The primary structure of MEK, a protein kinase that phosphorylates the ERK gene product. *Science* **258**, 478–480.

Dancey, J. and Sausville, E. A. (2003) Issues and progress with protein kinase inhibitors for cancer treatment. *Nat. Rev. Drug Discov.* **2**, 296–313.

Davidson, W., et al. (2004) Discovery and characterization of a substrate selective p38a; inhibitor. *Biochemistry* **43**, 11658–11671.

Diella, F., Gould, C. M., Chica, C., Via, A., and Gibson, T. J. (2008) Phospho.ELM: A database of phosphorylation sites update 2008. *Nucl. Acids Res.* **36**, D240–244.

Elia, A. E. H., Cantley, L. C., and Yaffe, M. B. (2003) Proteomic screen finds pSer/pThr-binding domain localizing Plk1 to mitotic substrates. *Science* **299**, 1228–1231.

Ellis, L. M. and Hicklin, D. J. (2008) VEGF-targeted therapy: Mechanisms of antitumour activity. *Nat. Rev. Cancer* **8**, 579–591.

Fabian, M. A., et al. (2005) A small molecule-kinase interaction map for clinical kinase inhibitors. *Natl. Biotechnol.* **23**, 329–336.

Fedorov, O., et al. (2007) A systematic interaction map of validated kinase inhibitors with Ser/Thr kinases. *Proc. Natl. Acad. Sci.* **104**, 20523–20528.

Giroux, V., Iovanna, J., and Dagorn, J.-C. (2006) Probing the human kinome for kinases involved in pancreatic cancer cell survival and gemcitabine resistance. *FASEB J.* **20**, 1982–1991.

Goldstein, D. M., Gray, N. S., and Zarrinkar, P. P. (2008) High-throughput kinase profiling as a platform for drug discovery. *Nat. Rev. Drug Discov.* **7**, 391–397.

Gumireddy, K., et al. (2005) A non-ATP-competitive inhibitor of BCR-ABL overrides imatinib resistance. *Proc. Natl. Acad. Sci.* **102**, 1992–1997.

Hanks, S. K., Quinn, A. M., and Hunter, T. (1988) The protein kinase family: Conserved features and deduced phylogeny of the catalytic domains. *Science* **241**, 42–52.

Harris, J. L., et al. (2000) Rapid and general profiling of protease specificity by using combinatorial fluorogenic substrate libraries. *Proc. Natl. Acad. Sci.* **97**, 7754–7759.

House, C. and Kemp, B. (1987) Protein kinase C contains a pseudosubstrate prototope in its regulatory domain. *Science* **238**, 1726–1728.

Hunter, T. (1982) Synthetic peptide substrates for a tyrosine protein kinase. *J. Biol. Chem.* **257**, 4843–4848.

Huss, K. L., Blonigen, P. E., and Campbell, R. M. (2007) Development of a TranscreenerTM kinase assay for protein kinase A and demonstration of concordance of data with a filter-binding assay format. *J. Biomol. Screen.* **12**, 578–584.

Hutti, J. E., et al. (2004) A rapid method for determining protein kinase phosphorylation specificity. *Natl. Methods* **1**, 27–29.

Kannan, N., Haste, N., Taylor, S. S., and Neuwald, A. F. (2007) The hallmark of AGC kinase functional divergence is its C-terminal tail, a cis-acting regulatory module. *Proc. Natl. Acad. Sci.* **104**, 1272–1277.

Kemp, B. E., Graves, D. J., Benjamini, E., and Krebs, E. G. (1977) Role of multiple basic residues in determining the substrate specificity of cyclic AMP-dependent protein kinase. *J. Biol. Chem.* **252**, 4888–4894.

Kennelly, P. and Krebs, E. (1991) Consensus sequences as substrate specificity determinants for protein kinases and protein phosphatases. *J. Biol. Chem.* **266**, 15555–15558.

King, A. J., et al. (2006) Demonstration of a genetic therapeutic index for tumors expressing oncogenic BRAF by the kinase inhibitor SB-590885. *Cancer Res.* **66**, 11100–11105.

Knighton, D., et al. (1991) Crystal structure of the catalytic subunit of cyclic adenosine monophosphate-dependent protein kinase. *Science* **253**, 407–414.

Krause, D. S. and Van Etten, R. A. (2005) Tyrosine kinases as targets for cancer therapy. *N. Engl. J. Med.* **353**, 172–187.

Linding, R., et al. (2007) Systematic discovery of in vivo phosphorylation networks. *Cell* **129**, 1415–1426.

Manley, P. W., Bold, G., Brüggen, J., Fendrich, G., Furet, P., Mestan, J., Schnell, C., Stolz, B., Meyer, T., Meyhack, B., Stark, W., Strauss, A., and Wood, J. (2004) Advances in the structural biology, design and clinical development of VEGF-R kinase inhibitors for the treatment of angiogenesis. *Biochim. Biophys. Acta.* **1697**, 17–27.

Manning, B. D. and Cantley, L. C. (2002) Hitting the target: Emerging technologies in the search for kinase substrates. *Sci. STKE* **2002**, pe49.

Manning, G., Whyte, D. B., Martinez, R., Hunter, T., and Sudarsanam, S. (2002) The protein kinase complement of the human genome. *Science* **298**, 1912–1934.

Marais, R., Wynne, J., and Treisman, R. (1993) The SRF accessory protein Elk-1 contains a growth factor-regulated transcriptional activation domain. *Cell* **73**, 381–393.

Martin, L., Catherinot, V., and Labesse, G. (2006) kinDOCK: A tool for comparative docking of protein kinase ligands. *Nucl. Acids Res.* **34**, W325–329.

Martinez, A., Alonso, M., Castro, A., Perez, C., and Moreno, F. J. (2002) First non-ATP competitive glycogen synthase kinase 3beta (GSK-3beta) inhibitors: Thiadiazolidinones (TDZD) as potential drugs for the treatment of alzheimer's disease. *J. Med. Chem.* **45**, 1292–1299.

Noble, M. E. M., Endicott, J. A., and Johnson, L. N. (2004) Protein kinase inhibitors: Insights into drug design from structure. *Science* **303**, 1800–1805.

Olive, D. M. (2004) Quantitative methods for the analysis of protein phosphorylation in drug development. *Expert Rev. Proteomics* **1**, 327–341.

Olsen, J. V., Blagoev, B., Gnad, F., Macek, B., Kumar, C., Mortensen, P., and Mann, M. (2006) Global, in vivo, and site-specific phosphorylation dynamics in signaling networks. *Cell* **127**, 625–648.

Pinna, L. A. and Ruzzene, M. (1996) How do protein kinases recognize their substrates? *Biochim. Biophys. Acta* **1314**, 191–225.

Posner, I., Engel, M., and Levitzki, A. (1992) Kinetic model of the epidermal growth factor (EGF) receptor tyrosine kinase and a possible mechanism of its activation by EGF. *J. Biol. Chem.* **267**, 20638–20647.

Russello, S. V. (2004) Assessing cellular protein phosphorylation: High throughput drug discovery technologies. *Assay Drug Dev. Technol.* **2**, 225–235.

Schauble, S., et al. (2007) Identification of ChChd3 as a novel substrate of the cAMP-dependent protein kinase (PKA) using an analog-sensitive catalytic subunit. *J. Biol. Chem.* **282**, 14952–14959.

Schlessinger, J. (2000) Cell signaling by receptor tyrosine kinases. *Cell* **103**, 211–225.

Schlessinger, J. (2003) Signal transduction: Autoinhibition control. *Science* **300**, 750–752.

Schroter, T., et al. (2008) Comparison of miniaturized time-resolved fluorescence resonance energy transfer and enzyme-coupled luciferase high-throughput screening assays to discover inhibitors of Rho-kinase II (ROCK-II). *J. Biomol. Screen.* **13**, 17–28.

Slice, L. W. and Taylor, S. S. (1989) Expression of the catalytic subunit of cAMP-dependent protein kinase in Escherichia coli. *J. Biol. Chem.* **264**, 20940–20946.

Songyang, Z. and Cantley, L. C. (1995) Recognition and specificity in protein tyrosine kinase-mediated signalling. *Trends Biochem. Sci.* **20**, 470–475.

Songyang, Z., et al. (1996) A structural basis for substrate specificities of protein Ser/Thr kinases: Primary sequence preference of casein kinases I and II, NIMA, phosphorylase kinase, calmodulin-dependent kinase II, CDK5, and Erk1. *Mol. Cell. Biol.* **16**, 6486–6493.

Taylor, S., Radzio-Andzelm, E., and Hunter, T. (1995) How do protein kinases discriminate between serine/threonine and tyrosine? Structural insights from the insulin receptor protein- tyrosine kinase. *FASEB J.* **9**, 1255–1266.

Toomik, R., Ekman, P., and Engström, L. (1992) A potential pitfall in protein kinase assay: Phosphocellulose paper as an unreliable adsorbent of produced phosphopeptides. *Anal. Biochem.* **204**, 311–314.

Turk, B. E., Hutti, J. E., and Cantley, L. C. (2006) Determining protein kinase substrate specificity by parallel solution-phase assay of large numbers of peptide substrates. *Natl. Protoc.* **1**, 375–379.

Verkhivker, G. M. (2007) Exploring sequence-structure relationships in the tyrosine kinome space: Functional classification of the binding specificity mechanisms for cancer therapeutics. *Bioinformatics* **23**, 1919–1926.

Wang, Y. and Klemke, R. L. (2008) PhosphoBlast, a computational tool for comparing phosphoprotein signatures among large datasets. *Mol. Cell. Proteomics* **7**, 145–162.

Wilhelm, S., et al. (2006) Discovery and development of sorafenib: A multikinase inhibitor for treating cancer. *Nat. Rev. Drug Discov.* **5**, 835–844.

Wong, Y.-H., et al. (2007) KinasePhos 2.0: A web server for identifying protein kinase-specific phosphorylation sites based on sequences and coupling patterns. *Nucl. Acids Res.* **35**, W588–594.

Wu, J., Ma, Q. N., and Lam, K. S. (1994) Identifying substrate motifs of protein kinases by a random library approach. *Biochemistry* **33**, 14825–14833.

Xue, Y., et al. (2008) GPS 2.0: Prediction of kinase-specific phosphorylation sites in hierarchy. *Mol. Cell. Proteomics* **7**, 1598–1608.

CHAPTER 8

FUNDAMENTAL PRINCIPLES OF CELL-BASED ASSAYS

CELL-BASED ASSAYS use live cells as the test systems to detect the effect of test molecules. The live cells can be either primary cells isolated from an organism or immortal cell lines grown in artificial laboratory conditions. In general, cell-based assays are more complex and are less defined when compared with biochemical assays. However, cell-based assays possess many unique properties that are not shared with biochemical assays. Cell-based assays are particularly useful with less defined assay targets or in the situation that the target protein is known but unknown cofactors are required for the proper function of the target protein. In cell-based assays, the target protein is in its native cellular environment with appropriate cofactors at physiological concentrations for its proper function. Even when assaying a known target, cell-based assay eliminates the need to develop expression and purification procedures for the target, which can save a lot of time and effort in assay development. Cell-based assays can distinguish between agonist and antagonist and can study the synergic effect of two or more test compounds. Thus, cell-based assays are indispensable parts of bioassays and they complement biochemical assays very well.

After cells are exposed to test molecules, many intracellular changes and cells' responses may occur if the test molecules have effects on the cells. Cell-based assay development is a process to evaluate many of the intracellular changes and the cell response after the cells are exposed to test molecules and then define the final readout for the assay. The assay signals may be continuously monitored with live cells or be detected only at the end of the assay after the cells are fixed or lysed. The fixed cells can be examined with flow cytometry or imaging-based methods, which will be discussed in Chapter 12. The exposed cellular components from the lysed cells can be assayed biochemically. One or a few of the detectable signals may be selected as the final readout based on the relevance to the biological questions, the nature of the signal, the reliability and ease to measure the signal, and many other

Assay Development: Fundamentals and Practices. By Ge Wu

criteria. A good understanding of the cellular signal transduction processes and the final cellular responses is critical to develop cell-based assays.

8.1 CELL SIGNALING, SIGNAL TRANSDUCTION, AND CELLULAR RESPONSES

8.1.1 Overview of Cell-Signaling Process

The structural and functional capabilities of multicellular organisms depend on the coordination of various cells that make up the organisms. The coordination of all the cells is achieved by the intercellular communications among various cells. A cell can affect other cells through (1) releasing chemical messengers that interact with the surface receptors of the target cells; (2) interacting directly with other cells through the binding of cell surface proteins; (3) forming gap junctions that are channels connecting neighboring cells to allow direct exchange of signaling molecules; and (4) emitting electrical impulses in the form of changes in membrane potential in presynaptic nerve cells that is received by postsynaptic cells at chemical synapse via neurotransmitters or at electric synapse via gap junction. The signaling molecules for the communication among cells are hormones, growth factors, cytokines, neurotransmitters, and the like. The signaling molecules vary greatly in terms of their chemical nature. They can be small organic molecules (e.g., amino acids and their derivatives, steroids, retinoids, nucleotides, and derivatives of fatty acids), peptides, and proteins. The communication among cells via signaling molecules can be local or long distance. With local signaling, the transmitting cell secretes molecules that influence only cells in the vicinity. Autocrine signaling and paracrine signaling are local signaling. In autocrine signaling, cells secrete signaling molecules that are received among themselves. This type of the signaling is common in immune responses. In paracrine signaling, cells secrete signaling molecules that defuse in a short distance and only affect cells in close proximity. The communication between nerve cells via neurotransmitters in the synapse is a good example of paracrine signaling. Long-distance signaling is accomplished by endocrine signaling. In endocrine signaling, the cells secrete signaling molecules (hormones) into blood or lymphatic fluid. The hormones are distributed throughout the body to reach receiving cells distant from the hormone secreting cells. A well-known example of endocrine signaling is the regulation of glucose levels by insulin. In addition to receiving stimulation by other cells, cells can also respond to signals from other sources. For example, cells involved in vision receive light stimulation, a variety of sensory cells receive heat/cold stimulation, and the like. The discussion of cell signaling here will focus on those transmitted by chemical molecules.

After the cells receive a signal, many intracellular events will occur along the signal transduction pathways. The propagation of the intracellular signal will finally lead to certain cellular responses. Figure 8.1 illustrates the three stages of the signal transduction process: signal reception and initiation, signal transmission through pathways, and cellular responses. Signal reception is the first stage of the signaling process when the extracellular signal molecule binds to a cellular protein, commonly

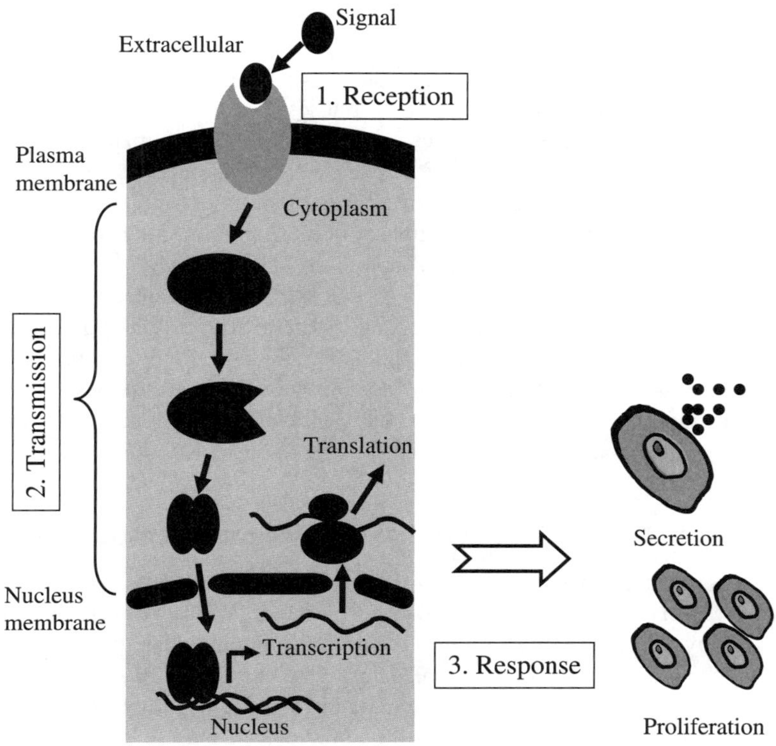

Figure 8.1 Three stages of cell-signaling process. The first stage of signaling process is the binding of extracellular signal molecule to the cell surface receptor. This is followed by the signal transduction stage at which the cell surface receptor changes its conformation and phosphorylation state that affected its interactions with other molecules in the signaling pathway. This subsequently results in a cascade of changes that involves many molecules. The final stage is the responses of the cells to the extracellular signal. Many responses may result including protein synthesis, secretion of molecules, proliferation, survival, apoptosis, and morphological changes.

referred to as a "receptor." The receptors can locate on the cell surface or inside the cells. Most receptors, such as epidermal growth factor receptors (EGFRs), fibroblast growth factor receptors (FGFRs), and insulin receptors (InsRs), are transmembrane proteins located in the plasma membrane. The signaling molecules that bind to receptors on plasma membrane can be lipophilic small molecules (e.g., prostaglandins, thromboxines, leukotrienes, and prostaclyclins) and a broad range of water-soluble molecules (e.g., epinephrine, histamine, insulin, grow factors, and glucagons). The receptors inside the cells can be in the cytosol or the nucleus. For example, steroid receptors reside originally in cytosol and translocate to the nucleus after binding to steroids. The signaling molecules that bind to the intracellular receptors are membrane-permeable lipophilic small molecules (e.g., steroids, thyroxine, and retinoic acid). These receptors transmit the initial signals by changing their conformations, changing their aggregation states (e.g., dimerization of receptors), or changing their interactions with different molecules. The signal reception and initiation stage is followed by the signal transmission stage. There are many pathways inside the cells that can transmit the initial signal to the points of cellular responses. The transduction pathways not only relay the initial signal downstream but also amplify the signal in the process. This is achieved through a cascade of activation/inactivation events of many molecules along the signal transduction pathway. Downstream of the signal transduction, transcription factors are activated and transported through the nuclear membrane to the nucleus to initiate transcription. The transcribed messenger RNAs (mRNAs) are then transported out of the nucleus to the cytoplasm where new proteins are synthesized. The final stage is the integrated systemwide response of the cells to the extracellular signal at the end of the signal transduction. Many cellular responses may occur that include, but are not limited to, protein synthesis and secretion to the extracellular surface, gluconeogenesis, cell proliferation, cell survival, apoptosis, and morphological changes.

8.1.2 Signal Reception and Initiation of Signal Transduction

The reception of an extracellular signal is the binding between the signal molecules and the receptors of the receiving cell. The receptors, which convert the physical stimulation by the signal molecule into their conformational changes, are the molecules responsible for signal initiation. The signal initiation may involve several other proteins that work together on the cytoplasmic side of the plasma membrane. Figure 8.2 shows the general scheme of the initial signal transduction process. After the binding of the ligand (L) to the receptor (R), the receptor is activated (R*). The activated receptor then activates an effector protein (E). The activated effector protein (E*) may directly generate diffusible intracellular signal (S) or activates a second effector protein (E2*) through an adaptor protein (A). The activated second effector protein E2* then generates diffusible signal (S). We discussed before that there are two types of receptors: transmembrane receptors on the plasma surface and soluble receptors in the cytosol or nucleus. The discussion here will focus on the cell surface receptors.

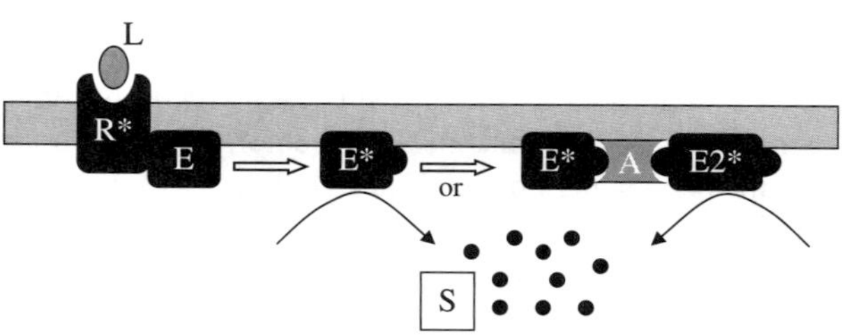

Figure 8.2 General signal reception and initiation process. After the binding of the ligand (L) to the receptor (R), the receptor is activated (R*). The activated receptor then activates an effector protein (E). The activated effector protein (E*) may directly generate diffusible intracellular signal (S) or activates a second effector protein (E2*) through an adaptor protein (A). The activated second effector protein E2* then generates diffusible signal (S).

Receptors on the plasma membrane can be categorized into four major classes: ion channels, G protein-coupled receptors (GPCRs), tyrosine kinase-linked receptors, and receptors with intrinsic enzymatic activities. Several types of enzymatic activities, such as GTPases, tyrosine kinases, serine/threonine kinases, and protein tyrosine phosphatases, are found linked to receptors. In this book, only the receptors with intrinsic tyrosine kinase activity, commonly referred to as receptor tyrosine kinases (RTKs), will be discussed.

Ion Channels These receptors are different from the other three types of receptors in that the signal reception, initiation, propagation, and response are all carried out by the same proteins. There are two major types of ion channel receptors: ligand-gated channels and voltage-gated channels. The ligand-gated channels open the channel to conduct ions after they bind to ligands. The voltage-gated channels will open the channel to conduct ions when appropriate voltage is applied across the membrane. The flow of ions across the membrane when the channels are open after receiving an initial signal (ligand binding or voltage changing) will alter the electric potential across the cell membrane. Ion channels will be discussed in more detail in Chapter 9.

G Protein-Coupled Receptors These receptors function together with G proteins. G proteins, short for guanine nucleotide binding proteins, are a family of proteins involved in second-messenger cascades. G proteins function as "molecular switches," alternating between an inactive and active state, to regulate downstream cellular processes. The G proteins are loosely attached to the cytoplasmic side of the plasma membrane and can be in the active or the inactive state depending on whether they bind to GTP or GDP. G proteins belong to the larger group of enzymes called GTPases. After GPCRs bind to ligands, the G proteins will associate with GPCRs and bind to GTP to become activated. Activated G proteins then activate an effector protein (adenylyl cyclase or phospholipase C_β) that in turn generates diffusible second-messenger signals (cAMP, DAG, IP_3, or Ca^{2+}). GPCRs comprise a large and diverse number of receptors (e.g., adrenergic receptors, olfactory receptors, muscarinic acetylcholine receptors, epinephrine receptors, and rhodopsin). GPCRs will be discussed in more detail in Chapter 10.

Receptor Tyrosine Kinases These receptors have ligand binding domains and intrinsic tyrosine-specific protein kinase domains on the same protein. The extracellular ligand binding domain is connected to the cytoplasmic domain by a single transmembrane helix. The cytoplasmic domain contains a conserved protein tyrosine kinase core and additional regulatory sequences that are subject to phosphorylation. The receptors for insulin and many soluble or membrane-bound growth factors are RTKs. Except for the insulin receptor family of RTKs, all known RTKs are monomers in cell membrane. Binding of ligand causes the dimerization of the RTKs and the subsequent activation of the intrinsic tyrosine kinases. This is followed by the autophosphorylation of the tyrosine residues in the cytoplasmic domain of each of the receptor monomers. The phosphorylated tyrosine residues serve as binding sites for the effector molecules (e.g., phospholipase C_γ, PI^{-3} kinases, tyrosine kinases, protein tyrosine phosphatases) or the adaptor molecules (e.g., Grb2, IRS, Shc). The

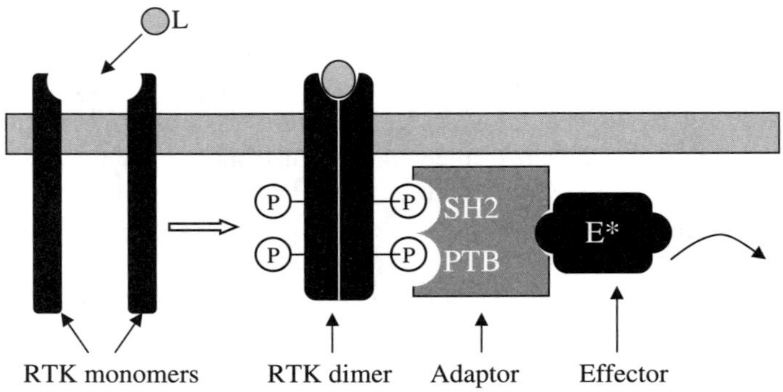

Figure 8.3 General signal reception and initiation process for RTKs. Binding of ligand L causes the RTKs to dimerize and the activation of the intrinsic tyrosine kinase. This is followed by the autophosphorylation of the tyrosine residues in the cytoplasmic domain of each of the receptor monomer. The phosphorylated tyrosine residues serve as binding sites for the effector molecules E (e.g., phospholipase Cg, PI_3 kinases, tyrosine kinases, protein tyrosine phosphatases), or adaptor molecules A (e.g., Grb2, IRS, Shc). The signals are then further propagated through various signaling pathways.

signals are then further propagated through different signaling pathways. The general signal reception and initiation process for RTKs is illustrated in Figure 8.3.

Tyrosine Kinase-Linked Receptors Similar to RTKs, these receptors have an extracellular ligand binding domain and the cytoplasmic domain connected by a single transmembrane helix. The cytoplasmic domain also contains regulatory sequences that are subjected to phosphorylation. However, they lack intrinsic catalytic activity. The signal reception and initiation of tyrosine kinase-linked receptors is similar to RTKs (see Fig. 8.3). Ligand binding causes the receptors dimerization. This is followed by binding of a cytosolic protein tyrosine kinases to the receptor and subsequent activation of the tyrosine kinase. Many cytosolic protein tyrosine kinases (e.g., Abl, Btk, Csk, Fak, Fes, Jak, Src, and Zap70) can phosphorylate these receptors. The activated kinase phosphorylates tyrosines in the receptor to provide binding sites for effector or adaptor proteins. The receptors for most cytokines and interferons belong to this type.

8.1.3 Pathways for Intracellular Signal Transmission

The signal transmission is a multistep process that involves many component molecules in the signal transduction pathways that eventually lead to signal amplification. Most of the molecules in the signal transduction pathways are proteins. Protein kinases and phosphatases are the major components in intracellular signal transduction pathways. The activities of the protein kinases in the pathways are regulated by upstream kinases or phosphatases. In addition to proteins, certain small molecules and ions (cAMP, DAG, IP_3, and Ca^{2+}) are also found to be key components in some pathways (e.g., pathways relaying extracellular signal binding to GPCRs). These small molecules are referred to as second messengers. The signal transduction pathways involving GPCRs and second messengers will be discussed in details in Chapter 10.

Mitogen-activated protein kinase (MAPK) pathways are the most important signal transduction pathways that involve a cascade of serine/threonine kinases. Figure 8.4 shows the general scheme of MAPK pathways and three actual MAPK pathways used to transmit different extracellular signals. After the reception of extracellular signals and the initiation of signal transduction, the signal is relayed to small regulatory GTPases (Ras, Rho, and Rac) or MAP4Ks (GCK, PAK) that activate MAP3Ks. Active MAP3Ks phosphorylate MAPKKs and activate them. The active MAPKKs then phosphorylate MAPK and activate them. The active MAPKs can phosphorylate factors affecting transcription (e.g., ATF2, Jun, Fos, Elk) or phosphorylate MAPK-activated protein kinases (e.g., ERK phosphorylates RSK and P38 phosphorylates MAPKAP kinases) that have cytosolic targets. Different extracellular signals may be carried through different MAPK pathways leading to different cellular responses. Many receptors, including RTKs and GPCRs, use Ras/MAPK pathways to transmit extracellular signals.

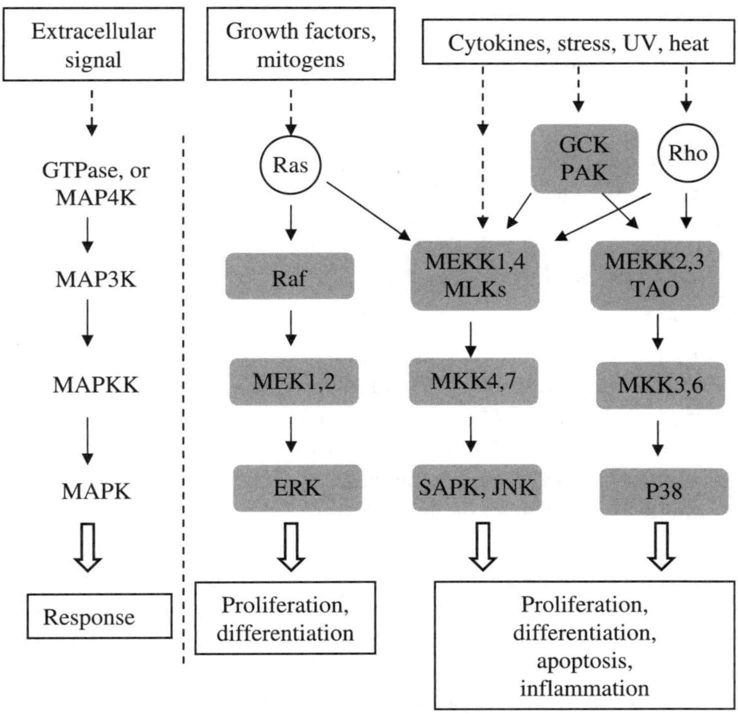

Figure 8.4 General scheme of MAPK pathways (left panel) and three actual pathways (right panel). After signal reception and initiation, the signal is relayed to small regulatory GTPases (Ras, Rho, and Rac) or MAP4Ks (GCK, PAK) that activate MAP3Ks. Active MAP3Ks phosphorylate and activate MAPKKs. The active MAPKKs then phosphorylate and activate MAPK. Different extracellular signals may be carried through different MAPK pathways leading to different cellular responses.

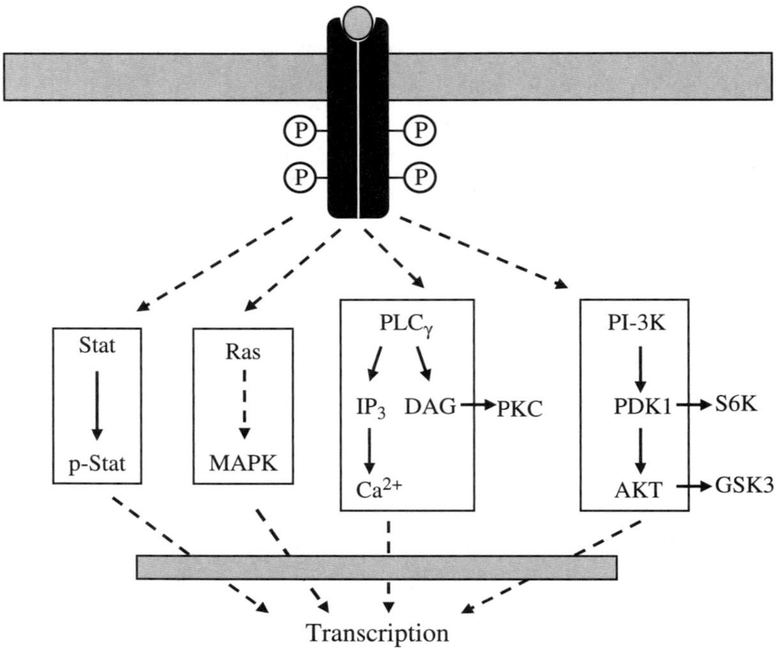

Figure 8.5 RTKs can pass the extracellular signal to many different signaling pathways. Four simplified pathways that are linked to the activated RTKs are shown here. All these four pathways lead to transcription activities. Some of pathways also generate signals that act on the targets outside of the nucleus.

After the binding of extracellular signal to cell surface receptors, the signal transduction can progress through many different pathways. Four simplified pathways that are linked to the activated RTKs are shown in Figure 8.5. The Ras/MAPK pathways are coupled with RTKs through adaptor proteins Grb2 and Sos, which were discussed earlier. The PI-3 kinases can directly bind to the phosphotyrosine group on activated RTKs through their SH2 domains or indirectly associated with RTKs by coupling to an adaptor protein (e.g., IRS) that binds to phosphorylated RTKs. Phospholipase C_γ binds to phosphotyrosine group on activated RTKs through their SH2 domains. The activated phospholipase C_γ catalyzes the production of several secondary messengers (IP_3, DAG, and Ca^{2+}). The Stat proteins are phosphorylated by activated RTKs. The phosphorylated p-Stat proteins form a dimer and then translocate to nucleus. All these four pathways finally lead to transcription. Some of pathways also generate signals that act on the targets outside of the nucleus.

8.1.4 Cell Responses to Extracellular Stimulation

The end of the signal transduction is the cell's responses to the extracellular signal. Many cellular responses can occur, including the ion channel opening/closing in the plasma membrane, cell metabolism change, specific genes' turning on/off for

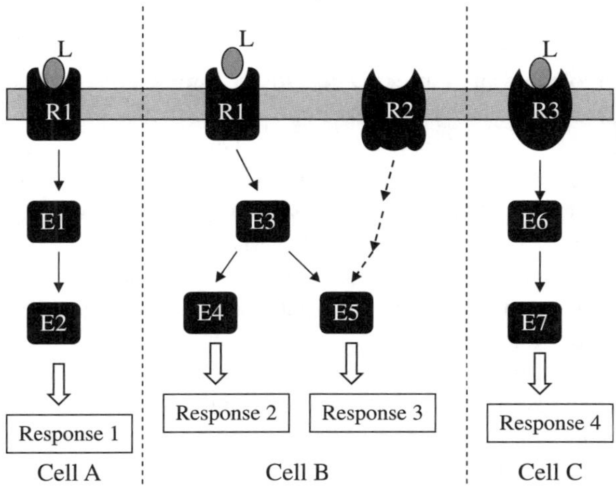

Figure 8.6 One signal molecule may produce many cellular responses. In cell type A, the ligand L binds to receptor R1 leading to cellular response 1 after the signal is propagated through a linear pathway. In cell type B, the ligand L binds to the same receptor R1. However, there exist branched signal pathways leading to two cellular responses, response 2 and response 3. Response 3 is further complicated because it is the result of two converging pathways involving different unrelated receptors, R1 and R2. In cell type C, the ligand L binds to a different receptor R3 leading to another cell response, response 4.

protein synthesis regulation, cell proliferation, cell survival, apoptosis, and morphological changes. The cell's responses to a particular signal can be different depending on the cell, the signaling molecule, and the receptor(s) with which the signaling molecule interacts. This is illustrated in Figure 8.6. In cell type A, the ligand L binds to receptor R_1, leading to cellular response 1 after the signal is propagated through a linear pathway. In cell type B, the ligand L binds to the same receptor R_1. However, there exist branched signal pathways leading to two cellular responses, response 2 and response 3. The response 3 is further complicated because it is the product of two converging pathways involving two different unrelated receptors, R_1 and R_2. In cell type C, the ligand L binds to a different receptor R_3, leading to another cell response, response 4.

8.2 GENERAL APPROACHES IN CELL-BASED ASSAYS

Cell-based assays use cells (primary cells or transformed cell lines) in an artificial environment to detect the effect of test molecules for their ability to change the intracellular events or final cellular responses. This artificial cell-based assay system is similar to the natural signal transduction process except that the extracellular environment of cells in cell-based assays is different from their natural environment (extracellular matrix, co-localized cells, components of fluids that the cells are exposed to). There are two general schemes to perform cell-based assays. One is to expose the unmodified cell directly to test molecules and then detect the changes in intracellular components or the final cellular responses. This scheme is primarily used to exam whether a test molecule may affect intracellular components, pathways, or cellular responses (e.g., proliferation differentiation and survival). The other scheme is to manipulate the cells in a way to boost the specific signal to be detected before exposing the cells to test molecules. The cells can be manipulated by overexpressing specific cell surface receptors or other components through recombinant technology or by exposing the cells to known activators that drive the cells to a specific state. The test molecules are assayed to determine whether they can affect the specific pathway or the final responses when the pathway is activated. This mode of assay is commonly used to investigate whether the test molecule can inhibit the elevated signal or whether there exists a synergic effect between the test molecule and the activating ligand. The test molecules may exert their effect by interacting with cell surface proteins or by directly interacting with intracellular proteins if it is permeable to the cell membranes.

The cell's responses to the extracellular signal could be detected at any point along the signal transduction pathway (see Fig. 8.1). Cell-based assays can be divided into four categories based on the where the signal is measured along the steps of the cellular transduction process:

1. Measure the initial changes of components associated with plasma membrane:
   a. The binding between test molecules and the cell surface receptors
   b. Receptor dimerization and phosphorylation (e.g., RTKs)
   c. Ion channel conductivity and membrane potential changes
   d. Binding among receptors, adaptors, and effectors
2. Measure intermediate changes involving components in the signal transmission pathways:
   a. Changes in effector protein activity by monitoring its association with activation molecules (e.g., Ras association with GTP) or the enzymatic reaction product catalyzed by effector protein (e.g., the amounts of phosphorylated Raf)
   b. Changes in the concentration of second messenger (e.g., cAMP, DAG, IP_3, and Ca^{2+} in GPCR activation)
   c. Changes in enzymatic activity along the signal transmission pathways (e.g., serine/threonine kinase activity and protease activity)
   d. Protein translocation within cells (e.g., steroid receptor movement from cytosol to nucleus and arrestin internalization after GPCR activation)
3. Measure cellular responses at transcriptional/translational level:
   a. Quantitation of mRNAs for those transcriptions that are linked to the signal transduction pathway with modern RNA quantitation technologies, such as quantitative PCR and branched DNA
   b. Signals from artificially constructed exogenous reporter gene product that is controlled by the activated transcription factors linked to signal transduction pathways

4. Measure integrated systemwide cellular changes:

a. Gene transcription pattern changes and protein expression pattern changes (genomics and proteomics)

b. Cell metabolic profile changes that affect the concentration of metabolites (e.g., oxygen, carbon dioxide, pH, and glucose)

c. Cell interaction with extracellular matrix (e.g., impedance changes in electrical field)

d. Cell morphology changes, cell membrane ruffling, organelle membrane movement

e. Cell proliferation, differentiation, viability, and cytotoxicity

The cell-signaling process is highly regulated to allow attenuation or termination of the intracellular signals or final cellular responses. This is usually achieved by feedback mechanisms. The feedback regulation will lead to different levels of modification and activation of component molecules and cellular response depending on the signaling molecules, the concentration of the signaling molecules, and the duration of exposure to the signaling molecules. Thus, it is important to study different ligand, ligand concentration, and the incubation time to determine the conditions for the optimal signal.

8.3 CONCEPT OF AFFINITY AND EFFICACY IN CELL-BASED ASSAYS

From discussions in the previous sections, it is clear that cellular signal transduction is a very complicated process. Because many changes in intracellular components may occur and several cell responses may happen after exposing to a test molecule, cell-based assays are inherently more complicated than the binding assays and enzymatic assays. However, the basic physical principle should still hold true for cell-based assays. As with enzymatic assays, the binding process is the first step in cell-based assays. After binding, the response of the enzymatic assay system is the enzymatic product, whereas the responses of the cell-based assay systems can be diverse. This analysis enables us to extract a simplified underlying physical model from the complicated system. With this simplified model, cell-based assays, enzymatic assays, and binding assays can be compared with a unified theory as shown in Figure 8.7. In binding assays, only one parameter, the association constant (K_a), is the focus of the experiment, which is obtained by measuring the concentration of the bound complex [LP]. In enzymatic assays, two parameters, the K_a and the k_{cat}, are focuses of the experiment. These two parameters are determined by measuring the initial velocity of the product formation at different substrate concentrations. In cell-based assays, the first step is again the binding between ligand and receptor. Different from enzymatic assays where the substrate only binds to one enzyme, the ligand in cell-based assays may bind to more than one type of receptors (R_1 and R_2) on the cell surface. In addition, each binding may generate several detectable signals instead of just one signal as in enzymatic assays (the enzymatic product). Each of the signals

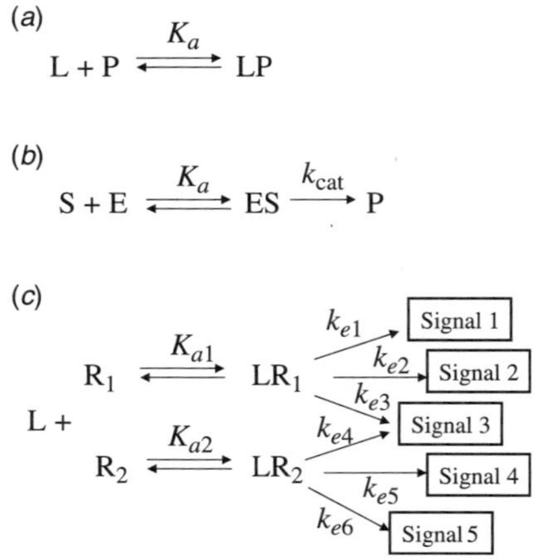

Figure 8.7 Comparison of binding assays, enzymatic assays, and cell-based assays. (*a*) Binding assays: Only one parameter, the association constant (K_a), is the focus of the experiment. (*b*) Enzymatic assay: Two parameters, the K_a and the k_{cat}, are focus of the experiment. (*c*) Cell-based assays. The first step is again the binding between ligand and receptor. The ligand in cell-based assays may bind to more than one type of receptors (R_1 and R_2) on the cell surface. In addition, each of the binding may generate several detectable signals. The measurement of a particular signal can be treated by two factors in a simplified model: K_a for the binding of the ligand to the receptor and k_e for an integrated efficacy leading to the particular signal.

in cell-based assays is an integrated function of all the cellular events leading to that signal. Borrowing the idea of the "k_{cat}" treatment in enzymatic assays, we will use k_e here as a parameter to represent the integrated efficacy at unit concentration of the corresponding receptor for a specific signal. This treatment separates the overall cellular response into the binding constant and maximum response analogs to the well-known pharmacological concepts of "affinity" and "efficacy." In ideal situations, the maximum response of a particular signal is the product of k_e and the receptor concentration. The sigmoid dose–response curves similar to that of enzymatic assays with increasing substrate–ligand concentration should be observed in cell-based assays. From this treatment, many K_a's and k_e's are determined in cell-based assays. To simplify the discussion, we only treat the situation where the ligand is specific and it only binds to one type of receptor on the cell surface. In this case, one affinity constant (K_a) and one or more efficacy constants (k_e's) may be determined depending on the number of signals that are measured in cell-based assays.

From the above analysis, the detected signals in cell-based assays are affected by affinity and efficacy. The interplay of affinity and efficacy that form the detected signal in cell-based assays is shown in Figure 8.8. In situation A, ligand A has unit affinity and unit efficacy (also the maximum efficacy). In situation B, ligand B has the same efficacy as ligand A but has higher affinity for the receptor. This situation can be the result of an enhancer molecule that acts on ligand A to increase its affinity. In situation C, ligand C has the same affinity to the receptor as ligand A. However, the efficacy of ligand C is much lower than ligand A so that it can only produce half of the maximum

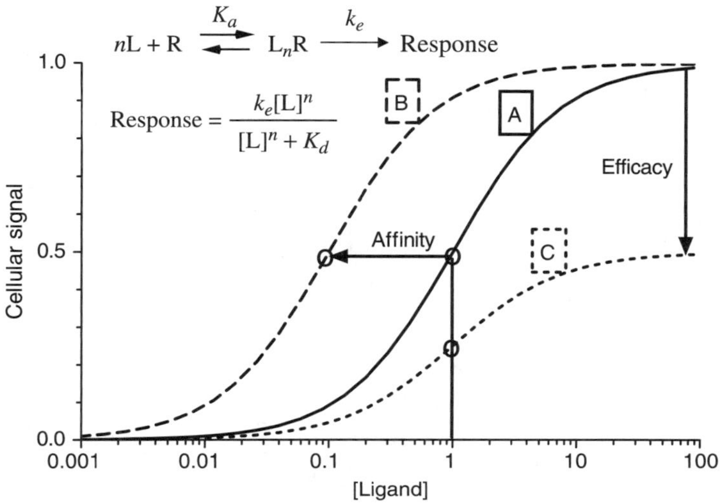

Figure 8.8 Interplay of affinity and efficacy in cellular signals for cell-based assays. Situation A: Ligand A has unit affinity and unit efficacy (also is the maximum efficacy). Situation B: Ligand B has the same efficacy as ligand A but has higher affinity for the receptor. This situation can also be the result of an enhancer molecule that acts on ligand A to increase its affinity. Situation C: Ligand C has the same affinity to the receptor as ligand A. However, the efficacy of ligand C is much lower than ligand A so that it can only produce half of the maximum signal at full receptor occupancy. This situation can also be the result of an inhibitor molecule that reduces the k_e for ligand A.

signal at full receptor occupancy. This situation can be the result of an inhibitor molecule that reduces the k_e of ligand A. The separation of affinity and efficacy also help explaining the agonism and antagonism phenomena. Agonism is defined as positive effects that are produced in the cells after a ligand is mixed with the cells. A ligand that can produce positive effects on the cells is called an agonist. Some ligands can produce maximum effect when they fully occupy the receptors on the cell surface and they are termed full agonists. Some ligands can only produce a fraction of the maximum effect and they are called partial agonists. On the other hand, some molecules are able to diminish the cellular response to agonist and they are called antagonists. In Figure 8.8, both ligand A and ligand B are agonists with maximum efficacy. Ligand C is a partial agonist with reduced efficacy as compared with ligand A and ligand B. Antagonists can totally diminish the efficacy of an agonist. This effect can be the result of its competition with the agonist for the same receptors or its effect somewhere in the signal transduction pathway.

Because the detected signals in cell-based assays are functions of affinity and efficacy, it is very important to be able to separate these two fundamental parameters from the detected signals when interpreting the results in cell-based assays. Direct interpretation of the detected signal at an arbitrarily set ligand concentration in cell-based assays often leads to questionable conclusions. The failure to correctly interpret the cell-based assay results with an arbitrarily set ligand concentration can be

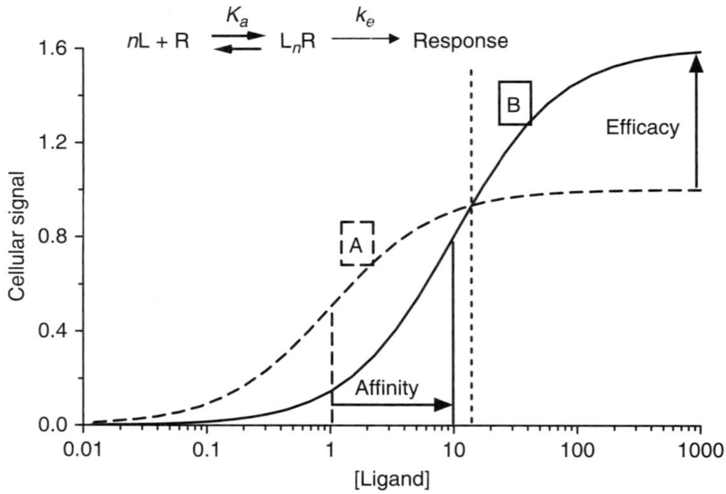

Figure 8.9 Importance of obtaining a full dose–response curve in cell-based assays. The amplitudes of detected signals versus the concentrations of ligand A (dotted line) and ligand B (solid line) are plotted. Ligand A has unit affinity and efficacy. Ligand B has lower affinity than ligand A (higher K_a value, right shift in the binding curve), but it has higher efficacy than ligand A (upward shift in the maximum response). The two dose–response curves of the two ligands intercept at the interception point. Ligand B appears more potent than ligand A when measured at any concentration lower than the interception point of the two curves. The opposite will occur when ligand A and ligand B are compared at the concentrations above the interception point of the two curves. If the concentration used to compare ligand A and ligand B is at the interception point; ligand A and ligand B will appear equally potent.

demonstrated in Figure 8.9. The amplitudes of detected signals versus the concentrations of two ligands A and B are plotted (agonist dose–response curves). Ligand A has unit affinity and efficacy. Ligand B has lower affinity than ligand A (higher K_a value, right shift in the binding curve), but it has higher efficacy than ligand A (upward shift in the maximum response). The two dose–response curves from the two ligands intercept at the interception point. If the detected signals at an arbitrarily set agonist concentration are used as a criterion to compare ligand A and ligand B, ambiguity will occur. As seen in Figure 8.9, ligand B appears more potent than ligand A when measured at any concentration lower than the interception point of the two curves. The opposite will occur when ligand A and ligand B are compared at the concentrations above the interception point of the two curves. If the concentration used to compare ligand A and ligand B is at the interception point, ligand A and ligand B will appear equally potent. Thus, comparing ligand A and ligand B in cell-based assays at arbitrarily set concentrations can lead to all scenarios (ligand A is more potent, equally potent, or less potent than ligand B), depending on the concentrations at which the two ligands are compared. The behavior of ligand A and ligand B demonstrated here is not just theoretical but exists in real situations for many pairs of agonists. For example, the partial agonists of β-adrenergic receptors, fenoterol and trimethoquilol, display the same dose–response curves as shown in Figure 8.9.

Many published works reached questionable conclusions when cell-based assays were interpreted using an arbitrarily set ligand concentration. A particular example is the determination of the FGF receptor binding specificity for FGFs using relative mitogenic activity as the readout. In this study, all the FGFs are potential agonists for a particular FGF receptor under study. To correctly rank the FGFs based on "their binding specificity," or affinity, the efficacy and affinity should be separated in the mitogenic readout unless it is known that all the FGFs have the same efficacy. The separation of affinity and efficacy can be done if the complete dose–response curves of each FGF are determined. In the published mitogenic activity experiment (Ornitz, 1996), the dose–response curves of each FGF against the FGF receptors were measured in the studies. However, the scattered data in the dose–response curves for each FGF does not allow determination of affinity and efficacy. The authors made a compromise to rank the binding specificity of all the FGFs against a chosen FGF receptor by using the average of the mitogenic activities at two FGF concentrations covering a wide range. Averaging cell responses at two arbitrarily set ligand concentrations is not much different from using just one concentration in terms of failing to separate affinity and efficacy. The ranking of FGFs based on this method will be questionable because of the ambiguity demonstrated with one arbitrarily chosen ligand concentration as shown in Figure 8.9. This example demonstrates the importance of obtaining the complete dose–response curves to separate affinity and efficacy when studying agonists in cell-based assays.

Because cells are very complex, many factors can cause the deviation from the simplified models discussed above. Cells may contain excess receptors so that the maximum of a particular signal does not increase with increasing receptor numbers on the cell surface when the number of receptors on the cell surface exceeds a certain value. The extra receptors are termed "spare" receptors. When cell-based assays are performed, the number of cell surface receptors should be reduced if there are spare receptors. For example, the vesicles prepared from the Torpedo electrical organ contain too many nicotinic acetylcholine receptors (nAcChRs; also see Chapter 9). When performing ion flux assays, the ions are emptied so fast through the many channels that the measured rate of ion flux remains constant even when some of the channels are inactivated by inhibitors. The situation is similar to oversaturating an instrument's detector. Only after eliminating "spare receptors" by α-bungalotoxin, can the ion flux kinetics be studied with accuracy. Feedback inhibition may also cause deviation from the normal sigmoid curve. In enzymatic assays, the reaction kinetics will deviate from a normal curve if the extent of reaction goes too far. The deviation is caused by substrate depletion and the accumulation of the reaction product, which may cause enzyme inhibition. Initial velocity measurement is used in enzyme kinetic studies to eliminate this effect. Similarly, we would like to introduce the concept of "initial response" in cell-based assays. Initial response in cell-based assays contains two aspects. One aspect is to pick the signal to be measured as close to the signal reception as possible. For example, the early measurable signals along the signal transduction process for RTKs are the dimerization and autophosphorylation of the receptors. These early signals are less prone to be affected by the cell's feedback response to ligands and are less affected by signal pathway crosstalk or branching. The other aspect of "initial response" is to start signal collection as soon as possible after the cells are exposed to ligands. Immediate measurement of signals after ligand exposure may reduce the cell's feedback inhibition, which may take time to accumulate.

8.4 DEVELOPMENT OF CELL-BASED ASSAYS

8.4.1 Selection of Cell Source

The cells used in cell-based assays are very critical for the development of an optimal assay. The selection of cells for a particular assay is based on the assay performance matrix common to other assays, that is, physiological relevance, availability of cells, signal-to-background ratio, signal-to-noise ratio, dynamic range of signal, reproducibility of the signal, and the feasibility of implementing the assay. Cell-based assays can be divided into two categories based on the cell source used: assays using primary cells and assays using cell lines. Because primary cells retain the phenotypic characteristics of the original tissue, primary cells of human origin are the most physiologically relevant to drug discovery. However, the use of human primary cells creates challenges throughout assay development. Scarce cell sources, donor variability, cell viability, and stromal environment are all critical variables that must be understood and controlled. For example, only a limited quantity of immune cells (such as natural killer cells) can be obtained from a single donor at one time. Immune cells from different donors will react with each other and cannot be pooled. Thus, primary cells from multiple donors have to be used separately during the study period. This may cause large assay variation due to genetic background and health status of the donor. In addition, freshly isolated primary cells can only remain functional in cell culture for a limited time. This imposes constrains for the assays that must be able to generate enough detectable signal within the cells' life span in culture. For example, freshly isolated primary adipocytes survive for less than 24 h in culture, which limits a primary adipocyte assay has to be executed within hours. Instead of using terminally differentiated primary cells directly, progenitor cells can be used to produce terminally differentiated cells. Progenitor cells are not terminally differentiated in terms of function or morphology. They can be expanded under mitogenic conditions. Progenitor cells can then be differentiated to the desired cell type under controlled conditions. For example, human adipocytes can be differentiated from human preadipocytes that are available from several commercial vendors.

Access to human tissues is often very difficult and time consuming. Other than human primary cells derived from blood that is readily available in large quantities from blood banks, very few primary human cells can be obtained because of the unavailability of human tissues. In addition, primary cells cannot be expanded to large quantities because they reach senescence within a limited number of cell divisions. Because mammalian tissues for many species, from rodents to bovine, are readily available, primary cells from nonhuman mammalian sources are often used as an alternative to human primary cells in bioassays. However, lack of cross-species reactivity can be an issue, particularly for proteins active in the immune system, which are generally poorly conserved between human and rodents. Thus, primary mammalian cells with functions known to be preserved in evolution should be chosen for a particular study.

Transformed cell lines of human origin are the most widely used cell sources in cell-based assays because they can be readily obtained in large quantity, can be maintained indefinitely, and can be modified genetically to express specific receptors or specific components along signal pathways. Different from primary cells that vary from donor to donor, cell lines are homogeneous and enable consistent assay results. Some cell lines can retain differentiated phenotypes, making them a good substitute for the primary cells to study biological functions. However, transformed cell lines are typically genetically abnormal and with profoundly perturbed biology. The results obtained from assays using cell lines may not represent the true biology, especially for apoptosis, cell cycle control, and cell proliferation studies. Human cell lines expressing receptors or other proteins of interest may serve as a good cell source if the cells contain a single pure population of the desired proteins. However, it is common that structurally and pharmacologically related receptors (receptor subtypes) or protein isoforms may be present in any given cell, which can complicate the assay. To obtain satisfactory assay signals, the cell lines are often further engineered to express or overexpress specific receptors that are missing or insufficient. In other situations, the cell lines may be engineered to express "reporter" proteins to facilitate assays. Commonly used reporter proteins are GFPs that have intrinsic colors and luciferase or β-galactosidase that has specific enzymatic activity. These reporters are not present in native cell lines. The transfection of cell lines can be transient or stable. In transient transfection, the transfected gene is only transiently expressed since the DNA introduced in the transfection process is not inserted into the genome. The foreign DNA is lost at a later stage when the cells undergo mitosis. Stably transfected cells have the exogenous DNA integrated into the cells' genome. The foreign DNA is passed to the daughter cells during mitosis. To obtain cell lines with stably transfected foreign genes, the cells must go through a selection process. The procedure to obtain stable cell lines is time consuming. However, the resulting cell lines can be used for a long time, although some drifting may occur after extended passages. A transient transfection is simple to perform and offers a rapid way to obtain cells expressing specific DNA. Transiently transfected cells have an advantage over stable transfected cells when the gene product to be expressed is toxic to the cells. Transiently transfected cells may still survive in this case.

Recent advances in embryonic stem cell research offered a new promising way to obtain novel cell types for assays. Since the isolation of mouse embryonic stem cells (mES) in the 1980s, numerous cell types, including adipocytes, cardiomyocytes, chondrocytes, glial cells, islet cells, and motor neurons, have been obtained by differentiation of mES. Cells differentiated from stem cells offer considerable advantages in assays compared with primary cells and cell lines. The cells differentiated from stem cells are genetically normal (diploid versus aneuploid in cell lines) and have uniform physiological responses (compared with variation in primary cells from different donors). Undifferentiated mES cells can grow almost indefinitely and remain totipotent. These properties enable them to be cultured for a long period of time and be expanded in large quantity. In addition, the genome of stem cells can be manipulated the same way as the cell lines to enable assay platforms that cannot be performed with primary cells, such as reporter gene assays. The ultimate cells for assays in drug discovery are of human origin. Since the establishment of human embryonic stem cells (hES) in 1998 in Thompson's lab, intensive efforts have been made to generate human cells using hES. In addition to obtaining normal human cells, derivatives of hES cell lines can be established from genetically abnormal embryos with a wide range of genetic disorders to provide various in vitro disease models. The ability to obtain suitable stem cell sources and to understand the differentiation processes in vivo is crucial for production of human cells in vitro. Technical, regulatory, and ethic issues have significantly hindered the application of hES. To date, only a handful of hES derivatives have been successfully isolated, passaged, and characterized to the extent that they can be considered similar to their in vivo counterparts. Recently, advances have been made to make inducible pluripotent stem cells (iPS). Researchers showed that fibroblasts can be induced to form ES-like cells by the forced expression of certain transcription factors, lately named iPS cells. The iPS cells behave like ES cells (e.g., they express ES cell markers, are diploid, can be induced to differentiation into various cell lineages, and can form taratomas in mice if implanted subcutaneously). Mouse iPS cells can transmit into germline if injected into blastocysts. Researchers have produced iPS cells from patients of various diseases. The iPS cells and their derivatives have the potential of providing a variety of cell types for cell-based assays.

8.4.2 Cell Culture and Cell Treatment During the Assay

Cells can behave differently depending on their environment, culture conditions, and assay conditions. Maintaining cells in consistent conditions during culture and treating the cells properly during the assay are very important for consistent and biologically relevant assay results. The important factors include the composition and the geometry of the surface that the cells grow on, the cell growth media, cell density, cell handling during passage, and the assay media to which the cells are subjected. Cells are commonly grown in cell culture flasks that provide a 2-D surface for adherent cells. It was found that many cells exhibit poor adhesion, poor growth characteristics, and even change of phenotypes when directly plated on unmodified plastic or polylysine-derived surface of cell culture flasks. The coating of the surface with extracellular matrix components, such as collagen (types I, II, IV, and V), fibronectin, laminin 1, proteoglycan, tenascin, thrombospondin, and vitronectin, has been shown to promote cellular adhesion and proliferation for many cell types. It is very important to test many coatings to find the optimal conditions for the cells under study. Even with the extracellular matrix coating on a 2-D surface, many cultured cells often have poor phenotypes because the system does not mimic the 3-D environment in vivo. For example, primary cardiomyocytes and chondrocytes will rapidly dedifferentiate into fibroblast-like cells after being plated on the 2-D surface of tissue culture flasks. It was shown that cells grown in 3-D environments, such as microcarriers, that mimic the ultimate cell-grown environment in the human body preserve cellular phenotypes of many cells. However, growing cells in 3-D environments imposes new operational challenges because it requires new cell culture techniques.

The media in cell culture may affect the cells significantly. If the formulation of the cell culture medium is changed, the cells may undergo selective adaptation or genetic drift leading to altered growth rate and even changes in gene expression. It is advisable to maintain the cells in the media in which the cells were originally isolated or in the media to which the cell line has been adapted. Common cell culture media contains many components, including amino acids, vitamins, inorganic salts, and glucose. The media are buffered with sodium bicarbonate. Constant pH is maintained with a proper fraction of CO_2 in the gas phase to balance with bicarbonate. In addition, growth factors, hormones, attachment factors, and other unknown factors are required to maintain the cells. Serum is commonly used in the cell culture media to supply these essential components. Much research has been done to substitute serum-containing media with defined media with known necessary medium components to maintain cells. However, defined media are often not as good as the one containing serum. In addition, it takes a lot of effort to obtain a media that is optimal in maintaining a specific cell line because media requirements are cell dependent. Thus, cell culture media containing fetal bovine serum (FBS) are used in most applications. Lots of FBS may vary considerably in the ability to maintain cells and promote growth. This is a significant uncontrollable variant in cell-based assays. It is advisable to use the same batch of FBS for the entire study by reserving aliquots of FBS from the same batch for the specific study. New FBS lots must be proven to be comparable to the old lot if changing the FBS lot is required. Mycoplama contamination may adversely affect many cell-based assays. Periodic analysis of the cell culture for mycoplama contamination is very important especially if changes in the cell's growth rate, morphology, or metabolism are observed.

Another factor that affects the cells in culture is the oxygen content. Regular cell culture incubators provide oxygen at about 20% of the gas phase content. This is fine with most cell lines that have adapted to this environment. However, 20% oxygen content in the gas phase may not be the optimal condition for isolated primary cells. It was estimated that most primary cells in internal organs are exposed to the gas phase with $\sim$5% oxygen content in vivo. There are limited studies in this area as to what percentage of oxygen in gas phase is optimal for culturing primary cells.

Cultured cells will become confluent if allowed to grow with sufficient nutrients. When this happens, the cells may undergo changes to adapt to the new environment or die. It is critical to subculture the cells before they reach 90% confluence to maintain the normal cell phenotype. However, primary human cells usually have a limited life span in vitro. For example, human skin fibroblasts reach senescence after about 50 passages. Many cell lines show a high mutation rate under normal culture conditions because they are inherently genetically unstable. This is especially true for many tumor-derived cell lines. The drifting of cells with successive passage may affect assay signals. When developing cell-based assays, potential signal variations with cell passages should be studied. It is advisable to keep the cell passages within a limited number to maintain consistent assay signal. In general, cells in culture should not be carried for longer than a month or two. Cell plating density during the assay is another important factor that affects assay signals, and it must be carefully studied during assay development.

Due to assay throughput, reagent usage, and operation easiness considerations, most modern assays are performed in microplates with various numbers of wells. There is always systematic variation in the assay signal between wells in the microplate because of differences in microplate geometry, moisture gradient, and other unknown factors. The variation is most significant with the wells on the edge of the microplates.

This is commonly referred to as "edge effects." Edge effects affect both biochemical assays and cell-based assays, but the amplitude is more significant in cell-based assays. The common remedy is to surround the microplate with a damp cloth to reduce the moisture gradient. It was found that the edge effects in cell-based assays could be further reduced by incubating the microplates at the ambient conditions (expose to air at room temperature) for about 1 h before placing them in a 37°C CO_2 incubator after cell plating.

The composition of the final assay media may affect the cells' viability and the assay quality. DMSO and serum are the two common components that may exert adverse effects on the assay quality. When small organic molecules are tested, the compounds are commonly stored in 100% DMSO that are diluted in the final assay media. Cells in general cannot tolerate more than 0.5% DMSO. Thus, DMSO's effects on the cell-based assays must be tested at the assay development stage. When cell-based assays are performed in high-density microplates with a final assay volume less than 60 μL (384-well or higher density plates), direct addition of compounds (in DMSO) requires the accurate delivery of less than a 0.3-μL solution to maintain final DMSO concentration at less than 0.5%. Special instruments are required to transfer such a small volume of solutions. In the absence of appropriate instruments, the compounds in DMSO are often diluted with aqueous solution first to create intermediate plates. The compounds from the intermediate plates are then transferred to the assay plates at higher volume. This manipulation may cause the compounds originally dissolved in DMSO to come out of solution in the intermediate plates. Since cells in general require serum for viability and growth, serum is often present in the final assay media throughout the assay. Because serum contains many growth factors and other unknown components that may activate selected cell pathways, this may cause high assay backgrounds, leading to false-negative measurements. If possible, it is preferable to use serum-free defined assay media to minimize assay uncertainty.

8.4.3 Selection of Final Assay Signals and Assay Format

Because there are so many potential detection points along the cell-signaling pathways, choosing which detection point after cell stimulation should be carefully studied. In cell-based assay development, many of the potential detectable signals are evaluated. One or a few of the detectable signals are selected as the final readout based on the relevance to the biological questions, the nature of the signal, the reliability and ease of measuring the signal, and many other criteria. As discussed before, the closer the detection point to the cell surface stimulation along the signaling pathway, the less interference from cell-feedback mechanism, signal crosstalk, and pathway branching will have on the signal. It is also easier to interpret the measured signals because of the lack of complications. The drawback of detecting the earlier changes along the signal pathway is that there is no or limited amplification of the stimulation signal, hence low detection signal. For example, if the binding between a growth factor and its receptor is measured, there is no amplification to the growth factor's signal because the binding between the receptor and the ligand is stoichiometric. If the phosphorylation of downstream kinases in the signal pathway, such as ERK, is measured, more than 100-fold of amplification of the original signal will be observed because one ligand binding may result in the phosphorylation of a significant number of ERK molecules. At the end of the phosphorylation cascade or at the transcription level, the signals will be amplified further. However, these signals at the end of the signal transduction pathway may be influenced by pathway crosstalk and feedback control. This may lead to unexpected assay results and the results may be difficult to interpret. Although in general the signal is amplified further down a signal transduction pathway, the final cells' integrated systemwide response to a stimulation may not be always higher than the earlier signal, depending on what is measured. In general, the final signal is expected to be further amplified if the final reading is at a molecular level (e.g., glucose measurement for glucogenesis). However, the final signal should be less amplified if the final reading is at a cellular level (e.g., cell number measurement in cell proliferation assays) because it takes many cellular components to assemble a few cells. This can be demonstrated by a hypothetical signaling pathway that leads to either glucogenesis or cell proliferation after the ligand binds GPCRs on the cell surface, as shown in Figure 8.10. At the signal reception, ligand binds to GPCRs on the cell surface, which is one-to-one binding, and thus there is no signal amplification (*A*); one ligand-bound GPCR activates 100 G proteins (*B*); the activated G proteins activate equal amounts of adenylyl cyclase (*C*); one activated adenylyl cyclase activates 100 cAMP (*D*); cAMP activates equal amounts of PKA (*E*); one activated PKA activates 10 phosphorylase kinase (*F*); one phosphorylase kinase activates 10 glycogen phosphorylase (*G*); and one glycogen phosphorylase generates 100 glucose-1-phosphate (*H*). At the molecular level, the signal is continuously amplified down the signal transduction pathways. However,

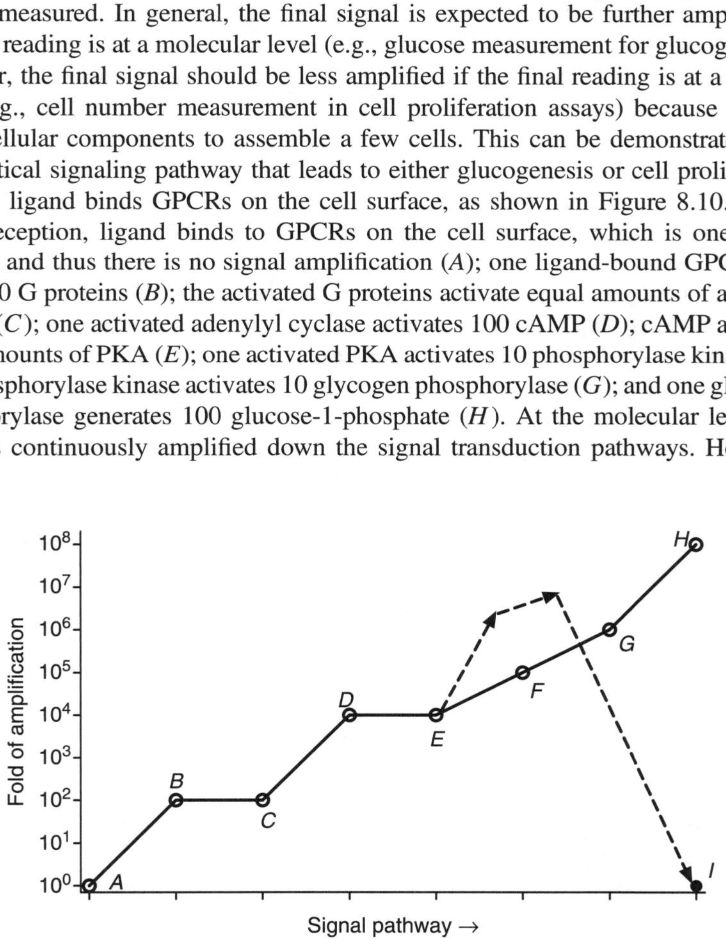

Figure 8.10 Illustration of signal amplification along hypothetical signal pathways. After ligands bind receptors, assuming there exist branched pathways in the middle of the signaling leading to glucose production (solid lines) and cell proliferation (dotted lines), respectively. *A*: ligand bound GPCR; *B*: activated G proteins; *C*: activated adenylyl cyclase; *D*: activated cAMP; *E*: activated PKA; *F*: activated phosphorylase kinase; *G*: activated glycogen phosphorylase; *H*: glucose-1-phosphate. A measurement of the number of new cells (*I*) will generate a measurable signal that is not amplified from the initial signal (number of ligand-bound receptors).

if a hypothetic pathway exists that leads to cell proliferation (dotted line), a measurement of the number of new cells (I) will generate a measurable signal that is not amplified from the initial signal (number of ligands). This is because that many newly synthesized cellular components will integrate together to make new cells at the end of the signaling process, and it takes days for mammalian cells to double in number after stimulation. If we use the total cell number as the final assay output, the signal amplification will be much lower than the measurement of earlier signals, such as the activation of PKA or transcription activity. In this case, a plot of signal amplification against the points down the signal pathway is a bell-shaped curve.

In addition to signal amplification and interpretability, several other factors must be considered as well when choosing final assays. These factors include the stability of the molecules to be measured inside the cells and the assay methods used to detect the molecules. One way of regulating cell signaling is the degradation of the molecules involved in the signaling pathway. Thus, the concentration of the molecules to be detected inside cells may not be constant and the life span of certain molecules along the cell-signaling pathways are limited. The optimal assay time after ligand exposure must be determined during assay development. For example, the concentration of cyclins fluctuates significantly during different cell cycle stages, and the life span of mRNAs ranges from minutes to 24 h. Different molecules require different detection methods that may affect the assay quality and may be an important factor in choosing the assay detection points. For example, detection of transcription activity requires the methods to detect mRNA and detection of kinase activation calls for the methods to detect phosphoprotein. These totally different detection methods have their inherent advantages and disadvantages. The time it takes for the cell to generate enough measurable signal will affect the assays too. For example, detection of mammalian cell growth by counting cell number demands that the incubation time must be several days because of the slow growth rate of mammalian cells. The long incubation time may result in high variations among wells in an assay plate. An assay developer should study the pattern of each measurable response along the signal pathway. The final assay can only be decided after careful evaluation of each assay with regard to biological relevance, signal amplification, the availability of proven assay technology for signal detection, the noise and sensitivity of the assay, the time for performing the assay, controllability of the variation of the assay, and the functional stability of the cells.

For being used as a test system, cells have many advantages over isolated proteins. The most obvious advantage is that cells are closer to the physiological state than isolated proteins. One of the most important properties of cell-based assay is that it is a system with "one treatment, multiple readouts." For example, binding between GTP and G_α, changes in cAMP concentration, changes in intracellular calcium, protein phosphorylation, changes in mRNA transcription, and other signaling events can be detected simultaneously after the stimulation of the GPCR by its ligand. This property offers an ideal platform for high-content screening or multiplexed assays. In addition, the preparation of the cell assay system is relatively standard and can be carried out easily with no need to develop the time-consuming protein purification procedure as in protein assay systems. Using cells as assay systems may also avoid the potential risk of loss of function for proteins that require multiprotein complexes. Because a cell-based assay is closer to physiology, some physiological processes can be measured, which is impossible with isolated proteins. For example, an effective molecule can be classified as "agonist" if it has a positive effect or "antagonist" if it has a negative effect on a subject in pharmacological studies. Such activities can be studied in cell-based assays by testing whether the molecules upregulate or downregulate certain detectable signals. This shows that cell-based assays can distinguish between agonists and antagonists, which cannot be achieved with assays using isolated proteins as the assay systems. Cells offer assay developers the flexibility to choose among many measurement points inside and outside of the cells. An assay developer can select among early change at receptor level, downstream change at amplified protein activation and mRNA level, or final cellular responses. This flexibility can be a big plus for an experienced assay designer but may be challenging for a novice who may pick an assay measurement point that is irrelevant to the proposed study. Because many processes often complicate the measurements in a cell-based assay, a novice assay developer may misinterpret the results.

There are some potential concerns when performing cell-based assays. In drug discovery, dissecting the point of action for a small cell-permeable molecule may be difficult if the detection is far downstream and the signaling pathways are complicated and not fully understood. Many cells, especially primary cells, may require undefined media supplements (such as serum) to maintain their proper function. It is difficult to control the consistency of these kinds of supplements from batch to batch and large variations in assay results may occur. Because cells are live and can be heterogeneous, cell-based assays on average have larger variations when compared with biochemical assays. Another problem inherent in the use of cellular assays for drug discovery is the cells' sensitivity to cytotoxic compounds, which can result in false positives in high-throughput screening applications. Thus, a separate cytotoxicity assay is usually required to follow up the hits from cell-based assays.

Useful Websites

http://www.atcc.org

http://ccr.coriell.org/

http://www.cellapplications.com/

http://www.promocell.com/us/homepage.htm

http://www.zen-bio.com/index.php

http://www.cambrex.com/

BIBLIOGRAPHY

Assil, I. Q., Qi, L. J., Arai, M., Shomali, M., and Abou-Samra, A. B. (2001) Juxtamembrane region of the amino terminus of the corticotropin releasing factor receptor type 1 is important for ligand interaction. **40**, 1187–1195.

Atkuri, K. R., Herzenberg, L. A., Niemi, A.-K., Cowan, T., and Herzenberg, L. A. (2007) Importance of culturing primary lymphocytes at physiological oxygen levels. *PNAS* **104**, 4547–4552.

Bashor, C. J., Helman, N. C., Yan, S., and Lim, W. A. (2008) Using engineered scaffold interactions to reshape MAP kinase pathway signaling dynamics. *Science* **319**, 1539–1543.

Bazley, L. A. and Gullick, W. J. (2005) The epidermal growth factor receptor family. *Endocr. Relat. Cancer* **12**, S17–S27.

Bjarnadottir, T. K. (2006) Comprehensive repertoire and phylogenetic analysis of the G protein-coupled receptors in human and mouse. **88**, 263–273.

Bjarnadottir, T. K., Fredriksson, R., and Schioth, H. B. (2005) The gene repertoire and the common evolutionary history of glutamate, pheromone (V2R), taste(1) and other related G protein-coupled receptors. **362**, 70–84.

Blum, G., et al. (2005) Dynamic imaging of protease activity with fluorescently quenched activity-based probes. **1**, 203–209.

Carron, C. (2003) Frizzled receptor dimerization is sufficient to activate the Wnt/[beta]-catenin pathway. **116**, 2541–2550.

Chen, Y. (2007) Interaction of novel positive allosteric modulators of metabotropic glutamate receptor 5 with the negative allosteric antagonist site is required for potentiation of receptor responses. **71**, 1389–1398.

Cherezov, V. (2007) High-resolution crystal structure of an engineered human [beta]2-adrenergic G protein coupled receptor. **318**, 1258–1265.

Cho, W. (2006) Building signaling complexes at the membrane. *Sci. STKE* **2006**, pe7.

Civelli, O., Saito, Y., Wang, Z., Nothacker, H. P., and Reinscheid, R. K. (2006) Orphan GPCRs and their ligands. **110**, 525–532.

Clark, S., et al. (1997) Novel insulinoma cell lines produced by iterative engineering of GLUT2, glucokinase, and human insulin expression. *Diabetes* **46**, 958–967.

Costa, M., et al. (2007) A method for genetic modification of human embryonic stem cells using electroporation. *Nat. Protocol* **2**, 792–796.

Dhundale, A. and Goddard, C. (1996) Reporter assays in the high throughput screening laboratory: A rapid and robust first look? *J. Biomol. Screen.* **1**, 115–118.

Dixon, R. A. (1986) Cloning of the gene and cDNA for mammalian [beta]-adrenergic receptor and homology with rhodopsin. *Nature* **321**, 75–79.

Dufayet de la Tour, D., et al. (2001) {beta}-Cell differentiation from a human pancreatic cell line in vitro and in vivo. *Mol. Endocrinol.* **15**, 476–483.

Foord, S. M. (2005) International Union of Pharmacology. XLVI. G protein-coupled receptor list. *Pharmacol. Rev.* **57**, 279–288.

Frank-Kamenetsky, M. (2002) Small-molecule modulators of Hedgehog signaling: Identification and characterization of Smoothened agonists and antagonists. *J. Biol.* **1**, 10.

Fredriksson, R., Gloriam, D. E., Hoglund, P. J., Lagerstrom, M. C., and Schioth, H. B. (2003a) There exist at least 30 human G-protein-coupled receptors with long Ser/Thr-rich N-termini. *Biochem. Biophys. Res. Commun.* **301**, 725–734.

Fredriksson, R., Lagerstrom, M. C., Lundin, L. G., and Schioth, H. B. (2003b) The G-protein-coupled receptors in the human genome form five main families. Phylogenetic analysis, paralogon groups, and fingerprints. *Mol. Pharmacol* **63**, 1256–1272.

Fredriksson, R. and Schioth, H. B. (2005) The repertoire of G-protein-coupled receptors in fully sequenced genomes. *Mol. Pharmacol* **67**, 1414–1425.

Gloriam, D. E., Fredriksson, R., and Schioth, H. B. (2007) The G protein-coupled receptor subset of the rat genome. *BMC Genomics* **8**, 338–405.

Guo, A. (2008) Signaling networks assembled by oncogenic EGFR and c-Met. *PNAS* **105**, 692–697.

Hallis, T. M., et al. (2007) An improved {beta}-lactamase reporter assay: Multiplexing with a cytotoxicity readout for enhanced accuracy of hit identification. *J. Biomol. Screen.* **12**, 635–644.

Hayflick, L. (1965) The limited in vitro lifetime of human diploid cell strains. *Exp. Cell Res.* **37**, 614–636.

Heldin, C. H. (1995) Dimerization of cell surface receptors in signal transduction. *Cell* **80**, 213–223.

Hewitt, W. (2004) *Microbiological Assay for Pharmaceutical Analysis.* Interpharm/CRC, London.

Jacoby, E., Bouhelal, R., Gerspacher, M., and Seuwen, K. (2006) The 7 TM G-protein-coupled receptor target family. *Chem. Med. Chem.* **1**, 761–782.

Jones, J. O. and Diamond, M. I. (2007) Design and implementation of cell-based assays to model human disease. *ACS Chem. Biol.* **2**, 718–724.

Klimanskaya, I., Rosenthal, N., and Lanza, R. (2008) Derive and conquer: Sourcing and differentiating stem cells for therapeutic applications. *Nat. Rev. Drug Discov.* **7**, 131–142.

Kost, T. A., Condreay, J. P., Ames, R., Rees, S., and Romanos, M. A. (2007) Implementation of BacMam virus gene delivery technology in a drug discovery setting. *Drug. Discov. Today* **12**, 396–403.

Kunz-Schughart, L. A., Freyer, J. P., Hofstaedter, F., and Ebner, R. (2004) The use of 3-D cultures for high-throughput screening: The multicellular spheroid model. *J. Biomol. Screen.* **9**, 273–285.

Lagerstrom, M. C. and Schioth, H. B. (2008) Structural diversity of G protein-coupled receptors and significance for drug discovery. *Nat. Rev. Drug Discov.* **7**, 339–357.

Lefkowitz, R. J. (2004) Historical review: A brief history and personal retrospective of seven-transmembrane receptors. *Trends Pharmacol. Sci.* **25**, 413–422.

Li, G. N., Livi, L. L., Gourd, C. M., Deweerd, E. S., and Hoffman-Kim, D. (2007) Genomic and morphological changes of neuroblastoma cells in response to three-dimensional matrices. *Tissue Eng.* **13**, 1035–1047.

Lundholt, B. K., Heydorn, A., Bjorn, S. P., and Pr[æ]stegaard, M. (2006) A simple cell-based HTS assay system to screen for inhibitors of p53-Hdm2 protein: Protein interactions. *Assay Drug Dev. Technol.* **4**, 679–688.

Lundholt, B. K., Scudder, K. M., and Pagliaro, L. (2003) A simple technique for reducing edge effect in cell-based assays. *J. Biomol. Screen.* **8**, 566–570.

Malnic, B., Godfrey, P. A., and Buck, L. B. (2004) The human olfactory receptor gene family. *PNAS* **101**, 2584–2589.

McCudden, C. R., Hains, M. D., Kimple, R. J., Siderovski, D. P., and Willard, F. S. (2005) G-protein signaling: Back to the future. *Cell Mol. Life Sci.* **62**, 551–577.

McNeish, J. (2004) Embryonic stem cells in drug discovery. *Nat. Rev. Drug Discov.* **3**, 70–80.

McNeish, J. D. (2007) Stem cells as screening tools in drug discovery. *Curr. Opin. Pharmacol.* **7**, 515–520.

Nauck, M. A. and Meier, J. J. (2005) Glucagon-like peptide 1 and its derivatives in the treatment of diabetes. **128**, 135–148.

Oikonomopoulou, K. (2006) Proteinase-mediated cell signalling: Targeting proteinase-activated receptors (PARs) by kallikreins and more. *Biol. Chem.* **387**, 677–685.

Olson, K. R. and Eglen, R. M. (2007) Galactosidase complementation: A cell-based luminescent assay platform for drug discovery. *Assay Drug Dev. Technol.* **5**, 137–144.

Olwin, B. and Hauschka, S. (1988) Cell surface fibroblast growth factor and epidermal growth factor receptors are permanently lost during skeletal muscle terminal differentiation in culture. *J. Cell Biol.* **107**, 761–769.

Onuffer, J. J. and Horuk, R. (2002) Chemokines, chemokine receptors and small-molecule antagonists: Recent developments. **23**, 459–467.

Ornitz, D. M., et al. (1996) Receptor specificity of the fibroblast growth factor family. *J. Biol. Chem.* **271**, 15292–15297.

Palczewski, K. (2000) Crystal structure of rhodopsin: A G-protein-coupled receptor. **289**, 739–745.

Parham, J., Iannone, M., Overton, L., and Hutchins, J. (1998) Optimization of transient gene expression in mammalian cells and potential for scale-up using flow electroporation. *Cytotechnology* **28**, 147–155.

Pawson, T. and Scott, J. D. (1997) Signaling through scaffold, anchoring, and adaptor proteins. *Science* **278**, 2075–2080.

Pazolli, E. and Stewart, S. A. (2008) Senescence: The good the bad and the dysfunctional. *Curr. Opin. Genet. Dev.* **18**, 42–47.

Pittenger, M. F., et al. (1999) Multilineage potential of adult human mesenchymal stem cells. *Science* **284**, 143–147.

Rasmussen, S. G. (2007) Crystal structure of the human [beta]2 adrenergic G-protein-coupled receptor. *Nature* **450**, 383–387.

Robbins, M. J. (2000) Molecular cloning and characterization of two novel retinoic acid-inducible orphan G-protein-coupled receptors (GPRC5B and GPRC5C). *Genomics* **67**, 8–18.

Rubin, H. (1997) Cell aging in vivo and in vitro. *Mech. Ageing Dev.* **98**, 1–35.

Schlessinger, J. (2000) Cell signaling by receptor tyrosine kinases. *Cell* **103**, 211–225.

Siehler, S. (2008) Cell-based assays in GPCR drug discovery. *Biotechnol. J.* **3**, 471–483.

Takahashi, K., Tanabe, K., Ohnuki, M., Narita, M., Ichisaka, T., Tomoda, K., and Yamanaka, S. (2007) Induction of pluripotent stem cells from adult human fibroblasts by defined factors. *Cell* **131**, 861–872.

Thomson, J. A., et al. (1998) Embryonic stem cell lines derived from human blastocysts. *Science* **282**, 1145–1147.

Tyndall, J. D. and Sandilya, R. (2005) GPCR agonists and antagonists in the clinic. *Med. Chem.* **1**, 405–421.

Umezawa, Y. (2002) Assay and screening methods for bioactive substances based on cellular signaling pathways. *J. Biotechnol.* **82**, 357–370.

Vassilatis, D. K. (2003) The G protein-coupled receptor repertoires of human and mouse. *PNAS* **100**, 4903–4908.

Vauquelin, G., Bottari, S., and Strosberg, A. D. (1980) Inactivation of beta-adrenergic receptors by *N*-ethylmaleimide: Permissive role of beta-adrenergic agents in relation to adenylate cyclase activation. *Mol. Pharmacol.* **17**, 163–171.

Zhang, X., et al. (2006) Receptor specificity of the fibroblast growth factor family: The complete mammalian FGF family. *J. Biol. Chem.* **281**, 15694–15700.

CHAPTER 9

FUNCTIONAL ION CHANNEL ASSAYS

9.1 INTRODUCTION TO ION CHANNELS

Ion channels are a diverse family of membrane proteins that are capable of allowing the passive diffusion of ions across the plasma membrane. They are found in most cell types and control a diverse range of physiological processes including proprioception, nociception, cell volume, secretion, and heart rate. Ion channels can exist in many different states, such as resting (closed) state, conducting (open) state, and desensitized (inactive) state. Only channels in the open state allow the passage of ions. Some ion channels are highly selective in conducting ions—they only allow the movement of a particular ion, such as Na^+, K^+, Ca^{2+}, or Cl^-. Ion channels are the fundamental excitable elements in the membranes of excitable cells such as neurons, cardiac and skeletal muscles, and sensory cells. These cells are able to produce and respond to electrical signals. The timing of the alternation of ion channels in different states and their ionic selectivity are the basic properties that allow the establishment of resting membrane potential and transmission of electrical signals. Because ion channels control a diverse range of physiological processes, malfunction of ion channels causes many diseases such as arrhythmia, hypertension, migraine, neuropathic pain, seizure, and cystic fibrosis. Thus, ion channels are important protein targets for drug discovery.

9.1.1 Ion Channel Classification

Early researches found that some ion channels were activated by voltage while others were activated by chemical ligands. It was further found that ion channels are often selectively permeable to specific ions, such as sodium, potassium, calcium, and chloride. Thus, the channels are conveniently named based on their ion selectivity, for example, sodium channels, potassium channels, calcium channels, and the like. Ligand-gated channels are often named after the ligands that activate them, followed by the word "receptor," as in nicotinic acetylcholine receptors, glutamate receptors, GABA (γ-aminobutyric acid) receptors, and so forth. Voltage-gated Na, K, and Ca channels have many functional similarities. Likewise, ion channels gated by chemicals, such as acetylcholine, glutamate, glycine, and GABA, seem similar. Molecular

Assay Development: Fundamentals and Practices. By Ge Wu

genetics has confirmed most of these relationships. These findings form the basis to broadly classify ion channels into voltage-gated ion channels and ligand-gated ion channels. The predicted amino acid sequences for ion channels reveal strong structural similarities among groups of ion channels, allowing for further classification of ion channels into families of homologous ion channel proteins that have evolved by processes of successive gene duplication, mutation, and selection from common ancestral ion channels. With the complete genomic sequence of human and several other organisms, several hundred genes encoding ion channels were discovered, which calls for a more systematic naming method for ion channels. Ion channels can be classified based upon their ion selectivity, gating mechanism, and sequence similarity. Among them, the system based on ion channel gating mechanisms is most successful and widely adopted. With this system, ion channels can be classified into three main groups: (1) the voltage-gated channels, (2) the extracellular ligand-gated channels, and (3) other gating mechanisms. It has been shown by sequence comparison that ion channels within the above groups show the greatest sequence similarity and are therefore most likely all descended from a common ancestor. The gating mechanism along with a combination of sequence similarity and ion selectivity further subdivide ion channels into several subtypes.

Voltage-Gated Channels This group contains several families: (1) The voltage-gated sodium channel family contains 9 functional members ($Na_V1.1$ to $Na_V1.9$). They are largely responsible for action potential creation and propagation. They contain pore-forming α subunits and auxiliary β subunits. The α subunits (up to 4000 amino acids) consist of four homologous repeat domains (I to IV) each comprising six transmembrane segments (S1 to S6) for a total of 24 transmembrane segments. Each β subunit spans the membrane once. (2) The voltage-gated calcium channel family contains 10 members belonging to 3 subfamilies ($Ca_V1.1$ to $Ca_V1.4$, $Ca_V2.1$ to $Ca_V2.3$, and $Ca_V3.1$ to $Ca_V3.3$). Voltage-gated calcium channels contain pore-forming α subunits and other regulatory δ, β, and γ subunits. These channels play an important role in both linking muscle excitation with contraction as well as neuronal excitation with transmitter release. The α subunits have an overall structural resemblance to those of the sodium channels. (3) The voltage-gated potassium channel family contains almost 40 members, which are further divided into 12 subfamilies (K_V1 to K_V12). These channels are known mainly for their role in setting resting membrane potential and repolarizing the cell membrane following action potentials. The pore-forming α subunits have an overall structural resemblance to a single repeat domain of the sodium channels. In addition, there are several other families of voltage-gated channels that include transient receptor potential (TRP) channels, hyperpolarization-activated cyclic nucleotide-gated channels, Catsper channels (cation channels of sperm), and voltage-gated proton channels.

Extracellular Ligand-Gated Channels This group of channels is also referred to as ionotropic receptors. They form intrinsic transmembrane ion channels that are opened in response to the binding of a chemical ligand. Many of these channels are located at synapses where they convert chemical signals of presynaptically released neurotransmitters into postsynaptic electrical signals. Ligand-gated ion channels are classified into three superfamilies: the cys loop receptors (including nicotinic acetylcholine receptors, $5HT_3$ receptors, glycine receptors, $GABA_A$ receptors, and $GABA_C$ receptors), the ionotropic glutamate receptors (including N-methyl-D-aspartate (NMDA) receptors, α-amino-3-hydroxy-5-methyl-4-isoxazolepropionate (AMPA) receptors, and Kainate receptors), and the ATP-gated P2X receptors. The nicotinic acetylcholine receptor is the most studied ligand-gated ion channel. It consists of five subunits (2α, β, γ, and δ) with two binding sites on the α-subunits for acetylcholine. Each of the subunits contains four transmembrane helices. Upon binding of acetylcholine, the receptor changes configuration, which causes the opening of its internal pore, which in turn allows cations, including Na^+, to go through. When many channels are open at the same time, the concentration of intracellular Na^+ rises, which causes significant depolarization of the cell and initiates the firing of action potentials.

Ion Channels with Other Gating Mechanisms This diverse group includes ion channels with activation mechanisms that are different from those for the voltage- and ligand-gated channels. They include inward-rectifier potassium channels, calcium-activated potassium channels, two-pore potassium channels, cyclic nucleotide-gated channels, light-gated channels, and mechanosensitive ion channels.

9.1.2 Electrical Conduction and Action Potential in Excitable Cells

Cells normally maintain a negative electric potential across the plasma membrane when they are in the resting state, meaning the electric potential inside the cells are more negative than the electric potential outside the plasma membrane. This is commonly referred to as the resting potential. For neurons, the resting potential is about −70 mV. The resting potential in neurons is mainly the results of selective permeability of ion channels to specific ions as illustrated in Figure 9.1. Three membrane proteins, a potassium channel, a sodium channel, and a sodium–potassium pump are shown. The concentrations of ions inside the cells are ~150 mM for potassium, ~15 mM for sodium, ~10 mM for chloride, and collectively ~100 mM of other anions, such as proteins, amino acids, phosphate, and sulfate, that do not cross the cell membrane. The concentrations of ions outside the cells are ~5 mM for potassium, ~150 mM for sodium, and ~120 mM for chloride. The calcium concentration is very low in the cytosol (<0.1 μM in resting cells) and is ~2 mM in extracellular media. Because potassium channels are relatively more permeable than other channels when the cell is at rest, the net result is that more potassium ions cross the membrane down the concentration gradient to the extracellular side. The accumulation of extra potassium ions outside the cell results in more positive charge outside the plasma membrane and more negative charge inside the membrane. The sodium channels are much less permeable when the cell is at rest, but some sodium ions can still leak into the cells due to the finite, albeit small, channel open probability and the favorable concentration gradient and electric forces. Sodium ions that leak into the cell are subsequently pumped out by the sodium–potassium pump to maintain the normal distribution of the ions. Thus, potassium ions are the major components responsible for the resting membrane potential.

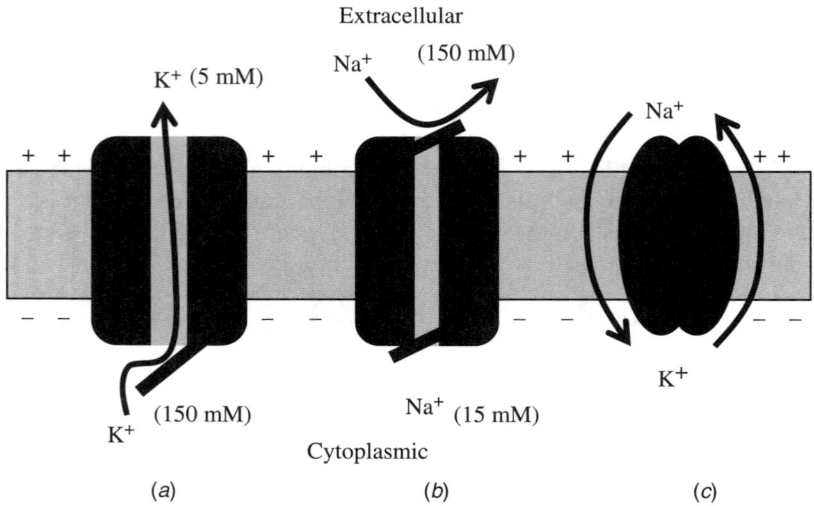

Figure 9.1 Basis of membrane resting potential. Three membrane proteins, (*a*) a potassium channel, (*b*) a sodium channel, and (*c*) a sodium–potassium pump are shown. The concentrations of ions inside the cells are: $[K^+] = 150$ mM, $[Na^+] = 15$ mM, $[Cl^-] = 10$ mM, and about 100 mM anions that do not diffuse across cell membranes. The concentrations of ions outside the cells are: $[K^+] = 5$ mM, $[Na^+] = 150$ mM, and $[Cl^-] = 120$ mM. Potassium channels are the major components responsible for the negative membrane resting potential. They are more permeable in the resting state than sodium channels, resulting in net negative charge inside and positive charge outside the membrane. Sodium–potassium pump helps to maintain the membrane resting potential by pumping three sodium ions out of and two potassium ions into the cell during each cycle.

The membrane potential of an excitable cell (such as a neuron) changes when the ion channels embedded in the membrane are activated. If a stimulus causes the potassium channels to open, more potassium ions will flow out of the cell, and the membrane potential will become more negative than the resting potential. This is referred to as hyperpolarization. If a stimulus causes the sodium or calcium channels to open, more sodium or calcium ions will flow into the cell and the membrane potential will become less negative inside. This is referred to as depolarization. When the depolarization reaches a critical point, referred to as the threshold potential, an action potential is triggered. This process is illustrated in Figure 9.2. The membrane potential is maintained at the resting potential (RP) level at time zero before activation of the sodium channels. After stimulation, a small number of sodium channels are open if the threshold potential (TP) is not reached. A stronger stimulation results in the membrane depolarization reaching the threshold potential, which in turn opens a large number of sodium channels. This triggers the action potential (AP), which subsequently opens voltage-gated potassium channels and hyperpolarizes the cell. At this stage, the sodium channels are inactivated and closed and the potassium channels are open. The potassium efflux results in a brief hyperpolarization before reverting back to normal resting potential. The whole process takes place on the millisecond scale.

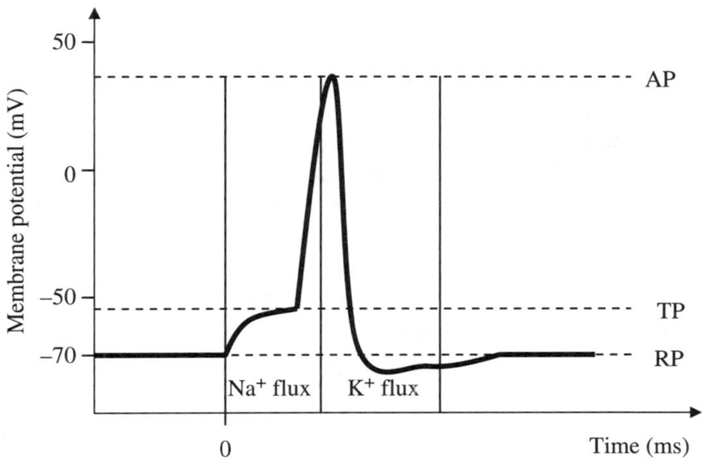

Figure 9.2 Basis of membrane action potential. The membrane potential is maintained at the resting potential (RP) level at time zero before sodium channel activation. After stimulation, a small number of sodium channels are open if the threshold potential (TP) is not reached. A strong stimulation will result in the membrane depolarization reaching the threshold potential, which in turn opens a large number of sodium channels. This triggers the action potential (AP) that subsequently activates potassium channels and hyperpolarizes the membrane while sodium channels are closed along the way back to the resting state.

9.2 STRATEGIES FOR ION CHANNEL ASSAYS

Early ion channel assays were performed using the radioligand competition method to identify compounds that interact with the channel. This method was discussed in Chapter 5. However, the value of this type of assay is limited because the hits from the assay depend on the binding site of the tracer ligand used in the assay. Compounds that bind to other (allosteric) sites on the ion channel protein could be missed. The application of this assay is particularly difficult for voltage-gated channels where no endogenous ligand exists. In addition, it is unclear as to the functional effect of a hit identified using this method since it is not a functional assay. Functional assays for ion channels are thus preferred to binding assays. From the discussion of the basic principles of ion channels, we understand that signal transduction in excitable cells is based on the changes in membrane potential and the ion flux through the ion channels embedded in the membrane. Thus, the function of ion channels can be measured either by monitoring the movement of specific ions through these channels or by monitoring the changes in the membrane potential. Currently, there are three common approaches to assay ion channels: electrophysiological methods, ion flux measurement, and membrane potential measurement with sensor dyes.

9.2.1 Electrophysiological Methods

Membrane potential can be measured by placing an electrode inside the cells and another electrode outside the cells. In addition to membrane potential, the ion flux through the membrane can also be measured in this configuration by monitoring the electric current. The advantage of this method is that the membrane potential is under control. This control is especially important for voltage-gated channels because activation of these channels is highly sensitive to the voltage across the membrane. Many early ground-breaking studies of ion channels were performed by placing a tiny electrode made from glass capillaries inside a giant squid axon or muscle fiber. However, limitations in this technology caused it to be subsequently replaced by the patch-clamp technology. Patch clamp allows the measurement of small current (<1 pA) with submillisecond temporal resolution while the membrane potential is under full control. It is the primary tool in modern electrophysiological studies and is the gold standard in studying ion channel function and pharmacology.

9.2.2 Ion Flux Methods

Measurement of the ionic movement between the cytoplasmic and extracellular media offers an alternative way to study ion channels. This method focuses on the ionic species conducted by ion channels similar to the study of enzymes by studying their substrate turnover. In this method, changes in the concentration of a specific ion on either side of the plasma membrane before and after ion channel activation are measured. There are three common ways to measure the concentration of an ion. The first method uses a radioactive tracer ion, such as $^{86}Rb^{+}$ (to mimic K^{+}), $^{22}Na^{+}$, [^{14}C]guanidinium (to mimic Na^{+}), $^{45}Ca^{2+}$, and $^{36}Cl^{-}$. The tracer ions are preloaded inside or outside the cells. After channel activation, the media with tracer ions inside or outside of the cells are removed and counted for radioactivity. The extent of movement of the tracer ion from one side of the cell to the other is a function of the activity of the channels. The second method uses atomic absorption spectroscopy to measure the concentration of ions. Different atoms have distinct absorption spectra. A vaporized atom will absorb at several distinct wavelengths. The extent of absorption is proportional to the concentration of the atom that is vaporized. Quantitation of ionic species with atomic absorption spectroscopy avoids the use of radioactive substance. Both of the above-mentioned methods require the disruption of the assay system before signal detection. The third method to measure the concentration of a specific ion is to use a dye whose excitation/emission wavelength changes or fluorescent intensity changes upon selective binding to the ion. There exist specific dyes for both monovalent and divalent cations. However, except for calcium ions, the dyes for most other ions are not suitable for quantitation due to a combination of factors, such as the poor ion specificity of the dye, the small fluorescence change upon binding to ions, and the small concentration changes of ions upon cell excitation. There are several excellent dyes whose fluorescence spectra or intensities change upon binding to Ca^{2+}. Figure 9.3 shows the molecular structure of some of the commonly used calcium-sensing dyes. These dyes all contain a fluorophore and a calcium-binding motif (inside the boxes). Many calcium-sensing dyes are not toxic and can be preloaded into cells. With these dyes, the intracellular Ca^{2+} concentration can be conveniently monitored continuously in real time without the need for offline calcium concentration measurement.

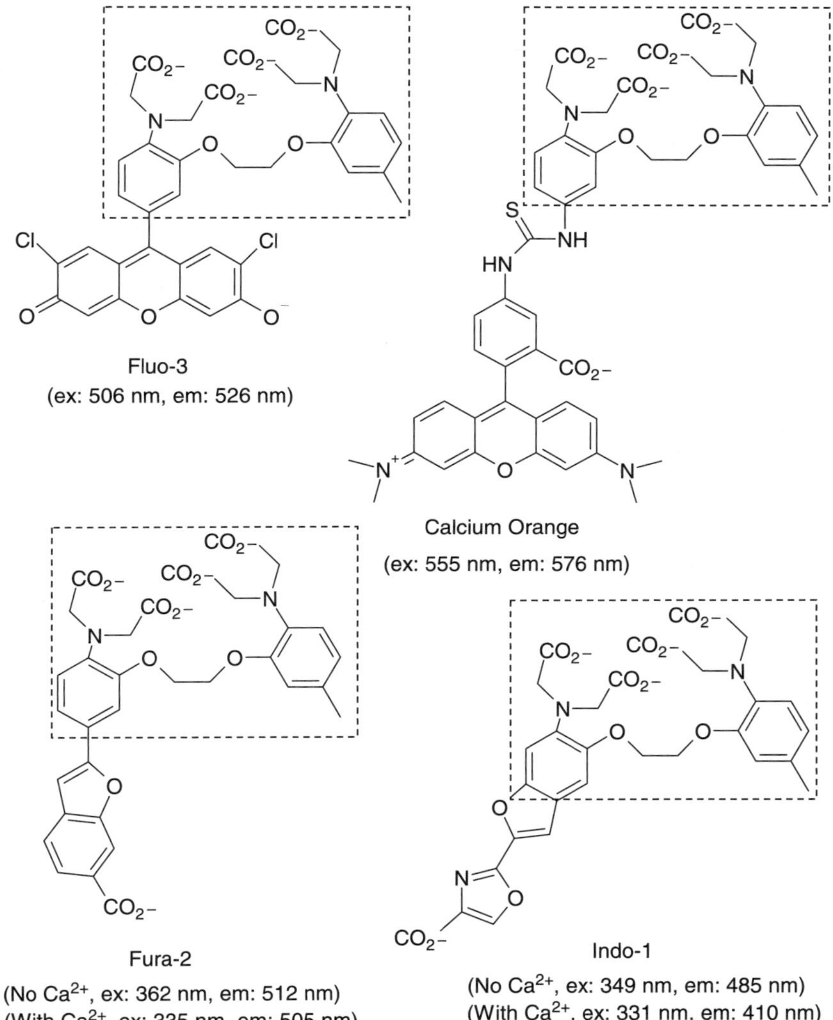

Figure 9.3 Structures and the excitation/emission maxima of some commonly used calcium-sensing dyes. The calcium binding motif is inside the boxes.

9.2.3 Membrane Potential Methods

Because the cell membrane potential changes when ion channels are activated, membrane potential measurement offers an alternative way to indirectly monitor the channel activity. There exist fluorescent voltage-sensitive dyes (VSDs) that can associate with membranes and change their fluorescence properties when membrane potential changes. The structures of some commonly used VSDs are shown in Figure 9.4. VSDs can be divided into two groups based on the speed of voltage sensing: fast VSDs and slow VSDs. The mechanisms of voltage sensing for a slow

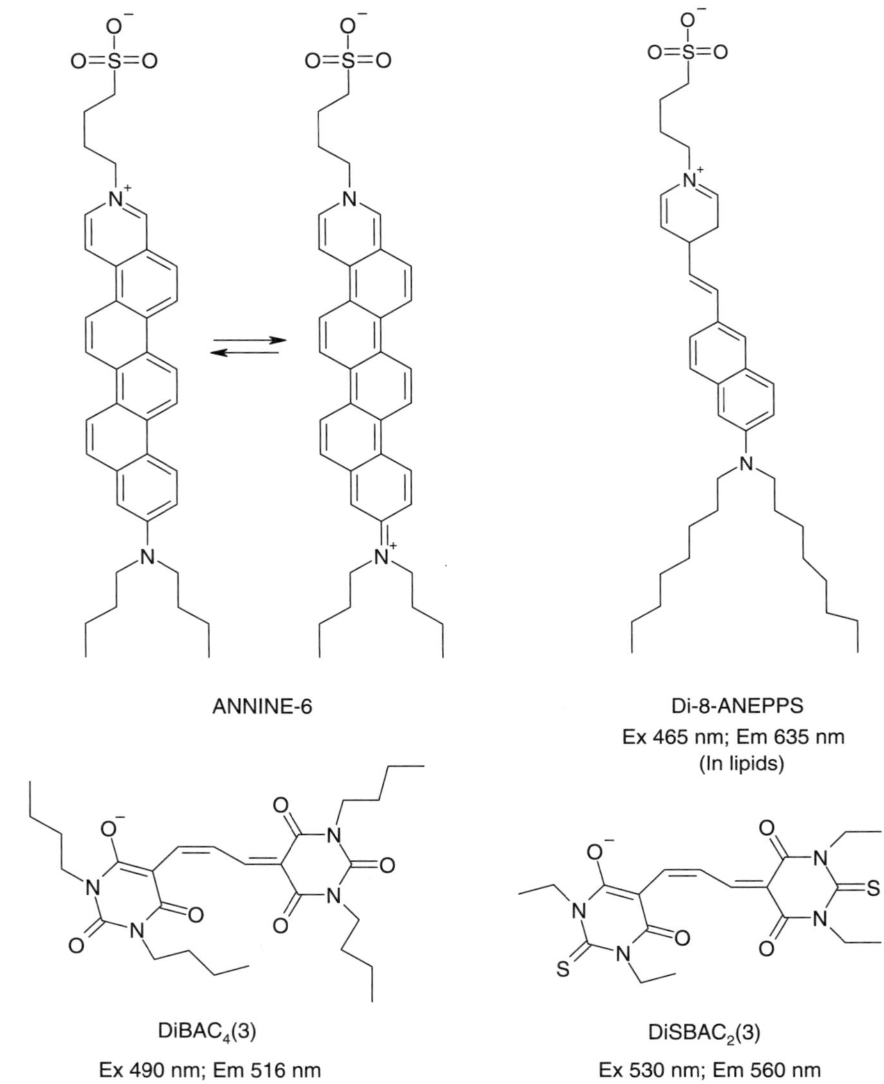

Figure 9.4 Structures and the excitation/emission maxima of some commonly used voltage-sensitive dyes.

VSD [e.g., DiBAC4(3)] and a fast VSD (e.g., Di-8-ANEPPS) are illustrated in Figure 9.5. Fast VSDs, such as Di-8-ANEPPS and ANNINE-6, partition into one leaflet of the plasma membrane where they are influenced by the membrane electric field. The fluorescence of fast VSDs are modulated by membrane potential through several mechanisms, such as potential-dependent spectral shifts, dipole reorientation, and change in the equilibrium between the monomer–dimer of the dyes. Fast VSDs have response times of milliseconds. They are used to detect transient membrane

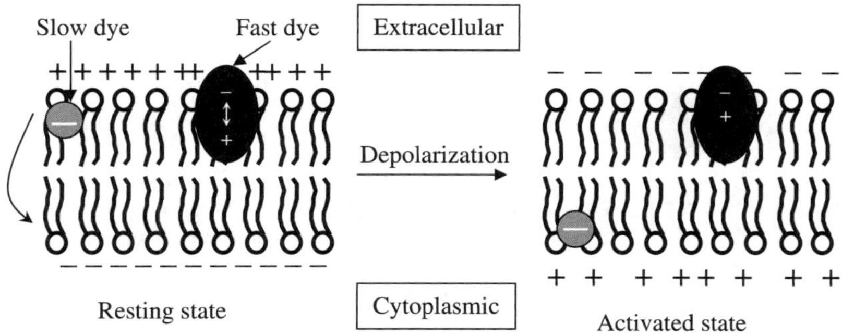

Figure 9.5 Mechanism of measuring membrane potential by Di-8-ANEPPS (fast VSD, black oval) and DiBAC4(3) (slow VSD, gray round). Di-8-ANEPPS inserted inside the membrane bilayer with its dipoles oriented along the electrical field in resting state. When the membrane is depolarized, the intramolecular charge separation is reduced leading to changes in its fluorescent intensity and emission spectra. The negatively charged hydrophobic DiBAC4(3) are associated with the outer leaflet of the membrane at resting state because of interaction with the positively charged surface. When the membrane is depolarized, DiBAC4(3) is attracted by the positive charge on the cytoplasmic side of the membrane surface and translocates to the inner leaflet of the membrane, causing redistribution of the dye between the two side of the membrane.

potential changes in excitable cells such as neurons and cardiac cells. However, the fluorescent intensity changes for fast VSDs are usually less than 0.2%/mV and are too small for routine assays. Slow VSDs, such as cationic carbocyanines, rhodamines, and anionic oxonols, change fluorescent intensity when changes in membrane potential trigger their redistribution between the two leaflets of the plasma membrane. As shown in Figure 9.5, DiBAC4(3) is initially associated with the outer leaflet of the membrane at the resting membrane potential because of its interaction with the positively charged outer surface. When the membrane is depolarized, the negatively charged DiBAC4(3) is attracted by the positive charge on the cytoplasmic side of the membrane and translocates to the inner leaflet of the membrane, causing redistribution of the dye between the two sides of the membrane. Depolarization exhibits enhanced fluorescence and a red spectral shift. The amplitude of fluorescence change for slow VSDs are severalfold larger than the fast response dye and can reach 1.5%/mV depending on the nature of the dye. Because the slow response time ranging from tens of milliseconds to tens of seconds, slow VSDs are only suitable for detecting changes in average membrane potentials of nonexcitable cells. Slow response dyes are commonly used in bioassays where the ion channels are expressed in nonexcitable cells such as CHO and HEK293.

9.3 ELECTROPHYSIOLOGICAL METHODS

9.3.1 Classic Patch-Clamp Methods

Early studies with ion channels were made possible by inserting a tiny electrode into giant axons and giant muscle fibers. These electrophysiological studies treat the lipid

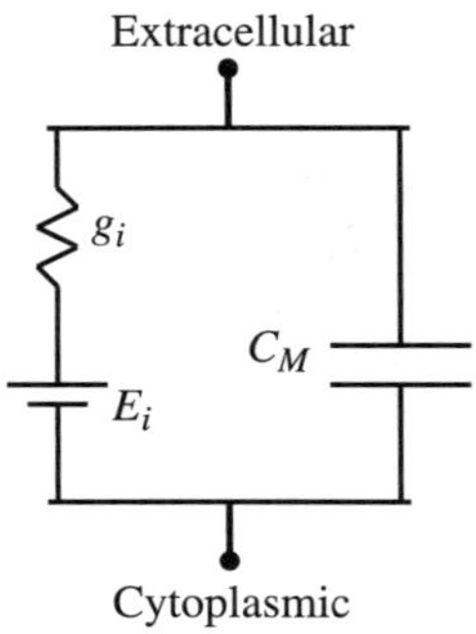

Figure 9.6 Representation of excitable membranes with an equivalent electrical circuit. The lipid bilayer of the membrane is treated as a capacitor (C_M), the ion channels embedded in the membrane are treated as conductors (g_i), and the concentration gradient of a specific ion is treated as a battery with electromotive force (E_i).

bilayers of the membrane as a capacitor (C_M) and the ion channels embedded in the membrane as conductors (g_i). The concentration gradient of a specific ion is treated as a battery that generated the electromotive force (E_i). The excitable membranes can be viewed as an electrical circuit as shown in Figure 9.6 where i can be any ionic species such as sodium, potassium, and chloride. When different channels permeable to specific ionic species coexist, they form parallel circuits with one capacitor and several conductors with electromotive forces. This simple model leads to most of the electrophysiological theories that are widely used to interpret the data from electrophysiological experiments. The voltage clamping is the most useful procedure in an electrophysiological experiment. With this procedure, the voltage of the membrane is first clamped at resting potential. The voltage is then rapidly changed to induce depolarization or hyperpolarization. The electrical current is measured in the millisecond to second scale in a series of different voltages. The information about the ion channels can be inferred by analyzing the time course and the pattern of the electric current. Figure 9.7 shows a schematic drawing of a typical voltage clamp experiment

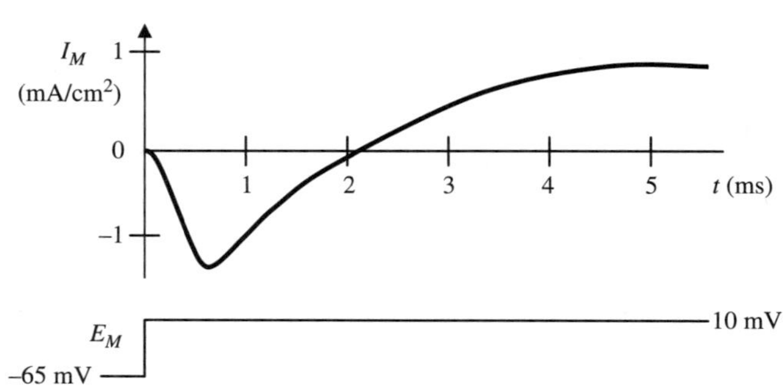

Figure 9.7 Schematic drawing of a typical voltage clamp experiment with the squid giant axon membrane. The membrane was first clamped at −65 mV, which was rapidly switched to 10 mV to induce depolarization. The electric current was monitored for ~5 ms. The inward current was assigned negative and the outward current was assigned positive. After depolarization, the inward current was first observed that peaks at about 0.7 ms. The inward current then became smaller and reached zero at about 2 ms. After this, the current is reversed and flowed outward and reached a plateau after 5 ms.

with the squid giant axon membrane. The membrane is first clamped at −65 mV and then rapidly switched to 10 mV to induce depolarization. The electric current is monitored for ~5 ms. The inward current is assigned to be negative and the outward current positive. After depolarization, the inward current is first observed, which reaches the maximum at about 0.7 ms. The inward current then becomes smaller and reaches zero at about 2 ms. After that, the current is reversed (moves outward) and plateaus after 5 ms. The biphasic pattern hints that the initial inward current is from sodium ions flowing into the cell through sodium channels, which is followed by the outward movement of potassium ions through potassium channel. At about 2 ms when the net current flow is zero, the influx of sodium ions into the cell and the efflux of potassium ions out of the cell are equal. This interpretation was further verified by performing the same experiment with selective sodium or potassium channel blockers and with selective removal of an ion (either Na^+ or K^+) from the test buffer.

Though early electrophysiological studies are very powerful, they have many limitations. The development of patch-clamp methods profoundly changed electrophysiology and they became the primary tools to study ion channels. Patch clamp is based on the observation that a glass pipette tip can fuse with the cell membrane to form a seal that is mechanically stable with high electric resistance (>1 GΩ). Because of the high resistant seal, the background current noise is very small in patch-clamp recording. Patch-clamp methods are the most sensitive means to measure ionic current flowing through the cell membrane while the voltage across the membrane is fully under control. Patch clamp can be configured in four different modes: on cell (or cell attached), whole cell, inside out, and outside out. Figure 9.8 shows the procedure to obtain different patch-clamp configurations. The electrode is first positioned to the cell surface under a microscope. After gentle suction, a gigaohm seal is formed and an on-cell patch is obtained. If an ion channel is in the patch surrounded by the glass electrode, single-channel activity can be measured. When the electrode is pulled from the cell after forming an on-cell patch, a piece of cell membrane is separated from the rest of the cell but is still attached to the electrode, forming an inside-out patch. Strong suction in the on-cell configuration will break the piece of cell membrane surrounded by the electrode, resulting in whole-cell patch clamp. Alternatively, pore-forming compounds, such as amphotericin B or nystatin, can be included in the pipette solution, which causes perforation of the membrane patch in the on-cell configuration. The perforated area of the cell membrane allows small molecules and ions, but not larger compounds, to cross the patch. The ion channel measurement in this mode is similar to the conventional whole-cell patch clamp in that the sum of all ion channel currents is measured. Pulling of electrode from the whole-cell configuration will carry a piece of cell membrane with the electrode tip that reseals to form an outside-out patch. The on-cell, inside-out and outside-out patch modes are often used for single-channel studies. The whole-cell patch-clamp mode is most commonly used in ion channel drug discovery.

Patch-clamp methods are extremely powerful in studying ion channel function and pharmacology. They enable analysis of ion channel function by direct measurement of ion current flowing through ion channels with full control of the membrane potential under physiological conditions. In addition, they offer the flexibility to

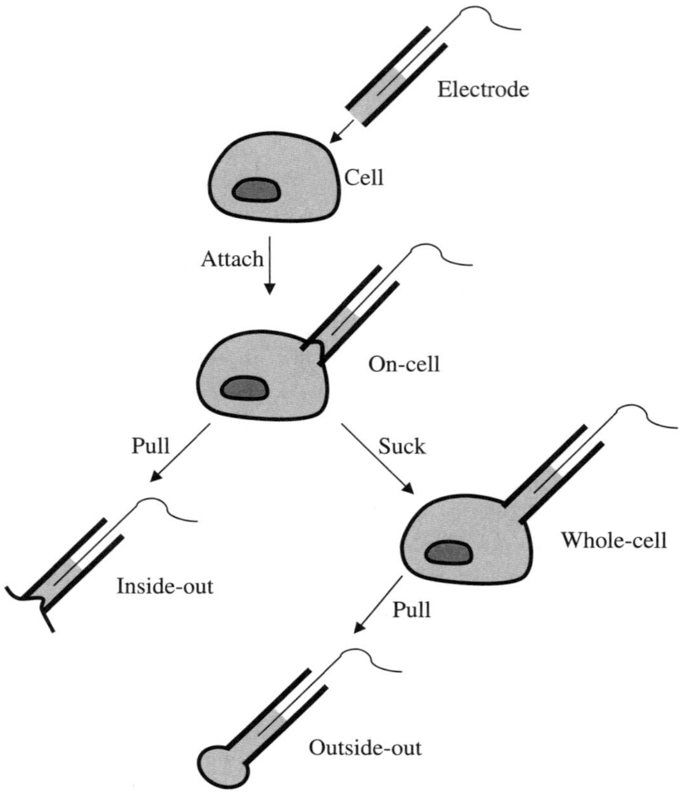

Figure 9.8 Procedures to obtain the four modes of patch clamp. The electrode is first positioned to the cell surface under a microscope. After gentle suction, the gigaohm seal is formed and an on-cell patch is obtained. When the electrode is pulled from the cell, a piece of cell membrane separates from the rest of the cell but is still attached to the electrode, forming an inside-out patch. Strong suction in the on-cell configuration breaks the piece of cell membrane interface with the electrode, resulting in the whole-cell patch-clamp configuration. Pulling of the electrode from the whole-cell mode carries a piece of cell membrane with the tip of the electrode that reseals to form an outside-out patch.

manipulate the solution composition on either side of the cell membrane. Patch-clamp methods are widely applicable to all channel types with high sensitivity enabling the detection of single-channel activity with fast temporal resolution in the micro- to millisecond time scales. The key to patch-clamp technology's sensitivity and accuracy is the seal formed between the glass and the membrane. To discriminate a real signal from background noise, the seal resistance must be high. The quality of the seal is measured in ohms of electrical resistance of the patch and good quality seals typically exceed 1 GΩ. The procedure to produce a good patch clamp is demanding and a good electrophysiologist may patch ~10 cells in a day. Patchclamp methods produce much fewer false-positive hits compared with VSD-based assays and flux assays. However, patch clamp is technically challenging, requires highly skilled electrophysiologists, and has very low throughput. The limited throughput of classic patch-clamp methods has significantly restricted their use to a few late-stage compounds in the drug discovery process.

9.3.2 Patch Clamping with Planar Electrodes

With high-throughput screening rapidly evolving and becoming an indispensable part of drug discovery, a lot of effort has been made to automate the patch-clamping process since the late 1990s. The initial efforts were focused on automating the traditional patch clamping, such as the Apatchi-1 system from Sophion Bioscience and AP1 AutoPatch from CeNeS Pharmaceuticals. However, these instruments offer little increase in throughput. Planar electrodes or patch-clamp chips made it possible to perform multiple automated recordings in parallel. These technologies typically employ a flat substrate with a small hole (1 to 2 μm in diameter) as replacement for the glass pipette tips (Fig. 9.9). The substrate separates the upper and the lower reservoirs, which are filled with extracellular and intracellular fluids, respectively. Cells are deposited in the upper reservoir, and a negative pressure is applied from the bottom reservoir to draw a cell tightly against the hole to form a seal. The result is essentially an upside-down version of the traditional whole-cell patch clamp. The flat substrate can be made on the bottom of a microplate and the test compounds are applied from the upper reservoir to multiple cells in parallel, significantly increasing the screening throughput. Several factors can affect the quality of assays with planar patch-clamp methods, such as the positioning of a cell over the hole on the substrate, adequate suction control, perfusion control of the test compounds, and seal resistance. A number of companies are engaged in developing planar chip patch-clamp technologies and several platforms are either commercially available or in development. Currently, the predominant instruments in the market are IonWorks and PatchXpress from Molecular Devices and QPatch from Sophion Bioscience.

IonWorks was originally developed at Essen Instruments and later acquired by Molecular Devices. It is the first marketed planar electrode instrument for ion

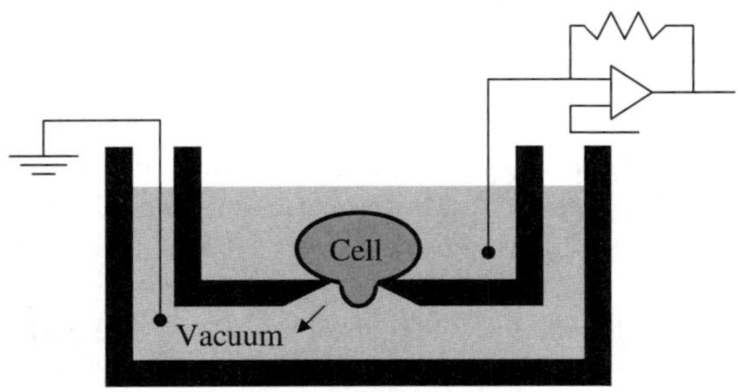

Figure 9.9 Ion channel assays with a planar electrode. The substrate separates the upper and the lower reservoirs that are filled with extracellular and intracellular fluid, respectively. Each of the reservoirs contains an electrode. Cells are deposited in the upper reservoir and a negative pressure is applied from the bottom to draw a cell tightly against the aperture forming a seal.

channel assays. IonWorks uses a 384-well PatchPlate and comes with 12 channel liquid dispensers. The original IonWorks HT uses microplate with only one hole per well. It can perform experiments with up to 3000 cells per day. While the success rate is respectable, many wells still fail to successfully patch a cell. In addition, variability from well to well (i.e., from cell to cell) is also relatively large. When only one cell is tested, biological variability among individual cells, such as cell health, cell size, and channel expression levels, can significantly affect the success rates and create variations in experimental results. The new IonWorks Quattro significantly increases the success rate by using a Population Patch Clamp (PPC) technology that records averaged ionic currents from a population of cells expressing a recombinant ion channel. With IonWorks Quattro, cells are plated on the substrate of a 384-well PatchPlate that contains 64 holes instead of just one per well so that the readout per well is the average current from all 64 cells. Because ionic currents are measured from a population of cells, the average current measured on the IonWorks Quattro system is more consistent from one well to the next as compared with IonWorks HT. For example, Molecular Devices reported that the CV for $K_v1.3$ currents is only 8% for IonWorks Quattro compared to 34% for IonWorks HT. A major drawback of the IonWorks is the low quality of the seal, typically ~100 MΩ. This makes IonWorks less sensitive than other patch-clamp methods that obtain gigaohm seals. Another drawback of IonWorks is that it is not fast enough to study fast ligand-gated channels that require fast ligand exchange and recording (typically within 20 ms). In IonWorks, the test compounds are added to the bulk solution, and it takes time for the compounds to defuse to the cells. In addition, the liquid handling head must be removed before the electrode can access the solution from the top of the plate, which creates a delay in recording.

PatchXpress 7000A was originally developed by Axon Instruments (now part of Molecular Devices) in collaboration with Aviva Biosciences. The instrument uses Aviva's SealChip 16, which is a chip with 16 wells and each well contains a hole made of glass substrate. Each well has independent pressure control, to increase the success rate of forming a gigaohm seal, and the test compounds are only introduced to wells that have a good seal. In comparison, IonWorks does not have this control, and test compounds are added to all wells including those that do not have a viable seal. PatchXpress also incorporates vessel wash, allowing faster perfusion of test compounds. Most of the bath solution can be removed before addition of test compounds. Thus, PatchXpress can handle fast ion channels.

Sophion Bioscience markets QPatch 16 and QPatch HT, which use planar chips with 16 and 48 wells, respectively. These instruments include individual low-noise patch-clamp amplifiers and pressure controllers. Independent pressure control on each individual cell greatly increases the success rate of forming a gigaohm seal. Sophion's instruments also incorporate microfluidics for rapid perfusion of test compounds that are driven by electroosmotic pumps. The laminar flow of solution over the cells allows for rapid exchange of solutions (~100 ms). QPatch is able to do simultaneous compound addition and recording as well as study both voltage-gated and ligand-gated channels. In addition, the microfluidic channels allow repeated perfusion of different solutions on the same cell. Thus, dose–response experiments of a test compound and screening of multiple compounds on the same cell is possible.

The microfluidic channel perfusion method requires only a small volume of solution (~2 to 5 μL) for each perfusion. The QPatch HT has eight pipettes attached to the fluid-handling robot, allowing up to 48 simultaneous patch-clamp experiments. In comparison, the QPatch 16 has four pipettes.

9.4 ION FLUX METHODS

9.4.1 Ion Flux Assays with Radioactive and Atomic Absorption Measurement

In these ion flux methods, a tracer ion is used and analysis is performed offline by either radioactivity measurement if the tracer ion is radioactive or by atomic absorption spectrum measurement if the tracer ion is nonradioactive. The cells expressing the ion channel or the vesicle containing the ion channel are used as the assay system. A tracer ion resembling the physiological ions permeable to the ion channel under study is employed. The concentrations of the tracer ion inside and outside of the cells or vesicles are measured before and after ion channel activation. The ion flux rate, the rate of the change of the tracer ion concentration inside or outside the cells or vesicles at a fixed time after ion channel activation, is directly proportional to the number of activated ion channels. Ion channel flux experiments can be configured in two ways, efflux and influx. An efflux experiment measures the rate of the tracer ion flowing out of a cell or vesicle. An influx experiment measures the rate of the tracer ion flowing into a cell or vesicle. In a typical efflux experiment, the tracer ion is first loaded into the cells or vesicles by co-incubation for a long period that allows the tracer ion to passively cross the membrane of the cells or vesicles to reach equilibrium. The tracer ion outside the cell or vesicle is removed just before the experiment by passing the co-incubation mixture through a size-exclusion column. The cells or vesicles loaded with the tracer ions are eluted out of the column first. The tracer-loaded cells or vesicles are then stimulated to open the ion channel, allowing the tracer ions to efflux. The cells or vesicles are then separated from the extracellular media by filtration. The concentration of the tracer ions remaining inside the cells or vesicles is analyzed. It is critical to include a control in the experiment that is subject to the same manipulation except for the activation of the ion channel.

A study of muscular nicotinic acetylcholine receptors (nAcChR) with the efflux of rubidium tracer ion is presented here to illustrate the ion flux method. The muscular nicotinic acetylcholine receptor (nAcChR) is a transmembrane protein with five subunits (2α, β, γ, δ). It is a ligand-gated channel with two asymmetric binding sites (one in each of the α subunits) for acetylcholine, the endogenous ligand. When the channel opens, it is nonselectively permeable to a variety of small cations, such as Na^+, K^+, Ca^{2+}, and Rb^+. Vesicles containing large quantities of nAcChR can be obtained from the electric organ of *Torpedo*. They are widely used as a model to study ligand-gated ion channels. Figure 9.10 shows the scheme for an efflux experiment. Rubidium ions are preloaded inside the vesicles containing nAcChR. After exposure to acetylcholine, some of the nAcChRs open and rubidium ions flux out

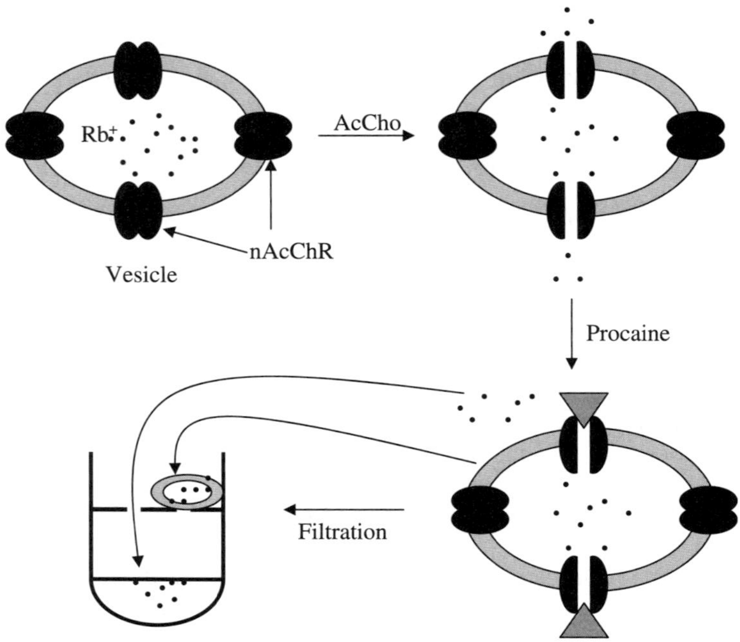

Figure 9.10 Illustration of ion flux experiment. Rubidium ions are preloaded inside the vesicles expressing nAcChRs on the surface. After exposure to acetylcholine, some of the nAcChRs open and rubidium ions flux out of the vesicle down their concentration gradient. The rubidium efflux is stopped by blocking the open nAcChRs. The mixture is then poured on to a filter that retained the vesicles on the membrane surface. The rubidium ions outside of the vesicle pass through the filter. The rubidium ions inside the vesicles are retained on the filter paper that can be further analyzed.

of the vesicle down their concentration gradient. The rubidium efflux is stopped by blocking the open nAcChRs. The solution is then poured on to a filter that retains the vesicles on the membrane surface together with the rubidium ions inside the vesicles. The rubidium ions outside of the vesicle pass through the filter. With radioactive tracer method, $^{86}Rb^+$ tracer ion is used in the above experiment. The filter paper or the filtrate is transferred to a vial containing liquid scintillant. The radioactivity is then counted with a liquid scintillation counter to analyze the amount of $^{86}Rb^+$. Because the nAcChRs are rapidly desensitized (within milliseconds) in the presence of acetylcholine, the flux experiment is best performed with the quenched-flow apparatus discussed in Chapter 3 (Fig. 3.4). In a quenched-flow experiment, nAcChR-expressing vesicles preloaded with $^{86}Rb^+$ are rapidly mixed with different concentrations of acetylcholine, and the mixture is incubated for a few milliseconds before quenching the open nAcChRs with procaine. The quenched solution is then transferred to a filtration station to separate the $^{86}Rb^+$ inside the vesicles from the $^{86}Rb^+$ outside of the vesicles. Because rubidium ions continuously leak out of the vesicle, it is very important to start the experiment immediately after the

$^{86}Rb^{+}$-loaded vesicles are separated from the rest of the media. Immediately separate the vesicle from the media after the quenching step is equally important.

The radioactive ion tracer methods discussed above can be performed similarly using atomic absorption when nonradioactive tracer, such as rubidium, is used. The tracer ion is analyzed with an atomic absorption spectrum instead of the scintillation counting. To obtain an atomic absorption spectrum, the samples have to be atomized by heating them to very high temperature. There are two methods for sample atomization: flame atomization and electrothermal atomization. In flame atomization, the sample is nebulized by a flow of gaseous oxidant (air or oxygen), mixed with a gaseous fuel (natural gas, hydrogen, or acetylene), and carried to a flame where atomization occurs. In electrothermal atomization, a few microliters of a sample are placed in a graphite tube that is electrically heated. The sample is first evaporated at a low temperature and then ashed at a higher temperature. After ashing, the temperature is rapidly increased to about 3000°C. Atomization of the sample occurs within a few seconds. Because the entire sample is atomized in a short period of time, relatively higher concentrations of the atoms are present in the optical path. This makes the electrothermal atomization several orders more sensitive than flame atomization. For example, the detection limits for Na are 2 ng/mL with flame atomization and 0.2 pg/mL with electrothermal atomization. However, the relative error associated with flame atomization is only about 1 to 2% while the relative error associated with electrothermal atomization is about a factor of 5 to 10 larger. Because of the relatively small error, flame atomization is more suitable for bioassays. Traditional atomic absorption spectrum with flame atomization only handles one sample at a time, which is not suitable for high-throughput applications. Aurora Biomed (Vancouver, B.C., Canada) markets an instrument (ICR12000) that can simultaneously read 12 samples and with throughput of more than 10,000 samples/day. The technology has been used in studying many potassium channels (Shaker, hERG, Kir, K_{ATP}, BK, and SK) and nAcChR.

The advantage of studying ion channel function using offline quantitation of tracer ions is the relative higher throughput compared with the patch-clamp method. The drawback of this method is that it requires two separation steps to obtain the loaded cells and to separate the tracer ions inside the cells from those outside the cells. With ion flux assays that do not employ quenched-flow technology, the assay is too slow to resolve fast ion channel kinetics. The signal-to-background ratio is strongly affected by incomplete removal of extracellular tracers after loading and the continuous leak of tracers out of cells during the experiment. Because of the lack of voltage control, the study of voltage-gated channels with this method has to rely on alternative means of channel activation, such as depolarization of the cell membrane with high concentrations of extracellular potassium or employing toxins to activate the channel or delay channel inactivation leading to an accumulation of channels in the open state. Discrepancies between the assay results obtained with the ion flux method and with the electrophysiological method are often attributed to the lack of voltage control and/or alternative activation methods used in flux assays. The original method of tracer quantitation relied on the radioactivity measurement, which is hazardous to both the environment and the operator. Quantitation of tracer ions with atomic absorption methods eliminated this problem.

9.4.2 Ion Flux Assay with Ion-Specific Indicator Dyes

In addition to radioactive and atomic absorption methods, some ions can be measured by fluorescence due to the existence of ion-specific indicator dyes that bind selectively to these ions. This method is widely used to detect the calcium ion flux because of the wide availability of several excellent calcium ion-specific dyes and the large range of intracellular calcium ion concentration changes when calcium-permeable ion channels are activated. With this method, the dye can be present in the assay solution, allowing real time measurement of the calcium concentration changes. The elimination of offline analysis also greatly increased the flexibility and quality of the assay. There are two types of calcium-sensing dyes: organic dyes and proteins dyes. Though this method is very successful with calcium ions, it has found only limited application with monovalent ions due to the lack of suitable selective dyes and the small changes in intracellular concentrations of monovalent cations and anions.

Organic Calcium Dyes The commonly used calcium-sensing organic dyes fall into two operational classes (see Fig. 9.3): single-wavelength dyes (i.e., Fluo-3, Fluo-4, and Calcium Orange) and dual-wavelength dyes (i.e., Fura-2, Fura-Red, and Indo-1). All the four dyes shown in Figure 9.3 have the same calcium binding motif. With single-wavelength dyes, the changes in calcium concentration result in changes in the fluorescence emission intensity without changing the excitation or emission maximum. Dual-wavelength dyes exhibit changes in either the excitation maximum or the emission maximum after binding with calcium. Fura-2 exhibits a shift in the excitation spectrum and Indo-1 exhibits a shift in the emission spectrum after Ca^{2+} binding. Because of these properties, the ratiometric measurements are used for dual-wavelength dyes to correlate with the concentration of calcium. Ratiometric measurements minimize the artifacts such as variations in dye loading and time-dependent concentration change. The advantages of organic dyes are their brightness, high specificity, large dynamic range, and relative insensitivity to temperature, buffer composition, and pH. When nontoxic organic calcium-sensing dyes are preloaded inside the cells, the calcium concentration changes inside the cells can be monitored in real time.

In a typical calcium flux experiment with organic dyes, the adherent cells are first plated on a microplate and allowed to grow to about 80 to 90% confluence. The acetoxymethyl (AM) form of a calcium dye is dissolved in dry DMSO and mixed with a solution of 15 to 20% Fluronic F-127 in dry DMSO. The mixture is then dispersed into the aqueous loading medium that is then added to the cells to allow the calcium dye to diffuse into the cells. The AM group is cleaved inside the cells, leaving the negatively charged calcium dye inside the cells. The negatively charged dyes cannot diffuse out of the cells due to their charges. However, some cells can actively extrude organic anions such as the calcium-sensing dyes. Lowering the temperature can significantly reduce the rate of dye extrusion. Inhibitors for the anion-exchange protein, such as probenecid and sulfinpyrazone, can also be used to inhibit the extrusion of the calcium-sensing dyes. Care must be taken when using high concentrations of inhibitors because they may induce cell stress such as blebbing. The dye in the extracellular media is washed away before experiment. The microplate is then placed in a fluorimeter with online liquid handling capability where the fluorescence is read immediately after the cells are treated with reagents that induce calcium flux. Calcium flux assays were greatly facilitated by the FLIPR (Fluorometric Imaging Plate Reader) family of instruments and associated kit reagents from Molecular Devices. The FLIPR family of instruments can handle microplates with different formats up to 1536 wells. Molecular Devices also markets no-wash kits for calcium ion studies that contain quencher solutions to quench the dyes outside the cells. The quenchers are not permeable to the cell membrane and remain outside the cells. Because the washing step is the major cause of the variation between different wells due to inconsistent removal of dyes and disturbance to the cells, experimental results with no wash kits have much smaller variations between wells. One potential concern is that the proprietary no-wash quenchers may change the behavior of the cells and the ion channels. The FLIPR instruments have an online liquid handling module that dispense reagents from the top of the microplate. The fluorescence signal from the whole plate is read from the bottom of the microplate continuously with a temporal resolution on the order of seconds. A typical calcium flux curve is shown in Figure 9.11. Because the time course of calcium concentration changes are collected, the calcium flux can be analyzed in several ways, such as maximum calcium signal, the slope of the calcium signal rise, or the area under the curve (integration of total amount of calcium).

Protein Calcium Sensors Aequorin is the most widely used protein calcium sensor. Aequorin is composed of two components, the apoprotein apoaequorin (MW ~22 kDa) and the prosthetic group coelenterazine (MW 472). The two components form functional aequorin spontaneously in the presence of oxygen. When aequorin binds to Ca^{2+}, it undergoes a conformational change and converts

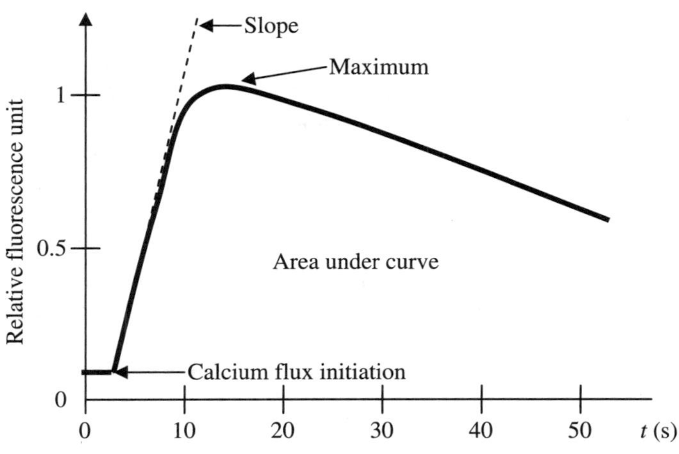

Figure 9.11 Typical calcium flux pattern. The baseline is prerecorded before the addition of reagents to initiate calcium flux. After addition of the reagents, the calcium concentration in the cytosol rapidly increases and reaches the maximum in a few seconds. The calcium concentration then decreases (maybe gradually or rapidly depending on the cells).

coelenterazine into excited coelenteramide and CO_2 through oxidation. As the excited coelenteramide relaxes to the ground state, blue light (469 nm) is emitted. Because the aequorin method detects chemiluminescence, the noise is very low compared with the fluorescence detection. This method also has a large dynamic range from under 0.1 μM up to 100 μM Ca^{2+}. In comparison, the organic calcium dyes normally have a dynamic range about 100-fold or less. For bioassay applications, the apoaequorin gene is stably transfected in target cells (i.e., CHO and HEK293) containing the calcium channels to be assayed. Because aequorin is expressed at specific cellular sites, usually mitochondria, it is less disturbing to the cells and does not have the dye leakage issue associated with organic dyes. The affinity of aequorin to Ca^{2+} is about 10 μM, about 5-fold lower than organic dyes. Lower affinity for Ca^{2+} produces less buffering effect (the reduction of Ca^{2+} concentration due to the depletion of Ca^{2+} by the Ca^{2+}-sensing dyes). The major disadvantage of the aequorin-based method is the low quantum yield. Highly sensitive instruments are required to detect the luminescent signal. Dedicated instruments, such as Hamamatsu's FDSS systems, PerkinElmer's LumiLux, and Molecular Devices's FLIPR^{tetra}, are required for high-throughput applications. In addition to luminescence measurements, FDSS and FLIPR^{tetra} also have added fluorescence modules that can detect organic calcium dyes as well.

In addition to aequorin, there are a few other proteins that have been used in calcium sensing. PerkinElmer's PhotoScreen (Photina) product line is a chimeric photoprotein generated from obelin and clytin. It was originally developed at Axxam. The mechanisms to generate luminescence are similar to aequorin: Coelenterazine is oxidized to excited coelenteramide, which relaxes to the ground state and emits blue light at 469 nm. While the mechanism of action of Photina is similar to other photoproteins, the total light released is greater and the reaction kinetics is slower. The Photina platform includes parental cell lines expressing mitochondrially targeted apophotoprotein and double transfected cell lines expressing both mitochondrially targeted apophotoprotein and a recombinant calcium-coupled ion channel. Invitrogen's Premo Cameleon calcium sensor is based on FRET between two GFP proteins connected by calmodulin and a calmodulin binding peptide. One of the GFP variant is a cyan fluorescent protein (CFP) with emission maximum at 535 nm and the other is a yellow fluorescent protein (YFP) with emission maximum at 485 nm. In the absence of Ca, the protein complex adopts a conformation that prohibits FRET. The binding of Ca^{2+} ions to the calmodulin-M13 moiety induces a conformational change that brings the CFP and YFP domains closer, allowing FRET to occur. A major advantage of this assay is the ratiometric measurement that significantly reduces assay variations due to compound or cellular autofluorescence, nonuniform cell plating, differences in expression levels among cells, instability of instrument illumination, and changes in illumination pathlength.

Sensor Dyes for Other Ions Changes in the intracellular concentration of monovalent ions, such as K^+, Na^+, and Cl^-, are not dramatic during cellular activation. Therefore, directly monitoring these ions during cell activation will generate a very small signal over a large background. It is preferable to use a tracer ion that resembles the ion to be monitored and selectively place the tracer ion in either side of the plasma membrane before cell activation. After channel activation, the concentration of the tracer ion on either side is analyzed. Up to now, this method has not been successful in detecting most monovalent ions except for thallium for which the high-throughput assay has been reported and there is a kit available in the market. The FluxOR thallium detection kits from Molecular Probes (now part of Invitrogen) utilize a thallium ion (Tl^+) specific fluorescent dye to detect thallium ion flux in live cells. It is well known that potassium channels are permeable to thallium ions. The dye is loaded inside the cells to be assayed. Thallium is added to the extracellular solution. When potassium channels open, an increase in fluorescence is detected as thallium ions flow down their concentration gradient into the cell. The FluxOR thallium detection assay has been demonstrated with CHO and HEK293 cells stably expressing the hERG channel.

9.5 MEMBRANE POTENTIAL SENSING METHODS

Since the fast VSDs only generate small fluorescence changes when the membrane potential changes ($<0.2\%$ per 1 mV), they are not suitable for robust assays for ion channels. The slow VSDs, such as oxonols, are commonly used in assays that measure membrane potentials. It must be noted that the relationship between the membrane potential and the ion channel activity is not linear since the gating of the ion channel is voltage dependent. In addition, the temporal resolution with slow VSDs reflects the movement of the dye rather than the gating kinetics of the ion channel. The loading characteristics, response times to voltage changes, and the types of applications among different VSDs are different even within the same family of VSDs such as oxonols. Most ion channel screening applications have been conducted with DiSBAC2(3) because it is more water soluble, sensitive, stable, and easier to load to the cell. The time response of DiSBAC2(3) is ~500 ms. In contrast, the more hydrophobic DiSBAC4(3) responds to membrane potential changes in ~20 ms. Because of its hydrophobicity, DiSBAC4(3) requires Pluronic®-127 surfactant for cellular loading and a washing step to remove excess dye. DiSBAC4(3) is useful for applications that require faster response times or no added dye in the extracellular solution. The performance of membrane potential measurements with slow VSDs is improved to some degree with the commercial kits that optimize the measurement of the dye between the two leaflets of the membrane bilayers. FLIPR and VIPR are the two most commonly used commercially available membrane potential assays with proprietary technologies or reagents. Though these VSD-based indirect ion channel measurements have low signal-to-background ratio and produce very high rates of false hits, they remain one of the major methods in primary screening of ion channels in the absence of alternative high-quality, high-throughput method.

Voltage sensor probes marketed by Invitogen was originally developed by Aurora Biosciences (merged with Vertex Pharmaceuticals). A voltage ion probe reader (VIPR) is required to perform the assay. The technology is based on FRET and uses a pair of dyes that are composed of a voltage-sensitive oxonol [(DiSBAC2(3) or DiSBAC4(3)] and a fluorescent, membrane-bound coumarin-modified dimyristoyl-phosphatidyl-ethanolamine (CC2-DMPE). The mechanism of the assay is illustrated in Figure 9.12. When the two dyes are loaded into the cellular

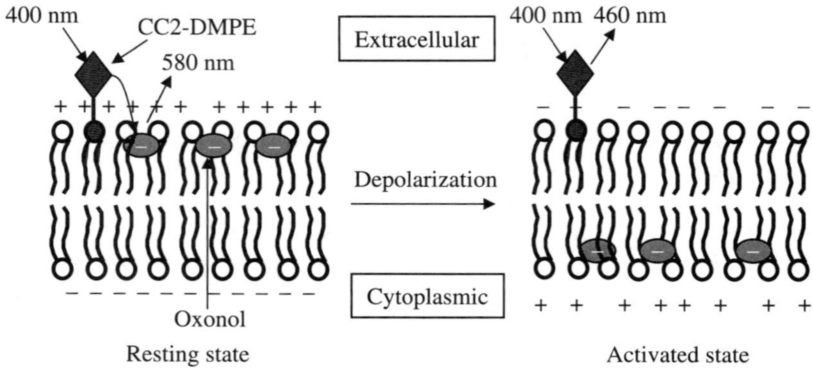

Figure 9.12 Membrane potential assay with VIPR. A voltage-sensitive acceptor oxonol and a fluorescent, membrane-bound donor coumarin-modified dimyristoyl-phosphatidylethanolamine (CC2-DMPE) are loaded into the cellular plasma membrane. The CC2-DMPE donor stays on the outer monolayer of the cellular membrane. The oxonol initially binds to the outer face of the plasma membrane but will diffuse to the inner side of the membrane in response to changes in the membrane potential to establish a new equilibrium. When the CC2-DMPE and oxonol dyes are on the same side of the membrane, excitation of CC2-DMPE at 400 nm will result in oxonol emission at 580 nm due to FRET. When the oxonol moves to the opposite side of the membrane, a decrease in emission at 580 nm is observed.

plasma membrane, the CC2-DMPE donor stays in the outer monolayer of the membrane while the oxonol dye can be associated with either side of the membrane, depending on the membrane potential. The oxonol initially binds to the outer face of the plasma membrane after loading but will diffuse to the inner side of the membrane in response to changes in the membrane potential to establish a new equilibrium. When the CC2-DMPE and oxonol dyes are on the same side of the membrane, excitation of CC2-DMPE at 400 nm will result in oxonol emission at 580 nm due to FRET. When the oxonol moved to the opposite side of the membrane, a decrease in emission at 580 nm is observed. This technology offers the potential of screening more than 32,000 compounds a day. Because the FRET-based measurement is restricted to the plasma membrane, it is more relevant than non-FRET (single-dye) based approaches that measure signals throughout the cell. When carrying out experiments with this technology, the optimal CC2-DMPE and DiSBAC2(3) loading concentrations fall between 0.5 and 20 μM for most cells. The optimal DiSBAC4(3) loading concentrations is between 2 and 3 μM.

The FLIPR membrane potential assay kit contains a proprietary (oxonol-based) voltage-sensing dye and a quencher. The mechanism of the assay is illustrated in Figure 9.13. The quencher is impermeable to the cell membrane. After loading the VSD into the cells, the VSD initially stays in the outer leaflet of the membrane bilayer due to its negative charge. After channel activation, the membrane potential changes, resulting in the redistribution of the VSD between the two sides of the cell membrane. The VSD on the outer side of the membrane is quenched by the quencher while the VSD on the inner side of the membrane is protected from quenching. Thus, channel activation results in an increase in fluorescence. This kit generates a large fluorescence

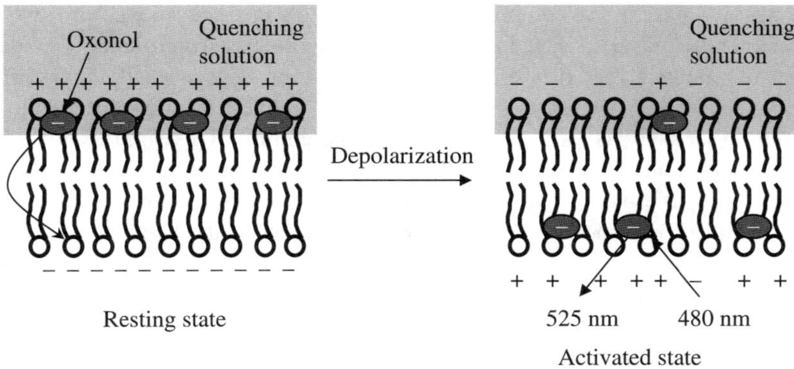

Figure 9.13 Membrane potential assay with FLIPR. The proprietary oxonol dyes are loosely attracted to the outer membrane surface in the resting state due to electrostatic interaction. A solution containing a membrane-impermeable quencher is added extracellularly that quenches all the oxonal molecules located on the extracellular surface. When the membrane is activated and depolarized, the negatively charged dyes are attracted by the positive charge inside the membrane and move inside where they are not accessible by the quencher. These dyes will emit fluorescence at 525 nm when excited at 480 nm.

response upon ion channel activation (as high as 1.5% per 1 mV). However, the quencher molecule, whose composition and structure are not disclosed, could have adverse pharmacological effects on the ion channel under study.

9.6 SELECTING SUITABLE ASSAYS FOR ION CHANNEL STUDIES

We have discussed patch-clamp, ion flux, and the voltage-sensing dye methods for measuring ion channel activities. For an assay developer, choosing an assay for a particular ion channel depends on many factors. Patch clamp is the only method that provides good sensitivity, temporal resolution, and physiological relevance. For high-quality studies with a handful of compounds, conventional and planar patch-clamp methods are the preferred choices. Conventional patch clamp is cheaper to set up. Planar patch clamp is new and requires a significant initial investment. However, the throughput could be potentially increased by up to a few hundred-fold over conventional patch clamp. Even with this increased throughput, planar patch clamp still cannot compare with the throughput of true high-throughput screening methods. To screen a large compound library, the ion flux or voltage-sensing dye method are usually used. Among different flux methods, assays with calcium-sensing dye is the most robust and is easier to perform than the methods that require offline analysis of tracer ions, such as radioactive analysis and the atomic absorption analysis. Voltage-sensing dye methods offer very high throughput. However, the quality of the assay usually is very poor. For example, the assay of $K_V1.5$ with VIPR only gives a signal-to-background ratio of about 2. Coupled with the low signal-to-noise ratio of the assay, screening of the compound library with this method produces a significant number of false positives and false negatives. In situations with small changes of membrane potential, the change of the dye distribution is very small and is difficult to measure with voltage-sensing dyes. The major problem of high-throughput flux and voltage-sensing assays is the lack of voltage control. This significantly affects the assays for voltage-gated channels where nonphysiological channel activation methods, such as toxins or high concentration of potassium, have to be used. In addition to the concerns of choosing appropriate assay methods, artificially created cell lines expressing the ion channel under study can lead to questionable assay results. For example, the resting membrane potential in neurons is about −70 mM whereas most cell lines heterologously expressing cloned channels have a more positive resting potential, typically around −20 mV. This may result in large discrepancies between native systems and artificial systems when studying voltage-gated channels.

Useful Websites and Vendors

http://www.iuphar.org/

http://www.aurorabiomed.com/

http://www.sophion.dk/

http://www.moleculardevices.com/pages/instruments/electrophys_main.html

http://www.axxam.com/

http://www.euroscreen.com/

http://jp.hamamatsu.com/sp/sys/fdss7_e.html

http://www.invitrogen.com/site/us/en/home/Products-and-Services/Applications/Cell-and-Tissue-Analysis/Cellular-Imaging/Cell-Based-Reporter-Assays/Promo_Cameleon_Calcium_Sensor.html

http://www.avivabio.com/

BIBLIOGRAPHY

Baker, B. J., et al. (2005) Imaging brain activity with voltage- and calcium-sensitive dyes. *Cell. Mol. Neurobiol.* **25**, 245–282.

Borst, J. W., et al. (2008) Structural changes of yellow cameleon domains observed by quantitative FRET analysis and polarized fluorescence correlation spectroscopy. *Biophys. J.* **95**, 5399–5411.

Bovolenta, S., Foti, M., Lohmer, S., and Corazza, S. (2007) Development of a Ca^{2+}-activated photoprotein, Photina(R), and its application to high-throughput screening. *J. Biomol. Screen.* **12**, 694–704.

Bullen, A. and Saggau, P. (1999) High-speed, random-access fluorescence microscopy: II. Fast quantitative measurements with voltage-sensitive dyes. *Biophys. J.* **76**, 2272–2287.

Conn, P. J. and Pin, J.-P. (1997) Pharmacology and functions of metabotropic glutamate receptors. *Annu. Rev. Pharmacol. Toxicol.* **37**, 205–237.

Dabrowski, M. A., et al. (2008) Ion channel screening technology. *CNS Neurol. Disord. Drug Targets* **7**, 122–128.

Doyle, D. A., et al. (1998) The structure of the potassium channel: Molecular basis of K^+ conduction and selectivity. *Science* **280**, 69–77.

Dunlop, J., Bowlby, M., Peri, R., Vasilyev, D., and Arias, R. (2008) High-throughput electrophysiology: An emerging paradigm for ion-channel screening and physiology. *Nat. Rev. Drug Discov.* **358**, 358–368.

Dutzler, R., Campbell, E. B., Cadene, M., Chait, B. T., and MacKinnon, R. (2002) X-ray structure of a ClC chloride channel at 3.0 Å reveals the molecular basis of anion selectivity. *Nature* **415**, 287–294.

Falconer, M., et al. (2002) High-throughput screening for ion channel modulators. *J. Biomol. Screen.* **7**, 460–465.

Finkel, A., et al. (2006) Population patch clamp improves data consistency and success rates in the measurement of ionic currents. *J. Biomol. Screen.* **11**, 488–496.

Frey, W., et al. (2006) Plasma membrane voltage changes during nanosecond pulsed electric field exposure. *Biophys. J.* **90**, 3608–3615.

Gill, S., et al. (2003) Flux assays in high throughput screening of ion channels in drug discovery. *Assay and Drug Dev. Technol.* **1**, 709–717.

Hille, B. (2001) *Ion Channels of Excitable Membranes*, 3rd ed. Sinauer Associates, Sunderland.

Jiang, Y., et al. (2003) X-ray structure of a voltage-dependent K^+ channel. *Nature* **423**, 33–41.

Kandel, E. R., Schwartz, J. H., and Jessell, T. M. (2000) *Principles of Neural Science*, 4th ed. McGraw-Hill Medical, New York.

Lü, Q. and An, W. F. (2008) Impact of novel screening technologies on ion channel drug discovery. *Comb. Chem. High Throughput Screen.* **11**, 185–194.

Mayer, M. L. and Armstrong, N. (2004) Structure and function of glutamate receptor ion channels. *Annu. Rev. Physiol.* **66**, 161–181.

Miller, C. (2000) An overview of the potassium channel family. *Genome Biol.* **1**, 1–5.

Molokanova, E. and Savchenko, A. (2008) Bright future of optical assays for ion channel drug discovery. *Drug Discov. Today* **13**, 14–22.

Morimoto, T., et al. (2007) Voltage-sensitive oxonol dyes are novel large-conductance $Ca2^+$-activated K^+ channel activators selective for beta1 and beta4 but not for beta2 subunits. *Mol. Pharmacol.* **71**, 1075–1088.

Palmer, A. E. and Tsien, R. Y. (2006) Measuring calcium signaling using genetically targetable fluorescent indicators. *Nat. Protocols* **1**, 1057–1065.

Park, K.-S., Yang, J.-W., Seikel, E., and Trimmer, J. S. (2008) Potassium channel phosphorylation in excitable cells: Providing dynamic functional variability to a diverse family of ion channels. *Physiology* **23**, 49–57.

Poul, E. L., et al. (2002) Adaptation of aequorin functional assay to high throughput screening. *J. Biomol. Screen.* **7**, 57–65.

Priest, B. T., Swensen, A. M., and McManus, O. B. (2007) Automated electrophysiology in drug discovery. *Curr. Pharm. Des.* **13**, 2325–2337.

Schapira, A. H. V., et al. (2006) Novel pharmacological targets for the treatment of Parkinson's disease. *Nat. Rev. Drug Discov.* **5**, 845–854.

Siegel, G. J., Albers, R. W., Brady, S., Price, D. L., and Neurochemistry, A. S. f. (eds.) (2005) *Basic Neurochemistry: Molecular, Cellular and Medical Aspects*, 7th ed. Academic, New York.

Sjulson, L. and Miesenbock, G. (2007) Optical recording of action potentials and other discrete physiological events: A perspective from signal detection theory. *Physiology* **22**, 47–55.

Solly, K., et al. (2008) Miniaturization and HTS of a FRET-based membrane potential assay for Kir channel inhibitors. *Assay Drug Dev. Technol.* **6**, 225–234.

Squire, L. R., Bloom, F. E., and Spitzer, N. C. (eds.) (2008) *Fundamental Neuroscience*, 3rd ed. Academic, New York.

Tang, W., et al. (2001) Development and evaluation of high throughput functional assay methods for hERG potassium channel. *J. Biomol. Screen.* **6**, 325–331.

Terstappen, G. C. (2004) Nonradioactive rubidium ion efflux assay and its applications in drug discovery and development. *Assay Drug Dev. Technol.* **2**, 553–559.

Trivedi, S., et al. (2008) Cellular HTS assays for pharmacological characterization of NaV1.7 modulators. *Assay Drug Dev. Technol.* **6**, 167–179.

Wang, X. and Li, M. (2003) Automated electrophysiology: High throughput of art. *Assay Drug Dev. Technol.* **1**, 695–708.

Weaver, C. D., Harden, D., Dworetzky, S. I., Robertson, B., and Knox, R. J. (2004) A thallium-sensitive, fluorescence-based assay for detecting and characterizing potassium channel modulators in mammalian cells. *J. Biomol. Screen.* **9**, 671–677.

Wolff, C., Fuks, B., and Chatelain, P. (2003) Comparative study of membrane potential-sensitive fluorescent probes and their use in ion channel screening assays. *J. Biomol. Screen.* **8**, 533–543.

Wu, G., Raines, D. E., and Miller, K. W. (1994a) A hydrophobic inhibitor of the nicotinic acetylcholine receptor acts on the resting state. *Biochemistry* **33**, 15375–15381.

Wu, G., Tonner, P., and Miller, K. (1994b) Ethanol stabilizes the open channel state of the Torpedo nicotinic acetylcholine receptor. *Mol. Pharmacol.* **45**, 102–108.

Xu, J., et al. (2003) A benchmark study with SealChip planar patch-clamp technology. *Assay Drug Dev. Technol.* **1**, 675–684.

Yellen, G. (2002) The voltage-gated potassium channels and their relatives. *Nature* **419**, 35–42.

Zheng, W., Spencer, R. H., and Kiss, L. (2004) High throughput assay technologies for ion channel drug discovery. *Assay Drug Dev. Technol.* **2**, 543–552.

CHAPTER 10

ASSAYS WITH GPCRs

10.1 INTRODUCTION TO GPCRs AND G PROTEINS

G Protein-Coupled Receptors are transmembrane proteins that can interact with G proteins and possess seven sequences with each of them containing 25 to 35 consecutive hydrophobic residues. They are found only in eukaryotes, including yeast, plants, choanoflagellates, and animals. Analysis of the complete human genome reveals that there are approximately 800 genes encoding GPCRs. Among them, approximately 400 are nonolfactory GPCRs. With the exception of slightly over 100 orphan GPCRs for which the ligands are unidentified, the ligands for most of the nonolfactory GPCRs are known. GPCRs are the receptors for hormones, growth factors, and other endogenous ligands. They are involved in a wide variety of physiological processes such as the visual sensing, smell sensing, behavioral and mood regulation, regulation of immune system activity and inflammation, autonomic nervous system transmission, and cell density sensing. A diversity of molecules, including ions, small volatile organic odorants, biogenic amines, amino acids, lipids, nucleotides, peptides, and proteins are found to be ligands of GPCRs. Several ligands for GPCR are top selling drugs. Malfunction of GPCR signaling pathways are involved in many diseases, such as diabetes, blindness, allergies, depression, cardiovascular defects, and certain forms of cancer. It is estimated that more than half of the modern drugs' cellular targets are GPCRs. This makes GPCRs one of the most pursued targets in drug discovery and in assay development.

GPCRs are integral membrane proteins that possess seven transmembrane helices (Fig. 10.1). The structure of human GPCRs was not resolved until 2007. However, the 7-helix structural model for GPCRs was proposed long ago based on a variety of biophysical and biological measurements. The seven transmembrane helices are numbered from helix 1 to helix 7 starting from the N-terminus to the C-terminus. The N-terminus resides on the extracellular side of the membrane, and C-terminus resides on the cytoplasmic side of the membrane. The seven helices are linked by three extracellular loops and three intracellular loops. The extracellular loops contain two highly conserved cysteine residues that form disulfide bonds to stabilize the receptor structure. The extracellular parts of the receptor can also be glycosylated. The structure of bacteriorhodopsin, which is weakly analogous to mammalian GPCRs, was determined by both electron diffraction (1996) and X-ray crystallography (1997). The first crystal structure of a mammalian GPCR, bovine

Assay Development: Fundamentals and Practices. By Ge Wu

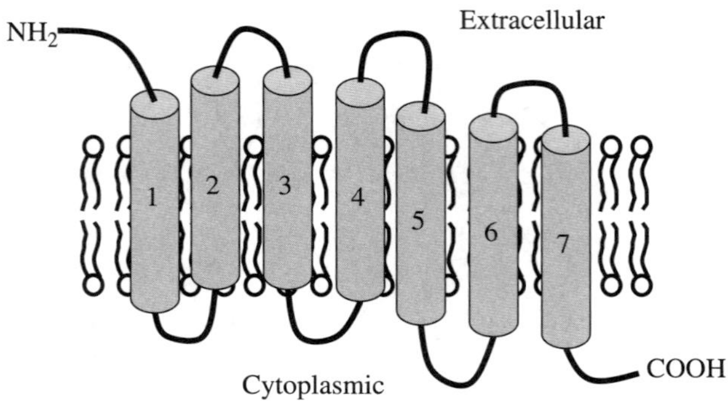

Figure 10.1 Structure of GPCRs. There are seven transmembrane helices that are numbered from helix 1 to helix 7 starting from N-terminal to C-terminal. The N-terminus resides on the extracellular side of the membrane, and C-terminus resides on the cytoplasmic side of the membrane. The seven helices are linked by three extracellular loops and three intracellular loops.

rhodopsin, was solved in 2000. It was found that the relative orientation of the helices in rhodopsin differs significantly from that of bacteriorhodopsin, though the seven transmembrane helices feature is conserved. When the structure of the first human GPCR, β_2-adrenergic receptor, was resolved, it was found to be highly similar to the bovine rhodopsin in terms of the relative orientation of the seven-transmembrane helices. However the conformation of the second extracellular loop that constitutes the "lid" covering the top of the ligand binding site is entirely different between the two structures. Though still controversial, there is evidence that some GPCRs can form homo- and/or heterodimers and possibly more complex oligomeric structures. It was found that heterodimerization is essential for the function of some GPCRs, such as metabotropic $GABA_B$ receptors and purinergic receptors.

The currently widely used classification of the GPCR superfamily was originally proposed by Kolakowski in 1994, who grouped GPCRs into six classes (A to F) based on their sequence homology and functional similarity. This system was adopted by the International Union of Basic and Clinical Pharmacology Committee on Receptor Nomenclature and Drug Classification (NC-IUPHAR) with some modifications (see Table 10.1). Based on phylogenetic criteria, the human GPCRs can be divided into five families; glutamate, rhodopsin, adhesion, frizzled, and secretin. The rhodopsin family (class A) is the largest family of GPCRs with more than 670 members. Each of the other families of GPCRs contains less than 35 members in the family. Most of the drugs targeting GPCRs interact with the rhodopsin family of GPCRs with the rest interacting with the secretin and glutamate families of GPCRs.

TABLE 10.1 Classification of GPCRs

Class A (or 1)	Rhodopsin-like
Class B (or 2)	Secretin receptor family
Class C (or 3)	Metabotropic glutamate
Class D (or 4)	Fungal mating pheromone receptors
Class E (or 5)	Cyclic AMP receptors
Class F (or 6)	Frizzled/smoothened

To understand the function and mechanism of GPCRs, it is essential to understand G proteins. G proteins are loosely associated with the cytoplasmic side of the cell membrane and function as "molecular switches" by alternating between an inactive GDP-bound state and an active GTP-bound state. There are two distinct families of G proteins, the "large" G proteins and the "small" G proteins. They belong to the larger group of enzymes called GTPases that can catalyze the hydrolysis of GTP. The large G proteins are associated with GPCRs and are heterotrimers containing G_α (38 to 52 kDa), G_β (35 kDa), and G_γ (8 to 10 kDa) subunits. The G_α subunit possesses the GTPase activity and can bind to both GTP and GDP. G_α is the main molecule that initiates downstream signal transduction. There exist approximately 16 members of G_α proteins that are classified into four main families: G_s, $G_{i/o}$, $G_{q/11}$, and $G_{12/13}$. They share a similar mechanism of activation but differ primarily in the recognition of downstream effector proteins. The G_α subunit is hydrophilic. Though some G_α subunits are covalently modified by the myristoyl group, the association of G_α with the cytoplasmic side of the membrane is primarily due to its association with the $G_\beta G_\gamma$ complex, which is anchored to the cytoplasmic side of the membrane. The G_β and G_γ subunits are always associated. There is evidence that they may initiate downstream signal transduction as well. The monomeric small G proteins (20 to 25 kDa) belong to the Ras superfamily of small GTPases, which are involved in the RTK signaling pathway. The small G proteins are homologous to the G_α subunit of heterotrimeric G proteins. In order to associate with the cytoplasmic side of the plasma membrane, many G proteins are covalently modified with lipid. The large heterotrimeric G proteins may be myristoylated, palmitoylated, or prenylated. The small monometric G proteins may be prenylated.

10.2 G PROTEIN-COUPLED RECEPTOR ACTIVATION AND SIGNAL TRANSDUCTION

The typical GPCR activation and signal transduction processes are illustrated in Figure 10.2. The GPCR activation starts with ligand binding. The ligand for a GPCR may bind to the receptor either in the extracellular domain or in the transmembrane region. In comparison, ligands with other types of receptors, such as RTKs, typically bind to extracellular domains. Small-molecule ligands for class A GPCRs typically bind within the transmembrane domain, whereas the ligands for class B and class C GPCRs bind to the extracellular domain. The detailed mechanism of GPCR activation by ligand has not been clearly defined. The induced-fit theory proposes that the ligand binding to GPCR induces a conformational change in the receptor, leading to receptor activation. A competing theory proposes that the GPCRs can preexist in a conformational equilibrium between the active and the inactive states. Binding of ligand to the GPCR shifts the equilibrium toward the active receptor states. Ligand binding may also induce GPCR to form homo- and heterodimmers, similar to the ligand-induced RTK dimerization. The activation of GPCR results in their association with inactive G protein heterotrimers (with the GDP-bound G_α subunit), which are loosely associated with the inner side of the cytoplasmic membrane. The conformational changes in the GPCR after ligand binding also induce the receptor

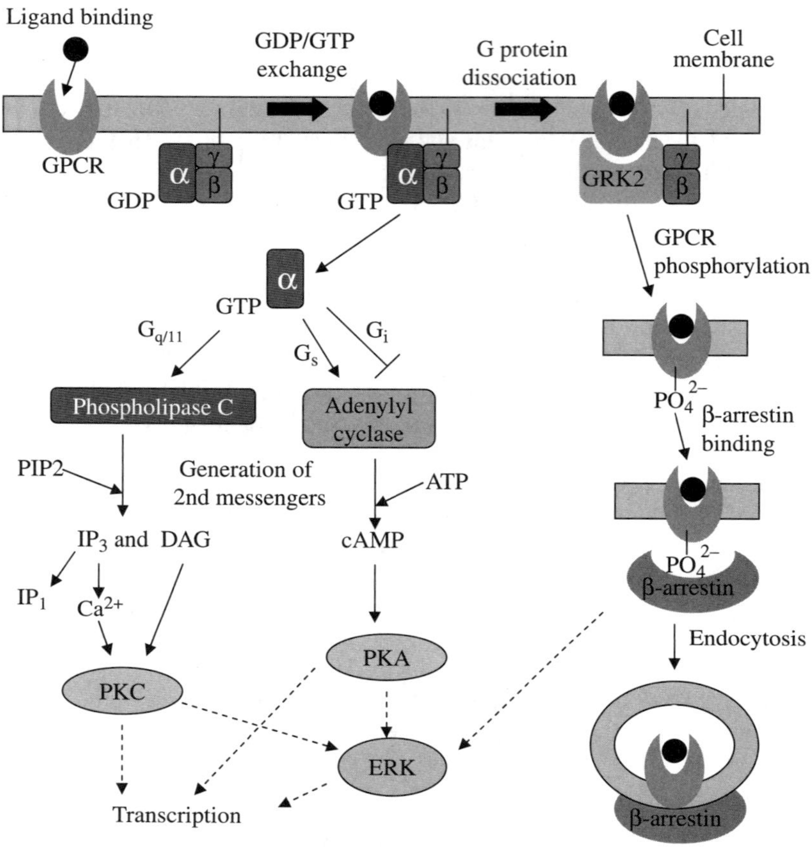

Figure 10.2 Signal transduction pathways of GPCRs. (See color insert.)

to function as a guanine nucleotide exchange factor (GEF) that facilitates the exchanges of bound GDP with GTP on the G_α subunit of the G proteins. This exchange in turn triggers the dissociation of the GTP-bound G_α subunit from the $G_{\beta\gamma}$ dimer and the GPCR. Both G_α-GTP and $G_{\beta\gamma}$ can then interact with different effector proteins, which lead to activation of different signaling cascades. At the site of GPCR activation on the cytoplasmic side of the membrane, G protein receptor kinases (GRKs) is recruited that phosphorylate the ligand-bound GPCR, leading to receptor desensitization. In response to GPCR activation and phosphorylation by GPCR kinases, β-arrestins move from the cytosol toward the cytoplasmic side of the membrane to bind the phosphorylated GPCR. The binding of β-arrestin targets the GPCRs for internalization via clathrin-coated pits and vesicles. The GPCR is then either degraded or recycled back to the plasma membrane. The binding of β-arrestin to GPCR also decouples the interaction between the GPCR and G proteins, providing a way to terminate the GPCR signaling. In fact, the desensitization of the majority of GPCRs is mediated by the binding of β-arrestins to the phosphorylated

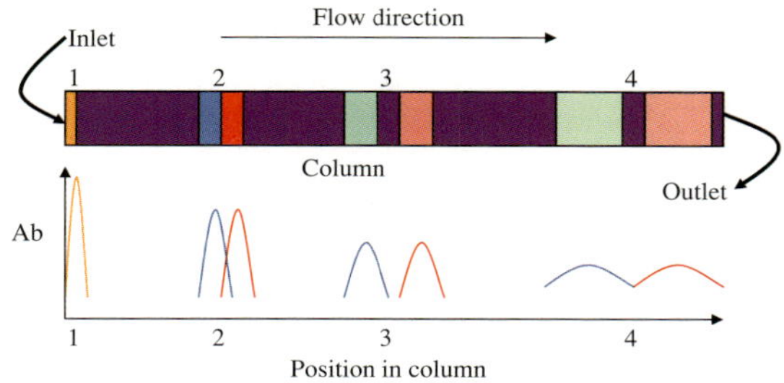

Figure 4.1 For caption see page 111.

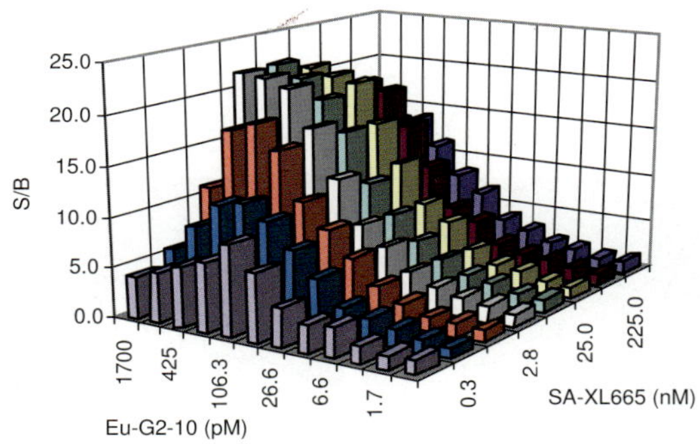

Figure 6.9 For caption see page 175.

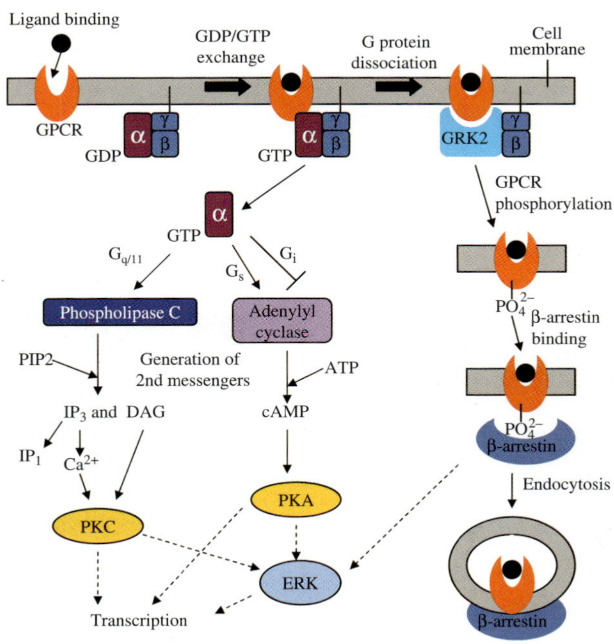

Figure 10.2 For caption see page 268.

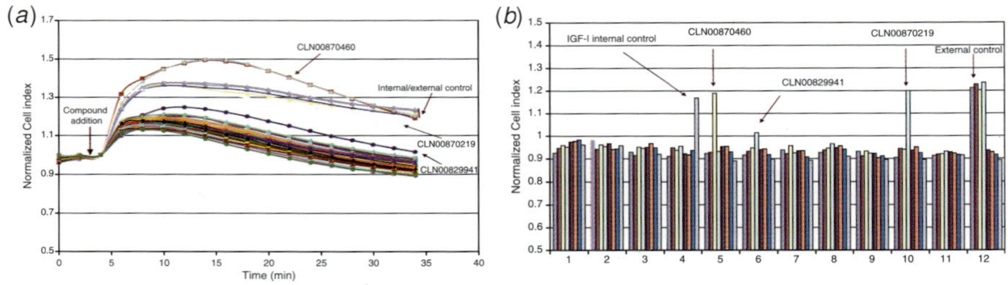

Figure 11.6 For caption see page 300.

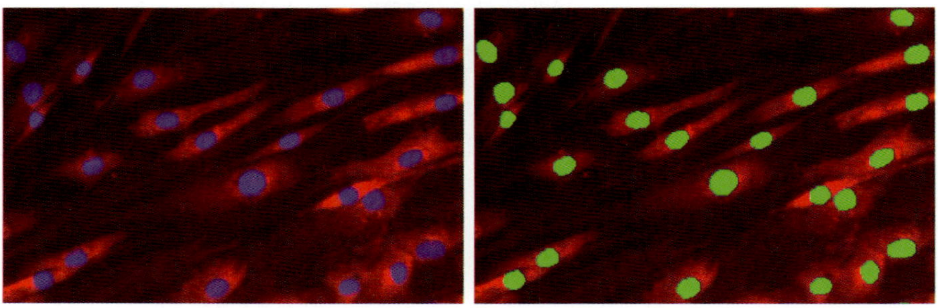

Figure 12.1 For caption see page 310.

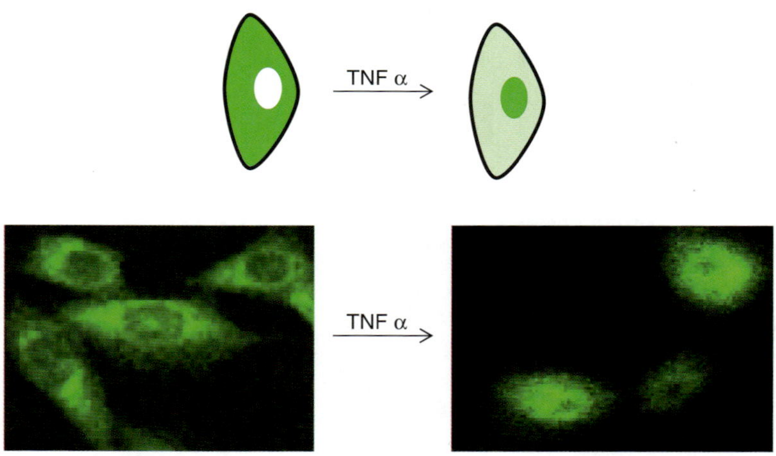

Figure 12.4 For caption see page 319.

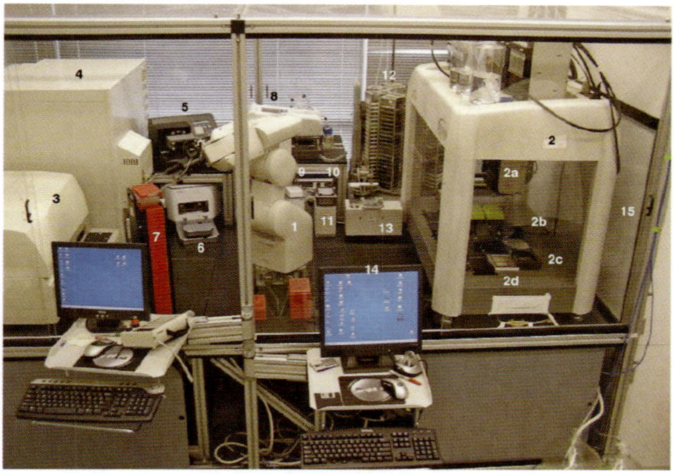

Figure 13.2 For caption see page 242.

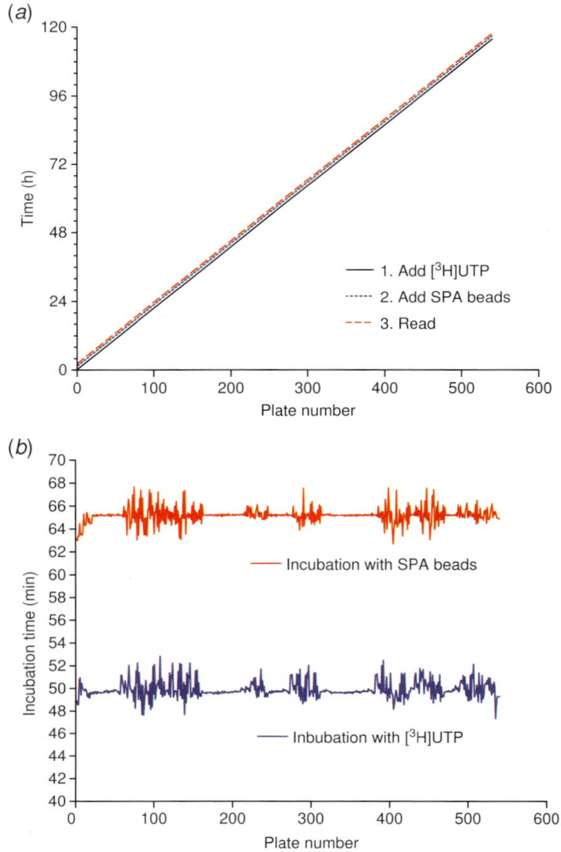

Figure 13.5 For caption see page 349.

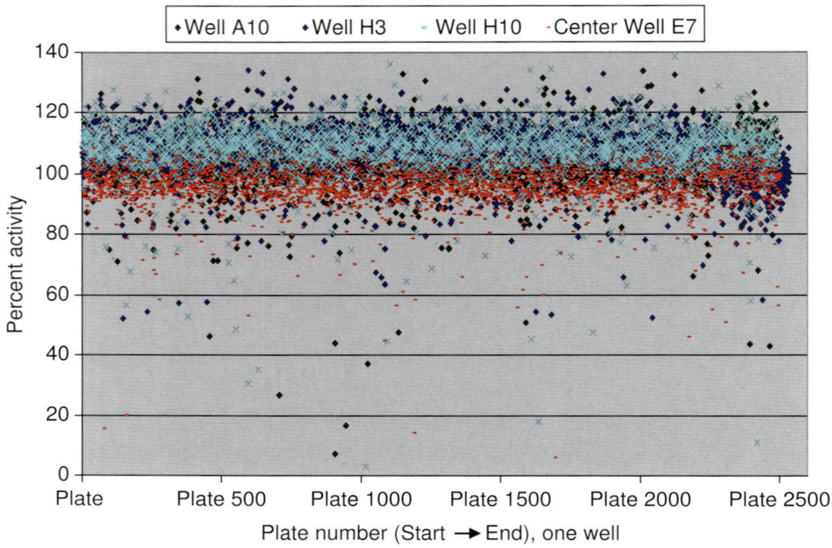

Figure 13.8 For caption see page 365.

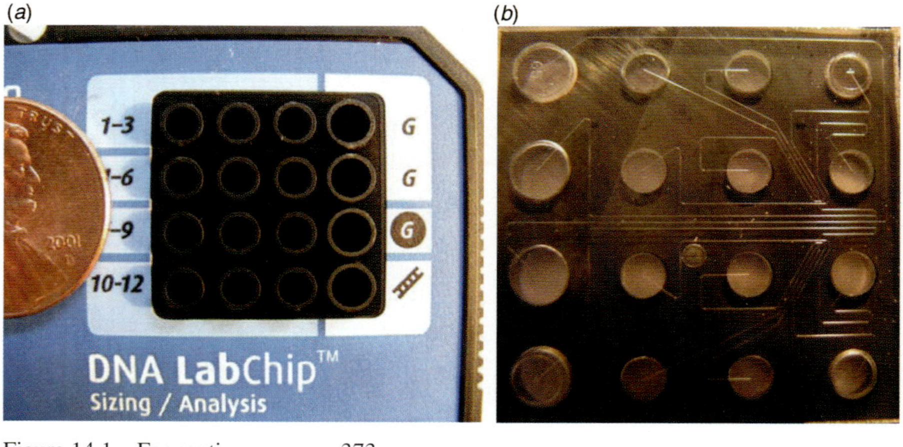

Figure 14.1 For caption see page 373.

GPCR. In addition to terminating GPCR signaling, β-arrestins also act as adapter proteins linking GPCR activation to the MAPK pathways.

Let us turn back to the activated GTP-bound G_α as a result of GPCR activation. The dissociated active G_α subunit can interact with different effector proteins depending on the nature of the G_α molecules. Both G_s and $G_{i/o}$ interact with adenylyl cyclases while $G_{q/11}$ interacts with phospholipase C_β (PLC_β). G_s stimulates adenylyl cyclases while $G_{i/o}$ inhibits them. The adenylyl cyclases catalyze the production of a second-messenger, 3′,5′-cyclic adenosine monophosphate (cAMP), through the hydrolysis of ATP. cAMP further activates protein kinase A (PKA). PKA then phosphorylates a myriad of downstream targets. $G_{q/11}$ activates membrane-bound phospholipase C_β that then cleaves phosphatidylinositol (4,5)-bisphosphate (PIP_2) to produce two second messengers, 1,4,5-inositol triphosphate (IP_3) and diacylglycerol (DAG). IP_3 is a signaling molecule that causes the opening of calcium channels in the endoplasmic reticulum, resulting in an increase in intracellular calcium ion concentration. IP_3 is not stable in a cell (with a half-life in seconds) and is further broken down to IP_1. The calcium ion is another second messenger that causes many downstream signal transduction events, including the activation of calmodulin (CaM). $G_{12/13}$ is involved in Rho family GTPase signaling and controls cell cytoskeleton remodeling, which in turn regulates cell migration. Though most downstream signal transductions are transmitted by the G_α subunit, $G_{\beta\gamma}$ sometimes may also show active functions (e.g., coupling to L-type calcium channels or activation of PI-3 kinase leading to MAPK activation). GPCRs are also known to activate MAPKs through other mechanisms including PKA-dependent phosphorylation of small G proteins (Rap1), PKC-dependent phosphorylation of small G proteins (Raf), and transactivation of RTKs (EGFRs). The activation of PKA, PKC, MAPK, and CaM resulting from GPCR activation can lead to the phosphorylation of several transcription factors (e.g., CREB, ELK-1, cFOS/cJUN, and NFAT) that regulate gene transcriptions. The G_α subunit will eventually hydrolyze the attached GTP to GDP by its inherent enzymatic activity, allowing it to reassociate with $G_{\beta\gamma}$ and starting a new cycle.

10.3 STRATEGIES OF GPCR ASSAY DEVELOPMENT

In the context of assay for GPCRs, several aspects of the GPCR signaling processes described above can be exploited in various assay formats. Many components involved in the GPCR activation pathways, from initial receptor binding all the way downstream to gene transcription and even whole-cell responses, offer potential read-out to assay GPCRs. Because of the demand for GPCR assays and the many different pathways involved in GPCR signaling, GPCRs offer the best example to showcase the power of modern cell-based assay technologies. Early assays for GPCR were mostly performed in radioligand competition format. However, many of the disadvantages of radioligand binding assays, which were discussed in Chapter 5, have limited their use in modern GPCR assays. Numerous technologies have been developed that offer functional assays for GPCRs over the years. Functional GPCR assays are currently the major format performed with GPCRs. Because of the large number of GPCR assays with a variety of technologies, it is difficult to classify the GPCR assays using the same methods as with the other assay targets employed throughout this book. Instead, we will discuss GPCR assays following the natural order of the assay targets along the signal transduction pathways from upstream to downstream. Since we already covered receptor binding assays in Chapter 5, ligand binding assays with GPCRs will not be discussed here. Though there are several assays for GPCR oligomerization with cell lines expressing heterologous GPCRs, especially in FRET formats, the GPCR oligomerization assays are not widely adopted in drug discovery and there is still controversy as to whether oligomerization happens in the native cells. We will not discuss oligomerization assays here. If interested, several references in this field are listed in the Bibliography at end of this chapter. Similar to RTK assay format through quantitation of phosphorylated RTKs after their activation, there are some GPCR assays to quantify the phosphorylation of GPCRs at C-terminals on the cytoplasmic side following GPCR activation. However, this method will not be discussed here due to the limited adoption in GPCR assays. We will start the discussion of GPCR assays with the readout that measures the association of radioactive GTP with G proteins following GPCR activation. This is one of the early GPCR assays, but it is still widely used today. The second messengers, such as cAMP, inositol phosphates (IP_1 and IP_3), and Ca^{2+}, are the most widely used readouts in modern GPCR assays due to the availability of many commercial kits. In general, cAMP assays are applied to GPCRs coupled to $G_{i/o}$ or G_s, whereas IP_3 formation and Ca^{2+} mobilization assays are applied to GPCRs coupled to $G_{q/11}$. Recently, MAPK activity readout has emerged to assay GPCR activation. Since GPCR activation results in increased transcription, reporter gene assays have long been a workhorse for GPCR assays, and there are many commercial kits available for reporter gene-based GPCR assays. The broad adoption of automated cell-imaging analysis (discussed in Chapter 12) in recent years also opened the door for image-based GPCR assays that can quantify the extent of β-arrestin-associated GPCR internalization. Further, electro label-free assay that measures the global cellular interaction with the cell matrix offers another assay platform for GPCR activation (will be discussed in Chapter 11).

10.4 G PROTEIN-COUPLED RECEPTOR ASSAYS BY MEASURING THE EXTENT OF GTP BINDING TO G_α

The G_α subunit of the heterotrimeric G proteins binds to GDP at the resting state. After GPCR activation, GDP is replaced by GTP. Thus, the extent of GTP bound to G_α is a measure of the GPCR activation. However, GTP itself cannot be used as a tracer since G_α is also a GTPase and will hydrolyze GTP. An ideal tracer would be a molecule that resembles GTP in terms of G_α binding and cannot be hydrolyzed by G_α. Guanosine 5′-γ-thiophosphate (GTPγS) is a GTP analog with the substitution of the oxygen by a sulfur atom at the γ-position of GTP. The rate of GTPγS hydrolysis catalyzed by GTPase is remarkably slower, making it an excellent tracer for the measurement of G_α binding. The radioactive [^{35}S]GTPγS is one of the most used GTP analogs for studying GPCR activation. The G_α-bound [^{35}S]GTPγS can be analyzed by scintillation counting of the radioactive ^{35}S. In theory, a competitive radiolabel [^{35}S]GTPγS assay can be configured with a cell line heterologously overexpressing the GPCR under study. However, the concentration of GPCR and G_α is not high enough for the detection of bound [^{35}S]GTPγS over baseline. Instead, the membrane preparation of the cell line overexpressing the GPCR under study is commonly used in this assay. To further increase the signal and the sensitivity of the assay, membrane preparation from cells expressing GPCR/G_α fusion protein can be used. The interaction between GPCR and G_α in GPCR/G_α fusion proteins is more efficient than that between independent GPCR and G_α. Because GTPγS binding assay depends on the exchange between GTP and GDP, it is applicable to studying GPCRs that have fast GDP/GTP exchange rates. Thus, this assay is ideal in studying the $G_{i/o}$ coupled GPCRs with high GDP/GTP exchange rate, but may not be able to detect signals from GPCRs that are coupled to G_s and $G_{q/11}$ with slow GDP/GTP exchange rates.

In a typical filtration-based GTP binding assay, the GPCR and G_α are co-expressed in a mammalian cell line (e.g., CHO and HEK293). The cells are then homogenized to make membrane preparations. Approximately 10 pmol receptor/mg of membrane protein in the final membrane preparation usually give a good assay signal. In a typical 96-well plate assay format, 10 μg/well of membrane proteins are used. The GTPγS binding to G_α is strongly affected by the presence of ADP and Mg^{2+}. Thus, an appropriate amount of ADP (usually 1 to 40 μM) must be added to the assay buffer to increase the signal-to-background ratio of the assay. The optimal concentration of Mg^{2+} in the assay buffer is about 10 mM. For some membranes, mild detergent (such as saponin) is required in the buffer to facilitate the exchange between GDP/GTP. After incubating membranes with the ligand and all other assay components for an appropriate time (which should be determined experimentally, but 30 min to 1 h is usually good enough), the assay mixture is filtered through a glass membrane. The radioactivity on the glass filter is then counted with liquid scintillation counting.

There are several variations of the GTPγS binding assays. One variation is the use of SPA technology that eliminates the filtration step involved in the filtration-based assays. With homogeneous SPA format, the same radioactive [^{35}S]GTPγS tracer is used to bind to G proteins associated with the cell membranes. The membrane containing the radiolabeled G_α is captured with wheat germ agglutinin (WGA)-coated SPA beads. This brings the radioactive [^{35}S]GTPγS in close proximity to the bead. The decay of the radioactive isotope stimulates the scintillant, resulting in an SPA signal. In a typical 96-well format, the SPA assays are performed in a 200 μL reaction volume with 10 μg membrane protein per well. The assay buffer contains 20 mM HEPES, pH 7.4, 100 mM sodium chloride, 10 mM magnesium chloride with 0.3 nM [^{35}S]GTPγS and 10 to 50 μM GDP. After incubation with the ligand for about 30 min, the plates containing the reaction mixture and SPA beads are centrifuged at ~1500 rpm for 5 min. The SPA signal is then counted with a Topcount or a MicroBeta microplate scintillation counter.

To eliminate the use of radioactivity, the GTP-G_α binding assay for GPCR can also be configured with DELFIA (PerkinElmer) technology, which uses europium-labeled GTPγS (Eu-GTPγS) as the tracer. The basic steps are similar to the methods discussed above except for the tracer substitution. This technology still requires the filtration step to separate the G_α-bound tracer from the tracer in the solution. The proprietary tracer used in this assay is autofluorescent and no enhancement reagent is needed.

An important advantage of GTPγS binding assay using membrane preparation is that false-positive reactions are rare, whereas other cell-based GPCR assays may be affected by false-positive responses. Different from receptor ligand binding assays, the GTPγS binding assay is a functional assay that can distinguish between agonism and antagonism, which cannot be done with ligand binding assays. When mammalian systems are not required in the assay, a large amount of membrane expressing the GPCR-G_α fusion protein can be easily prepared by using the baculovirus-sf9 systems. The disadvantage of GTPγS binding assay using membrane preparation is that the sensitivity is not as high as that of the cell-based assays. In addition, GTPγS binding assay may not work with some G_s- and $G_{q/11}$-coupled GPCRs with slow GDP/GTP exchange rates.

10.5 G PROTEIN-COUPLED RECEPTOR ASSAYS BASED ON MEASUREMENT OF cAMP

Because cAMP is the second messenger downstream of the G_s- and $G_{i/o}$-coupled receptors, the change in intracellular cAMP concentration after ligand binding offers a way to assay the G_s- and $G_{i/o}$-coupled receptors. As with all signaling molecules, the levels of intracellular cAMP are tightly regulated. The production of cAMP is controlled by adenylyl cyclases, while the degradation of cAMP is controlled by the cAMP phosphodiesterases (PDE), which catalyze the hydrolysis of the 3′-ester bond of cAMP to form 5′-adenosine monophosphate (AMP). Both adenylyl cyclases and PDEs are in turn regulated by the concentration of cAMP through a negative-feedback system. There are many commercially available kits for a cAMP assay. The majority of them are based on the measurement of the displacement of a tracer cAMP by cellular cAMP that is released into the assay media after the cells are lysed. There are very few assays that can directly measure cAMP levels in live cells. In a typical 96-well format assay, a few hundred thousand cells expressing GPCRs may be used. In typical nonstimulated cells, the total cAMP released is about 1 pmol or less after lysing 100,000 cells. If the total assay volume is 100 μL, there will be less than 1 nM cAMP in the assay solution, which is approximately at the detection limits of most cAMP assays (0.1 to 1 nM). When assaying the cAMP level involved in G_s pathways, the number of cells in the assay is a major factor that affects the assay sensitivity. Cell numbers and assay volumes should be tested in advance to make the baseline cAMP concentration just above the lower detection limits of the assay kits so that any increase in cAMP would fall in the assay detection linear range. When assaying the cAMP level involving G_i pathways, the number of cells and known cAMP stimulating factors (e.g., forskolin, PGEs, Isoproterenol, etc.) must be tested to raise the cAMP level to an appropriate level but must be below the upper detection limits of the assay (~100 nM cAMP in assay solution) to ensure a good detection window.

10.5.1 Cyclic AMP Enzyme Immunoassay (EIA), RIA, DELFIA, and ECL Assays

The cAMP EIA assay is illustrated in Figure 10.3. Anti-cAMP antibody (Ab) is coated on the surface of the wells in a microplate. The cAMP tracer (En-cAMP) is a cAMP covalently linked to an enzyme (En), such as HRP and AP. In the absence of cellular cAMP, the En-cAMP tracer binds to Ab. After washing the plate to remove the unbound En-cAMP, an enzyme substrate is added to the wells. The enzymatic product can be colorimetric, fluorescent, or chemiluminescent depending on the substrate used. In the presence of cellular cAMP, the En-cAMP tracer is displaced from binding to Ab. The displaced En-cAMP in solution will be washed out resulting in a reduced enzymatic product. A variety of assay kits based on this method are offered by different vendors. In general, the most sensitive assays are those that generate a chemiluminescence signal. One of the advantages of EIA is that there are many affordable competing commercial kits available. The assay format is well known and not much training is required to perform the assay. No special instrument or new plate reader is required. The major disadvantages of this assay are those commonly found in heterogeneous assay format, such as many washing steps, a large volume of waste buffer generated, and higher assay variation.

Quantitation of cAMP based on RIA employs a similar scheme to the EIA assay discussed above, but with a ^{125}I-labeled cAMP ([^{125}I]cAMP) substituting for En-cAMP as the tracer. In the absence of cAMP, the [^{125}I]cAMP will bind to the Ab that is coated on the surface of the wells in a microplate. After washing the plate, scintillant is added to the plate and the radioactivity in the wells is counted with a TopCount or MicroBeta microplate counter. The presence of cellular cAMP will compete with [^{125}I]cAMP for the binding to Ab. The displaced [^{125}I]cAMP in solution is

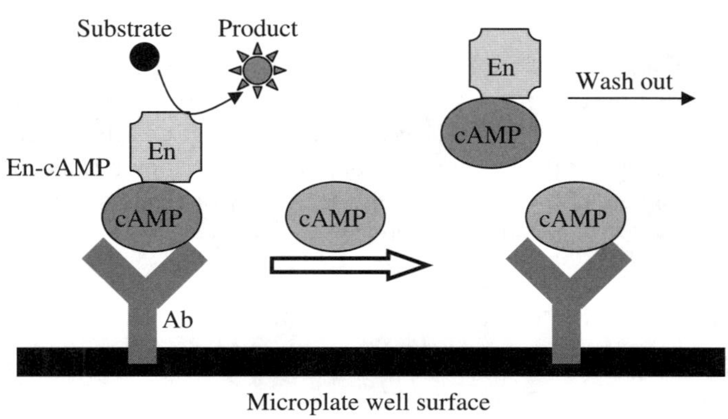

Figure 10.3 Illustration of cAMP EIA. Anti-cAMP antibody (Ab) is coated on the surface of a well in a microplate. In the absence of cellular cAMP, the En-cAMP tracer binds to Ab. After washing the unbound En-cAMP, an enzyme substrate is added to the wells. The enzyme product is then formed that gives rise to a detectable signal. The cellular cAMP competes with En-cAMP in binding to Ab. The displaced En-cAMP in solution will be washed out, resulting in reduced enzymatic product.

washed away, resulting in reduced radioactivity in the well. A variation of this assay is the use of FlashPlate or SPA beads. With these assays, the Ab is coated on the surface of SPA beads or the surface of FlashPlate. The [^{125}I]cAMP that binds to the Ab will give a signal because of proximity to the scintillant coated on the beads or the plate. The displaced [^{125}I]cAMP in solution will not give a signal because it is far away from the scintillant. The major advantage of the SPA and FlatePlate assay is that the assay is homogenous with no washing step required.

The DELFIA cAMP assay (marketed by PerkinElmer) uses the same general format as EIA and RIA but with a europium-labeled cAMP (Eu-cAMP) as the tracer. After exposure to cellular cAMP that is followed by a wash step, an enhanced solution is added to the wells that releases the europium from the Ab-bound Eu-cAMP. An instrument capable of measuring time-resolved fluorescence is then used to detect the signal. DELFIA assays are usually used to replace the radioimmunoassay to eliminate the hazardous radioactive waste.

A similar cAMP assay strategy was also adopted in ECL format (the technology was discussed in Chapter 2). Meso Scale Discovery marketed an ECL-based cAMP kit. The anti-cAMP antibody is coated at the bottom of the special ECL plate. A ruthenium-labeled cAMP (Ru-cAMP) is used as the tracer. The anti-cAMP antibody binds to Ru-cAMP and brings the ruthenium close to the plate surface. The ruthenium close to the ECL plate surface is excited electronically to generate an ECL signal. The cellular cAMP can displace the Ru-cAMP from binding to the antibody. The displaced Ru-cAMP in solution is far away from the plate surface and thus does not generate an ECL signal.

10.5.2 Cyclic AMP Assays in FP Format

Several vendors (e.g., PerkinElmer and GE Health Science) offer cAMP assay kits based on FP. In this assay, a fluorescently labeled cAMP (Fl-cAMP) tracer is used. In the absence of cellular cAMP, the Fl-cAMP tracer binds to an anti-cAMP antibody. Because of the large size of the antibody, the bound Fl-cAMP will have a slow rotational correlation time and give a high FP value. In the presence of cellular cAMP, the Fl-cAMP tracer is displaced from the antibody. The free Fl-cAMP will give a low FP value. The advantage of this assay is the simplicity of the format (mix and read, less assay components involved). However, the sensitivity of FP assay kits in general is lower than the RIA-based and AlphaScreen-based assays.

10.5.3 Cyclic AMP Assay in EFC Format

DiscoverX markets HitHunter cAMP assay kits that are based on the EFC technology (discussed in Chapter 7). Figure 10.4 illustrates the principle for EFC-based cAMP assays. cAMP covalently linked to ED (ED-cAMP) is used as a tracer. In the absence of cellular cAMP, the ED-cAMP tracer binds to anti-cAMP antibody, preventing it from binding to EA. The cellular cAMP competes with ED-cAMP for the binding to the anti-cAMP antibody. The displaced ED-cAMP in solution combines with EA to form a functional enzyme. Upon addition of the enzyme substrate, a detectable signal will be generated.

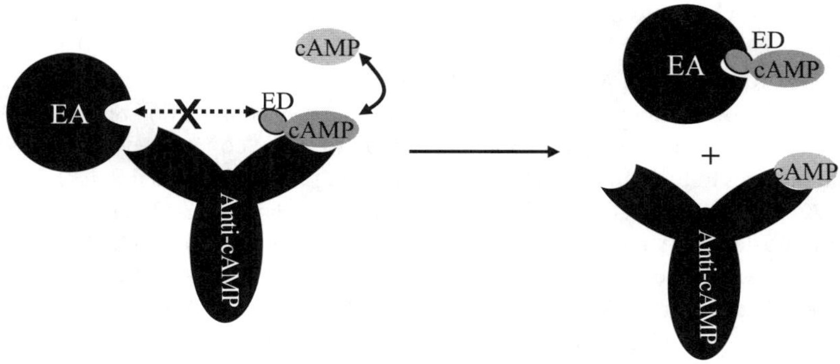

Figure 10.4 Illustration of cAMP EFC assay. cAMP is covalently linked to ED (ED-cAMP) that serves as a tracer. In the absence of cellular cAMP, the ED-cAMP tracer binds to anti-cAMP antibody, thus preventing it from binding to EA. The cellular cAMP competes with ED-cAMP in binding to the anti-cAMP antibody. The displaced ED-cAMP in solution combines with EA to form a functional enzyme. Upon addition of enzyme substrate, a detectable signal is generated.

10.5.4 Cyclic AMP Assay in TR-FRET and AlphaScreen Format

There are two commercially available cAMP assay kits that are developed based on TR-FRET (Lance from PerkinElmer and HTRF from Cisbio). The difference between the two technologies is discussed in Chapter 2. The principle of cAMP assay in HTRF format is illustrated in Figure 10.5. The assay employs a biotin-labeled cAMP as the

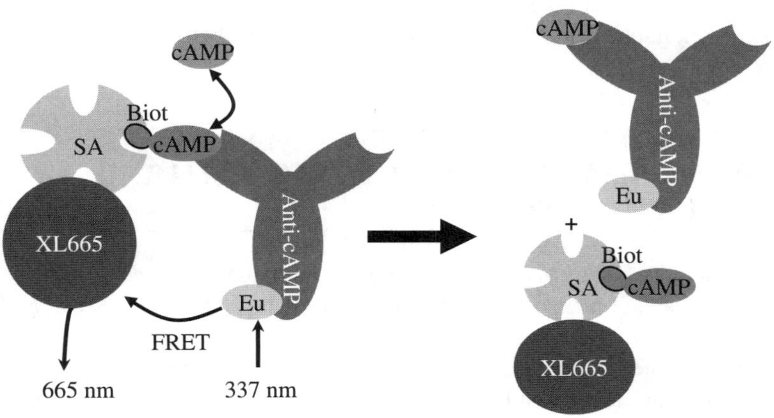

Figure 10.5 Illustration of cAMP assay in HTRF format. The assay employs a biotin-labeled cAMP as the tracer (biot-cAMP). A europium-labeled anti-cAMP antibody is used to bind to the cAMP and a XL665-labeled streptavidin is used to bind to the biotin group on the biot-cAMP. The complex brings the europium chelates and the XL665 together to enable TR-FRET. The emission signal is detected at 665 nm when europium is excited at 337 nm. Cellular cAMP competes with biot-cAMP in binding to the anti-cAMP antibody, resulting in reduced TR-FRET signal.

tracer (biot-cAMP). A europium-labeled anti-cAMP antibody is used to bind the cAMP moiety in biot-cAMP while XL665-labeled streptavidin is used to bind to the biotin group in the biot-cAMP. The binding complex brings the europium chelates and the XL665 together to enable TR-FRET. The emission at 665 nm is measured when the europium is excited at 337 nm. Cellular cAMP competes with biot-cAMP for the binding to the anti-cAMP antibody, resulting in a decreased TR-FRET signal.

A similar strategy for cAMP assay can also be adapted in AlphaScreen format. The XL665-labeled streptavidin is substituted by the donor bead attached to streptavidin while the europium-labeled anti-cAMP antibody is substituted by the acceptor bead attached to anti-cAMP. The presence of the biot-cAMP tracer will bring the donor beads and acceptor beads in close proximity and enable AlphaScreen signal. Cellular cAMP competes with biot-cAMP for binding to anti-cAMP antibody, resulting in decreased AlphaScreen signal.

10.5.5 Cyclic AMP Assay with ACTOne Kit from BD Biosciences

All the cAMP assays discussed above are based on the competition between cellular cAMP and a tracer, and they are single-point assays that require the lysis of cells. BD Biosciences has developed the ACTOne cAMP assay kit that can directly and continuously monitor the intracellular cAMP changes in live cells in real time. The assay is based on using a mutated cyclic-nucleotide-gated ion channel (CNGC) as a cAMP biosensor that is co-expressed with the GPCR under study. When the cAMP level changes as a result of GPCR activation, the CNGC channel opens or closes accordingly. This in turn results in changes in transmembrane potential. Thus, the intracellular level of cAMP is coupled to the change in transmembrane potential that can be measured by a fluorescent membrane potential sensitive dye.

10.6 G PROTEIN-COUPLED RECEPTOR ASSAYS BASED ON MEASUREMENT OF INTRACELLULAR INOSITOL PHOSPHOLIPIDS

The stimulation of $G_{q/11}$-coupled receptors can induce the activation of PLC_{β}, which catalyzes the breakdown of PIP_2 to generate IP_3 and DAG. IP_3 can be further broken down in the inositol phosphate (IP) cascade. Though many of the metabolites in this pathway are potential targets for GPCR assays, IP_3 and IP_1 are the two commonly used assay targets, and there are several commercial assay kits available to assay IP_3 and IP_1.

The strategies to measure IP_1 and IP_3 levels are similar to the measurement of cAMP with the competition method. Here the competition between IP_1 or IP_3 and its corresponding tracer to the binding site of a protein is measured. The binding between IP_3 and specific intracellular receptors provides the basis for competition binding assays without the need of antibodies. For example, IP_3 can be measured either in fluorescence polarization format (assay kit from DiscoverX) and

AlphaScreen format (assay kit from PerkinElmer). Because IP_3 has a short half-life in cells after GPCR stimulation (peak at about 20 s and totally disappear into baseline in about 4 min), the assay should be performed in a short time window when IP_3 is at its peak to obtain optimal signals. The time constraint makes it difficult to perform the assay. In addition, rapid changes in IP_3 level make it difficult to be accurately quantified. In comparison, IP_1 can be assayed with no such time constraint when LiCl is used in the assay buffer. LiCl can block inositol monophosphate phosphatases to prevent the breakdown of IP_1. Thus, the presence of LiCl in the assay buffer causes accumulation of IP_1 in cells following $G_{q/11}$ receptor activation. CisBio offers IP-ONE HTRF-based assay kit and IP-ONE ELISA-based assay kit for quantitation of IP_1. Both assays are based on competition between cellular IP_1 and IP_1 tracers in binding to anti-IP_1 antibodies. Since the general assay principle is similar to what has been discussed in competitive cAMP assays, it will not be discussed further. An important property of IP_1 and IP_3 assays is that they are highly selective to the molecule to be assayed (IP_1 or IP_3) over other inositol polyphosphates because of the highly selective binding protein used in the assays.

10.7 G PROTEIN-COUPLED RECEPTOR ASSAYS BASED ON MEASUREMENT OF INTRACELLULAR Ca^{2+}

The activation of $G_{q/11}$-coupled receptors induces the transient increase of IP_3, which in turn releases calcium from intracellular stores. The measurement of intracellular free calcium concentration thus can be used to study the functional activity of GPCRs. Calcium ion concentration can be readily detected with a variety of organic fluorescent dyes and calcium sensing protein (see Chapter 9). Given the transient nature of Ca^{2+} flux, it should be measured with instrumentations that have 'inject and read' capability. The kinetics of Ca^{2+} flux is usually measured immediately following GPCR activation. The intracellular calcium sensing technologies will not be discussed here since they have been discussed in Chapter 9. The major advantage of measuring Ca^{2+} flux to assay GPCR is the real-time kinetic measurement in live cells. The methods are well established. However, the assay requires expensive instrumentation, and Ca^{2+} is a second messenger further downstream after IP_3. As always, the measurement of a signaling event closer to the receptor is a more direct measure of the functional activity of the receptor if enough signals can be obtained. When screening a compound library, more false positives may occur with calcium flux methods because some compounds may have nonspecific effects on Ca^{2+} flux.

Because calcium flux measurement is well established and it enables kinetic measurement in live cells, much effort has been spent to broaden the assay to cover G_s- and $G_{i/o}$-coupled receptors in addition to $G_{q/11}$-coupled receptors. It was found that Ca^{2+} flux is also induced by the activation of the promiscuous G proteins $G_{15/16}$, which can couple to G_s- and $G_{i/o}$-coupled receptors. Co-expression of G_s- and $G_{i/o}$-coupled receptors and $G_{15/16}$ thus can link the activation of these receptors to calcium flux. Alternatively, the C-terminus of G_q proteins can be replaced with that from G_s or $G_{i/o}$. The substitution enables the coupling between G_s- and $G_{i/o}$-coupled receptors and G_q protein, leading to the activation of $G_{q/11}$ pathways and calcium flux.

10.8 G PROTEIN-COUPLED RECEPTOR ASSAYS BASED ON MEASUREMENT OF MAPK ACTIVITY

The GPCR assays discussed so far all have their limitations and no single assay format is suitable for all situations. This creates demands for new methods to assay GPCRs. Recent studies showed that the activation of many GPCRs, regardless of their G protein-coupling patterns, leads to the modification of ERK1/2 activity. Available data showed that all $G_{q/11}$-coupled receptors and about 70% of $G_{i/o}$-coupled receptors tested so far lead to changes in ERK1/2 activity. However, much fewer G_s-coupled receptors tested so far lead to changes in ERK1/2 activity. Though the mechanism is not clear, it is evident that several GPCR pathways converge on ERK1/2. Thus, measurement of ERK1/2 activity offers a broad assay format for many GPCRs with different coupling mechanisms. Several studies evaluating the ERK1/2 activity as a generic readout for GPCRs were performed with the three main classes of G proteins: G_s, $G_{i/o}$, and $G_{q/11}$. GPCR-mediated ERK activity assay compared favorably against other assay formats, such as GTPγS binding assay, cAMP assays, and calcium ion flux assays. Because of the high intracellular concentration of ERK1/2, the activity of ERK1/2 can be readily detected with endogenous GPCRs (not overexpressed) in various cells, including primary cells. The GPCR assay results obtained by measuring ERK1/2 activity are more physiologically relevant with no need for transfection, dye loading, or prestimulations that are required in other GPCR assay formats.

We mentioned in Chapter 7 that the activity of a kinase in live cells can be measured by either the cleavage of its substrate(s) or the phosphorylation of the kinase itself. However, neither assays can be easily accomplished in live cells because there exist many active kinases inside the cell. Due to substrate promiscuity, it is difficult to pinpoint whether the phosphorylation of a substrate is due to the action of a particular kinase in the cell. Similarly, there is no technology that can directly detect phosphorylation of a kinase in live cells. Thus, the activity of a kinase in live cells can only be assessed by analysis of the cell lysate. The major concern with the analysis of the cell lysate is whether the modification of the kinase and the activity of the kinase are preserved during the lysing process. To minimize the perturbation of the native system, a cocktail of phosphatases inhibitors and protease inhibitors are commonly included in the lysing buffer. For the purpose of monitoring GPCR activation through the analysis of ERK1/2 activity, quantitation of phosphorylated ERK1/2 is more direct and less prone to interference from cell lysate than measuring phosphorylation of a ERK1/2 substrate because it is difficult to identify a substrate that is only phosphorylated by ERK1/2 among the many active kinases present in the cell lysate.

The SureFire (developed by TGR BioSciences and PerkinElmer) cellular active ERK1/2 assay is based on AlphaScreen to detect phosphorylated ERK1/2 (Fig. 10.6). Two antibodies are used in the assay with one of them recognizing the Thr202/Tyr204 and the other recognizing nondisclosed ERK1/2 sequences. The two antibodies are linked to donor beads and acceptor beads, respectively. The phosphorylated ERK1/2 as a result of ERK1/2 activation brings the two antibodies and their attached beads in close proximity to enable the AlphaScreen signal. When performing SureFire assays, cultured cells (typically CHO or HEK293) are treated with

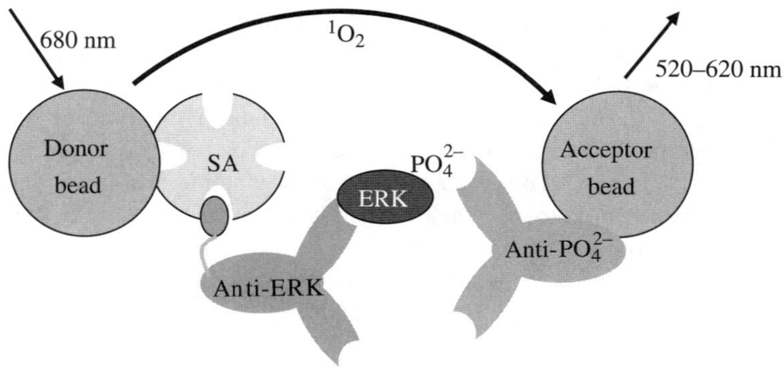

Figure 10.6 Illustration of GPCR assay with Surefire platform. Phosphorylated ERK is detected with AlphaScreen technology. The donor bead is labeled with streptavidin and the acceptor bead is labeled with anti-phosphoERK antibody. The anti-ERK antibody is labeled with biotin. The presence of phosphorylated ERK brings the two beads together. When excited at 680 nm, the emission signal between 520 and 620 nm is measured, whose intensity is proportional to the concentration of phosphorylated ERK.

a ligand for an appropriate time (usually 5 to 10 min). The cells are then lysed. The extent of ERK1/2 phosphorylation in the cell lysate is determined by incubating the lysate with the activation buffer (supplied with the kit) that is followed by a reaction buffer (supplied with kit) containing the AlphaScreen beads, for 2 h at room temperature. The AlphaScreen signal is then measured. This is a homogenous assay with no wash step. The assay can be performed in 96-, 384-, or 1536-well microplates.

10.9 G PROTEIN-COUPLED RECEPTOR ASSAYS WITH REPORTER GENE

10.9.1 Introduction to Reporter Gene Assays

A reporter gene is a gene possessing special characteristics that can be easily measured qualitatively or quantitatively when they are expressed in cell cultures or in organisms. Reporter genes are generally used to determine whether the gene of interest has been taken up by or expressed in the cell or organism. For applications in bioassays, genes expressing either a protein that is autofluorescent or expressing an enzyme that can turn a substrate into a fluorescent or luminescent product are commonly used as reporter genes. Reporter gene-based assays can be formatted into promoter assays and gene expression assays. In promoter assays, reporter genes are linked to a particular promoter in a cell or organism. The reporter gene is placed under the control of the target promoter and the reporter gene product's activity is measured. In gene expression assays, the reporter genes are fused, in frame, to the coding sequence of the gene under study. The two genes are under the control of the same promoter and are transcribed into a single messenger RNA molecule. The mRNA is then translated into a fused protein. It is important that both proteins be properly folded into their active conformations and function independently despite being fused. To minimize the interference between the two fused proteins, a segment of DNA coding for a flexible polypeptide linker region is usually included when building the DNA construct. Since the expression of the reporter gene is used to obtain a ubiquitous signal, it is important to use a reporter gene that is not natively expressed in the cell or organism under study. Thus, reporter genes originating from nonmammalian species are commonly used to study mammalian systems. Further, most of the reporter genes have been modified to enhance their expression in mammalian cells and to change their characteristics for better signal detection and broad applications. The genes encoding β-galactosidase, secreted alkaline phosphatases (SEAP), GFPs (including many variants), luciferases, and β-lactamase are commonly used as reporter genes.

The *LacZ* gene from *E. coli*, which encodes β-galactosidase, is the earliest reporter gene that is still widely used in many applications, especially in histological analysis. There is a wide variety of β-galactosidase substrates from many vendors that can generate colorimetric, fluorescent, or luminescent products. For example, the chromogenic substrate, 5-bromo-4-chloro-3-indolyl-*b*-D-galactoside (X-gal), is cleaved by β-galactosidase into a stable blue insoluble precipitate that can be observed under microscope. The *LacZ* reporter gene is typically used with fixed cells or cell lysate. The gene encoding SEAP is another widely used reporter gene to analyze the activity of promoters and to trace gene expression in cell culture or animals. A unique property of using SEAP as a reporter gene is that it is secreted into culture medium or serum, making it convenient to kinetically monitor the gene expression. Experiments with both β-galactosidase and SEAP can be performed economically without the complicated patent issues surrounding some other reporters. GFP and its variants have been extensively used in recent years. A key advantage of using GFPs as the reporter genes is that they are autofluorescent. No substrate is required to monitor the fluorescence signal, eliminating the need to deliver a substrate to the cells or to lyse the cell in order to expose the reporter enzyme to its substrates. However, the signal from GFP is weak compared with other reporter genes because the signal in a GFP-based reporter assay is not amplified. In addition, it is very expensive to use GFP as a reporter gene because of the complicated patents surrounding GFPs. There are several luciferase reporter genes, such as *Firefly* luciferase, *Renilla* luciferase, and *Gaussia* luciferase. We have discussed *Firefly* luciferase in Chapter 7 when discussing ATP quantitation. Different from *Firefly* luciferase that uses luciferin as the substrate, *Renilla* luciferase and *Gaussia* luciferase use coelenterazine as their substrate. The oxidation of coelenterazine to coelenteramide generates light at ~480 nm. However, coelenterazine may also emit light from enzyme-independent oxidation, thus increasing the background of the assay. Because of much higher quantum yields, a reporter assay based on *Gaussia* luciferase can generate much larger signals than assays based on *Firefly* luciferase and *Renilla* luciferase. In recent years, reporter gene systems using β-lactamase (GeneBLAzer marketed by Invitrogen) have become very popular because of the ratiometric substrate, CCF2, which was originally developed by Aurora Biosciences. The structure of CCF2 is shown in Figure 10.7. It contains a coumarin moiety and a fluorescein moiety. The two dyes are joined by a penicillin moiety

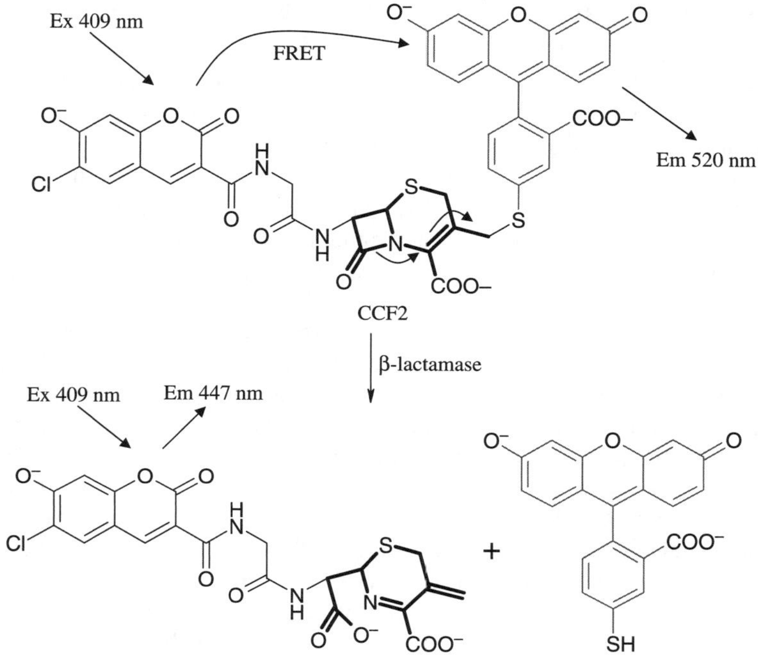

Figure 10.7 Molecular structure of CCF2 and its cleavage reaction catalyzed by β-lactamase. CCF2 contains a coumarin moiety and a fluorescein moiety, those are joined together by a penicillin moiety (bold). Upon excitation at 409 nm, CCF2 emits fluorescent signal at 520 nm through FRET. After cleavage of CCF2 catalyzed by β-lacatamase, the coumarin moiety and fluorescein moiety are separated from each other. Excitation of the separated coumarin at 409 nm gives fluorescent signal at 447 nm. [After Zlokarnik et al. (1998), *Science*, 279: 84.]

(shown in bold) containing a β-lactam bond that can be hydrolyzed by β-lactamase. When excited at 409 nm, the energy absorbed by the coumarin moiety transfers to the fluorescein moiety through FRET, resulting in green light emission at 520 nm. Cleavage of CCF2 by β-lactamase results in separation of the coumarin and fluorescein moieties. When excited at 409 nm, the free coumarin part in solution gives an emission at 447 nm. Thus, the ratio of 520-nm/447-nm emission is proportional to β-lactamase activity. The ratiometric nature of CCF2 emission as a function of its cleavage can reduce the experimental errors from many different sources between samples, such as variations in cell number, substrate concentration differences, excitation pathlength difference, and fluorescence detector fluctuations or changes in instrument settings. Because CCF2 is highly negatively charged, it is modified into CCF2/AM with acetyl, butylyl, or acetoxymethyl groups for cell-based assay applications. CCF2/AM is a neutral, cell-membrane-permeable, and nontoxic molecule. The modifying groups on CCF2/AM are cleaved by esterases inside the cells to release CCF2 that cannot diffuse out of the cell because of its negative charges.

10.9.2 Application of Reporter Gene Assays in GPCR

It is well known that GPCR activation leads to changes in transcription activities of many genes, particularly the genes that are controlled by transcription factors CREB, ELK-1, cFOS/cJUN, and NFAT. Reporter genes, such as β-lactamase, β-galactosidase, luciferases, and GFPs, can be placed behind the responsive elements to these transcription factors to provide a direct readout of the transcription activity. For example, GPCR coupled to cAMP (G_s- and $G_{i/o}$-coupled receptors) can be assayed by placing the reporter genes behind CRE (cAMP response element). Similarly, GPCRs coupled to intracellular calcium ($G_{q/11}$-coupled receptors) can be assayed by placing the reporter genes behind AP1 (activator protein 1) or NFAT (nuclear factor of activated T cells) elements. There are many commercial reporter gene assay kits available for GPCRs that provide cell lines transfected with GPCRs and reporter genes. The principal advantages of reporter gene assays are the wide linearity and sensitivity of the technique, making them very suitable for detection of weak GPCR signaling. Reporter gene-based assays are readily scalable. For example, reporter gene assays based on β-lactamase have been shown to perform in 96-well microplate all the way down to 3456-well microplates with extremely low assay volume (~2 μL). A major concern with reporter gene-based assays is the relative high occurrence of false positives resulting from off-target action in the assay because the final readout is at the bottom of the signal transduction pathways. Many intermediate proteins along the pathways may be affected by the test compound. To minimize the long signal transduction path from GPCR activation to reporter gene transcription, Invitrogen recently developed the Tango GPCR assay using β-lactamase as the reporter gene. With this method, the GPCR under study is linked through a protease-recognizing site to a nonnative transcription factor at the C-terminus of the GPCR. Further, β-arrestin in the cell is modified by linking it to a protease. Upon GPCR activation, which is followed by desensitization, the protease-linked arrestin is recruited to bind to the C-terminus of GPCR. The protease then cleaves the protease site on the C-terminus of the GPCR, releasing the nonnative transcription factor. The transcription factor immediately enters the nucleus to regulate the transcription of β-lacatamase. This approach directly links receptor desensitization to reporter gene activation, bypassing many signaling intermediates involved in normal GPCR signaling.

10.10 G PROTEIN-COUPLED RECEPTOR ASSAYS BY MONITORING EVENTS LEADING TO GPCR INTERNALIZATION

After GPCR activation, the receptor is phosphorylated and desensitized. This is followed by GPCR internalization that is mediated by a family of proteins called β-arrestins. There are two forms of β-arrestins: β-arrestin 1 and β-arrestin 2. β-arrestins bind to the phosphorylated GPCRs, inhibiting its ability to interact with G proteins. The complex of β-arrestin and GPCR are then internalized via clathrin-coated pits and vesicles. This process proceeds in parallel with the pathways involving the generation of second messengers. β-arrestin 2 is relatively promiscuous and can couple to many GPCRs while β-arrestin 1 binds to a limited number of GPCRs. There are two modes of interaction between GPCRs and β-arrestins. Hence, GPCRs can be divided into two different groups based on the mode of their interaction with β-arrestins. Class A GPCRs interact primarily with β-arrestin 2 and dissociate from β-arrestin 2 at or near the cell surface prior to internalization. Class B GPCRs remain associated with both β-arrestins during internalization. Several processes in this pathway, such as the association between GPCR and β-arrestin and the internalization of GPCR and/or β-arrestins, can be used to assay GPCR activation.

10.10.1 Measurement of Association Between GPCR and β-arrestin

The association between two separate proteins inside the cells can be studied in many ways. Here we focus discussions on the commonly used FRET, bioluminescence resonance energy transfer (BRET), and EFC methods to study the interaction between GPCR and β-arrestin. In FRET assays, two GFP variants with overlapping excitation and emission spectra are used to label GPCRs and β-arrestin, respectively. The association between GPCR and β-arrestin brings the two GFPs together, and FRET occurs when the donor GFP is excited. Though FRET-based assays give off bright fluorescence, they are limited by a high background from cellular autofluorescence and from excitation of the acceptor fluorophore due to their broad absorption spectra.

The BRET-based assay can reduce the high background problem with FRET assays because there is no need for external illumination of the sample. The original BRET system uses *Renilla* luciferase as the donor, a derivative of coelenterazine as its substrate and a yellow fluorescent protein (YFP) as the acceptor. The light emission of *Renilla* luciferase-catalyzed reaction peaks at 480 nm that overlap poorly with the YFP excitation spectrum (maximum absorption at 513 nm). In addition, the emission spectrum of *Renilla* luciferase-catalyzed reaction is broad, extending to the YFP emission region (peak at 527 nm). The *Renilla* emission therefore contributes to a high assay background. The new version of the assay, $BRET^2$, uses a new coelenterazine derivative, DeepBlueC, which emits light with a maximum at 395 nm. DeepBlueC is nontoxic and can rapidly (in seconds) penetrate cell membranes. A new version of GFP acceptor with its excitation spectra that overlaps well with the donor emission spectra is used in $BRET^2$ assay. The new GFP emits at 510 nm that is far away from the DeepBkueC emission, resulting in a much reduced assay background. In addition to substrate modification and GFP improvement, using a mutated β-arrestin that dissociates slowly from the GPCR further enhances the $BRET^2$ signal. The $BRET^2$ signal is a ratiometric measurement and inherits all the advantages associated with ratiometric measurement.

In addition to BRET and FRET assays, EFC assays are developed to examine the interaction between GPCR and β-arrestin (marketed by DiscoverX). In this assay, GPCR and β-arrestin are fused to two separate fragments of β-galactosidase, respectively. Neither of the two β-galactosidase fragments is functional. The binding of GPCR and β-arrestin brings the two fragments of the β-galactosidase together, which recombines to form an intact functional β-galactosidase.

10.10.2 Visualization of Internalization of GPCR and/or β-arrestin

Since GPCR desensitization only occurs with an activated receptor, monitoring the GPCR/β-arrestin complex translocation and subsequent receptor recycling provides a method to detect the activation of any GPCR. If either GPCR or β-arrestin is labeled, the internalization process can be visualized under a microscope. Using modern automated image collection instruments and image analysis algorithms, GPCR activation can be readily assayed by following the GPCR and/or β-arrestin internalization (image-based assays are discussed in Chapter 12). This assay is universally applicable to receptors coupled to all G proteins (G_s, $G_{i/o}$, and G_q). One potential issue with this assay is the high levels of internalized receptor under basal conditions due to constitutive recycling of GPCRs in some cells.

Transfluor technology, which was originally developed by Norak Biosciences and is now marketed by MDS Analytics, is one of the early image-based assays for GPCRs and has been validated with more than 100 GPCRs spanning all classes. With this assay, a GFP is attached to β-arrestin. The location of the GPCR–arrestin complex can be visualized by taking the fluorescence image of the sample under a microscope. Before GPCR activation, the fluorescence is uniformly distributed in cytosol. After GPCR activation, the fluorescence is seen as bright aggregated pits within seconds that is followed by the movement of fluorescence to endocytic vesicles within minutes. This technology requires the use of cell lines that are genetically engineered to express both the GFP-labeled β-arrestin and the GPCR of interest. This assay can be performed in live cells and kinetic data can be obtained.

Instead of monitoring GPCR internalization by GFP-labeled β-arrestin, GPCR can be fused to GFP directly and the GPCR internalization process can be monitored fluorescently. Upon GPCR activation, the GPCR-GFP fusion protein moves from the plasma membrane to internalized recycling compartments, forming bright aggregated pits. This assay can be performed in live cell and kinetic data can be obtained. However, the GFP tag may affect the properties of the GPCR under study and the fluorescent signal of GFP in live cells is not bright. Alternatively, the native GPCR internalization process can be visualized in fixed cells by immunostaining. A fluorescently labeled antibody is used to recognize the native GPCR after fixing the cells. No modification of the native GPCR is required. Fluorescent signals from the immunostaining procedure are more sensitive than visualization of the GFP signal in live cells. However, no kinetic data is obtained with an immunostaining assay and the additional immunostaining procedures may not be easy to automate.

CypHer5 (marketed by GE Healthscience) is a pH-sensitive cyanine dye that is nonfluorescent at pH 7.4 but is fluorescent at pH 5.5. It is therefore ideally suited for monitoring internalization of any cell surface receptors, such as GPCRs and RTKs, via acidic endosomal vesicles upon agonist stimulation. In the assay, the receptor is recognized with an antibody that is labeled with CypHer5. The antibody can directly recognize the receptor sequences exposed to the cell surface or recognize a tag that is engineered to the receptor surface. Binding of the antibody to cell surface receptors does not produce a fluorescence signal. Following ligand binding, the receptor, together with the bound antibody, internalizes to endosomal compartments, producing bright fluorescence because of low pH environment in the endosomal vesicles.

Useful Websites

http://www.gpcr.org/7tm/

http://www.discoverx.com/

http://www.meso-scale.com/

http://www.htrf.com/splash.asp

http://www.promega.com

http://las.perkinelmer.com/Catalog/FamilyPage.htm?CategoryID=SteadyLite+Plus

BIBLIOGRAPHY

Adie, E. J., et al. (2003) CypHer 5: A generic approach for measuring the activation and trafficking of G protein-coupled receptors in live cells. *Assay Drug Dev. Technol.* **1**, 251–259.

Barak, L. S., Ferguson, S. S. G., Zhang, J., and Caron, M. G. (1997) A beta-arrestin/green fluorescent protein biosensor for detecting G protein-coupled receptor activation. *J. Biol. Chem.* **272**, 27497–27500.

Bertrand, L., et al. (2002) The BRET2/arrestin assay in stable recombinant cells: A platform to screen for compounds that interact with G protein-coupled receptors (GPCRS). *J. Recept. Signal Transduct. Res.* **22**, 533–541.

Casadó, V., et al. (2007) Old and new ways to calculate the affinity of agonists and antagonists interacting with G-protein-coupled monomeric and dimeric receptors: The receptor-dimer cooperativity index. *Pharmacol. Ther.* **116**, 343–354.

Cassutt, K. J., Orsini, M. J., Abousleiman, M., Colone, D., and Tang, W. (2007) Identifying nonselective hits from a homogeneous calcium assay screen. *J. Biomol. Screen.* **12**, 285–287.

Cherezov, V., et al. (2007) High-resolution crystal structure of an engineered human beta 2-adrenergic G protein coupled receptor. *Science* **318**, 1258–1265.

Conway, B. R., et al. (2001) Quantitative analysis of agonist-dependent parathyroid hormone receptor trafficking in whole cells using a functional green fluorescent protein conjugate. *J. Cell. Physiol.* **189**, 341–355.

Eglen, R. M., Bosse, R., and Reisine, T. (2007) Emerging concepts of guanine nucleotide-binding protein-coupled receptor (GPCR) function and implications for high throughput screening. *Assay Drug Dev. Technol.* **5**, 425–452.

Elster, L., Elling, C., and Heding, A. (2007) Bioluminescence resonance energy transfer as a screening assay: Focus on partial and inverse agonism. *J. Biomol. Screen.* **12**, 41–49.

Fang, Y. (2006) Label-free cell-based assays with optical biosensors in drug discovery. *Assay Drug Dev. Technol.* **4**, 583–595.

Ferguson, S. S. and Caron, M. G. (2004) Green fluorescent protein-tagged beta-arrestin translocation as a measure of G protein-coupled receptor activation. *Methods Mol. Biol.* **237**, 121–126.

Ferguson, S. S. G. (2001) Evolving concepts in G protein-coupled receptor endocytosis: The role in receptor desensitization and signaling. *Pharmacol. Rev.* **53**, 1–24.

Franco, R., et al. (2007) G-protein-coupled receptor heteromers: Function and ligand pharmacology. *Br. J. Pharmacol.* **153**, S90–S98.

Fredriksson, R., Lagerstrom, M. C., Lundin, L.-G., and Schioth, H. B. (2003) The G-protein-coupled receptors in the human genome form five main families. Phylogenetic analysis, paralogon groups, and fingerprints. *Mol. Pharmacol.* **63**, 1256–1272.

Gales, C., et al. (2005) Real-time monitoring of receptor and G-protein interactions in living cells. *Nat. Methods* **2**, 177–184.

Gether, U. (2000) Uncovering molecular mechanisms involved in activation of G protein-coupled receptors. *Endocr. Rev.* **21**, 90–113.

Gether, U., Fazila, A., Meinild, A. K., and Rasmussen, J. G. F. (2002) Structural basis for activation of G-protein-coupled receptors. *Pharmacol. and Toxicol.* **91**, 304–312.

Goetzl, E. J. (2007) Diverse pathways for nuclear signaling by G protein-coupled receptors and their ligands. *FASEB J.* **21**, 638–642.

Haasen, D., Schnapp, A., Valler, M. J., and Heilker, R. (2006) G protein-coupled receptor internalization assays in the high-content screening format. *Methods Enzymol.* **414**, 121–139.

Hammer, M. M., Wehrman, T. S., and Blau, H. M. (2007) A novel enzyme complementation-based assay for monitoring G-protein-coupled receptor internalization. *FASEB J.* **21**, 3827–3834.

Hanson, B. J. (2006) Multiplexing Fluo-4 NW and a GeneBLAzer(R) transcriptional assay for high-throughput screening of G-protein-coupled receptors. *J. Biomol. Screen.* **11**, 644–651.

Hill, S. J. (2006) G-protein-coupled receptors: Past, present and future. *Br. J. Pharmacol.* **147**, S27–S37.

Hollinger, S. and Hepler, J. R. (2002) Cellular regulation of RGS proteins: Modulators and integrators of G protein signaling. *Pharmacol. Rev.* **54**, 527–559.

Kenakin, T. (2005) New concepts in drug discovery: Collateral efficacy and permissive antagonism. *Nat. Rev. Drug Discov.* **4**, 919–927.

Kocan, M., See, H. B., Seeber, R. M., Eidne, K. A., and Pfleger, K. D. G. (2008) Demonstration of improvements to the bioluminescence resonance energy transfer (BRET) technology for the monitoring of G protein-coupled receptors in live cells. *J. Biomol. Screen.* **13**, 888–898.

Kolakowski, L. F. J. (1994) GCRDb: A G-protein-coupled receptor database. *Recept. Channels* **2**, 1–7.

Kostenis, E., Waelbroeck, M., and Milligan, G. (2005) Techniques: Promiscuous Galpha proteins in basic research and drug discovery. *Trends Pharmacol. Sci.* **26**, 595–562.

Lagerstrom, M. C. and Schioth, H. B. (2008) Structural diversity of G protein-coupled receptors and significance for drug discovery. *Nat. Rev. Drug Discov.* **7**, 339–357.

Lee, S., Howell, B., and Kunapuli, P. (2006) Cell imaging assays for G protein-coupled receptor internalization: Application to high-throughput screening. *Methods Enzymol.* **414**, 79–98.

Leifert, W. R., Aloia, A. L., Bucco, O., Glatz, R. V., and McMurchie, E. J. (2005) G-protein-coupled receptors in drug discovery: Nanosizing using cell-free technologies and molecular biology approaches. *J. Biomol. Screen.* **10**, 765–779.

Leroy, D., Missotten, M., Waltzinger, C., Martin, T., and Scheer, A. (2007) G protein-coupled receptor-mediated ERK1/2 phosphorylation: Towards a generic sensor of GPCR activation. *J. Recept. Signal Transduct.* **27**, 83–97.

Litschig, S., et al. (1999) CPCCOEt, a noncompetitive metabotropic glutamate receptor 1 antagonist, inhibits receptor signaling without affecting glutamate binding. *Mol. Pharmacol.* **55**, 453–461.

Liu, B. and Wu, D. (2004) Analysis of the coupling of G12/13 to G protein-coupled receptors using a luciferase reporter assay. *Methods Mol. Biol.* **237**, 145–149.

Lohse, M. J., et al. (2008) Optical techniques to analyze real-time activation and signaling of G-protein-coupled receptors. *Trends Pharmacol. Sci.* **29**, 159–165.

Luttrell, D. K. and Luttrell, L. M. (2003) Signaling in time and space: G protein-coupled receptors and mitogen-activated protein kinases. *Assay Drug Dev. Technol.* **1**, 327–338.

Ma, L. and Pei, G. (2007) Beta-arrestin signaling and regulation of transcription. *J. Cell Sci.* **120**, 213–218.

Milligan, G. (2004) Applications of bioluminescence- and fluorescence resonance energy transfer to drug discovery at G protein-coupled receptors. *Eur. J. Pharm. Sci.* **21**, 397–405.

Milligan, G. (2006) G-protein-coupled receptor heterodimers: Pharmacology, function and relevance to drug discovery. *Drug Discov. Today* **11**, 541–549.

Milligan, G. and Kostenis, E. (2006) Heterotrimeric G-proteins: A short history. *Br. J. Pharmacol.* **147**, S46–S55.

Monastyrskaia, K., et al. (1999) Effect of the umami peptides on the ligand binding and function of rat mGlu4a receptor might implicate this receptor in the monosodium glutamate taste transduction. *Br. J. Pharmacol.* **128**, 1027–1034.

Oakley, R. H., et al. (2002) The cellular distribution of fluorescently labeled arrestins provides a robust, sensitive, and universal assay for screening G protein-coupled receptors. *Assay Drug Dev. Technol.* **1**, 21–30.

O'Dowd, B. F., et al. (2007) Using ligand-induced conformational change to screen for compounds targeting G-protein-coupled receptors. *J. Biomol. Screen.* **12**, 175–185.

Osmond, R. I. W., et al. (2005) GPCR screening via ERK 1/2: A novel platform for screening G protein-coupled receptors. *J. Biomol. Screen.* **10**, 730–737.

Palczewski, K., et al. (2000) Crystal structure of rhodopsin: A G protein-coupled receptor. *Science* **289**, 739–745.

Pebay-Peyroula, E., Rummel, G., Rosenbusch, J. P., and Landau, E. M. (1997) X-ray structure of bacteriorhodopsin at 2.5 angstroms from microcrystals grown in lipidic cubic phases. *Science* **277**, 1676–1681.

Perez, D. M. and Karnik, S. S. (2005) Multiple signaling states of G-protein-coupled Receptors. *Pharmacol. Rev.* **57**, 147–161.

Peters, M. F., et al. (2007) Evaluation of cellular dielectric spectroscopy, a whole-cell, label-free technology for drug discovery on Gi-coupled GPCRs. *J. Biomol. Screen.* **12**, 312–319.

Pfleger, K. D. G. and Eidne, K. A. (2005) Monitoring the formation of dynamic G-protein-coupled receptor-protein complexes in living cells. *Biochem. J.* **385**, 625–637.

Pierce, K. L. and Lefkowitz, R. J. (2001) Classical and new roles of [beta]-arrestins in the regulation of G-protein-coupled receptors. **2**, 727–733.

Pin, J.-P., et al. (2007) International Union of Basic and Clinical Pharmacology. LXVII. Recommendations for the recognition and nomenclature of G protein-coupled receptor heteromultimers. *Pharmacol. Rev.* **59**, 5–13.

Prinster, S. C., Hague, C., and Hall, R. A. (2005) Heterodimerization of G protein-coupled receptors: Specificity and functional significance. *Pharmacol. Rev.* **57**, 289–298.

Rasmussen, S. G. F., et al. (2007) Crystal structure of the human beta 2 adrenergic G-protein-coupled receptor. *Nature* **450**, 383–387.

Rees, S., et al. (2001) Development of a homogeneous MAP kinase reporter gene screen for the identification of agonists and antagonists at the CXCR1 chemokine receptor. *J. Biomol. Screen.* **6**, 19–27.

Rodgers, G., et al. (2003) Development of displacement binding and GTPrS scintillation proximity assays for the identification of antagonists of the μ-opioid receptor. *Assay Drug Dev. Technol.* **1**, 627–636.

Rohere, D. K. and Kobilka, B. K. (1998) G protein-coupled receptors: Functional and mechanistic insights through altered gene expression. *Physiol. Rev.* **78**, 35–52.

Rozengurt, E. (2007) Mitogenic signaling pathways induced by G protein-coupled receptors. *J. Cell. Physiol.* **213**, 589–602.

Russello, S. V. (2004) Assessing cellular protein phosphorylation: High throughput drug discovery technologies. *Assay Drug Dev. Technol.* **2**, 225–235.

Scheerer, P., et al. (2008) Crystal structure of opsin in its G-protein-interacting conformation. *Nature* **455**, 497–502.

Selkirk, J. V., et al. (2006) A novel cell-based assay for G-protein-coupled receptor-mediated cyclic adenosine monophosphate response element binding protein phosphorylation. *J. Biomol. Screen.* **11**, 351–358.

Siehler, S. (2008) Cell-based assays in GPCR drug discovery. *Biotechnology J.* **3**, 471–483.

Tang, Y., Li, X., He, J., Lu, J., and Diwu, Z. (2006) Real-time and high throughput monitoring of cAMP in live cells using a fluorescent membrane potential-sensitive dye. *Assay Drug Dev. Technol.* **4**, 461–471.

Tate, C. G. and Grisshammer, R. (1996) Heterologous expression of G-protein-coupled receptors. *Trends Biotechnol.* **14**, 426–430.

Thomas, S. M., et al. (2006) Cross-talk between G protein-coupled receptor and epidermal growth factor receptor signaling pathways contributes to growth and invasion of head and neck squamous cell carcinoma. *Cancer Res.* **66**, 11831–11839.

Thomsen, W., Frazer, J., and D. U. (2005) Functional assays for screening GPCR targets. *Curr. Opin. Biotechnol.* **16**, 655–665.

Tobin, A. B. (2008) G-protein-coupled receptor phosphorylation: Where, when and by whom. *Br. J. Pharmacol.* **153**, S167–S176.

Urwyler, S., et al. (2001) Positive allosteric modulation of native and recombinant gamma-aminobutyric acid B receptors by 2,6-di-*tert*-butyl-4-(3-hydroxy-2,2-dimethyl-propyl)-phenol (CGP7930) and its aldehyde analog CGP13501. *Mol. Pharmacol.* **60**, 963–971.

Vauquelin, G. and von Mentzer, B. (2007) *G Protein-Coupled Receptors: Molecular Pharmacology.* Wiley, West Sussex, England.

von Degenfeld, G., Wehrman, T. S., Hammer, M. M., and Blau, H. M. (2007) A universal technology for monitoring G-protein-coupled receptor activation in vitro and noninvasively in live animals. *FASEB J.* **21**, 3819–3826.

Warne, T., et al. (2008) Structure of a beta 1-adrenergic G-protein-coupled receptor. *Nature* **454**, 486–491.

Williams, C. (2004) cAMP detection methods in HTS: Selecting the best from the rest. **3**, 125–135.

Wise, A., Jupe, S. C., and Rees, S. (2004) The identification of ligands at orphan G-protein coupled receptors. *Annu. Rev. Pharmacol. Toxicol.* **44**, 43–66.

Xiao, S. H., et al. (2008) High throughput screening for orphan and liganded GPCRs. *Comb. Chem. High Throughput Screen.* **11**, 195–215.

Xing, H., Tran, H. C., Knapp, T. E., Negulescu, P. A., and Pollok, B. A. (2000) A fluorescent reporter assay for the detection of ligands acting through Gi protein-coupled receptors. *J. Recept. Signal Transduct. Res.* **20**, 189–210.

Xu, Y., Piston, D. W., and Johnson, C. H. (1999) A bioluminescence resonance energy transfer (BRET) system: Application to interacting circadian clock proteins. *Proc. Nat. Acad. Sci. U.S.A* **96**, 151–156.

Yan, Y.-X., et al. (2002) Cell-based high-throughput screening assay system for monitoring G protein-coupled receptor activation using {beta}-galactosidase enzyme complementation technology. *J. Biomol. Screen.* **7**, 451–459.

Zhao, X., et al. (2008) A homogeneous enzyme fragment complementation-based {beta}-arrestin translocation assay for high-throughput screening of G-protein-coupled receptors. *J. Biomol. Screen.* **13**, 737–747.

Zhong, H. and Neubig, R. R. (2001) Regulator of G protein signaling proteins: Novel multifunctional drug targets. *J. Pharmacol. Exp. Ther.* **297**, 837–845.

Zlokarnik, G., et al. (1998) Quantitation of transcription and clonal selection of single living cells with {beta}-lactamase as reporter. *Science* **279**, 84–88.

CHAPTER 11

ASSAYS BASED ON INTEGRATED CELL SYSTEM PROPERTIES

ALL THE cell-based assays we discussed in previous chapters are based on the analysis of individual cellular components that reside in the signal transduction pathways. Individual pathways transmit signals along linear tracts that relay the signal from the cell surface all the way to transcription and translation activities. However, cellular signaling is not a simple linear process but a network of many individual pathways. These pathways are not free standing but are interconnected through integration, interactions, and branching to regulate discrete cellular functions. After stimulation, cells have the ability to coordinate systematically their intracellular activities to cope with the new environment, resulting in systemwide changes in the cells. In this chapter, we will discuss the assays that measure systemwide cellular changes (or cellular global response to stimulation), such as cell proliferation, apoptosis, metabolic changes, interaction with extracellular matrix, protein secretion, changes in optical properties, and migration. One caveat for these assays is that the measurements sometimes were only made with the final global cellular responses, and it may not possible to definitively identify which component inside the cells has changed. The systemwide changes in cells can also be detected with the analysis of the transcription of large set of DNAs (genomics) or the translation and modification of a large set of proteins (proteomics). Due to the limited scope of this book, we will not discuss genomic and proteomic approaches to studying systematic cellular changes.

11.1 CELL VIABILITY, PROLIFERATION, AND CYTOTOXICITY ASSAYS

One of the most common assays to study integrated cellular changes is the measurement of cell viability. Cell viability is the determination of the total number of live cells in a sample. Since all live organisms are made up of cells, cell viability assays have broad applications, such as the evaluation of the effectiveness of certain treatments to human cells (i.e., cancerous cells), the evaluation of the effectiveness of a

Assay Development: Fundamentals and Practices. By Ge Wu

pesticide or insecticide, the evaluation of environmental damages of chemical reagents to living species, and the evaluation of the rejection of implanted organs. The outcomes of a treatment to the cells may lead to cell proliferation or cell death. Both of these outcomes can be readily assayed by the determination of the change in the total number of cells before and after the treatment. In addition to simply counting the cell numbers, cell proliferation and cell death can also be determined by biomarkers. Cell death can be the result of apoptosis or necrosis. Though the end results of cell death are the same (24 h or longer after treatment), the processes leading to cell death are different. There are many biomarkers involved in apoptosis and necrosis that can usually be measured between 0.5 and 5 h after treatment. Thus, many assays can be configured to measure the biomarkers to study these processes.

11.1.1 Methods to Count the Number of Viable Cells

Testing cell viability involves looking at cell population in a sample under the microscope with the help of staining (applying special chemicals to the sample) to distinguish between the live and dead cells. There are numerous methods for measuring viability of cells. Trypan blue is the most widely used molecule that can selectively stain dead cells with blue color. Live cells with intact cell membranes are not stained because live cells are very selective in permitting compounds to pass through their membranes. The Trypan blue assay is performed under a microscope. The dead cells are shown as distinctive blue dots and the live cells appear as white dots. Since live cells are excluded from staining, this staining method is also referred to as a dye exclusion method. When performing the experiment, cells are first suspended in appropriate media and then mixed with an equal volume of 0.4% Trypan blue. After incubation for about 5 min at room temperature, 10 μL of the mixture is transferred to a hemocytometer. The number of viable (unstained) and dead (stained) cells can be counted under the microscope manually. Manual measurement puts a lot of strains on human eyes and is error prone, resulting in miss counting. Recently, Nexcelom Bioscience developed a line of instruments that automated the cell counting process, eliminating the need for human reading. Beckman also markets Vi-CELL series cell analyzers based on monitoring the size of the particles (cells) flowing through a sensing gate to count the number of cells. Though the measurements from both companies are automated, they are not interfaced to microplates and have seen limited use in high-throughput assays. The expensive high-end cell imaging instruments (discussed in Chapter 12) enable high-throughput counting of cell numbers and provide much more information about individual cells. In addition to directly counting cells, there exist methods that can measure cellular parameters that are proportional to the number of cells, providing high-throughput methods to count number of cells.

ATP as Surrogate Marker for Cell Number It was found that the ATP concentration in a live cell under normal condition is constant. Thus, the measurement of total ATP in the cell lysate is an indirect measure of the number of live cells in the sample. ATP can be readily measured by the luminescence produced in the luciferase/luciferin system (see Chapter 7). There are many commercial cellular

ATP measurement kits, such as CellTitor-Glo from Promega and ATPlite from PerkinElmer. These kits typically contain cell lysing reagents, ATPases inhibitors, and luciferase/luciferin. Because ATP detection with the luciferase/luciferin system is very sensitive, these kits can detect as few as 10 cells. After mixing CellTitor-Glo reagents with cells, the luminescence signal is immediately detectable (<2 min after mixing) and the signal is stable for several hours. Thus, ATP assays offer a quick snapshot of the ATP content in the cells. In comparison, other enzymatic substrate reduction assays, such as the AlamarBlue and XTT, can only detect the signal after a period of incubation time that is dependent on gradual turnover of substrate to product. In normal high-throughput screening mode with ATP assays, the signal is usually measured between 20 min and 1 h after mixing the cells and the reagents for ease of scheduling of the screening process with an integrated screening platform.

Total DNA as Surrogate Marker for Cell Number Because cellular DNA content is constant except when the cells undergo mitosis, the total DNA measurement provides another way to measure the total cell number. Molecular Probes (now part of Invitrogen) markets the CyQUANT assay kit to measure cell numbers. The kit employs a dye that gives intense fluorescence upon binding to DNA. The intensity of the fluorescence is proportional to total DNA and thus the number of cells in the sample. The kit requires no washes, extractions, growth medium changes, or long incubations.

Metabolic Activities as Surrogate Markers for Cell Number The number of cells in a sample can also be determined by the metabolic activities of the sample. Tetrazolium salts, such as MTT [3-(4,5-dimethylthiazol-2-yl)-2,5-diphenyl tetrasodium bromide], MTS[3-(4,5-dimethylthiazol-2-yl)-5-(3-carboxymethoxyphenyl)-2-(4-sulfophenyl)-2H-tetrazolium], and XTT (sodium 3′-[1-(phenylaminocarbonyl)-3,4-tetrazolium]-bis (4-methoxy-6-nitro) benzene sulfonic acid hydrate), are widely used for the quantification of viable cells. Terazolium salts are reduced by viable cells to formazan dyes, which can be detected by measurement of their absorption at specific wavelength. Figure 11.1 shows the reaction scheme for the reduction of MTT. The reaction mostly happens in mitochondria, and so the assays are largely a

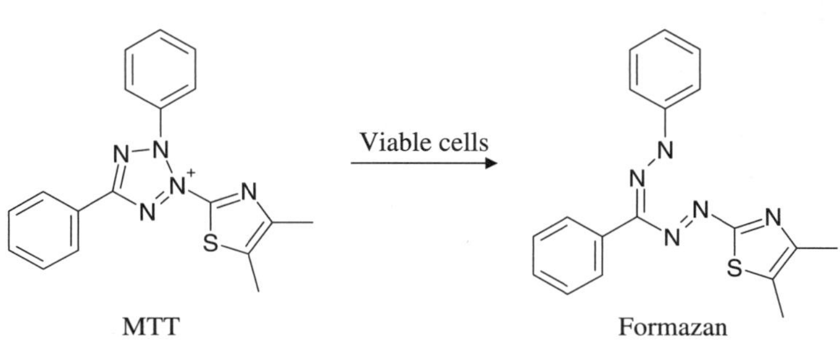

Figure 11.1 Reaction scheme of MTT reduction to formazan by live cells. Yellow MTT is reduced to insoluble purple formazan in the mitochondria of live cells. A solubilization solution is added to the sample to dissolve the insoluble formazan product into a colored solution.

measure of mitochondrial activity. At death, cells rapidly lose the ability to reduce tetrazolium salts. The production of the colored formazan product, therefore, is proportional to the number of viable cells in the culture. When performing the assays, cells are incubated 1 to 4 h at 37°C with the tetrazolium salts and then the absorbance is measured. MTT assay is the oldest among the three. The product from MTT reduction is insoluble in normal cell culture media. The insoluble formazan product has to be solubilized by a solubilization solution (such as DMSO) before the sample can be read. The absorbance of the colored solution can be quantified by absorption at a certain wavelength (usually between 500 and 600 nm). The absorption maximum is dependent on the solvent employed. MTS is a newer alternative to MTT. It is used together with phenazine methosulfate (PMS). The product of MTS/PMS is a water-soluble formazan product that has an absorbance maximum at 490 to 500 nm in phosphate-buffered saline. MTS/PMS has an advantage over MTT because MTS/PMS is reduced more efficiently than MTT, and the product is water soluble, making the assay less toxic to cells as compared with the insoluble product in MTT. XTT is another variation of the MTT assay. XTT can be reduced into soluble formazan product by the mitochondria of viable cells as well. The product of XTT can be read at 450 nm. The metabolic activity assays with tetrazolium salts can detect as low as 200 cells, which is an order less sensitivity than measuring ATP with luciferase.

The metabolic activities of live cells can also be measured by the oxidoreduction indicator dye resazurin. The principle of resazurin-based cell metabolic activity assay is shown in Figure 11.2. Resazurin can penetrate cells, where it is reduced to resorufin, probably as the result of the action of several different redox enzymes in mitochondria, cytosol and microsome. The fluorescent resorufin then diffuses from cells and back into the surrounding medium. The ability of different cell types to reduce resazurin to resorufin varies depending on the metabolic capacity of the cell line and the length of incubation with resazurin. The number of cells/well and the length of the incubation should be empirically determined. For most cells, 1- to 4-h incubation is adequate. There are many commercial kits available for this assay, such as AlamarBlue from Trek Diagnostic Systems. Resazurin in normal assay solution is dark blue in color. Reduction of resazurin by viable cells produces red fluorescent resorufin. The amount of resazurin conversion into resorufin in solution can be measured either fluorimetrically (detection of resorufin) or spectrophotometrically

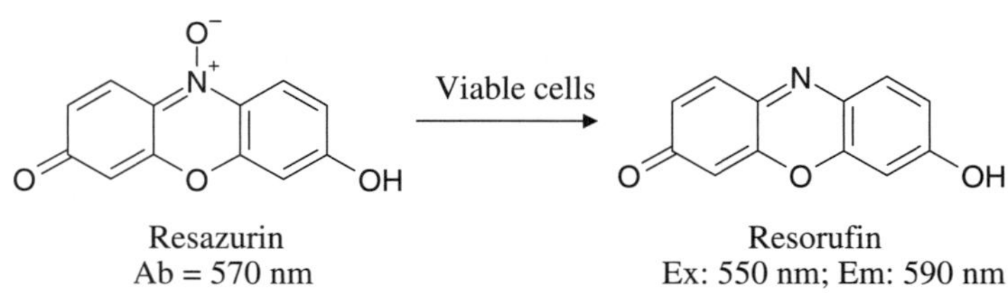

Figure 11.2 Reaction scheme of resazurin reduction to resorufin by live cells. The substrate resazurin absorbs light at 570 nm but does not fluoresce. The product resorufin is a fluorescent molecule that emits at 590 nm when it is excited at 570 nm.

(detection of resazurin). The absorbance maximum of resazurin is at 605 nm and that of resorufin is at 573 nm. The fluorescent signal is monitored at 590 nm when the sample is excited at 560 nm. The resazurin method is simple to perform and is inexpensive because the major component in the assay kit is resazurin. Since the fluorescent resorufin can be further reduced to nonfluorescent hydroresorufin with some cell types, the plot of assay signal versus cell number will be nonlinear (or even decrease, producing the so-called hook effect) at a high concentration of cells. Therefore, it is important to conduct a cell number titration assay for the particular cell line under study to determine the linear range of the assay to avoid this potential problem. Resazurin-based assays are usually performed by detecting signal at a particular incubation time when the reduction reaction is still going on. Alternatively, the reaction can be stopped by addition of 3% SDS and the sample can be read within 24 h. In general, resazurin-based assays with fluorescence measurement can detect as few as 40 cells, which are less sensitive than the ATP/luciferase-based assays (luminescence) but is more sensitive than the tetrazolium-based assays (absorption).

11.1.2 Measurement of Cell Proliferation

Cell proliferation can be measured by counting and subtracting the total number of cells before and after a treatment. Any one of the methods described above can be used to count cells. However, there is a caveat in this method to study cell proliferation because cell death and cell growth may happen simultaneously. For example, isolated primary human cells (such as T cells) will continuously die in cultured media and they grow slowly as well. To measure the effect of a treatment on T-cell proliferation, the cells must be treated with a test molecule and then incubated for a few days for them to double their number. If half of the T cells died and half of the remaining doubled through mitosis during the incubation period, the net gain of cell number will be zero. Even with no cell death, the signal-to-background ratio of the assay will only be 2 if the cells doubled after a few days of incubation. Thus, assays capable of measuring only the proliferative cells without the interference from nonproliferative cells will be desirable. It is known that DNA synthesis must happen before cell growth. Thus, measurement of DNA synthesis is a marker for cell growth. Measurement of [^{3}H]thymidine incorporation as cells enter S phase has long been the traditional method for the detection of cell proliferation. With this method, radioactive thymidine is added to the cell growth media. After the treatment of the cells with test compounds, the cells are transferred to a filtration membrane where the cells are retained on the membrane and the cell growth media will pass through the membrane. After washing the membrane, the radioactivity remaining on the membrane is counted. The radioactivity is proportional to the DNA in the newly formed cells, and hence is proportional to the number of newly formed cells. In theory, the signal-to-background ratio of this assay should be large because the background signal should be close to zero. However, the noise of the assay is very large because of the wash steps involved. Taking account of both the signal-to-background and the signal-to-noise ratios, the radioactive thymidine incorporation assay may not be as good as the other assays despite its obvious advantages in theory. Thus, experimental evaluation of different assay technologies should be performed for a particular project.

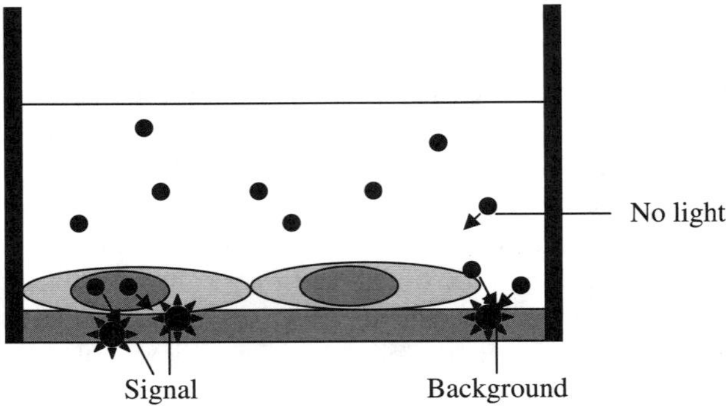

Figure 11.3 Illustration of Cytostar-T technology. The bottom of the wells in Cytostar-T microplate is coated with scintillant. The cells under study grow on the bottom of the well (adherent cells) or are loosely attached to the bottom (nonadherent cells). The [^{14}C]thymidine (black dots) are added to the growth media that will not give an SPA signal due to the long distance from the bottom of the plate. The [^{14}C]thymidines incorporated into the cells will give an SPA signal. Some [^{14}C]thymidines that are loosely associated with the cells or close to the bottom of the microplate by random motion produce assay background.

Cytostar-T is a technology from GE Healthscience that can perform the radioactive thymidine assays without the filtration step. This technology is based on scintillation proximity using scintillating microplates for the real-time analysis of a broad spectrum of cell-associated phenomena (Fig. 11.3). The bottom of the wells in Cytostar-T microplate is coated with scintillant. The cells grow on the bottom of the microplate (adherent cells) or are loosely attached to the bottom of the microplate (nonadherent cells). The [^{14}C]thymidines are added to the growth media, which will not give an SPA signal when they are in solution due to the long distance from the bottom of the plate. However, the [^{14}C]thymidines incorporated into the cells will give an SPA signal. The [^{14}C]thymidines that are loosely associated with the cells or are close to the bottom of microplate by random motion produce assay background. This technology permits real-time analysis of a diverse range of dynamic cellular activities. Though widely used, one should be cautious in interpreting data from experiments using the radioactive tracer to determine the rates of DNA synthesis and cell replication. There is accumulating evidence demonstrating that metabolic incorporation of a low-energy ß emitter can globally influence a diverse set of cellular activities that can, in turn, affect the outcome of many experiments by altering the cell cycle, metabolism, signaling, or redox status of the cell. Alternatively, stable isotope-labeled thymidines coupled with chemical reaction interface mass spectrometry have been demonstrated in detecting thymidine incorporation in cells. However, the technology requires expensive instrument and the throughput is too low for high-throughput applications.

Bromodeoxyuridine (BrdU), a thymidine analog, has been demonstrated capable of replacing [^{3}H]thymidine for thymidine uptake assay. BrdU is incorporated into newly synthesized DNA strands of actively proliferating cells. Following partial denaturation of double-stranded DNA, BrdU can be detected by an anti-BrdU antibody. The assay can be formatted by ELISA or formatted as image-based assays.

There are many commercial ELISA BrdU assay kits available. In a typical ELISA-based assay, BrdU is added to the wells of a microplate containing the cells in the final 2 to 24 h of culture. The cells are then fixed and permeabilized. The DNA is denatured with acid or alkali. After washing, anti-BrdU monoclonal antibody is added into the wells and allowed to bind to any incorporated BrdU for about 1 h. The unbound antibody is washed away. A second antibody with conjugated horse-radish peroxidase, such as goat antimouse antibody, is added to the wells. The unbound secondary antibody is then washed away. When a substrate of HRP is added to the microplate, the signal can be detected that is proportional to the amount of BrdU in the well. One drawback of this approach is the need to fix and permeabilize the cells and denature cellular DNA by acid or alkali to allow access to BrdU, which may give inconsistent results. Alternatively, antibody access can be achieved by nuclease digestion of DNA under mild condition (GE Healthscience's cell proliferation kit).

11.1.3 Measurement of Cell Death

Cell death can occur by two different mechanisms, necrosis and apoptosis. Necrosis is the premature or unnatural cell death that results from acute cellular injury caused by external factors, such as infection, toxins, or trauma. When cells go through necrosis, they swell and brust, releasing their intracellular contents that can damage surrounding cells and cause inflammation. On the other hand, apoptosis is programmed cell death in multicellular organisms that follows a well-defined sequence of biochemical changes leading to characteristic cell morphology changes, including blebbing, cell membrane changes (e.g., loss of membrane asymmetry and attachment), cell shrinkage, nuclear fragmentation, chromatin condensation, and chromosomal DNA fragmentation. In contrast to necrosis, the contents of the apoptotic cells are taken by phagocytes and are not damaging to neighboring cells. Apoptosis, in general, confers advantages during an organism's life cycle. For in vitro studies of cell death, the cell membrane will leak, and the cellular content will be released at the end of the process no matter whether the cells go through necrosis or apoptosis. Thus, measurement of cellular content releasing to the media has been widely used to study cell death. The release of ^{51}Cr radioactivity of cells labeled by radioactive chromate is a classic method to measure short-term cytotoxicity. With this method, the cells under study are first labeled with ^{51}Cr by incubating with [^{51}Cr]sodium chromate for about 1 h at 37°C. The labeled cells are then separated from the radioactive chromate by centrifugation/washing. After treatment of the labeled cells with test reagent, the cells are separated from the media. The radioactivity released in the media is then measured. This assay is easy to perform, highly sensitive, with low spontaneous release of radioactivity, and the labeling is nontoxic to the cells. Furthermore, this classic assay is applicable to a wide range of effector and target cells. However, radioactivity and the short half-life of the label hamper the use of this method. Alternatively, europium chelate can be used to label the cells. The release of europium into solution, where it forms a highly fluorescent chelate, can be measured by time-resolved fluorescence with high sensitivity. In addition to the cell-labeling methods, the release of many enzymes originally in cytosol into the media offers another way to measure cell death. For example, lactate dehydrogenase (LDH) released into the media is widely used as a cytotoxicity assay and many commercial kits are available to perform this assay. LDH is a stable cytosolic enzyme that is released upon cell lysis in much the same way as [^{51}Cr] is released in radioactive assays. The LDH in culture supernatants can be measured with coupled enzymatic assays that covert a tetrazolium salt into a red formazan product with absorbance at 490 nm. The extent of absorption is proportional to the amount of LDH and hence the number of lysed cells.

Because apoptosis is a programmed process, many events involved in the process can be used to assay apoptosis in addition to the cell leakage assay discussed above. When cells commit to apoptosis, the caspase family of cysteine proteases is activated. These ubiquitous enzymes exist as inactive zymogens in cells and are cleaved before forming active heterotetramers that drive apoptotic events. Caspases are the central mediators of the proteolytic cascade leading to cell death. Assays that directly measure caspases activities can provide valuable information about the mechanism of cell death. There are many commercial kits available that employ fluorescent and luminescent substrates to assay the activity of many caspases. For example, Ac-DEVD-AMC is a cell-permeable fluorogenic substrate for caspase-3 (it was discussed in Chapter 6). The loss of membrane asymmetry is another apoptotic event that can be measured. Normally, eukaryotic cells maintain asymmetric distribution of phospholipids between the inner and outer leaflets of the cell membrane with a majority of phosphatidylserine (PS) on the inner leaflet. During apoptosis, PS becomes abundant on the outer leaflet. Annexin V is a protein that has a high affinity for PS. Annexin V does not bind to normal intact cells but will bind to the PS in the outer leaflet of the apoptotic cell membrane. With this assay, Annexin V is usually conjugated to a dye to label apoptotic cells. The labeled cells can be studied with an imaging-based method (see Chapter 12) or flow cytometry. DNA fragmentation is another characteristic of apoptotic cells that can be employed to assays cells going through apoptosis. During apoptosis, the genomic DNA in apoptotic cells is cleaved into 180 to 200 bp fragments. TUNEL (TdT-mediated dUTP nick end labeling) is a common method for detecting DNA fragmentation. The presence of nicks in the DNA can be recognized by terminal deoxynucleotidyl transferase, an enzyme that catalyzes the addition of dUTPs to fragmented DNA. The dUTPs can be labeled fluorescently with dyes, such as fluorescein or Alexa Fluor dyes. Alternatively, 5-bromo-2′-deoxyuridine 5′-triphosphate (BrdUTP) can be used to label the fragmented DNAs. Once incorporated into DNA, the BrdU moiety can be detected by an anti-BrdU antibody. Nonapoptotic cells do not incorporate much of the labeled dUTP because of the absence of exposed 3′-hydroxyl DNA ends. The labeled cells can be studied with image-based assays.

11.2 MEASUREMENT OF EXTRACELLULAR INDICATORS OF CELLULAR METABOLISM

Cellular metabolism is the chemical reaction in live cells by which energy is generated/utilized for vital processes and activities and new material is assimilated. The chemical reactions are executed through a series of intracellular biochemical processes (e.g., glycolysis, TCA cycle, electron transport, and oxidative phosphorylation). The process involves substrate uptake (e.g., oxygen and glucose), production/utilization of energy (ATP), generation of chemical by-products (proton, CO_2, and lactate), and generation of heat. In response to extracellular signals, cells rapidly change their metabolism to cope with the perturbations. The change in cellular metabolism leads to specific molecules moving in and out of the cells and are indicators of changes in cellular physiology. McConnell and colleagues (1992) built a device called a microphysiometer that can measure the extracellular pH with light-addressable potentiometric sensor. (The commercial version of the microphysiometer, Cytosensor, was marketed by Molecular Devices). They showed that bioactive ligands for cell surface receptors (e.g., GPCRs and RTKs) can be identified by analysis of the rate of proton excretion. Oxygen utilization is another important measure of cell metabolism because it is often a limiting substrate for cell growth owing to its low solubility in water. There are two common methods to measure oxygen in bioassays. The Clark-type electrode is the early and commonly used sensor for dissolved oxygen. A second and newer method to measure oxygen uses fluorescence probes sensitive to quenching by dissolved oxygen. The electrochemical sensors perform better at high oxygen concentrations. However, they actively consume oxygen, which may influence some measurements. Optical sensors are more sensitive at low oxygen concentrations (generally $<50\%$ of air saturation) and do not consume oxygen during measurement. Under typical in vitro cell culture conditions, the rate of oxygen consumption is an indicator of mitochondrial respiration, and the rate of proton efflux is predominantly a measure of lactic acid formed during glycolytic energy metabolism. It has been shown that glucose utilization by the H9C2 rat heart myoblast cell line measured by oxygen consumption and acid extrusion rates are comparable with standard radiometric assay. Using modified electrodes in the standard Cytosensor plunger, Cytosensor was shown to be able to measure changes in extracellular oxygen, glucose, and lactate in addition to pH. Glucose and lactate was measured indirectly at platinum electrodes by amperometric oxidation of hydrogen peroxide, which is produced from catalysis of glucose and lactate at films containing their respective entrapped oxidase.

Though the measurement of extracellular molecules for analysis of cell metabolism has a long history, currently only Seahorse Bioscience markets high-throughput Extracellular Flux (XF) assay platform that can read both oxygen consumption and the acidification rate simultaneously in 96-well microplates. The XF96 Extracellular Flux Analyzer is a fully integrated instrument that simultaneously measures aerobic respiration and glycolysis. The XF Assay Kit consists of a disposable dual-analyte sensor cartridge. Two fluorophores for analyte detection are embedded in a polymer spotted on the end of each of the sensor sleeves. The oxygen is measured by the fluorescence quenching method. The proton concentration is measured when the protons diffuse across the fluorophore-based polymetric substrate. The XF Analyzer's fiber-optic waveguide is inserted into the sleeves of the sensor cartridge. Excitation light and emission light is conducted through the fiber-optic bundles. Each sensor cartridge also contains 96 compound storage wells. The multichannel drug delivery system can inject up to four different drug compounds into each well allowing multiple measurements with a single population of cells. XF assays enable label-free, time-resolved analysis of the cell metabolic activities. Adherent cells, primary cells, suspension cells, and isolated mitochondria can be analyzed with the instruments.

BD Biosciences markets an oxygen biosensor system in 96-well format (Oxoplate) capable of kinetically monitoring dissolved oxygen. An oxygen-sensitive dye embedded in a gas-permeable silicone matrix is permanently attached to the bottom of each well in the microplate. Oxygen in the cell media can freely diffuse through the dye-embedded matrix. Oxygen concentration is detected by the extent of fluorescence quenching. When oxygen is depleted from the media as a result of cell growth that consumes oxygen, the concentration of oxygen in the matrix decreases and the fluorescence increases.

11.3 MEASUREMENT OF CELL'S EFFECT ON ELECTRICAL IMPEDANCE

When discussing ion channels in Chapter 9, we introduced the concept that cell membranes can be modeled as a resistance and a capacitance in parallel. When a layer of cells is placed in the electric field, the same concept can be applied though the situation is more complicated. With alternating current, the impedance (Z) is calculated as the ratio of applied voltage to measured current as described by Ohm's law ($Z = V/I$), the same way as the resistance is calculated in direct current. Figure 11.4 illustrates the current flow and resistance with a cell layer in the electric field. The current can flow either through the space between cells (extracellular current) or flow though the cell membrane by capacitive current (transcellular current). The resistance to extracellular current comes from the media between the bottom of the cell surface and the electrode plus the media filling the space between the cells. The measured impedance in the presence of a cell layer depends on many factors, such as cell-substrate adherence, cell shape and volume, and cell–cell interactions. These factors individually or collectively affect the flow of extracellular and transcellular current, influencing the magnitude and characteristics of the signal measured.

Giaever and Keese (1984) originally described impedance measurement of cells in externally applied electric fields. The spreading and motion of fibroblasts cultured on gold electrodes was observed as the electrical impedance changes because the cells

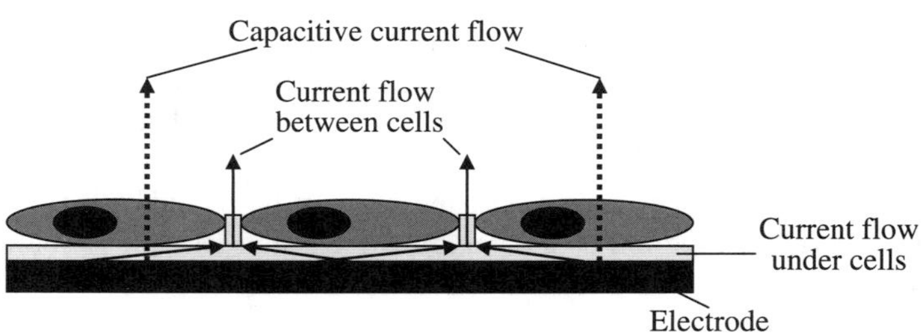

Figure 11.4 Current flow and resistance with a cell layer in electric field. The current can flow either through the space between cells (extracellular current, solid lines) or flow though the cell membrane by capacitive current (transcellular current, dotted lines). The resistance to extracellular current comes from the media between the bottom of the cell surface and the electrode plus the media filling the space between the cells.

constrain the current flow. As the fibroblasts attach and spread on these surfaces, the measured impedance between the electrodes increases, reflecting the amount of area blocked by the spreading cells. Applied BioPhysics was subsequently formed to market this technology. In addition to Applied BioPhysics, ACEA Biosciences and MDS Analytical Technologies also market their cell-based assay systems based on the measurement of impedance but with different designs that offer higher throughput and higher sensitivity in many applications. In addition to measurement of cell attachment and spreading, cellular impedance measurements have been shown to be able to sense many other biological phenomena, such as cell migration, wound healing, invasion of cancer cells, cell membrane conductivity, monolayer permeability, cell morphology changes, cellular micromotion, and receptor (GPCR and RTK) activation. The weak electrical field used in impedance assays has no noticeable effect on the cells. The method requires no labeling of cells and significantly reduced the assay development time. The Cellkey system from MDS Analytical Technologies can perform assays in standard 384-well microplate format. Remarkably, the system is capable of distinguishing the main subset (G_s, G_i, and G_q) of GPCR pathways when applied to GPCR studies. Because of the unique needs at Five Prime Therapeutics (a former employer) to screen for secreted factors using primary cells, our attention was immediately drawn to the label-free impedance system (RT-CES) from ACEA which was the first marketed instrument capable of high throughput. The system uses E-plate that conforms to the normal 96-well microplate standard. The bottom of each well in the E-plate is covered with more than two dozen interlaced microelectrodes that are used to sense the impedance (Fig. 11.5). The dense array of

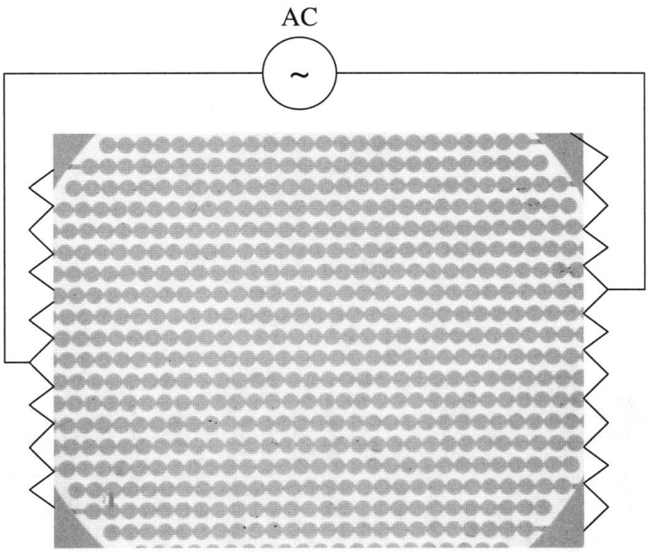

Figure 11.5 Microscopic view of the bottom of a well in the 96-well ACRA E-plate from the top. The whole well is extensively covered with interlaced arrays of microelectrode. AC is applied to the interlaced electrodes to obtain the impedance value. When cells are attached to the surface of the electrodes, the impedance changes from the baseline in the absence of the cells.

microelectrode offers much higher sensitivity than the system using only one or a few pairs of electrodes. We obtained the first six-station 96-well plate assay system from ACEA for screening applications before the Cellkey system was marketed. Below we will show an example of the application of impedance assays using the ACEA system.

We were trying to identify secreted proteins that act on primary muscle cells from a library of more than 2000 secreted proteins. Before using the impedance system, a scientist in our company had spent more than half a year trying to develop an assay system using other established methods with both primary muscle cells and

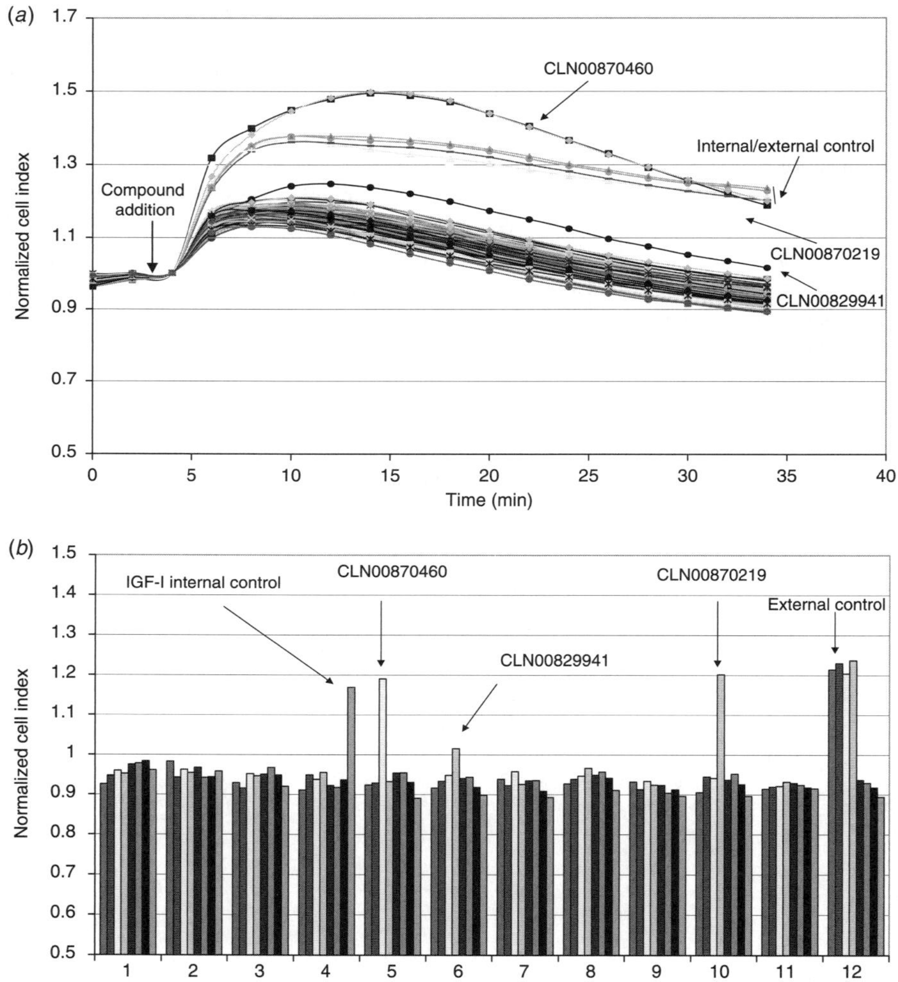

Figure 11.6 Screening of test compounds in 96-well format impedance assays with (*a*) primary muscle cells, and (*b*) raw data from ACEA RT-CES containing 96 traces. A single point in the kinetic traces in (*a*). In addition to the controls, three secreted proteins are identified as hits in this plate. (See color insert.)

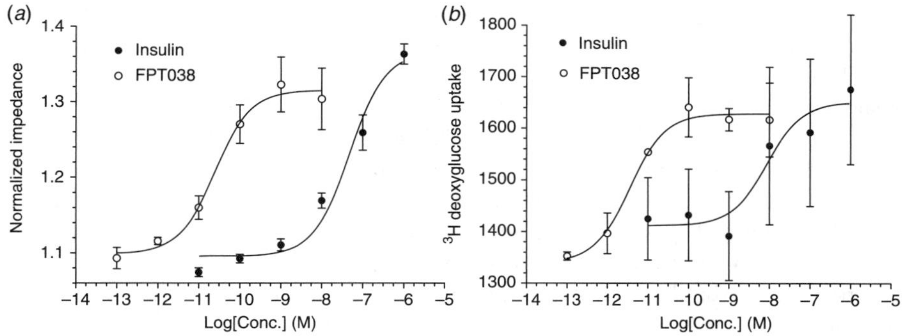

Figure 11.7 Comparison of data generated from (*a*) ACEA RT-CES impedance assay and (*b*) from radioactive glucose uptake assay with primary muscle cells. A known active protein (insulin) and a newly discovered secreted protein (FPT038) are tested with both methods. The data quality with RE-CES is significantly better than the established glucose uptake assays.

cell lines (L6 and C2C12) without much success. With the ACEA system, it takes less than one month to develop an assay system for screening. The assay development in electrical impedance format only requires the microplate coating reagent (e.g., poly-D-lysine, collagen, gelatin, BSA, etc.), pretreatment of the cells (e.g., starvation or not and the timing between starvation and treatment), and the media to maintain the cells after treatment. After assay development, screening of secreted proteins was performed in 96-well microplates. The primary cells were first allowed to grow in the E-plate and the baseline data were collected. The secreted proteins were then added to the E-plate and the data collection is continued. The normalized kinetic data with a particular 96-well microplate is shown in Figure 11.6*a*. From the kinetic traces in Figure 11.6*a*, the data at a single time point can be obtained, and are shown in Figure 11.6*b*. The assay signal-to-background ratio is only about 1.5. However, the kinetic nature of the assay enabled high-quality screening that would otherwise be difficult with end-point assays. To further illustrate the data quality in impedance-based assays, two secreted proteins were tested with both RT-CES assays and radioactive glucose uptake assays. The data are shown in Figure 11.7. Insulin is a known protein that induces glucose uptake in primary muscle cells, and FPT038 is a newly discovered secreted protein in our screening effort. FPT038 was found capable of inducing glucose uptake with EC_{50} at about 500-fold lower than insulin.

11.4 MEASUREMENT OF PROTEIN SECRETION FROM CELLS

The regulated release of proteins from cells is a common phenomenon. For example, insulin is secreted from β cells and cytokines are secreted from white blood cells when activated. Thus, measurement of the secretion of proteins from a specific cell offers a way to measure the activation of the cell. The secreted proteins are released to the cell growth media that normally contains many proteins, especially with undefined media that usually contains bovine serum. To quantify a particular secreted protein mixed with a large number of other proteins with varying quantity is not straightforward. The common method is to take the supernatant and run SDS-PAGE to separate the proteins into different bands. The band corresponding to the secreted protein of interest can be visualized by staining (coomassie blue or silver stain) or by Western blot with antibodies against the secreted protein. However, this method is labor intensive and the throughput is too low for applications with a large number of samples. Quantification of secreted proteins based on ELISA is the most common method for identifying and quantifying secreted proteins though the cell media sometimes may interfere with the assay. The assay is high throughput and multiple analytes can be assayed simultaneously (multiplexed assays). Multiplexing technologies are particularly important when studying the secretion of cytokines in a cell culture or patient sample. With more than 100 cytokines identified so far, individually testing for each cytokine is time consuming.

There are many multiplexed cytokine-specific assay kits available commercially. In general, they can be divided by whether the assay is performed on the surface of a microplate or on the surface of beads from Luminex. With microplate-based assays (such as SearchLight from Thermo Scientific), antibodies for multiple cytokines are spotted at the bottom of a microplate. The supernatant of the cell growth media are added to the microplate and the cytokines are allowed to bind selectively to the antibodies immobilized on the surface. The assay procedure is the same as traditional sandwiched ELISA assays. The final readout is chemiluminescence with SearchLight. Similar assays can also be carried out in ECL format (Meso Scale Discovery multiplexed cytokine kits). In addition, Luminex bead-based multiplexed cytokine assay kits are available from several vendors (e.g., BeadLight from Millipore). With Luminex technology, each bead is individually addressable and up to 100 beads can be distinguished form each other. Thus, up to 100 proteins can be simultaneously analyzed with Luminex technology in theory. Our lab at Five Prime Therapeutics has extensively used the Luminex bead-based multiplex assays to screen for proteins that can activate white blood cells by monitoring the secretion of a panel of cytokines. The antibody specific for each cytokine was attached to the beads with the same code. The beads with different antibodies attached to them were mixed together. The mixed beads were treated with supernatant from the cells that had been treated with a library of secreted proteins. The released cytokines from the activated cells were allowed to bind to the antibodies on the beads. After washing the beads to remove the unbound substances in the supernatant, the beads were treated with fluorescently labeled antibodies that bind the cytokines on the beads. After washing away excessively labeled antibodies, the mixed beads were analyzed with Luminex 100 instruments that read the fluorescence intensities for each coded bead. The median value of the fluorescent intensity from the beads with the same code was used to calculate the concentration of the cytokine.

11.5 MEASUREMENT OF DISCOLORATION OF MELANOPHORE CELLS

In amphibians, such as *Xenopus laevis*, skin coloration is controlled by melatonin through an action on melanin-containing pigment granules (melanosomes) in melanophores. In these cells, very low concentrations of melatonin activate the Mel(1c) receptor subtype, triggering the movement of granules toward the cell center and causing skin discoloration. Functional assays based on the discoloration of melanophore cells have been developed for GPCR, RTKs, and cytokine receptors. In GPCR assays, the GPCR is cloned into the melanophore cells. Activation of GPCR causes pigment redistribution within the cells resulting in discoloration. The assay is applicable to all subset of GPCRs (G_s, G_i, and G_q). Thus, this assay can be used to screen orphan GPCRs for novel ligands. The melanophore-based assay was also demonstrated with PDGF receptors. In this assay, the murine PDGFβ receptor was transiently expressed in melanophore cells. The homodimeric ligand PDGF-BB activated the receptor and led to dose-dependent pigment dispersion, whereas it did not induce pigment dispersion in wild-type cells. In addition, PDGF-AA had no ability to induce pigment dispersion in melanophore cells transiently expressing the PDGFβ receptor. The PDGF-BB-induced pigment dispersion could be blocked by protein kinase C inhibitors. When the full-length EGFR was expressed in melanophores, it was shown that EGF could mediate pigment dispersion in a time- and dose-dependent manner. A functional assay for the erythropoietin receptor (a cytokine receptor) was demonstrated using frog melanophore cells. In this assay, a chimeric receptor that comprised the extracellular portion of the murine erythropoietin receptor (EPOR) and the transmembrane and intracellular domains of the human EGFR was subcloned into the expression vector. When the chimeric EPOR/EGFR was expressed in melanophore cells, EPO but not EGF stimulated pigment dispersion in a time- and dose-dependent manner. Neither EGF nor EPO had any effect on pigment dispersion in wild-type melanophores. The EGF- and EPO-mediated pigment dispersion can be blocked by protein kinase C inhibitor too.

11.6 MEASUREMENT OF CELL MOTILITY

The migration of specialized cells is essential in diverse physiological and pathological processes including embryogenesis, immunity, and diseases such as cancer and chronic inflammatory disease. The movement of many cell types, such as leukocytes, fibroblasts, stem cells, and cancer cells, is directed by extracellular gradients of diffusible chemicals. The phenomenon of cells, bacteria, and multicellular organisms directing their movements to certain chemicals in their environment is referred to as chemotaxis. A large family of small proteins (chemokines) serves as the extracellular signals. The chemokine receptors (a family of GPCRs) detect the gradients of chemokines and guide cell movement in vivo. Evaluating the chemotactic activity of cells offers a way to screen for active molecules that act on these specialized cells. The assays for chemotaxis require the following three components: the establishment and maintenance of the concentration gradients of the attractant during the assay; a barrier that can initially separate the cells from the chemical but allows the movement of cells toward or away on the axis of the concentration gradient; and the detectable signal as the result of the active migration of cells. Early mobility assay was performed in an agar plate. Small wells are cut into the layer of the semisolid. Cells and test substance are then introduced into separate wells. Cells can migrate toward the chemical gradient in the semisolid layer. The movement of the cells can be observed using a microscope in native condition or with stained cells. The agar assay is easy to perform with a small number of samples. However, the throughput is low. When screening a large number of samples, two chamber systems, such as the Boyden chamber, are widely used. In this design, the two chambers are isolated by filters with varying pore size (e.g., 3, 5, and 8 μm). The pore size of the filter is determined by the size of the motile cells under study. It is essential to choose the pore size that allows active transmigration of the cells through the filter. The motile cells are placed in the upper chamber and the solution containing the test substance is placed in the lower chamber. The filter can be coated with extracellular matrix substances (e.g., collagen and elastin) to mimic in vivo conditions. The Boyden chamber design adapted into standard microplate format can be obtained from Millipore and Neuro Probe. Up to 96 samples can be evaluated in parallel. The migrated cells can be counted either in the lower chamber (long incubation time) or on the filter (short incubation time).

Useful Websites

http://www.nexcelom.com/

http://www.beckman.com/products/instrument/partChar/pc_vicell.asp

http://www.trekds.com/products/alamarBlue/alamarblue.asp

http://www.seahorsebio.com/

http://www.biophysics.com

http://www.cellkey.com/

http://www.aceabio.com/

http://www.neuroprobe.com/

http://www.millipore.com/

BIBLIOGRAPHY

Brunner, K. T., Mauel, J., Cerottini, J. C., and Chapuis, B. (1968) Quantitative assay of the lytic action of immune lymphoid cells of 51Cr-labelled allogeneic target cells in vitro; inhibition by isoantibody and by drugs. *Immunology* **14**, 181–196.

Carrithers, M. D., Marotti, L. A., Yoshimura, A., and Lerner, M. R. (1999) A melanophore-based screening assay for erythropoietin receptors. *J. Biomol. Screen.* **4**, 9–14.

Castellano, F. N. and Lakowicz, J. R. (1998) A water-soluble luminescence oxygen sensor. *Photochem. Photobiol.* **67**, 179–183.

Ciambrone, G. J., et al. (2004) Cellular dielectric spectroscopy: A powerful new approach to label-free cellular analysis. *J. Biomol. Screen.* **9**, 467–480.

Ciobanu, M., Taylor, D. E., Wilburn, J. P., and Cliffel, D. E. (2008) Glucose and lactate biosensors for scanning electrochemical microscopy imaging of single live cells. *Anal. Chem.* **80**, 2717–2727.

Eklund, S. E., Kozlov, E., Taylor, D. E., Baudenbacher, F., and Cliffel, D. E. (2005) Real-time cell dynamics with a multianalyte physiometer, in *NanoBiotechnology Protocols* (eds. S. J. Rosenthal and D. W. Wright), Humana, Totowa, NJ, pp. 209–223.

Eklund, S. E., Taylor, D., Kozlov, E., Prokop, A., and Cliffel, D. E. (2004) A microphysiometer for simultaneous measurement of changes in extracellular glucose, lactate, oxygen, and acidification rate. *Anal. Chem.* **76**, 519–527.

Entschladen, F., et al. (2004) Analysis methods of human cell migration. *Exp. Cell Res.* **307**, 418–426.

Fang, Y. (2006) Label-free cell-based assays with optical biosensors in drug discovery. *Assay Drug Dev. Technol.* **4**, 583–595.

Ferrick, D. A., Neilson, A., and Beeson, C. (2008) Advances in measuring cellular bioenergetics using extracellular flux. *Drug Discov. Today.* **13**, 268–274.

Giaever, I. and Keese, C. R. (1984) Monitoring fibroblast behavior in tissue culture with an applied electric field. *Proc. Natl. Acad. Sci. U.S.A.* **81**, 3761–3764.

Giaever, I. and Keese, C. R. (1991) Micromotion of mammalian cells measured electrically. *Proc. Natl. Acad. Sci. U.S.A.* **88**, 7896–7900.

Giaever, I. and Keese, C. R. (1993) A morphological biosensor for mammalian cells. *Nature* **366**, 591–592.

Gonzalez, R. J. and Tarloff, J. B. (2001) Evaluation of hepatic subcellular fractions for alamar blue and MTT reductase activity. *Toxicol. In Vitro* **15**, 257–259.

Graminski, G. F. and Lerner, M. R. (1994) A rapid bioassay for platelet-derived growth factor beta-receptor tyrosine kinase function. *Biotechnology (NY)* **12**, 1008–1011.

Hafner, F. (2000) Cytosensor microphysiometer: Technology and recent applications. *Biosens. Bioelectron.* **15**, 149–158.

Hanson, M. A., et al. (2007) Comparisons of optical pH and dissolved oxygen sensors with traditional electrochemical probes during mammalian cell culture. *Biotechnol. Bioeng.* **97**, 833–841.

Harms, P., Kostov, Y., and Rao, G. (2002) Bioprocess monitoring. *Curr. Opin. Biotechnol.* **13**, 124–127.

Hootman, S. R., Hobbs, E. C., and Luckie, D. B. (2005) Direct measurement of acid efflux from isolated guinea pig pancreatic ducts. *Pancreas* **30**, 363–368.

Hu, V. W., Black, G. E., Torres-Duarte, A., and Abramson, F. P. (2002) 3H-thymidine is a defective tool with which to measure rates of DNA synthesis. *FASEB J.* **16**, 1456–1457.

Hynes, J., Floyd, S., Soini, A. E., O'Connor, R., and Papkovsky, D. B. (2003) Fluorescence-based cell viability screening assays using water-soluble oxygen probes. *J. Biomol. Screen.* **8**, 264–272.

Jin, T., Xu, X., and Hereld, D. (2008) Chemotaxis, chemokine receptors and human disease. *Cytokine* **44**, 1–8.

Lash, G. E., et al. (2006) Comparison of three multiplex cytokine analysis systems: Luminex, SearchLight™ and FAST Quant. *J. Immunol. Methods* **309**, 205–208.

Lerner, M. R. (1994) Tools for investigating functional interactions between ligands and G-protein-coupled receptors. *Trends Neurosci.* **17**, 142–146.

Liebsch, G., Klimant, I., Frank, B., Holst, G., and Wolfbeis, O. S. (2000) Luminescence lifetime imaging of oxygen, pH, and carbon dioxide distribution using optical sensors. *Appl. Spectrosc.* **54**, 548–559.

Lo, C. M., Keese, C. R., and Giaever, I. (1995) Impedance analysis of MDCK cells measured by electric cell-substrate impedance sensing. *Biophys. J.* **69**, 2800–2807.

Marko, N. F., et al. (2003) Does metabolic radiolabeling stimulate the stress response? Gene expression profiling reveals differential cellular responses to internal beta vs. external gamma radiation. *FASEB J.* **17**, 1470–1486.

McConnell, H. M., et al. (1992) The cytosensor microphysiometer: Biological applications of silicon technology. *Science* **257**, 1906–1912.

O'Brien, J., Wilson, I., Orton, T., and Pognan, F. (2000) Investigation of the alamar blue (resazurin) fluorescent dye for the assessment of mammalian cell cytotoxicity. *Eur. J. Biochem.* **267**, 5421–5426.

Owicki, J. C. and Parce, J. W. (1990) Bioassays with a microphysiometer. *Nature* **344**, 271.

Owicki, J. C. and Parce, J. W. (1992) Biosensors based on the energy metabolism of living cells: The physical chemistry and cell biology of extracellular acidification. *Biosens. Bioelectron.* **7**, 255–272.

Pancrazio, J. J., Whelan, J. P., Borkholder, D. A., Ma, W., and Stenger, D. A. (1999) Development and application of cell-based Biosensors. *Ann. Biomed. Eng.* **27**, 697–711.

Rolfe, D. F. and Brown, G. C. (1997) Cellular energy utilization and molecular origin of standard metabolic rate in mammals. *Physiol. Rev.* **77**, 731–758.

Solly, K., Wang, X., Xu, X., Strulovici, B., and Zheng, W. (2004) Application of real-time cell electronic sensing (RT-CES) technology to cell-based assays. *Assay Drug Dev. Technol.* **2**, 363–372.

Thomas, C. A. J., Springer, P. A., Loeb, G. E., Berwald-Netter, Y., and Okun, L. M. (1972) A miniature microelectrode array to monitor the bioelectric activity of cultured cells. *Exp. Cell Res.* **74**, 61–66.

Van Haastert, P. J. M. and Veltman, D. M. (2007) Chemotaxis: Navigating by multiple signaling pathways. *Sci. STKE* **2007**, pe40.

von Zons, P., et al. (1997) Comparison of europium and chromium release assays: Cytotoxicity in healthy individuals and patients with cervical carcinoma. *Clin. Diagn. Lab. Immunol.* **4**, 202–207.

Watanabe, M., et al. (2006) Bile acids induce energy expenditure by promoting intracellular thyroid hormone activation. *Nature* **439**, 484–489.

Weidmann, E., et al. (1995) Lactate dehydrogenase-release assay: A reliable, nonradioactive technique for analysis of cytotoxic lymphocyte-mediated lytic activity against blasts from acute myelocytic leukemia. *Ann. Hematol.* **70**, 153–158.

Wiley, C. and Beeson, C. (2002) Continuous measurement of glucose utilization in heart myoblasts. *Anal. Biochem.* **304**, 139–146.

Wodnicka, M., et al. (2000) Novel fluorescent technology platform for high throughput cytotoxicity and proliferation assays. *J. Biomol. Screen.* **5**, 141–152.

Wu, G. and Doberstein, S. (2006) High throughput screening technologies in biopharmaceutical discovery. *Drug Discov. Today.* **11**, 718–724.

Wu, G. and Hubbell, W. L. (1993) Phospholipid asymmetry and transmembrane diffusion in photoreceptor disc membranes. *Biochemistry* **32**, 879–888.

Wu, M., et al. (2007) Multiparameter metabolic analysis reveals a close link between attenuated mitochondrial bioenergetic function and enhanced glycolysis dependency in human tumor cells. *Am. J. Physiol. Cell. Physiol.* **292**, C125–136.

Xi, B., Yu, N., Wang, X., Xu, X., and Abassi, Y. A. (2008) The application of cell-based label-free technology in drug discovery. *Biotechnol. J.* **3**, 484–495.

Yu, N., et al. (2006) Real-time monitoring of morphological changes in living cells by electronic cell sensor arrays: An approach to study G protein-coupled receptors. *Anal. Chem.* **78**, 35–43.

CHAPTER 12

HIGH-CONTENT CELL-BASED ASSAY WITH OPTICAL IMAGING TECHNIQUES

MOST OF the bioassays we discussed so far are based on measuring one or more signals of the test system, and these signals are collected from a well in a microplate or from arrays of spots in a well in a microplate. In these studies, the signals are from the bulk sample with no *spatial* resolution within the well or within the spot. These signals are one-dimensional and contain limited information about the test system. In these studies, we selectively observe what we want to see with predefined labels and ignore all the other changes that are usually not detectable with the predefined assays. When performing biochemical assays, the components in the assay systems are usually less than 10 nm in scale, which is too small to be visualized with reasonable throughput with current technology. Hence the whole test system can only be treated as a homogeneous unit, and the integrated signal from the bulk is used to represent the microenvironments. In cell-based assays, however, the fine structures of the cells are in the 10-μm scale, which can be readily observed with conventional optical microscopes. When the images of the individual cells are collected in a cell-based assay, the signal is two dimensional with a conventional microscope and three-dimensional with a confocal microscope. This extra spatial information is very valuable and many cellular changes, intended and unintended for detection, are observed after the cells are stimulated. Thus, the analysis of the cellular images opens up many opportunities for new assays that were not possible with traditional assays lacking spatial resolution. This is not a new concept and using a microscope to collect cellular information has been practiced in pathology for a long time. However, the old image-collecting processes involve manually making tissue slices, manually staining the slices, and manually mounting the slices to the microscope stage. The images were recorded with film and were interpreted by pathologists. These processes are tedious and are not suitable for bioassays that require reasonable throughput.

Assay Development: Fundamentals and Practices. By Ge Wu

Technology advances in instrument hardware and computer software changed the whole field in the late 1990s. In 1997, together with about a dozen colleagues from Merck, I had a chance to visit a small biotechnology company called BioDx (later changed to Cellomics) in a Pittsburgh suburb. BioDx was just formed about a year ago. They offered integrated solutions to automate all the processes involved in imaging cells, such as cell staining, sample mounting, image collection, and image analysis. The technology was implemented in 96-well microplates and the cells in each well in the microplate were imaged with subcellular spatial resolution using multiplexed fluorescence microscopy. The automated subcellular image analysis platform offered by BioDx was called high-content analysis (HCA) or high-content screening (HCS), and it opened a new frontier for bioassays. Since there are many other high-content assay technologies that collect multiple readouts from the test system (e.g., flow cytometry), I will use a different name, image-based cellular high-content analysis (icHCS), to describe the platform by limiting its scope back to what it really does. Basic components of icHCS are sample preparation, image collection, and data handling (analysis, output, achieving, and retrieval). The imaging data collected in icHCS contain both the intended information and additional information that was not analyzed or could not be analyzed because of lack of knowledge to interpret them. However, the extra information may be valuable and may lead to new discoveries when the image is reanalyzed in the future. The rapid development in the computer industry allows the affordable storage of all original images that can be reanalyzed in the future.

12.1 SAMPLE PREPARATION

The purpose of sample preparation is to obtain fluorescently labeled cells that are ready for imaging with a microscope. In icHCS, there are three basic requirements for the cells: (1) The selected cells should be responsive to the stimuli, that is, contain surface receptors or other elements that will interact with the stimuli; (2) the responses of the cells to the stimuli should be detectable with fluorescence by incorporating fluorescent molecules inside the cells; and (3) the fluorescently labeled cells must be positioned (or arrayed) in a way to allow suitable image collection and automated image analysis. Though some endogenous cellular molecules may fluoresce, such as NAD(P)H, their fluorescence are not specific and can be interfered with by autofluorescence from the biological samples. Thus, most imaging studies employ exogenously introduced fluorescence probes.

The cellular components can be labeled with one of the commonly employed methods discussed below. There exist fluorescent probes that can selectively associate with a particular subcellular structure of the cell. For example, 4′,6-diamidino-2-phenylindole (DAPI) and Hoechst stains can specifically label nuclei. Fluorescent probes can also be made by covalently linking two molecules with one of them specifically associating with the target subcellular components and the other being fluorescent. For example, fluorescent molecule-conjugated paclitaxel, colchicine, and vinblastine can specifically label tubulin. Fluorescently labeled antibodies can specifically label any protein after the cells are fixed and permeabilized. GFPs are widely used probes to study proteins of interest through co-expression with the target protein. Thus, the location and movement of the target protein in live cells can be studied by following the fluorescence in real time. We also discussed in Chapter 9 that there are nonfluorescent molecules that gain intense fluorescence after binding selectively to intracellular ions (e.g., Ca^{2+} sensing dyes). These probes can help to image the distribution of the ionic species inside the cells.

Because multiplexed cell imaging studies are performed with at least two colors, choosing a fluorescence probe with nonoverlapping spectra is very important. Fluorescein is usually the choice of the first fluorescent label because of its bright fluorescence, solubility in water, conjugates are stable, and its excitation peak (495 nm) closely matches the argon laser line at 488 nm. However, fluorescein has a relatively high rate of photobleaching, is pH sensitive, and its fluorescence emission spectrum is relatively broad, which limits its utility in some multicolor applications. Tetramethylrhodamine (TMR) is often used as the second dye for double labeling. However, there are some overlapping spectra between the fluorescein and TMR. Photobleaching can be reduced by addition of antifade reagents, such as *p*-phenylenediamine and DABCO (1,4-diazabi-cyclo-2,2,2-octane) to protect fluorescein, *n*-propylgallate to protect rhodamine, and 2-mercapto-ethylamine to protect propidium iodide. Most current imaging studies often use recently developed improved dyes, such as the cyanine series of fluorescent molecules from Amersham (now part of GE Healthcare and Life Sciences) and Alexa Fluor series of molecules from Molecular Probes (now part of Invitrogen). These two classes of dyes cover a wide spectrum of wavelengths with higher quantum yield, lower photobleaching, and narrower excitation and emission spectra.

There are two distinct icHCS experimental modes: end-point fluorescence detection and live-cell kinetic fluorescence detection. As the name implies, the end-point icHCS only can take the image after fixing the cells at a particular end point after stimulation. In contrast, live-cell kinetic icHCS can constantly take the image of the cells during an experiment. This mode of operation requires nontoxic fluorescent probes to be present in the live cells during the time of study. In addition to gaining the temporal resolution, some cellular properties, such as membrane potential and ion concentration changes, can only be studied in live cells. End-point and live-cell icHCS require different treatments of the cells, different fluorescent probes, and different instrumentations.

12.1.1 End-Point icHCS

In this mode of operation, the cells are treated similarly as with other cell-based assays. Adherent cells are preferred in imaging studies because they can attach to the flat bottom of the microplate ready for imaging. The bottom of the microplate is usually coated with poly-D-lysine, collagen, or other polymers derived from extracellular matrix. The plating density and the distribution evenness of the cells are very critical for downstream automated image analysis. Because nuclei are the most easily recognized subcellular structures that give a dense blue round shape after staining with DAPI or Hoeschst 33258, the nuclei are usually used as the reference points to locate each cell and help mark the cell boundaries with the help of a second cell membrane dye (Fig. 12.1). Thus, the density of the plated cells in the microplate before imaging should be adequate to allow clear separation between nuclei. After the cells are plated, they are treated with stimuli. At a predefined time after stimulation, the cells are fixed and may be permeated to allow access of fluorescent molecules to the cell's interior when necessary. After removing the fixing and permeation reagents by a plate washer, fluorescent labels targeting specific subcellular components or fluorescently labeled antibodies targeting specific epitopes are applied to label the fixed cells. After removing the excess dyes, the microplates are ready for imaging with automated fluorescence image reader. The end-point icHCS offers many advantages. Fixed cells

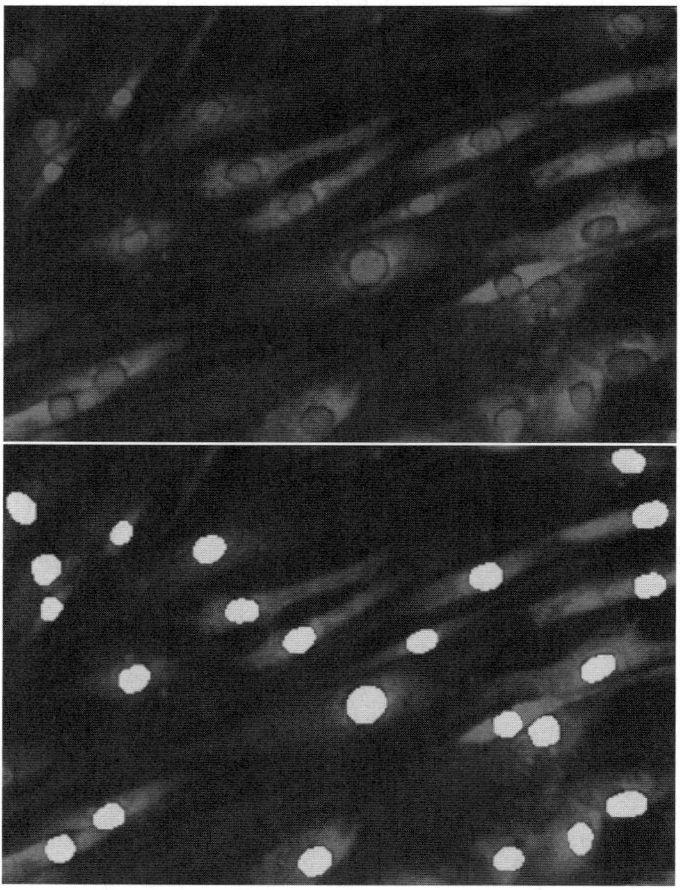

Figure 12.1 Identification of cells with the help of nuclei. The cells are stained with two dyes. The cell nuclei is labeled with DAPI, which gives a distinct blue color. The cell interior is labeled with a red dye. Top panel shows the raw image of the cell with overlaying two colors. Bottom panel shows the automatically processed data that locates the position of each nucleus and marks the boundary of each nucleus based on predefined fluorescent density criteria in the blue channel. The automated image processing was performed with Metamorph software from Molecular Devices. (See color insert.)

are flatter and the resulting images are sharper in the x and y dimensions. The stained samples can be stored for a longer time while waiting for the acquisition of the image, which helps the automation process. A large selection of fluorescent probes, especially many protein-specific fluorescently labeled antibodies, can be used in this mode since there is no cell toxicity constraint. Since no modification of the cells to incorporate exogenous fluorescent dyes is required, end-point icHCS is especially powerful in studying primary cells. In addition, most antifade reagents can be used on fixed cells. The disadvantage of end-point icHCS is the loss of temporal resolution because only one end point can be studied per sample. Thus, the test system must be carefully studied during assay development to identify the most critical end point of the assay.

12.1.2 Live-Cell Kinetic icHCS

The challenges of live-cell kinetic icHCS are to find a nontoxic probe and to deliver the probe inside the live cells. There exists a relatively small number of dyes that can be employed in live-cell studies as compared with the large number of dyes that can be employed in end-point studies. Hoeschst 33258 is a widely used nuclei stain in live-cell studies. In addition, fluorescent phalloidin was used as actin filament stain, fluorescent phospholipids were used as cell membrane stain, fluorescent ceramides were used as Golgi apparatus stain, rhodamine 123 was used as mitochondria stain, 3,3′-dihexyloxacarbocyanine iodide ($DiOC_6(3)$) was used as endoplasmic reticulum stain, and sulforhodamine was used as lysosome and vesicle stain. In addition to these fluorescent probes that associate with specific proteins or organelles, many fluorescent probes responsive to specific ions inside the cell upon binding to the ions are commonly used in live-cell studies. For example, Fluo3, Fura2, and indo-1 can probe the concentration of Ca^{2+}, and SNARF-1 (seminaphtharhodafluor-1) can be used as a pH indicator inside the cells. The delivery of these ion probes inside cells was originally performed with microinjection, which was not practical for screening because of the limited throughput. To solve the problem, these dyes were modified chemically to remove the charges from the carboxyl groups by forming ester bonds with them. The ester form of the probe is permeable to the cell membrane. After being inside the cells, the ester bond of the probe is cleaved by an intracellular enzyme. This chemical reaction converts the ester-modified probes back to their original charged form, which cannot passively defuse out of cell because of the cell membrane barrier. This strategy allows the loading of the fluorescent probes at high throughput.

Another method of introducing fluorescent probes into the cells is to genetically engineer a new stable cell line or transiently transfect the target cells with foreign genes that are fused to the gene of the protein of interest. GFPs are the first and the most used fluorescence probes for this purpose and they are well tolerated in many cells. Because commercial uses of GFPs are complicated with many intellectual property issues, several other gene constructs have been developed recently that can substitute GFP to produce fluorescence in live cells. For example, HaloTag interchangeable labeling technology from Promega genetically fuses a derivative of 33 kDa bacterial hydrolase to the gene of the target protein. The derived hydrolase is active to allow the covalent linking of fluorescence or other tags with the aliphatic halogen functional group. The introduced exogenous HaloTag has a similar size with GFP (~28 kDa). Another development is the use of fluorescein arsenical hairpin (FlAsH), which can selectively bind to tetracysteine that is genetically introduced —CCPGCC— residues to the protein of interest. This exogenous tag is much smaller than GPF and HaloTag and thus has less perturbation to the native system. Another technology takes advantage of *E. coli* biotin ligase, which is known to ligate biotin to a lysine side chain within a specific 15-amino-acid substrate sequence. By genetically fusing the 15-amino-acid acceptor peptide to proteins of interest, co-expressing the biotin ligase, and providing a ketone analog of biotin, proteins in live cells could be specifically labeled with a variety of small-molecule probes, including fluorophores. The location and translocation of target protein inside a cell can be readily monitored by studying these exogenously introduced fluorescent labels. The disadvantage of this approach is the need to make foreign gene constructs and introduce them into native cells. The resulting cells used in the assay may not truly reflect what happened in the native cells.

12.2 CELLULAR IMAGE COLLECTION

After fluorescently labeled cells are properly arrayed on the surface of a sample holder, the image of the cells is ready for collection. Cell image collection has the following components: the sample holder, the instrumentation for imaging, the interface between the sample holder and the imaging instruments, and the liquid handling and environment (for live-cell kinetic icHCS). Microscope slides were the most commonly used sample holdes for traditional microscopic studies. A high-quality clear and flat-bottom microplate with solid walls separating each well is the choice of sample holder for high-throughput studies. The best quality microplates for imaging are those with low autofluorescence solid plastics forming the sidewalls of the wells and thin flat high-quality glass forming the bottom of the microplate. Some high-quality microplates for imaging are made of transparent plastic at the bottom of the microplate. However, these microplates have the limitations when the excitation wavelength approaches the UV region. In addition, a plastic bottom usually is thicker than the glass bottom because too thin plastics may make the bottom of the microplate uneven because of the physical properties of the plastics. However, microplates with plastic bottoms are cost effective, and in many cases they are good enough for icHCS. Black adherent tapes are usually used to shield the top of the microplate to prevent the photobleaching from ambient lights before the microplates are delivered to the image reader.

Most automated cellular image readers have all the components enclosed in the dark and only allow the microplate being read to enter the enclosed environment through a narrow automatic shuffling door. A few image readers are not enclosed, which requires the ambient light be kept relatively dim (e.g., the Discovery-1 from Molecular Devices, as shown in Fig. 12.2). In this case, the top of the microplates must be shielded to prevent ambient light from interfering with the image collection. A black microplate lid or a black adherent tape can be used to block the ambient light. For the automated loading of the microplate, black microplate lids have proven to be very useful for automated operation. In addition, the black microplate lids can be

Figure 12.2 Interface between the sample stage (microplate) and the microscope in Discovery-1. The microplate stage is built on top of an inverted Nikon microscope. A microplate is placed on the sample stage that can move along the x and y axes to bring each well in the microplate to the viewing area of the inverted microscope. The lenses of the microscope can be selected automatically. The focus is performed automatically by the movement of the lens in the z axis.

reused many times before being discarded. The interface between the microplate and the imageing device is the plate stage above the inverted microscopes objective lenses with motorized xyz movement capability to allow automated sampling from well to well (x and y axis). Figure 12.2 shows the motorized microplate stage in Discovery-1. The stage is built on top of an inverted Nikon microscope, and it can move the microplate in the x and y axes. The movement in the z axis for autofocusing is made possible by the movement of the lenses of the microscope. For reasonable throughput applications, it is desirable to have a robotic arm and a microplate rack integrated together so that each microplate can be fed to the microscope continuously by a robot arm. Figure 12.3 shows the integrated system built at Five Prime Therapeutics with a CRS robot and a Discovery-1 imaging system.

The imaging device in icHCS usually is an inverted microscope similar to what was used in any pathology lab. The major difference is that icHCS employes automated sample loading to a microplate stage, automated lens and filter switching and the automated capture of the image. The sample stage in icHCS holds a microplate instead of a tissue slide. The light sources are usually high-powered xenon lamps or lasers covering at least two wavelengths. The objective lenses can be selected from 4× through 40×, which is controlled by software. At least two sets of filters are available and may go as many as to six color sets, which are more than enough for most applications. The selection of filters for excitation and emission are also controlled

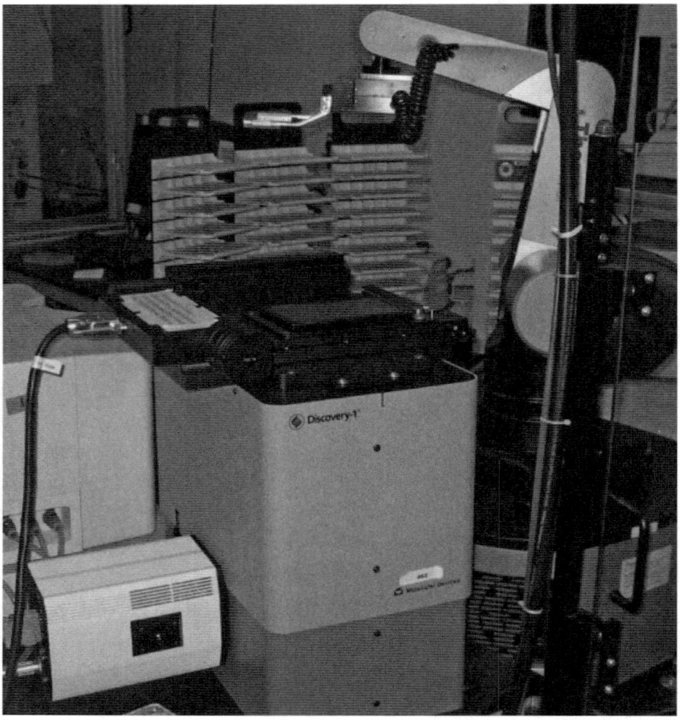

Figure 12.3 Automated loading of microplate with samples to be imaged to the sample stage on top of the microscope in Discovery-1. A Thermo CRS Catalyst Express system is employed in this automation process.

through software. The autofocus can be performed either with the image from the fluorescent sample or with a separate optical focusing light that projects a small spot on the plate bottom and adjusts the focus based on the reflected light intensity. Image-based focusing is more accurate and is very powerful when high magnification is used. However, image-based focusing is very slow compared with optical focus with infrared light. In most applications, infrared focusing is adequate and offers high-speed focusing. The focused images are usually captured with high-quality cooled CCD cameras because of their ability to capture two-dimensional images and other properties that were discussed in Chapter 2. Most of the commercially available imaging systems can resolve down to about 0.5 μm with 40× magnification objective lens. This is adequate enough for most applications. Though a light microscope's resolution is being pushed much lower than 0.5 μm, as discussed in Chapter 2, currently those technologies have not been adopted in icHCS. The drawback of high resolution in icHCS is the reduced field of view. The consequence of the reduced field of view is fewer cells per field of view, and thus many views are required to analyze enough cells to achieve statistical significance. This will significantly reduce the throughput of the assay. In addition to the objective lens, the CCD camera's sampling resolution also contributes to the overall resolution of the image. However, the sampling resolutions are often more than enough for most applications and a binning process is used to increase the throughput. Binning is a process that combines the adjacent pixels into one larger pixel. For example, the 4×4 binning process will reduce 16 pixels into 1. However, the sampling resolution is reduced from 16 small spots into 1 big spot in the same process. In practice, a scientist must balance the throughput and resolution based on particular assay requirements by choosing the objective lens (magnification and numerical aperture) and level of binning.

For live kinetic icHCS, an enclosed environment to support cell survival is needed. This requires the regulation of the enclosure to have air with 5% CO_2 and in an ~95% humidity environment. In addition, liquid handling instruments are needed inside the enclosure to allow rapid addition of reagents to initiate the kinetic studies. Thus, imaging systems with live-cell kinetic studies are more expensive than the end-point imaging systems.

There are two methods to collect the image of the sample. The most common method is to illuminate the whole sample simultaneously with lamp and collect the two-dimensional image simultaneously with a CCD camera. The other method is using a small light spot to scan through the whole imaging area sequentially and collect the signal from each spot with a PMT. The advantage of the scanning method is that it can achieve high resolution and can even break the diffraction limits (discussed in Chapter 2). In addition, PMT detection is more sensitive than CCD. However, a scanning method is usually slower than a simultaneous illumination and collection method. Because of the unique properties of a laser, it is the preferred light source in scanning especially in confocal microscopy, which requires strong excitation since most of the lights from out-of-focus planes are removed.

Most of the early icHCS systems employed wide-field imaging technology that has very poor resolution in the z axis. The resulting images are often blurred by out-of-focus light. Confocal (fluorescence) scanning microscopy is a microscopic technique that provides three-dimensional imaging and resolution. The 3-D resolution is obtained by blocking all the light that is not coming from an in-focus plane. The effect of blocking out the out-of-focused lights' contributions is also known as optical sectioning. It permits the imaging of separate (axial) slices within the specimen. Two confocal techniques have been developed to deliver light to every point of a specimen within the focal plane. The first is the laser-based point scanning confocal system, where the excitation laser source is scanned across the specimen in a point-by-point raster pattern. This technique offers the highest level of confocality and the ability to do very thin optical sections. However, there are several drawbacks to traditional laser-based confocal systems. Because it uses very small apertures and rejects so much light, it requires very bright emissions from the specimen. This in turn requires intense laser light sources that may cause other serious repercussions: photobleaching of the fluorophore and phototoxicity in the cell. In addition, the raster scan's point-by-point acquisition of the image is time consuming. The second is the spinning-disk confocal system used for high-throughput applications. Instead of scanning a diffraction-limited laser spot across the sample, a spinning disk with multiple pinholes is placed in front of the detector. All the light originating from an in-focus plane will pass through the pinhole, whereas light coming from out-of focus planes will be blocked by the pinhole. The size of the pinhole determines how much background reduction can be realized. The spinning-disk systems make possible far better speed, full frame imaging, and higher transmission rates than laser scanning systems. However, because of their fixed large pinhole or slit openings, this type of confocal system cannot deliver the same thinness of optical sectioning as the laser-based point scanning system.

12.3 IMAGE ABSTRACTION, ANALYSIS, AND DATA MANAGEMENT

After collection of images, the next challenge is to abstract useful information from the image, analyze the abstracted data, visualize the data, and archive the data. The collected images contain a wealth of information. Some of the information can be abstracted with the naked eye. However, it is not practical to analyze the large set of the screening images with human eyes. In addition, much information is beyond human vision and can only be obtained through the use of computational analyses. The development of automated algorithms to analyze images from icHCS is especially crucial when a large volume of images is collected via automated microscopy in live-cell kinetic studies and when the complexity and subtlety of patterns in the images are analyzed. There are two very distinct methods for image analysis: directed algorithms and pattern recognition.

The most common method is directed algorithms that abstract and analyze the images based on biological knowledge. The data collected from icHCS can be first divided based on which collection channel the data are from. Each channel collects data from one fluorophore. There is no interaction between the data in the difference channel except the physical locations where the pixels are related. Thus, data from each channel can be treated separately. Within each channel, the fluorescent intensity on a physical position is represented by a relative number in the corresponding pixel. Thus, a set of numbers in gray scale is assigned to each spot in a two-dimensional map. The next task is object recognition that will locate a specific cellular structure or a protein in the two-dimensional map. The most common method for object recognition is thresholding such that a histogram of all the values in the two-dimensional map can be generated and threshold (or background) values are set. The pixels with values significantly different from the threshold values form objects. Biological knowledge plays a significant role in algorithm development. A good example is to define the nuclei objects (Fig. 12.1). When DAPI is used to stain the nuclei, the pixels obtained from the blue channel contain information to recognize the nuclei. After defining the threshold in the histogram, a algorithms can be specified to look for a cluster of pixels forming round shapes, and the boundary of the clusters should be within a certain diameter range (e.g., $\sim 10\ \mu m$). This boundary condition will reject the artifact from a few high-intensity pixels or from a contaminant that may give a cluster of high pixels but may not have the round shape. Because of the distinct character of nuclei and the likelihood that they are separated from each other in most images, nuclei are the most used reference objects in directed algorithms. After the identification of nuclei, the cell boundary can be determined with the aid of other fluorescence channels and special algorithms such as thresholding, edge detection, watershed, and so forth. Other labeled objects within the cell boundary are also identified using data from corresponding fluorescent channels. Analysis of each object is then performed to obtain quantitative information, such as number of cells, total intensity of the object, size of the object, and shape of the objects.

The other method is based on pattern recognition, which will generate a set of parameters that may or may not be related to known biology. In pattern reorganization methods, the images of the same sample before and after stimulation are compared. The changes between the two are abstracted, and a set of parameters is derived mathematically to describe the changes. The difference between pattern recognition and direct algorithms can be demonstrated with the following example. In an icHCS assay, a stimulation of the cells induces cell proliferation and elongation. When analyzing the data in this experiment with a directed algorithm, the number of nuclei will be identified and are used as the parameter of the number of cells. The length of the cells will be measured too to study whether the stimulation also caused cell elongation. In comparison, a pattern recognition algorithm will analyze the changes in the collected image before and after the stimulation and generate a set of parameters to describe the changes. In the set of parameters generated, there may not exist a single parameter that corresponds to the cell number or the cell length. Instead, a number of parameters may be generated that may not correspond to any known biological parameters at all. Thus, the disadvantage of pattern recognition algorithms is that the abstracted data is difficult to relate to known biology. The advantage of pattern recognition is that it may detect some changes that are novel and may lead to new discovery. In some situations when the assay has no clear hypothesis, pattern recognition algorithms are very valuable. For example, a screening is aimed at finding secreted proteins that interact with B cells. Because there is no prior knowledge of how the B cells will respond to the unknown protein and there is no requirement in the assay on how the B cells should respond to the unknown protein, pattern recognition algorithms will be very valuable to identify which proteins in the library produce a change in the cells.

For large-scale high-throughput icHCS, the image processing algorithms must be robust to handle a variety of nonideal situations and must be fully validated. It takes a lot of effort and time to develop fully validated algorithms. The end users of icHCS systems are usually not expected to develop their own algorithms. Instead, vendors of icHCS systems usually have a team of experts to develop validated packages for specific biological applications using software development tools such as Metamorph from Molecular Devices, ImagePro from MediaCybernetics, and Axiovision from Carl Zeiss. There are many prepackaged turnkey solutions from several vendors that can handle most common biological applications such as cell proliferation (nuclei counting), cell viability (live/dead ratio), cell cycle (mitotic index), angiogenesis (tube formation), GPCR activation (receptor internalization), and nuclear receptor activation (translocation to nuclei). These packages usually limit the user's input by only allowing a few user-adjustable parameters because too many options may overwhelm inexperienced users or may significantly slow the assay development because more variables require more validations.

Data management in icHCS is very challenging because of the huge number of raw image data and the corresponding derived data from image abstraction and image analysis. An ideal data management tool should be able to visualize the primary and derived data (such as heat map at plate level, data tables, line graphs, and drill down from plate view to images), offer statistical tools to abstract higher level information (such as histograms, curve fitting, statistics across cells, wells, plates, and screens), and offer easy ways for achieving and retrieving data. A few vendors, such as Molecular Devices and Cellomics, offer complete data management solutions.

12.4 APPLICATIONS OF icHCS

More and more biological processes have been studied in icHCS format. The icHCS assays can be totally novel or can be performed to replace the existing assays because of cost, easy of use, or simply providing more information. A good example is the assays currently performed with flow cytometry that can be replaced with icHCS. Some icHCS assays enable measurement of biological processes that are either very difficult or impossible to do with conventional cell-based assay technologies (e.g., intracellular protein translocation and cell elongation in angiogenesis). Most icHCS applications are based on the measurement of the location, the area, and the intensity in a defined area of labeled proteins. The measurement can be divided into two broad areas: (1) fluorescence intensity level within specific subcellular compartment or structure. This application category includes intracellular translocation of proteins (e.g., cytoplasm to cell membranes and cytoplasm to nucleus), protein co-localization (e.g., protein–protein interaction), and GPCR activation (e.g., β-arrestin-GFP fusion protein aggregation and internalization). (2) The morphology of the object defined by the specific labeled protein. This application category includes discrete phases in cell cycle (e.g., morphological of actin or tubulin), cell motility (e.g., cells moving toward or away from stimulus), and angiogenesis assays by measuring the tube formation (e.g., the number of connected tubes and tube length and width) and neurite outgrowth (e.g., the neurite counts, the number of branches, and the lengths of neurite branches). For oncology applications, the the nuclei and mitochondria structures are usually evaluated. This includes counting the number of cells that are alive or dead based on its ability to uptake specific dyes and thus being stained, cell proliferation (by counting number of nuclei as shown in Fig. 12.1), and apoptosis (fragmentation of nuclei or the damage of mitochondria).

The nuclei are almost always stained with dyes that can associate with DNA in icHCS. Except in a few applications, such as with apoptosis, in which nuclei fragmentation is studied, the major purpose of staining the nuclei is to locate the position of nuclei to help define cell position and boundaries. Figure 12.4 shows an example of visualization of nuclear factor-κB (NF-κB) translocation from cytoplasm to nuclei. This application was among the earliest experiments to demonstrate the power of icHCS by BioDx (Cellomics). The NF-κB is a transcription factor that translocates into nuclei to initiate transcription in response to stimulation on the cell surface by inflammatory signaling molecules such as interleukin-1α (IL-1α) and TNFα. This process can be assayed by icHCS as shown in Figure 12.4. After treating the cells with TNFα at different concentrations, the cells were fixed and stained with two

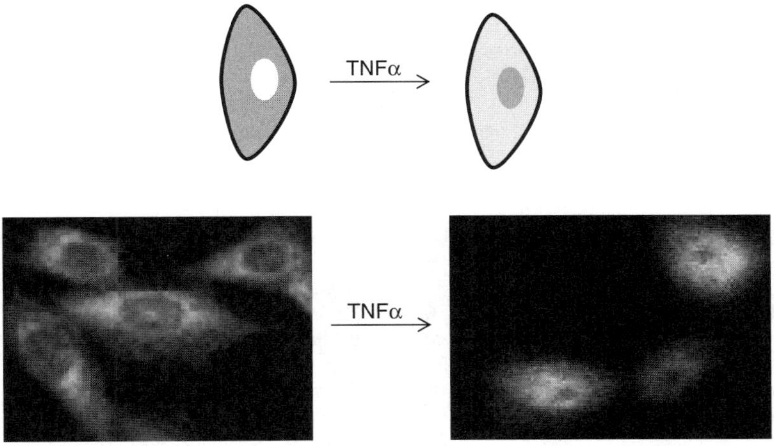

Figure 12.4 Translocation of proteins protein from cytoplasm to nucleus. Top: schematic illustration of the process. Bottom: experimental images. The nucleus region is marked by DAPI dye in blue channel (not shown for image clarity). The green channel measures fluorescein-labeled antibody against NF-κB. Before stimulation, NF-κB was mostly in cytoplasm. After stimulation, NF-κB translocates inside the nucleus. (See color insert.)

dyes, DAPI and anti-NF-κB antibody, which is labeled with a dye that emits at different wavelengths from DAPI. DAPI was used to locate the position of the nucleus and marks the position of each cell. Before the treatment with TNFα, NF-κB was located in the cytoplasm and the green fluorescence from the NF-κB channel marks the boundary of the cells. The nucleus region marked by DAPI did not show green fluorescence in the NF-κB channel. After TNFα treatment, the nucleus region showed intense fluorescence from the NF-κB channel because some NF-κB translocated into the nucleus. The total fluorescence in the cytoplasm can be integrated over the cytoplasm region, and the same can be done within the nucleus region. The degree of NF-κB translocation can be quantified by comparing the green fluorescence intensities in the two regions.

Useful Websites

http://www.cellomics.com/

http://www.moleculardevices.com/pages/instruments/imaging_main.html

http://www4.gelifesciences.com/APTRIX/upp01077.nsf/Content/incell_site2

http://www.atto.com/

http://las.perkinelmer.com/Catalog/default.htm?CategoryID=Cellular+Imaging+and+Analysis

http://www.zeiss.com/C12567BE0045ACF1/Contents-Frame/668C9FDCBB18C6E2412568C10045A72E

BIBLIOGRAPHY

Alivisatos, A. P., Gu, W., and Larabell, C. (2005) Quantum dots as cellular probes. *Annu. Rev. Biomed. Eng.* **7**, 55–76.

Bates, M., Huang, B., Dempsey, G. T., and Zhuang, X. (2007) Multicolor super-resolution imaging with photo-switchable fluorescent probes. *Science* **317**, 1749–1753.

Bullen, A. (2008) Microscopic imaging techniques for drug discovery. *Nat. Rev. Drug Discov.* **7**, 54–67.

Carpenter, A. E. (2007) Image-based chemical screening. *Nat. Chem. Biol.* **3**, 461–465.

Chen, I. and Ting, A. Y. (2005) Site-specific labeling of proteins with small molecules in live cells. *Curr. Opin. Biotechnol.* **16**, 35–40.

Chen, I., Howarth, M., Lin, W., and Ting, A. Y. (2005) Site-specific labeling of cell surface proteins with biophysical probes using biotin ligase. *Nat. Methods* **2**, 99–104.

Giepmans, B. N. G., Adams, S. R., Ellisman, M. H., and Tsien, R. Y. (2006) The fluorescent toolbox for assessing protein location and function. *Science* **312**, 217–224.

Goldman, R. D. and Spector, D. L. (eds.) (2005) *Live Cell Imaging: A Laboratory Manual*. Cold Spring Harbor Press, Cold Spring Harbor, NY.

Haney, S. A. (ed.) (2008) *High Content Screening: Science, Techniques and Applications*. Wiley, Hoboken, NJ.

Hell, S. W. (2007) Far-field optical nanoscopy. *Science* **316**, 1153–1158.

Heydorn, A., Lundholt, B. K., Praestegaard, M., and Pagliaro, L. (2006) Protein translocation assays: Key tools for accessing new biological information with high-throughput microscopy. *Methods Enzymol.* **414**, 513–530.

Huang, B., Wang, W., Bates, M., and Zhuang, X. (2008) Three-dimensional super-resolution imaging by stochastic optical reconstruction microscopy. *Science* **319**, 810–813.

Jacquier, V., Prummer, M., Segura, J.-M., Pick, H., and Vogel, H. (2006) Visualizing odorant receptor trafficking in living cells down to the single-molecule level. *PNAS* **103**, 14325–14330.

Korn, K. and Krausz, E. (2007) Cell-based high-content screening of small-molecule libraries. *Curr. Opin. Chem. Biol.* **11**, 503–510.

Lang, P., Yeow, K., Nichols, A., and Scheer, A. (2006) Cellular imaging in drug discovery. *Nat. Rev. Drug Discov.* **5**, 343–356.

Marks, K. M. and Nolan, G. P. (2006) Chemical labeling strategies for cell biology. *Nat. Methods* **3**, 591–596.

Murphy, D. B. (2001) *Fundamentals of Light Microscopy and Electronic Imaging*. Wiley-Liss, New York.

Ntziachristos, V. (2006) Fluorescence molecular imaging. *Annu. Rev. Biomed. Eng.* **8**, 1–33.

Rudin, M. and Weissleder, R. (2003) Molecular imaging in drug discovery and development. *Nat. Rev. Drug Discov.* **2**, 123–131.

Shaner, N. C., Patterson, G. H., and Davidson, M. W. (2007) Advances in fluorescent protein technology. *J. Cell Sci.* **120**, 4247–4260.

Shorte, S. L. and Frischknecht, F. (eds.) (2007) *Imaging Cellular and Molecular Biological Functions*. Springer, Berlin.

Stephens, D. J. and Allan, V. J. (2003) Light microscopy techniques for live cell imaging. *Science* **300**, 82–86.

Taatjes, D. J. and Mossman, B. T. (eds.) (2006) *Cell Imaging Techniques*. Humana, Totowa, NJ.

Taylor, D. L., Haskins, J. R., and Giuliano, K. A. (eds.) (2007) *High Content Screening*. Humana, Totowa, NJ.

Wang, Y., Shyy, J. Y.-J., and Chien, S. (2008) Fluorescence proteins, live-cell imaging, and mechanobiology: Seeing is believing. *Annu. Rev. Biomed. Eng.* **10**, 1–38.

Wolff, M., Wiedenmann, J., Nienhaus, G. U., Valler, M., and Heilker, R. (2006) Novel fluorescent proteins for high-content screening. *Drug Discov. Today.* **11**, 1054–1060.

Zhang, J., Campbell, R. E., Ting, A. Y., and Tsien, R. Y. (2002) Creating new fluorescent probes for cell biology. *Nat. Rev. Mol. Cell. Biol.* **3**, 906–918.

CHAPTER 13

HIGH-THROUGHPUT SCREENING

13.1 INTRODUCTION

High-throughput screening (HTS) is an operation that enables testing a large number of samples (e.g., small synthetic molecules, natural products, or recombinant proteins) according to a fixed assay protocol at high speed. HTS plays a pivotal role in modern drug discovery processes allowing rapid identification of initial "hit" molecules for many novel drug target. HTS was made possible by the advances in many other scientific disciplines including robotics, liquid handling devices, microplate readers, databases, and project managements. The practice of high-throughput testing of a large number of samples has existed for a long time in the diagnostic industry in centralized labs where a large number of blood samples are tested routinely. However, a majority of the tests performed in diagnostic industry are based on test tubes. Though the concepts and practices existed, the term HTS was not used until scientists working in drug discovery started practicing HTS to find "hit" compounds within a large number of compounds for drug targets. Thus, HTS here narrowly refers to the operation practiced in drug discovery. It is generally agreed that the evolution of HTS started from the need to test increasingly larger numbers of natural product samples by pharmaceutical companies. However, the majority of screening campaigns performed nowadays are with synthetic compounds. The size of the compound library can range from hundreds of thousands to millions. Though there is no clear definition, generally HTS refers to a screening operation that can test more than 10,000 compounds/day. Operations that can screen more than 100,000 compounds/day are sometimes referred to as ultra-HTS (uHTS).

13.1.1 Moving from Test Tube to 96-Well Microplate

Before the widespread adoption of microplates, most bioassays were performed with test tubes. A bench scientist can only test tens to hundreds of samples a day with test tubes. To facilitate the testing process and to increase the throughput of the assay, some rudimental automation instruments for handling test tubes were developed. For example, a carousal that can hold a few dozen test tubes was used to facilitate the automatic introduction of test samples sequentially into HPLC, and parallel filtration

Assay Development: Fundamentals and Practices. By Ge Wu

devices were used to filter more than 10 samples under vacuum simultaneously. To save the test samples and associated reagents, test tubes were made smaller, and gradually many tests were performed in smaller eppendorf tubes. The volume of bioassay shrinks from milliliter scale to microliter scale in the process. Small size allows the test tubes be arranged in smaller footprints that in turn allow parallel introduction of test reagents to many tubes simultaneously using handheld multichannel pipettor. Thus, the throughput of testing is increased. It is true that within the limits of existing technology, miniaturization results in both increased throughput and decreased cost of test/sample. Similar phenomenon has existed in the microprocessor industry where the famous Moore's law still holds.

Though miniaturization within the scope of test tubes saves test samples and associated testing reagents, the basic test unit is still an individual sample. This testing paradigm allows the flexibility to arrange samples in a different geometry. For example, the test samples can be arranged in a round carousal as often found in HPLC applications. Alternatively, the test tubes can be arranged in a square matrix in tube rack at 1×16, 2×8, 3×10, or whatever the researcher likes. The samples arranged in different arrays can still be handled with handheld multichannel one-dimensional pipettors. However, it will be difficult to design a liquid-handling instrument that can simultaneously handle samples arranged in a matrix if the tube matrix is not predefined. The same obstacle also prevents the introduction of parallel multisample measurement instruments. To increase throughput, the basic test unit must change from single tube to multiple tubes to allow parallel processes of all the samples in the test unit simultaneously. In this new paradigm of testing, a predefined block of samples is treated as an individual test unit, and all the samples in the block are subjected to the same treatment in parallel. This operation mode requires the standardization of the sample holders. Sample holders arranged in a 12×8 matrix emerged as the standard. It is conceivable to have a rack that holds 96 small tubes arranged in as 12×8 matrix as the standard test unit. However, for many reasons (such as the ease of manufacturing and ruggedness of the final product), microplates molded from plastics with 96 wells arranged in a 12×8 matrix emerged as the standard for a test unit.

Normal 96-well microplates used today can hold up to an ~400-μL sample in each well. There are special 96-well microplates that hold less (e.g., half-well microplates) or more (e.g., deep-well microplates). The most widely used microplates are made from polystyrene or polypropylene. The microplate can be made black (by addition of carbon to the plastic), white (by addition of titanium dioxide), or transparent (polystyrene or polycarbonate with no added pigments). Microplates made from polypropylene are more resistant to chemicals and are commonly used to store chemicals dissolved in DMSO. The wells in each microplate can adopt one of the three different shapes: flat bottom, U bottom, or V bottom. The U- and V-bottom microplates are usually made opaque and the measurement of the sample is made from the top of the microplate. The flat-bottom microplate can be made opaque (for top reading) or transparent (for bottom reading and absorbance reading). Polystyrene microplate can be made transparent to visible light for absorbance measurement and fluorescence/luminescence measurement from the bottom of the plate. Early microplates from different manufacturers are not exactly the same, causing some compatibility issues with the microplate liquid-handling instruments and microplate readers. To resolve this issue, a microplate standard committee in SBS (Society for Biomolecular Screening/Sciences) was formed to issue the microplate standard. Changing test unit from single sample in a tube to multisamples in a microplate is a dramatic operational change that allows parallel testing of multisamples in a single operation procedure. The throughput is gained in the parallel process while the assay volume is reduced because of the size of the well. These advances are made possible by the new generation of microplate liquid-handling instruments that can simultaneously transfer liquid into or out of microplates and microplate readers that can directly measure samples in the microplates.

13.1.2 Beyond the 96-Well Microplate

When biological testing units changed from a single tube to a microplate, the fields of genomics and combinatorial chemistry were rapidly advanced. The development of genomics provides more drug targets for screening and also enabled the production of large quantities of purified (or partially purified) target proteins and cell lines expressing target receptors. The production of large quantities of test systems (proteins or cells) for bioassays is a crucial requirement to carry large-scale screening. The large number of new targets demands increasingly higher throughput screening to support many drug discovery projects simultaneously. The advancement of combinatorial chemistry provides a large number of new chemicals to be screened that further demand higher throughput. After optimizing all existing operating components and procedures in the 96-well microplate format, the next logically way to increase throughput is to increase the number of test samples in a test unit, that is, to increase the number of wells in a microplate.

Higher density microplates with 384 wells and 1536 wells per microplate emerged as the most common format after the 96-well microplate. While the density of the wells on the microplate increases, the wells on the same sized microplate decreases from ~300 μL in 96-well microplates to ~100 μL in 384-well microplates and to <10 μL in 1536-well microplates. In the early 1990s, most screens were performed in 96-well microplates. By the end of 1990s, approximately equal numbers of screens were performed in 96-well microplates and 384-well microplates. Currently, only a small portion of all screens is carried out in 96-well microplate. Majority of the screens are now carried out in 384-well microplates and 1536-well microplates. Transition from 96-well microplates to 384-well microplates is relatively smooth. For example, the microplate multichannel pipettors for 96-well microplates can be modified to handling 384-well microplates by simply moving the liquid-dispensing head or tips four times in a quadrant. The same modification can be applied to most of the 96-well microplate detectors as well. There is no significant concern considering the physical and chemical properties of the reactions taking place in the wells in the transition.

However, the transition from 384-well microplates to 1536-well microplates had imposed greater challenges in both instrumentation and in the biological reactions. Because the wells in 1536-well microplates are small in diameter, the liquid-handling instruments must be able to precisely position the nozzles or tips at the center of the well. In addition, the tips that will be inserted inside the wells must have a small diameter relative to the diameter of the wells. Because the depth of the liquid in the well is small, the positioning of the tips in the z axis must be very precise too. Similarly, the detectors for 1536-well microplates must have precise positioning, and the diameter of the light beam for fluorescence excitation must be relatively small compared with the diameter of the wells. If a single detector is used to sequentially scan through a 1536-well microplate, it may take long time to finish reading a plate. For example, if it takes 1 s to read a well and reposition the detector head to the next well, it will take 1536 s (25.6 min) to finish reading one plate. If only one reader is present in the screening, the throughput in 24 h will be limited to 56 plates, no matter how fast the other steps in the whole-screen process are. To increase the speed, parallel reading of multiple wells in a plate should be adopted. CCD-based detection that can image the whole microplate (simultaneous reading of all 1536 wells) is of great advantage in this aspect. In addition to the challenge in instrumentation, the high surface-to-volume ratio in 1536-well microplates may change the properties of the biological transformation in the small well as compared to what happened in 96-well or 384-well microplates. Assays performed in microplates with higher than 1536-well density require highly specialized instruments (e.g., the Aurora screening platform uses 3456-well microplates). Though it is debatable, the author sees little benefit to using a microplate with a density higher than the 1536-well plate because of the associated high cost and other issues. In addition to performing bioassays in the wells in a microplate, there exist other screening formats, such as the Micro Arrayed Compound Screening (μARCS) developed in Abbott Laboratories and the microfluidic-based assays developed by Caliper Technologies. In this chapter, the discussion will focus on microplate-based screening.

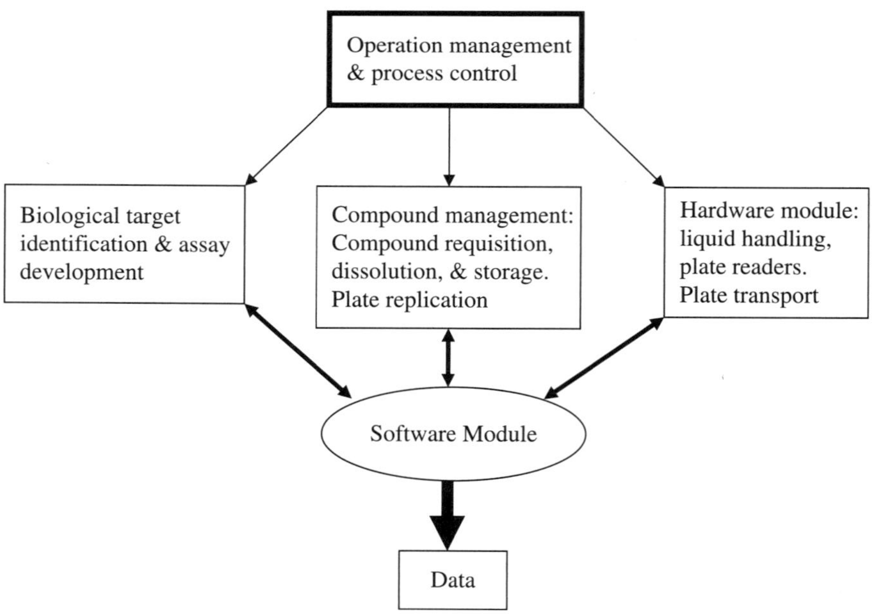

Figure 13.1 HTS operation modules and work flow. Infrastructures that include compound management module, hardware module, and software module must be built before carrying out HTS. After identification of biological target and the successful development of assays, the screening operation is performed with the protocol defined in the assay development. The final product of HTS operation is data.

13.1.3 High-Throughput Screening Operation Components

The HTS operation is composed of several components, as shown in Figure 13.1. In general, the HTS operation contains the following components: (1) assay development, (2) acquisition and maintenance of compound library, (3) hardware module that includes liquid-handling devices, microplate readers, and microplate transporters, (4) software module that includes scheduling of plate movement and handling of screening data, and (5) operation management that coordinates the HTS process. After identification of biological target and the successful development of assays, the screening operation is performed with the protocol defined in the assay development. The final product of HTS operation is data. Because of the huge amount of data generated, the old way of handling data with lab notebook or even in Excel spreadsheet is not sufficient. A database is required to handle the data generated by the HTS operation. In addition to handling screening data, software is also required to handle the communication between different instruments, the scheduling of the transport of microplates from different instruments, and tracking of assay protocols and compound library. Below we will discuss each of the HTS operation modules in detail.

13.2 MOLECULAR OR CELLULAR TARGETS AND ASSAY DEVELOPMENT

Choosing a molecular or cellular target for screening is a critical first step. In large pharmaceutical companies, there are many departments specializing in different therapeutic areas. The scientists in the therapeutic departments will identify the molecular or cellular target for specific diseases. In small biotechnology companies, the screening targets are often decided collectively with many scientists from different disciplines. After the identification of the screening target, an assay is developed against the target. We have discussed many assay formats in this book that may be applicable to the selected molecular or cellular target. It is often observed that scientists specialized in a therapeutic area may not be able to develop an assay suitable for HTS operation because they do not necessarily possess in-depth knowledge of the capability of the instruments in HTS, the special requirement for assay in HTS mode, and the cutting edge assay technologies. On the other hand, scientists specialized in HTS assay development may not possess in-depth knowledge of the therapeutic target and may design an HTS assay that may not be biologically relevant. Thus, close collaboration between the scientists from the two disciplines is crucial to develop a robust HTS assay that is also relevant to the biological target. In assay development, many factors common for HTS operation must be tested. These factors include the stability of the reagents during the assay, the effects of the media that contains the test sample on the assay, the signal-to-background ratio and the signal-to-noise ratio of the assay, the performance of the assay in the actual microplate wells (especially in the 1536-well format), and the performance of liquid-handling instrument in specific assay conditions. Initial assay developments are usually done manually in 96-well microplates and the liquid transfer is made with a handheld pipettor. With the assay development progresses, more assay procedures should be performed with the instruments and sample holders that will ultimately be used in HTS. In the final stage of assay development, all the liquid handling should be performed with the same instruments and the same sample holders (e.g., the same 1536-well microplate) that will be used in actual screening. In addition to following the common assay development principles, HTS assay development emphasizes the robustness of the assay (i.e., tolerance to many uncontrollable factors), minimazing the use of reagents (reduce reagent cost), the reagent's stability to fit to the time frame with continued robotic operation, and reducing reagent addition steps to the AssayPlate in the assay (by pooling different reagents in one reservoir). The outcome of the assay development is the assay protocol that will be carried in HTS.

13.3 COMPOUND LIBRARY MANAGEMENT

13.3.1 Compound Collection

For small-molecule drug discovery, the compound library is a collection of organic molecules with molecular weights typically less than 500 Da. One of the early obstacles for HTS was to obtain enough molecules to build a compound library for screening. Most big pharmaceutical companies in the United States evolved from chemical companies, and they have existed for a long time with many of them having over 100 years of history. These companies already possessed a large number of compounds ranging from a few hundred thousands to a million before HTS took off. However, these compounds were usually located in the individual scientist's bench inside glassware with different shapes. These compounds were either in powder form or were dissolved in various organic solvents and they were not readily accessible to HTS operation. To make compounds available for HTS, companies set up centralized compound management operations. This new function is responsible for collection, storage, management, and distribution of the compounds. Organic chemists in these companies are asked to make their compounds into powder form and then submit them to the compound collection. The compound powders are weighed and then dissolved into DMSO at fixed concentrations (usually $\sim$10 mM depending on the individual company). The dissolved compounds are then made into 96-well microplates (referred to as Masterplate or Motherplate). From a literature search and information from other sources, the initial compound libraries ready for HTS in the big pharmaceutical companies in the late 1990s had between 100,000 and 400,000 compounds. These in-house compounds usually are of high quality (with known identity and highly pure). In addition, many of these compounds are druglike molecules because they were possibly synthesized for specific therapeutic targets. While big pharmaceutical companies have relatively large and high-quality compound collections, smaller biotechnology companies must purchase compounds from the open market or by synthesizing them through combinatorial chemistry. Because of high demand, many companies were formed to sell compound libraries. These commercial compounds were typically collected from eastern European countries with no stringent quality control. The compounds typically were of low quality (high impurity, low diversity, and sometimes the compound purchased is not as was shown on paper). Even worse, some companies' compound libraries were on the beads directly from combinatorial synthesis, which might (or might not) have the supposed compounds attached to the bead. As the phrase "garbage in and garbage out" goes, screening using a low-quality library produced many failures in the early HTS operations. The costly failures prompted many companies to reevaluate their compound libraries. At the same time, companies engaged in providing compounds tightened their quality control when purchasing compounds. Nowadays, a relatively high quality compound library can be purchased. The compound libraries in big pharmaceutical companies have change dramatically too. In the early days, almost all accessible compounds were deposited in the library. Nowadays, many of the problematic compounds (such as those that may interfere with an assay) and compounds obviously of no use for drug discovery purposes are eliminated from the library. In the meantime, many new high-quality compounds are added to the library. Currently, many of the big pharmaceutical companies have compound libraries exceeding one million compounds.

There are many considerations in building a chemical library. Because of solubility, metabolism, and other issues, typical drugs are organic molecules with molecular weight less than 500 Da. It was estimated that the chemical space assembled from C, H, O, N, S, and P atoms contain at least 10^{30} molecules with molecular weights less than 500 Da. Assuming a screening cost of \$0.2/sample for one drug target, screening 5 million compounds will cost \$1 million. Thus, it is operationally prohibitive to screen a library with 5 million compounds. Even with 5 million compounds, it is still a tiny fraction of the possible chemical space. Thus, it is impossible to cover the whole chemical space in a screening campaign. With this in mind, the compound collection efforts nowadays are not focused on getting as many compounds as possible but are focused on selecting diverse druglike compounds. Drug-likeness criteria are based on the analysis of existing drugs. Some rules, such as Lipinski's rule of 5, were proposed based on the analyses of existing drugs. However, because of the limited number of drugs available and our limited understanding of the interaction between a chemical and the cellular target in vivo, it is very difficult to define a druglike molecule. Chemical diversity in a library is another important factor to decide what compounds should be included in the library. Early approaches to diversity analysis were based on traditional descriptors such as two-dimensional fingerprints. Current emphasis is on assessing scaffold coverage to ensure the inclusion of a variety of different chemotypes. It should be noted that the molecules obtained from combinatorial synthesis tend to be clustered in a smaller chemical space because only a limited number of reactions can be performed in combinatorial synthesis.

13.3.2 Compound Dissolution and Storage

The compounds must be dissolved in a solvent before they can be used in HTS. Though different compounds prefer different solvents, it is not practical to dissolve different compounds in different solvents for HTS operation. In addition, the residues of different solvents (usually 0.1 to 1%) in the final assay buffer may affect the assay differently and cause variations. Thus, a single solvent or mixed solvents with fixed composition must be used to dissolve all the compounds in the library to keep uniformity. The selected solvent must have the following properties: (1) It has strong solvating capacity for most of the compounds in the library; (2) it is miscible with water for effective delivery of the compounds to the drug target because all assays are performed in aqueous solution; (3) it must be compatible with in vitro assays at final diluted concentration (usually 0.1 to 1%); (4) it has a relatively low evaporation rate (same as high boiling points) so that compound concentrations remain the same during storage; and (5) it is inert to the compounds it dissolves. Common organic solvents can be divided into three groups: nonpolar (such as hexane, benzene, toluene, and tetrachloromethane), polar aprotic (such as THF, acetone, acetonitril, DMF, and DMSO), and polar protic (such as methanol, ethanol, and formic acid). The properties of some solvents related to this discussion are listed in Table 13.1. Nonpolar solvents are not miscible with water and they are ruled out. Many polar aprotic solvents (such as THF and dichloromethane) are ruled out too because they are not miscible with water. The polar protic solvents with short aliphatic chains, such as methanol and ethanol, are usually miscible with water. However, methanol and ethanol have very low boiling points and they can only dissolve a few classes of compounds. They are ruled out. Formic acid and acetic acid are not inert and are ruled out. Acetone and acetonitrile have low boiling points and are ruled out. This leaves only DMSO and DMF as the possible final candidates as the universal solvents. DMSO has a slightly higher boiling point and higher dielectric constant than DMF. While DMF is highly toxic, DMSO is much less toxic. In fact, DMSO is the first Food and Drug Administration (FDA) approved drug to treat interstitial cystitis. These favorable properties made DMSO the solvent of choice for compound storage across the industry. However, DMSO still has several shortcomings that need to be dealt with. DMSO is hygroscopic and rapidly absorbs water in the air. This results in the change of compound concentration, decreasing the melting point of the solution and increasing the rate of compound degradation. Thus, the compound stored in DMSO solution should be kept dry. The melting point for DMSO is 18.5°C. However, 5% water in DMSO will change the melting point to below 10°C. Usually, samples in frozen state have slower degradation rate. Thus, the compound library in DMSO can be stored at 4°C or −20°C while they remain frozen. Further lowering of the storage temperaure, such as to −80°C, may have some benefits because the reaction kinetics slows down with lower temperature. However, the added cost for −80°C storage may not justify the marginal benefits gained. Though DMSO has low toxicity, it can readily penetrate skin and bring potential toxic compounds inside the body. DMSO can penetrate gloves as well. Thick rubber gloves should be worn when handling compounds dissolved in DMSO. DMSO is a weak oxidant and can transfer its oxygen atom (oxidation) to some classes of compounds. Though mixed solvents may increase the solvating capacity to broader compounds, there is no report to store compound library in mixed solvents.

In addition to compound collection and storage, the centralized compound management operation in a company is also responsible for compound management and distribution of the compounds in the collection. Thus, in addition to the many copies of the MasterPlate (one copy of 1 million compounds will require more than 10,000 96-well plates), the compound management operation has to prepare and temporally store many copies of the whole library or sublibrary in microplates ready for distribution. Some compound management operations also store dry compounds. The dry compounds are usually stored in same-sized tubes that are addressable with two-dimensional barcodes. It became close to impossible to manage these tubes

TABLE 13.1 Comparison of Organic Solvents

Solvent	Boiling Point (°C)	Dielectric Constant	Density (g/mL)	Water Miscible	Low Evaporation	Inertness	Broad Solubility
Nonpolar Solvents							
Hexane	69	2	0.655	×	×	✓	×
Benzene	80	2.3	0.879	×	×	✓	×
Toluene	111	2.4	0.867	×	×	✓	×
Polar Aprotic Solvents							
Tetrahydrofuran (THF)	66	7.5	0.886	×	×	✓	×
Dichloromethane (DCM)	40	9.1	1.326	×	×	✓	×
Acetone	56	21	0.786	✓	×	✓	×
Acetonitrile (MeCN)	82	37	0.786	✓	×	✓	×
Dimethylformamide (DMF)	153	38	0.944	✓	✓	✓	✓
Dimethyl sulfoxide (DMSO)	189	47	1.092	✓	✓	✓	✓
Polar Protic Solvents							
n-Butanol	118	18	0.810	×	✓	✓	×
Isopropanol (IPA)	82	18	0.785	×	×	✓	×
Ethanol	79	30	0.789	✓	×	✓	×
Methanol	65	33	0.791	✓	×	✓	×
Formic acid	100	58	1.21	✓	✓	×	×
Water	100	80	1.000				

and microplates manually. To solve this problem, several companies provide fully automated compound storage solutions. One example is the REMP automated sample management system from Tecan. The system provides fully automated storage and retrieval of millions of chemical compounds or biological samples. Samples both in dry form or in solution can be managed by the REMP system. In addition to storing liquid samples in microplates, the system can also handle liquid samples in specially designed microtubes that reside in racks conforming to microplate format. Storing compounds in this type of tubes allow rapid cherry picking of desired compounds without the need to thaw the whole microplate. In addition to REMP, several other companies (e.g., RTS, Matrical, TTP Labtech, and The Automation Partnership) also provide fully integrated compound storage solutions with each of them having unique properties. All the large compound management solutions cost several million dollars and careful evaluation of different systems must be made before settling with one.

13.3.3 Compound Plate Manipulation and Distribution

The compounds in the MasterPlate may go through several intermediate plates (InterPlates) before they are transferred to the final assay plate (AssayPlate) where the actual assay takes places. In routine compound management operations, the MasterPlates may go through replication (making many copies of the MasterPlates), dilution (making InterPlates for assays requiring lower compound concentration), and reformatting (making InterPlates for assays that are performed in different plate formats, e.g., 96 to 384 wells). InterPlate is necessary because an assay may happen in different plate formats or the assay may require a lower compound concentration (or a lower DMSO content). For example, a cell-based assay usually requires DMSO concentrations at less than 0.1%. If the final assay volume is 50 μL and if the compounds are directly transferred from the MasterPlate to the AssayPlate, only less than 50-nL compounds should be transferred to satisfy the 0.1% DMSO requirement. It is difficult to transfer such a low volume accurately without special low-volume liquid transfer instruments. If the compounds are first diluted in an assay buffer by a factor of 100 into InterPlate, 5 μL of the diluted compounds in aqueous solution can be transferred to the final AssayPlate using common microplate multichannel pipettors. In early HTS, the compounds are delivered from a compound library operation to a screening operation as an InterPlate. The compound concentration in the InterPlate is usually lower than that in the MasterPlate and the compound can be in either 100% DMSO or in an aqueous solution containing DMSO. If the compounds are in an aqueous solution, they must be immediately used because the compounds may not be stable in an aqueous solution.

When transferring compounds from one microplate to another, instruments that can individually access each well in the microplates must be used. Because there may be 96 or 384 wells in the source plate, instruments with multiple tips that can transfer compounds in microplates in parallel are required to obtain adequate speed. Multichannel microplate pipettors with 96 tips arranged in the same way as 96-well plates (such as MultiMek from Beckman and MiniTrak from PerkinElmer) or arrays of 96 syringes arranged in the same way as 96-well plates (such as Robbins Hydra) are commonly used to transfer compounds with volumes in the microliter scale. Pin arrays with 96 pins arranged in the same way as 96-well microplates are used to transfer compounds with volume in the nanoliter scale. Multichannel microplate pipettors/syringes are expensive, especially the ones that can handle submicroliter volumes and the ones that can handle 384-well microplates. The operation is more tedious when reformatting is required. To avoid cross contamination, the pipette tips should be either disposed of or extensively washed. Disposing used tips guarantees zero contamination at the cost of a large number of tip boxes that collectively are expensive when thousands of microplates are involved. Washing pipette tips or pins is time consuming and generates a lot of toxic waste. A bigger risk is that incomplete washing due to the malfunction of the instrument might contaminate the whole library. The compound plate manipulation had been a major obstacle for HTS operations. This obstacle was tackled by both the development of novel liquid transfer methodologies and the innovation in screening processes. Novel liquid transfer methodologies to handle compound libraries will be discussed in the next section. Here we will discuss screening process innovations.

With the advance of new liquid transferring technologies, compounds can now be efficiently transferred from one plate to another at the nanoliter scale with great precision [$\sim$5% coefficient of variation (CV)]. However, it is expensive to install these instruments in every screening lab. Because these new technologies allow efficient and accurate transfer of submicroliter volumes of compounds directly from MasterPlate into AssayPlate, new screening process can be adopted that eliminate the need to install these instruments in every screening lab. In a typical screening, most reagents in the assay are the same in every well in the AssayPlate except the compounds. While the other reagents can be delivered from reservoirs to the AssayPlate using microplate dispensers, the compounds can only be transferred with instruments that can address individual wells in the source plate. There is a trend in the HTS industry to move the step of adding a compound to the AssayPalte from the screening operation to the compound management operation. The compound management operation, with the help of the cutting edge instruments designed to handle the compound, places the compounds directly into the specified AssayPlate for any screening project and distributes the plates to the screening operation. These AssayPlate-containing compounds are sometimes referred to as AssayReadyPlates (or ReadyPlate). After receiving the ReadyPlate, the screening operation only needs to add controls to specific wells (in one or two columns or rows) and then dispense other assay reagents uniformly into the whole plate. No expensive compound-handling instruments are required in the screening lab. This mode of operation not only streamlined the operation by eliminating many levels of InterPlate but also reduced the cost by eliminating expensive instruments in multiple screening sites. In addition, there is a huge savings with consumables, such as microplates and pipette tips.

13.3.4 Screening Single Compound per Well or Multiple Compounds per Well

Early in HTS when the screening throughput is relatively low, many screens were performed with one well containing a mixture of compounds (pooling). Pooling of multiple compounds in one well can increase the throughput and reduce the cost of screening. Matrix pooling is a strategy that mixes the pooled compounds in a two-dimensional matrix so that one compound can be tested twice to increase the confidence of the assay and also speed up identification of a hit compound. However, it was found that a significant number of hits were missed (false negatives) in screening pooled compounds compared with screening a single compound in a well. The reason for this discrepancy was not fully understood. There are some obvious issues with screening multiple compounds per well. (1) The effective compound concentration with multiple samples in one well will affect the assay outcome. If 10 compounds are mixed in one well at 10 μM each in the final assay, the effective total concentration of all the compounds will reach 100 μM. This effect can be better appreciated by considering all the 10 compounds as similar, such as the case we discussed when searching for kinase substrates in Chapter 8. Some compounds may have lower solubility and precipitate at 10-fold higher concentrations than normal screening concentration. (2) The pooled compounds may react or form a complex with each other leading to reduced activity in the assay. (3) Wells with many synergic but weakly active compounds may be identified and cannot be de-convoluted, leading to false positive; and the opposite may happen leading to false negative. (4) For many reasons, a small percentage of the compounds in the library may interfere with the assay. If there are 0.1% problematic compounds in the library, only 0.1% of the assay results will be affected when screening one compound/well. After mixing 10 compounds per well, 1% (10-fold higher) of the screening results will be affected. Some compounds in the library can affect the assay by interfering with the optical detection, such as fluorescent compounds and light quenching compounds. Recently, some compounds were found to be able to inhibit a large number of enzymes nonspecifically. These compounds are referred to as promiscuous inhibitors. They cannot be excluded from the library and they will affect the wells that contain it as a component in the mixture.

There is no argument that screening a single compound will result in better screen quality. Screening with a single compound or with multiplex depends on balancing the quality and economics of the screen. The major reasons for screening pooled compounds in early HTS were that the screening throughput was too low to handle the large number of compounds (on average about 500,000 compounds in a library) and the reagent costs were high (in 96-well microplates). With the rapid advance of HTS, throughput and reagent costs are not a major issue to handle a compound library with less than 1 million compounds. With most screening performed with even smaller focused libraries, most companies in small-molecule drug discovery now perform HTS with a single compound per well only.

13.4 HARDWARE MODULE

13.4.1 Microplate Liquid Transfer Instruments

The liquid transfer instruments in HTS perform the task of transferring solutions from a container (bottle or reservoir) or a source microplate to a designated microplate.

Except when employing acoustic droplet ejection (ADE) technology that will be discussed later, all other methods of transferring liquids must make contact with the source liquid. The microplate liquid transferring instruments can be divided into two types, depending on whether they make contact with the receiving microplate or not. If the liquid transferring instrument does not make contact with the receiving microplate, the liquid transferring process is called noncontact pipetting or noncontacting dispensing. If the liquid transferring instrument makes contact with the receiving microplate, the liquid transferring process is called contact pipetting or contact dispensing. Noncontact dispensing can add the same reagents across a microplate and between microplates without the need to change any part in the liquid path because there is no issue of cross contamination. It is a preferred method to transfer a single liquid to many microplates. Contact dispensing has two modes: dry dispensing and wet dispensing. Dry dispensing describes the process of contact dispensing to a dry plate. Wet dispensing describes the process of contact dispensing to a plate containing liquid in the wells. When dispensing small volumes, dry dispensing may create large variations because the dispensing tips may not touch the plate surface uniformly. Wet dispensing will cause less variation because the dispensing tips are inside the liquid in the receiving wells. However, the liquids already in the receiving wells may contaminate the dispensing tips. Microplate dispensers (noncontact dispensing) and multichannel pipettors (contact dispensing) are the two major types of liquid transferring instruments commonly used in HTS and they will be discussed below.

Microplate dispensers usually have a row of nozzles that are connected to the solution reservoir through a liquid path. The solution in the reservoir is pushed into the liquid path and then dispensed into a microplate through the nozzles. Microplate dispensers are used primarily for dispensing bulk solutions (such as enzymes, substrates, cells, and other reagents except compounds) across a whole microplate. Because the nozzles do not make contact with the well in the microplate, the advantage of using a microplate dispenser is the low chance of cross contamination. In addition, microplate dispensers are usually inexpensive. Multidrop microplate dispensers (now marketed by Thermo Scientific) were the most widely used microplate dispenser in the early days of HTS and are still widely used today. In the Multidrop design, eight separate tubings form the liquid paths that connect eight nozzles to a common solution reservoir. All the tubings share one peristaltic pump that compresses the tubings and drives the solution through all the nozzles. With a peristaltic pump, the fluids being pumped only contact the interior of the tubing, which is easy to sterilize and clean. Peristaltic pumps are inexpensive to manufacture and do not break easily because there is no moving part in contact with the fluid. Peristaltic pumps also minimize shear forces experienced by the fluid, which is good for dispensing cells. Another type of dispenser uses a syringe to draw liquid from the reservoir and then push the liquid through a shared path to the nozzles (e.g., the MicroFill microplate dispenser from BioTek). In this design, all the nozzles share the same liquid path to the reservoir. The dispensing volume is controlled by the syringe movement based on the sum of the volumes to be dispensed from all the nozzles. There is a potential issue with this design because there is no control of individual nozzle. If one nozzle is clogged, the same total volume of solution pushed by the syringe must be dispensed through the other nonclogged nozzles resulting in dispensing more volume to those wells. In comparison, a clogged tubing will not affect the dispensed volume from other tubings in the Multidrop because the liquid path is not shared. However, with proper maintenance, both types of liquid dispensers work well in my lab. Multidrop and MicroFill are low-end inexpensive instruments that perform well when the dispensing volume is larger than 5 μL. They can dispense liquid into 96-well and 384-well microplates. The high-end expensive FlexDrop microplate dispenser (marketed by PerkinElmer) can dispense four different solutions. In FlexDrop, a positive pressure is applied to the reservoir to push the liquid through an array of control valves inserted in the liquid path in front of each nozzle. It can precisely deliver liquid spanning from 0.2 μL to 2 mL. It can dispense liquid into 96-, 384-, and 1536-well microplates. However, the valves controlling the liquid dispensing in each nozzle have small inner diameters and they are prone to clogging. Flexdrop is limited to dispensing homogeneous solutions and cannot be used for solutions containing particles and cells.

When using microplate dispensers, a large void volume may be of concern. The reservoir has to maintain a certain volume to totally cover the tubing (~5 to 10 mL for MicroFill and Multidrop) and the liquid path from the liquid inlet to the outlet must be filled with liquid (another 5 to 10 mL). In addition to filling the liquid pathway, the solution must be purged through the tubing for at least 2 volumes of the liquid path for uniform dispensing. This leads to close to 30 mL of void solution. If dispensing 10 μL solution per well to ten 384-well microplates, the total volume delivered to the microplate is 38.4 mL. If a microplate dispenser is used, a total of 68 mL of solution is required with close to half of it (30 mL) being wasted. The void volume becomes negligible when the solution is dispensed to hundreds of microplates. It is observed that the first microplate in a screen, which usually results with a gradient pattern following the order of the wells being dispensed. The most likely cause of this pattern is the incomplete purging of the solution in the liquid path of microplate dispensers.

Another type of liquid transferring instruments commonly used in HTS is the multichannel pipettor. Multichannel pipettors aspirate and dispense liquid by the movement of the tightly sealed plungers. The pipette tips or needles (or cannula) are first inserted inside the source solution. A small volume of the solution is transferred inside the tips or needles under vacuum created by pulling the plungers. After positioning the tips or needles inside the receiving microplate, the solutions are expelled by the positive pressure created by pushing the plungers. Multichannel pipettors can be arranged in a single row (usually up to 8 tips) or in an array (96 or 384 tips). A good example of a single-row multichannel pipettor is the Genesis series of liquid transferring instruments from Tecan, which were very popular in early HTS operations. The instruments can be configured to have up to 8 cannulae arranged in a row. Each cannula can move independent of the others in the z axis so that samples can be addressed individually with one cannula. The spacing between the cannulae can change so that multiple samples in both the microplate and in the tube racks can be addressed with multiple cannulae simultaneously. In this design, the plunger is far away from the tip of the cannula, creating a large volume between the plunge and the cannula tip. To increase the pipetting precision, a relatively incompressible system liquid (water) is used to fill the large volume instead of leaving the space filled with compressible air. When using the instruments, a small air gap must be created before taking in samples to prevent the mixing of samples with system liquid. Because they can transfer samples from both tube racks and microplate to individual wells in a microplate, the instruments are typically used for placing controls in specific columns of the AssayPlate or for cherry picking selected compounds in a microplate. Though capable of transferring liquid from a whole microplate to another, the operation will be too slow with this type of instruments because it can only transfer a maximum of 8 wells at a time. Instead, another type of pipettors with tips or needles arranged in a two-dimensional array confirming to 96- or 384-plate formats should be used for high-speed transfer of liquid to a whole microplate. In this type of multichannel microplate pipettors, the tip/needle array is attached to fixed pipetting heads for dispensing liquid to 96- or 384-well microplates simultaneously. Tips/needles in the array are not individually addressable. Multichannel microplate pipettors can be found in both stand-alone instruments containing multiple deck positions for microplates (e.g., Agilent's Vertical Pipetting Station, Apricot Design's multichannel pipetting stations, and Robbins Hydra) and in integrated workstations (e.g., Beckmann's FX 2000, PerkinElmer's JANUS, Caliper's Sciclone, and Tecan's Freedom Evo). Multichannel pipettors with 96 or 384 tips/needles are commonly used in HTS to replicate, dilute, and reformat compound microplates. When performing reformatting microplates, the pipetting head can shift in quadrants to fit the new plate format. Another application of these multichannel pipettors is the addition of radioactive reagents to AssayPlates using disposable tips. Dispensing radioactive reagents across a microplate with microplate dispensers will contaminate the whole liquid path of the dispenser, which should be avoided. In special cases when dispensing bulk valuable reagents, it may be too costly to use microplate dispensers because of large void volume and the volume wasted in purging. On the other hand, multichannel pipettors can save the reagents using special low dead volume reservoirs. A major disadvantage of multichannel pipettors is the contact dispensing that carries the risk of cross contamination of samples when the needles or used tips are incompletely washed. In addition, multichannel pipettors are more expensive than microplate dispensers. The use of disposable tips adds to the cost too.

Piezoelectric-based pipettes are commercially available to dispense small volumes of liquid. This dispensing technology is based on the phenomenon that piezoelectric crystals will deform when a voltage is applied to it. Thus, a piezoelectric crystal can be placed in contact with a glass capillary that is attached to a syringe pump to form a piezoelectric pipettor. The sample to be dispensed is first drawn into the reservoirs in the pipettor by the syringe. When a voltage is applied to the crystal, the crystal will deform and squeeze the capillary resulting in a small amount of liquid in the pipettor's reservoir being ejected from the tip. The drop volume as a result of small deflection of the crystal is on the order of hundreds of picoliters. However, large volumes can be dispensed because the fast response time of the crystal allows a fast dispensing rate on the order of several thousand drops per second. Piezoelectric pipettors are not widely adopted because there have known reliability and robustness issues, such as the air bubbles in the sample resulting in reduced reliability in dispensing, relatively large sample volume required to dispense a small volume of samples, and the difficulties involved to wash the pipettors when changing dispensing samples.

Pin arrays are devices containing 96 or 384 rigid floating pins (usually made of stainless steel) that are arranged to fit a microplate format. When the pins are dipped into a sample solution, a small volume of liquid is transferred onto the tips of the pins. The samples loaded on the pins are transferred by touching the pins onto the surface of a dry microplate or by dipping the pins inside the solutions in wet microplates followed by agitation. The pins are then removed from the microplate and they must be washed thoroughly before transferring the next samples. Normal pins with round or flat tips can transfer liquid in the range of high picoliter to low nanoliter. To increase the transfer volume, a fine slot can be machined into the end of the pin creating a pin tool called a split pin. When a split pin is dipped into the sample solution, the sample is loaded into the slot. Tapping the pin onto the solid surface of the microplate with sufficient force can release the sample into the wells. There are several other pin tools with different patterns at the tip of the pin. Pin arrays are used in HTS primarily to transfer compounds between microplates in a similar way to a multichannel pipettor but with much smaller volume. The transferred volume depends on the diameter of the pin and the pattern of the pin tip. The errors in the volumes transferred with pin arrays are relatively high ($\sim$10%) compared with more expensive pipettors, such as piezoelectric pipettors and the instruments discussed below. However, the cost of a pin array is much lower. Pin arrays are usually not disposable and must be washed between transferring different samples. Pin arrays are especially useful in transferring small volumes of compound solution (usually less than 100 nL) into 1536-well microplates in the absence of more expensive instruments.

For transferring compounds between microplates, an instrument that makes contact with neither the source liquid nor the receiving liquid will be ideal to eliminate the cross-contamination issue. Acoustic droplet ejection (ADE) is a technique that allows the transfer of liquid between microplates with the instruments touching none of the liquid. The technology has existed for a long time before it was applied in transferring compounds in 100% DMSO between microplates. Currently, two companies (Labcyte and EDC Biosystems) offer instruments based on ADE for HTS applications. ADE focuses acoustic energy (a pulse of ultrasound) into a fluid sample to eject droplets of fluid (typically nanoliters or picoliters) without any physical contact. The diameter of the droplet is inversely proportional to the frequency of the acoustic energy—higher frequencies produce smaller droplets. ADE technology is a very gentle process, and it can be used to transfer not only compounds but also proteins, high-molecular-weight DNA, and even live cells without damage or loss of viability. Another major advantage of transferring liquid with ADE is that the CV of a transferred volume is independent of how small the transferred volume is. In comparison, liquid transfer methods using pipette tips, capillary nozzles, or pins rely on droplet formation through an orifice, which will lose precision as the transfer volume decreases. For the application in transferring compounds from microplate to microplate with ADE, the destination plate is placed inverted on top of the source plate. The energy source (transducer) is placed at the bottom of the source plate. The transducer travels in the z axis to focus the acoustic pulse at the fluid meniscus in order to eject droplets. Liquids ejected from the source are captured by the inverted receiving dry plates due to surface tension. The fluid in any well in the source plate can be transferred to any well or position of the destination. The transfer volume per ejection can be adjusted by changing the ultrasound frequency. If a larger volume is required in a well, multiple droplets can be ejected from the source (typically 100 to 500 droplets/s). However, the overall throughput will decrease. One caveat of ADE is that the transferred volume depends on the composition of the liquid solution. Whenever the liquid composition changes, the instrument's parameters need to be adjusted to eject the same volume. This is not a significant problem for dealing with compound libraries because they are uniformly dissolved in DMSO. However, if the DMSO solution absorbed significant amounts of water, the instrument has to be adjusted. This property of the ADE imposes a challenge in adapting the technology to dispense aqueous solution where the conductivity of the solution may change dramatically. The Echo series of instruments from Labcyte were installed in most compound management centers in major pharmaceutical companies to transfer small volumes of compounds. However, when transferring larger volumes of compounds in the microliter scale, traditional pipetting is still the method of choice.

TTP LabTech's Mosquito liquid pipettor has a novel design that is different from traditional multichannel pipettors. Each Mosquito micropipettor is built with a plastic tube wrapped around a stainless steel piston. The piston moves up and down to aspirate and dispense liquid with no dead volume. Because there is no air gap or system liquid, the effect of the liquid type, viscosity, surface tension, and environmental conditions are minimum and the solution can be transferred more accurately. The Mosquito pipettors are fed into the Mosquito by a continuous reel with all the pipettors equally spaced. A one-dimensional array of pipettors is available at one time for aspirating and dispensing liquid into 96-, 384-, or 1536-well microplates. The low-cost micropipettes are disposable, which is a huge advantage in preventing cross contamination between samples. Mosquito pipettors can transfer between 25 nL and 1.2 μL of liquid, which covers the volume commonly required for ReadyPlate.

13.4.2 Microplate Manipulation Instruments

In addition to liquid transfer instruments, several other types of instruments are required in HTS operations. Plate hotels (racks or carousals) are used to store hundreds of AssayPlates and InterPlates. The microplates in the hotel can be accessed randomly through either built-in shuttle or by an external robotic arm. The plate hotels can be placed inside incubators to gain cell-based assay capability. Cytomat (now part of ThermoFisher) and Liconic Instruments are the two major providers for microplate hotels. To track the microplate and to link the final assay results obtained from the AssayPlate to the InterPlate, and ultimately to MasterPlate, each microplate must be bar-coded. It is difficult to manually apply barcode labels at the same position on thousands of microplates. Automatic barcode appliers are commonly used in HTS operations (e.g., Vcode from Velocity11, now part of Agilent). Barcode scanners are important devices in HTS operation. For heterogeneous assays (e.g., ELISA), an automatic microplate washer is indispensable. The most widely used Elx405 plate washer from Biotek has two sets of liquid paths for each well arranged in 96 arrays of metal tubes. One set of the liquid path is connected to the washing solution and the other liquid path is connected to a vacuum. The microplate wells are washed by alternating suction and liquid dispensing. An automated filtration station is often present in HTS operations to handle microplate filtration steps. This operation requires the filter plates. The filtration is carried on either by applying a positive pressure on top of the filtration plate or by applying a negative pressure at the bottom of the filtration plate. For cell-based assays, centrifugation of microplates is sometimes required to spin down cells (e.g., changing cell media). Other common microplate manipulation instruments include microplate shaker, microplate sealer, and microplate piercer.

13.4.3 Microplate Readers

Most microplate readers were developed within the past 20 years as a result of the demands for high-speed reading of samples in microplates in HTS. Since instruments for detecting single samples in different detection modes already exist, early microplate readers were mostly developed by fitting the original design into the microplate format. However, this is not a simple task. For example, measurement of fluorescence with a single sample was done by detecting the emission light at 90° from the excitation light. Such arrangement cannot be adapted for the detection of fluorescence in the microplate format. Compromise must be made to have the excitation and detection on one side of the microplate (top or bottom). This compromise causes much higher background when detecting fluorescence in microplates. Early microplate readers read only a single mode. For example, BMG Labtech's FluoStar only reads a fluorescence signal. For the need to increase flexibility and to reduce the number of readers placed on the integrated HTS screening platform, instrument developers started to make multimode readers (one instrument can measure many modes of signals). Molecular Devices' Analyst instrument is one of the early multimode readers that can read absorption, fluorescence, FP, and FRET. Currently, many companies build multimode readers, such as Envision from PerkinElmer, PolarStar Omega from BMG, Synergy from BioTek, Infinite 200 from Tecan, and Mithras LB 940 from Berthold Technologies. These instruments all read microplates by repositioning the detector from well to well in sequence.

Microplate detectors based on CCD cameras were not evolved from single-sample detection technology but were developed primarily for the need of high-speed measurement of samples in high-density microplates in HTS. Because of the relative large footprint of the microplate (compared with a single well in a microplate), expensive telecentric optical lens must be used to image the whole microplate. Even with telecentric optical lens, the signals from the wells away from the center of the plate are gradually decreased because of the gradual increase of the distance from the lens. The signal distortion is profound for the wells at the edge of the microplate. This systematic distortion of signals can be mathematically corrected by applying a multiplication factor for each well. The multiplication factor for each well is obtained by measuring a standard microplate with uniformed signals from each well. The CCD camera is cooled to less than $-70°C$ to suppress the dark current to reduce the electric noise. LEADseeker from Amersham (now part of GE Heathcare) was the first whole-plate reader based on the telecentric lens and the cooled CCD. It was originally developed for high-speed measurement of the red luminescence signal from excited YSi SPA beads. The current LEADseeker is a multimode whole-plate imaging system that can measure fluorescence, luminescence, FRET, and FP. The competing

ViewLux whole-plate imager from PerkineElmer has similar functionality. Because these instruments read an entire plate in one exposure, they have an advantage over the other types of detectors, especially when used with high-density microplates. For example, ViewLux can measure the fluorescence signal in a 1536-well microplate with a throughput of more than 200,000 samples per hour under continuous operation. However, these instruments are expensive with a price tag close to or exceeding \$1 million.

13.4.4 Microplate Transport Instruments

We introduced all the instruments in HTS to handle individual microplates, that is, liquid transfer instruments, microplate manipulation instruments, and microplate readers. With these instruments, a scientist can perform screening by manually placing each microplate into an instrument and manually transferring each microplate or a stack of microplates between instruments. However, the throughput is low and it is labor intensive to have a person performing these tasks. Microplate transport instruments automate the microplate loading and transportation using two types of instruments: microplate conveyors and robotic arms.

Microplate conveyors move microplates from one position to another with belts or tracks. The early conveyor systems used in HTS are the simple linear tracks that transport microplates from microplate stackers to the interface position of an instrument for processing. Microplate stackers are microplate holders with the microplates stacked on top of each other. A pair of stackers are usually attached to an individual instrument to allow automated loading of a large number of microplates into the instrument at high speed (e.g., the MiniTrak system and Topcount reader from PerkinElmer with attached microplate stackers). Because the interface between the stacker and the instruments is at the bottom of the stacker, the microplates in a stacker cannot be accessed randomly. Thus, at least two stackers are required in any operation with one stacker loaded with the microplate to be processed and the other empty stacker receiving the processed microplates. The stacking of microplates in a stacker is a first-in-last-out operation, that is, the first microplate transferred to the empty stacker will be the last microplate coming out in subsequent stacking operation. Restacking of the microplates after each operation may be required to maintain the proper order of operation on each microplate in some assays to maintain the same incubation time for each microplate. If each instrument involved in screening is attached to stackers, an HTS operator can sequentially perform each step of the assay at a time with a stack of microplates. The stack of processed microplates is manually moved from one instrument to the next. This mode of screening is called batch processing because a batch of microplates is processed for one step of the assay at a time. The next step of the assay only starts after all the microplates in the batch are done in the previous step. In this mode, the bottleneck of screening is how fast the instrument can handle the microplate but is not how fast the microplate can be delivered into the instruments. An instrument that can carry out one or a few screening steps (but not all) is referred to as a workstation. Screening operations with screening operators transferring a batch of microplates between workstations is commonly referred to as semiautomation.

The HTS operation carried out in semiautomation mode with batch processing has the advantage of high throughput, flexibility to handle many assays, and rapid response to operation errors because the HTS operator is present at every step of the assay. The same advantage can be a disadvantage as well because human error may occur during long repeated operations. The morale may be low too for the operator to perform the tedious repeated work for a long term. Some assays that require a short time interval between adjacent steps cannot be carried out in a batch process mode if it takes longer to finish the whole batch of microplates than the time interval required for the execution of the next step. In this case, continuous parallel operation on each microplate in a conveyor system or arm system has the advantage.

In addition to loading microplates from stackers into an instrument, conveyors are also found in some large fully integrated HTS systems to transport microplates between instruments. In contrast to semiautomated operation that relies on humans to transport microplates between instruments, fully integrated HTS operation carries every step of the assay in one system with no need for an operator to touch the microplates during the screening except for loading the microplates on the system. Fully automated HTS systems usually process one microplate continuously through each step of the assay in parallel with several other plates at different steps of the assay. Conveyor-based integrated HTS systems usually carry out screening by transporting microplates on a linear track/belt. PlateTrak from PerkinElmer and UHTSS system from Aurora Biosciences are good examples of conveyor-based fully integrated HTS systems. The Allegro HTS system from Zymark (now Caliper) transports microplates continuously in predefined sequence between instruments along a linear path using a series of simple robot arms instead of a belt/track. However, its operation process is similar to the operation in conveyor-based systems. In conveyor-based HTS systems with a single lane, the microplates are moved linearly in the direction following the assay sequence. Such systems are less flexible to handle different types of assays. To alleviate this limitation, the Allegro HTS system was built in modules. The sequence of the modules can be changed according to the assay steps. Because of the ability to continuously send the microplates on the belt/track, the throughput of conveyor-based HTS system does not depend on the speed of the microplate transport between instruments but depends on the speed of the slowest instrument module on the track. This bottleneck can be widened by placing multiple slow instruments on the track to increase the speed of the assay steps involving the slow instruments. A fully integrated HTS system based on conveyor transporter (or similar mechanisms) can achieve very high throughput exceeding 600 microplates/day. Dimension 4 (Thermo CRS) uses the combination conveyors (linear plate transport) and many simple robot arms (Flip mover) to transport microplates to achieve both high throughput and flexibility. However, conveyor-based fully automated HTS systems are expensive to set up and to maintain.

Robot arms are instruments that can mimic the function of human arms. They were used widely in many industries to perform tedious, repeated, precision work. In HTS applications, robot arms can grip microplates with the fingers attached to the arm and move the microplate between instruments within the reach of the arm. A group of instruments can be placed in a circle around the robot arm to form a fully integrated HTS system. Because the arm can randomly access any instrument within its reach, the HTS operation with a robot arm as the microplate transporter is flexible to handle many different assays. Because the microplates are almost always held horizontally with upside-up during screening, the robot arms do not need to perform sophisticated movement when handling the microplate. Some HTS instrumentation companies built their own low-cost robots capable of movement in four axes. However, these low-cost robots proved unreliable for heavy HTS use. Currently, most HTS system integrators use proven industrial robots (e.g., Staubli, Thermo CRS, Mitsubishi, and Motoman) because of their reliability, though these instruments offer extra sophisticated movements (five or six axes) that are not neccesary for the application. When choosing a robot arm, it is important to known whether the arm is built with absolute encoder or not. The arm without absolute encoder (such as Catalyst 5 from Thermo CRS) only knows its current position by comparing to the zero position. If the robot crashed in the middle of an experiment, the robot will not know its current position. The robot has to be placed back to the zero position and restarted. This limitation dictates that the screening cannot continue where the crash happened but has to restart from the beginning. On the other hand, the arm with absolute encoder (such as F3 from Thermo CRS) always knows its position. After a crash, the robot can continue from where the crash happened. Early robot arm-based integrated HTS systems (such as the Robocon's HTS system and the SAGIAN Core system from Beckman) placed the robot arm on a track so that the robot arm can reach more instruments. Current robot arm-based integrated systems tend to use one or a few fixed robots with long-reach to access more instruments (such as the integrated systems from RTS and Kalypsys). The robot arm-based integrated HTS system without track can increase the system's reliability and reduce the maintenance cost. Figure 13.2 shows a fully integrated robotic system with one Mitsubishi robot arm (with absolute encoder) to transport microplates between instruments. This system was built to handle screening for biopharmaceutical discovery. The flexibility to handle cell-based assays was the primary goal instead of throughput when I designed this system. The reader table is modular and can be swapped for another reader if necessary. For example, an identical table was built with a PerkinElmer Envision multimode reader placed at the same position as the Fusion. These two readers were swapped later with the Envision as the primary reader. In the event that one reader fails, it can be rolled out and the other reader can be immediately rolled in to reduce the down time.

Integrated HTS systems with only one robot arm generally have lower throughput compared with the conveyor systems because one arm has to carry all the assay steps. The throughput in an arm-based integrated HTS system is limited by the slowest instrument on the system or by the total time for the arm to carry all the necessary assay steps for one microplate. This can be illustrated by a hypothetical RNA polymerase assay with SPA technology. The assay scheme is outlined in Figure 13.3. The biotin-labeled substrate is mixed with radioactive tracer ([^{3}H]UTP). After the addition of polymerase, the radioactive tracer is attached to the biotin-labeled substrate. After the addition of streptavidin-labeled SPA beads, the radioactive tracer is brought close to the beads through the binding between biotin and streptavidin. The SPA signal is generated and it is detected with a TopCount reader. The robotic movements

Figure 13.2 A fully integrated robot arm-based HTS system. The system is fully enclosed and comprises the following components: (1) Mitsubishi MELFA RV2A 6-axis robot; (2) Caliper's Sciclone ALH3000 with interchangeable 96- or 384-tip pipetting head, an independent 8-channel pipettor, two bulk-reagent dispensers, and plate gripper (2a). The following accessories are integrated into the Sciclone: microtiter plate shaker (2b); positive-pressure filtration system (2c); and ultrasonic tip-wash station (2d). (3) PerkinElmer's Fusion with 11-mode detection, which includes absorbance, fluorescence, fluorescence polarization, time-resolved fluorescence, time-resolved fluorescence–resonance energy transfer, AlphaScreen, etc. (4) Kendro Cytomat6001 with humidity, temperature, and CO_2 controls, and 189 normal microtiter plate storage capacity. (5) Biotek's ELX-405 plate washer, which can be used for 96-well and 384-well plates. (6) Volecity11's Vspin centrifuge can be used for normal- and deep-well plates. (7) Thermo CRS's high-capacity stacker is used to store up to 32 stacked tip boxes. (8) PerkinElmer's Flexdrop equipped with four individual dispensing heads that can dispense four bulk reagents in a broad volume range for each head (from 200 nL to 2 mL). (9) Volecity11's Vcode automatic barcode labeler. (10) MicroScan's MS-3 barcode reader. (11) Caliper's plate regrip station that changes the plate orientation to facilitate the interaction between the robot arm and individual components. (12) Kendro's room temperature incubator that stores 189 regular microtiter plates. (13) Caliper's plate-lid-handling station. (14) Six Variomag shaker station that provides an independent plate-shaking operation (behind the monitor, not visible). (15) Liberty Industry air purifier provides ultra-dust-free conditions for the enclosed system and prevents the introduction of contaminants, from the surrounding air, to the work area. (See color insert.)

that carry the assay steps for one microplate are listed below with the robotic time for each step shown in parenthesis:

> Take an AssayPlate from Hotel1 that holds all AssayPlates (20 s).
>
> Place the AssayPlate in Multidrop1 (20 s).

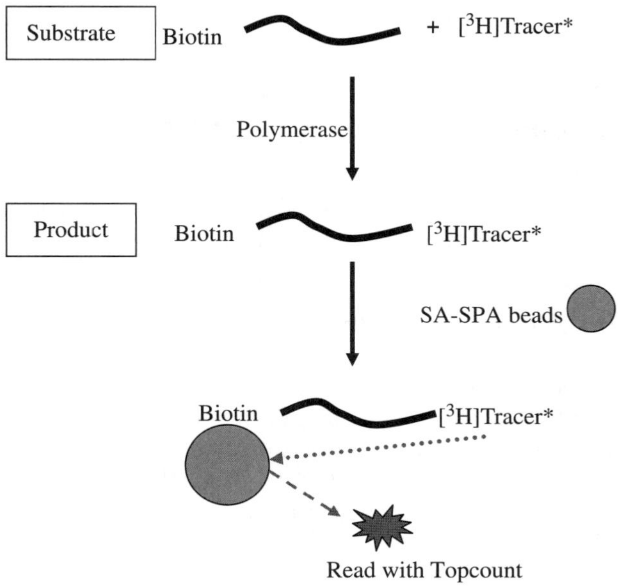

Figure 13.3 Hypothetic RNA polymerase assay in SPA format. The biotin-labeled substrate is mixed with radioactive tracer ([^{3}H]UTP). After addition of polymerase, the radioactive tracer is attached to the biotin-labeled substrate. After addition of streptavidin-labeled SPA beads, the radioactive tracer is brought close to the beads through the binding between biotin and streptavidin. The SPA signal is generated and it is detected with a TopCount reader.

The polymerase is dispensed into the microplate (no time for the arm, 0 s).
Take the AssayPlate to Tecan Genesis to add controls (20 s).
Take one CompoundPlate from Hotel2 that holds all CompoundPlates (20 s).
Place the CompounPlate in Hydro (20 s).
Hydra aspire compounds from CompoundPlate(0 s).
Put ComoundPlate back to Hotel2 (20 s).
Take the AssayPlate to Hydra (20 s).
Hydra dispense the compounds into the AssayPlate (0 s).
Take AssayPlate to shaker (20 s).
Shaking for 10 min (0 s).
*1 Take the AssayPlate to MultiMek to add substrate and [^{3}H]UTP (20 s).
Take the AssayPlate to shaker (20 s).
Shake for 1 min (no time for the arm).
Take the AssayPlate to incubator (20 s).
Incubate for 48 min.
*2 Take the AssayPalte to Multidrop to add stop solution together with SPA beads (20 s).

Take the AssayPalte to shaker (20 s).

Shake for 1 min.

Take the AssayPlate to incubator (20 s).

Incubate for 63 min.

*3 Take the AssayPlate to Topcount for detection (20 s).

Reade the microplate for 10 min (the time it takes a Topcount with 12 heads to read a 96-well plate with 1 min read time/well).

Take the AssayPlate from Topcount to Dump (20 s).

The execution time of three steps marked with an asterisk (*) is monitored to calculate the incubation time for the two critical reactions (will be discussed in next section).

The total robotic time to handle one microplate is about 320 s or rounded to 6 min (by addition of all the time in parenthesis). The incubators and shakers all have multiple positions and are not rate-limiting instruments. If there is only one TopCount reader in the HTS system, it is the slowest instruments (10 min) and it is slower than the total robotic time (6 min). Thus, the maximum theoretic throughput during a 24-h run will be about $24 \times 60/10 = 144$ plate/day. If two TopCount readers are present in the HTS system, the robotic time (6 min) becomes the bottleneck. The maximum theoretical throughput will be $24 \times 60/6 = 240$ plate/day. The only way to further increase the throughput is to add another robot arm to the system.

13.5 SOFTWARE MODULE

The software module in HTS operation contains the software that controls the HTS work flow and the software that handles the screening data. Each instrument in an integrated HTS system has an instrument control program (ICP) supplied by the manufacturer. In stand-alone operation, the user directly interacts with the ICP. In integrated systems, a central software that manages the work flow sends the command to each instruments through ICPs. Because every instrument is different, there is no single software that can communicate with all ICPs. The system integrator must closely work with each instrument vendor to develop a software (sometimes referred to as adaptor) to communicate with the ICP. It may be a long process to obtain a reliable adaptor to enable the central software to communicate with the ICP of a specific instrument, and the cost may be high, especially when the system integrator has never dealt with the specific instrument. When selecting system integrators to build an integrated system, it is very important to check whether the integrator has already integrated all the instruments you had in mind. If the integrator already has the adaptor for all instruments, the cost of building the system will be lower since no new software is required. In addition to interfacing with all the instruments, the central software also directs the sequence and timing of the transport of microplates between instruments carried out by the robot arm or the conveyor. Optimal timing of the execution of each assay steps for all the microplates are made by a scheduler. When screening many microplates in parallel, the scheduler may not always be able to schedule the execution of each step at a specified time due to conflicting requests for the robot arm when two or more microplates may request the robot arm at the same time. To eliminate conflicts, the scheduler is allowed to shift the operation timing slightly (within the range specified by the operator). A good scheduler should be able to optimize the utilization of the robot arm, allow parallel screening of different assays, allow dynamic run-time rescheduling, and allow on-fly recovery of the operation in case of error (require a robotic arm with an absolute encoder).

The advance of HTS has significantly increased the volume and complexity of the data. This requires a sophisticated database management system (DBMS) to handle a myriad of tasks including but not limited to (1) store the screening data, (2) facilitate quality control of the screening data, (3) manage the link among the screening data, the identity of the compounds, and the screening protocols, (4) facilitate the analysis of the screening data, (5) facilitate the hit picking using predefined criteria, (6) facilitate the retrieval of screening data, (7) link to existing database that contains compound information, and (8) offer user-friendly interface for presentation of the screening result. There are several commercial HTS database suites available (e.g., ActivityBase from IDBS, Assay Explorer from Symyx, and Screener from Genedata) with Activitybase commanding the largest market share. These application-specific databases are created for HTS using Oracle relational DBMS (RDBMS). The advantage of using commercial databases in HTS is the rapid deployment (usually within one month) and the confidence of a working database that satisfies a majority of the operational needs (though it may not be exactly what is required). The downside of purchasing a commercial HTS database is the high cost (initial lower six-figure purchasing price for a small organization plus annual maintenance fees running at $\sim$20% of the purchasing price) and less flexibility to add/modify the database to fit the needs of a unique situation.

Some HTS operations prefer to develop their own HTS databases. Currently, there are three RDBMS platform (Oracle, IBM DB2, and MS SQL Server) upon which a HTS database can be built. However, Oracle RDBMS is the dominant one. The RDBMS allows users to create a new application-specific database and specify their schema, support the storage of a large amount of data, allow the end user to query and modify the data, and control the access of the data. To develop an HTS database, the database designer/architect must fully understand the HTS operation work flow and can use the selected RDBMS to model the HTS operation. The database development process is the same as the general software development that follows eight phases as shown in Figure 13.4. The process begins with a specification phase when the database designer decides what operation should be modeled. This is followed by the design phase during which the designer decides how each operation should be modeled and the relationship/interaction between the operation modules. After the design stage, the database developer is involved and together they modify and verify the design. After verification of the design, the database developer starts coding. After coding and debugging, the database is tested (alpha test) with test data. In the alpha test, issues are identified and the database is further refined. The refined database then goes through a beta test. After the beta test, the production database is placed in use in the real world. After database production, database administrator (DBA) is required to maintain the database.

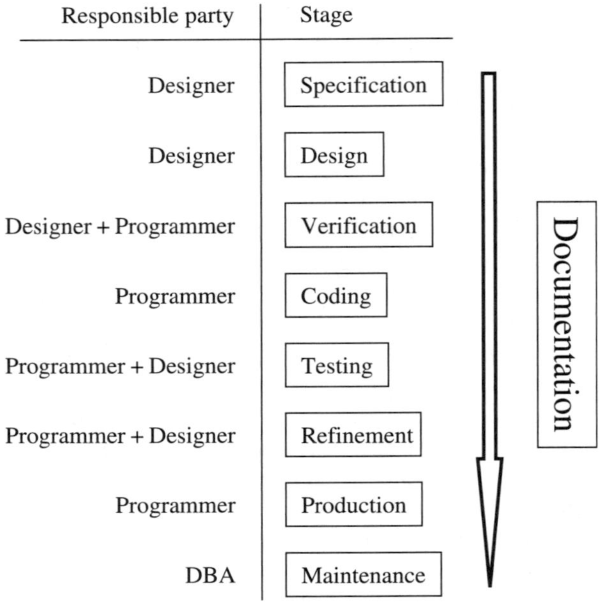

Figure 13.4 Eight phases of software (Database) life cycle. For HTS application-specific RDBMS, the process begins with specification phase when the database designer decides what operation should be modeled. This is followed by the design phase when the designer decides how each operation should be modeled and the relationship/interaction between operation modules. After design stage, database developer is involved and together they modify and verify the design. After verification of the design, the database developer starts coding. After coding and debugging, the database is tested (alpha test) with test data. In the alpha test, issues are identified and the database is further refined. The refined database then go through the beta test. After the beta test, the production database is placed in use in the real world. After database production, the database administrator (DBA) is required to maintain the database. Throughout all these eight phases, clear documentation must be made.

Throughout all these eight phases, clear documentation must be made. The most basic element in a database is the table used to model all the objects in the operation. Each object in the real world has its attributes and so do tables. For example, a "person" object may be associated with the following attributes: "name, birthday, SSN, ..." A table must have a primary key and the table should be made as simple as possible (opposite to the spreadsheet approach). For example, a table to model "assay protocols" may contain the assay name as the primary key and the assay developer, assay title, assay submission date, and so forth as the attributes. A table to model "assay jobs" can have the assay "job name" as the primary key and the assay protocol, plate ID, assay date, and so forth as the attributes. After building all the necessary tables, the relationship between tables are built. The relationship allows query data across tables. On top of table layer, forms are built to facilitate input data into tables and query specific information in the database. Since most biological scientists do not know SQL language, a user-friendly graphic user interface (GUI) must be built. Usually a Web interface is a good choice since most users are familiar with using the Web. In addition, many applications are built to allow scientists run specific applications using the data in the database.

Before releasing the newly built HTS database to users, it is very important to build access control in the database. Without access control, novice users may accidentally change the value in the table. For example, an operator is creating an assay job with assay protocol X associated with the job. All the screening data in this job is associated with assay protocol X. If there is no access control and every user is a superuser, a person may accidentally change the assay protocol from X to Y from a protocol pull-down menu that lists all available protocols. This single click can compromise the whole job and render the whole database useless if the action is undetected. Documentation is one of the most important parts in developing HTS databases and any other software. The database developer should document in detail each implemented module. Improper documentation often leads to the software improperly calculating some parameters that is very hard to detect if the calculated values are not obviously different from the expected values. This can be caused by miscommunication between the database designer and the database developer. In addition, proper documentation ensures the smooth transition if either party involved in the project leaves the company. I have participated in building one in-house HTS database and led in building another. Depending on the scope of the database, it may take anywhere from half a year to a year to get a nonpolished basic system running with the help of one to two Oracle developers. The database is then constantly modified/improved while it is running. It takes about 2 years for the database to remain relatively steady without major improvements. The advantages of developing an in-house HTS database are that you get what you need, the ease and flexibility to add new improvements, and the final overall cost is lower compared with purchasing commercial software. However, a mastermind (database designer) who knows both database and HTS very well is required to undertake this task. The downside of building an in-house HTS database is the risk of failure and the relatively long period before a minimum functional database can be deployed.

13.6 HTS OPERATION MANAGEMENT

The most important aspect of running HTS is to treat this function as an operation, that is, it is an engineering process rather than scientific research. The goals of HTS operation are to produce reliable data at an acceptable speed and at a minimum cost. Some early HTS operations built massive robotic systems from scratch. Not only is it expensive to build novel integrated robotic systems with in-house expertise, but also it takes a long time to finish the project and delays the data generation that is the main goal of a HTS operation. In addition, the knowledge gained in the process of building the integrated HTS system is of little value to the pharmaceutical companies because their product is the drug not the instrument. Experience shows that efficient HTS operations are usually built by choosing among commercially available instruments and making the best use of them through in-house process innovation. Process innovation is the key attribute to gain a winning edge against competitors when the same instruments and technologies are available to everyone. How can McDonald's and

Burger King establish a winning edge and make a huge profit in making hamburgers while everyone knows the technique of making a hamburger? The answer is process innovation. Building and managing a successive HTS operation also depends on process innovation.

13.6.1 Building HTS Operation Infrastructure

Before building an HTS operation, the manager must first study the unique situation in a particular organization because there is no one-fits-all model. After assessing the short-term goal and long-term goal of the organization, a corresponding plan is developed based on the needs but not what is fashionable. Short-term goals usually dictate rapid solutions to existing problems. Too much emphasis on short-term goals may result in an operation that is not scalable and will be obsolete within a short period, resulting in wasted investments. Long-term goals usually dictate to build an operation with more functionality than is needed currently. It is expensive to build something that is not useful today, and it may (or may not) be useful in the future because companies often change goals voluntarily or involuntarily. The manager must strike a balance between the short-term goal and long-term goal when building the operation. There is a lot of debate on whether semiautomation or full automation should be built. Since everything has the Ying-and-Yang parts, there will never be an agreed-upon decision as to which operation is better. It all depends on the unique situation of the organization and the experience of the manager who runs the operation. The best practice is to keep an open mind and be ready to perform both types of operation. Because it takes time to build a functional fully integrated screening system, many HTS labs started operation with semiautomation to process microplates in a batch mode. The instruments used in this operation can be used for both screening and assay development. The integrated HTS system is gradually built while the operation generating data in semiautomated mode of operation. Even after the fully integrated system is built, the instruments used routinely for assay development should be made ready (by attaching stackers to them) for executing semiautomated screening in case the big robotic system breaks down or in case the screen cannot fit into the parallel screening mode in fully integrated systems.

A semiautomation operation is much easier to build since most vendors provide turnkey solutions for individual instrument and workstations. It is inexpensive to build and it is flexible to carry a broad range of assays. The chance of failure to set up the semiautomated operation is low and the manager can sleep better without the worry of losing his/her job. If the organization has limited financial resources or if the manager is inexperienced, sticking with semiautomation is a good decision. However, many of the benefits of fully automated operation will be missed and the screening operators will sweat in the lab, causing a higher chance of making mistakes during screening. In addition, the morale of the screening operators will be low because of the routine running of the instruments for an extended period of time.

A fully automated system has the advantage of operating 24 h a day and 7 days a week with minimum user intervention if the system is set up correctly. Because fully integrated systems can work three times longer than the typical 8-h working day for a human operator, the throughput of fully integrated systems with only one robot arm can match or exceed the throughput with semiautomated batch processing. The fully integrated system can execute each assay step consistently and leave no chance for human errors. Figure 13.5 shows the execution of three critical steps

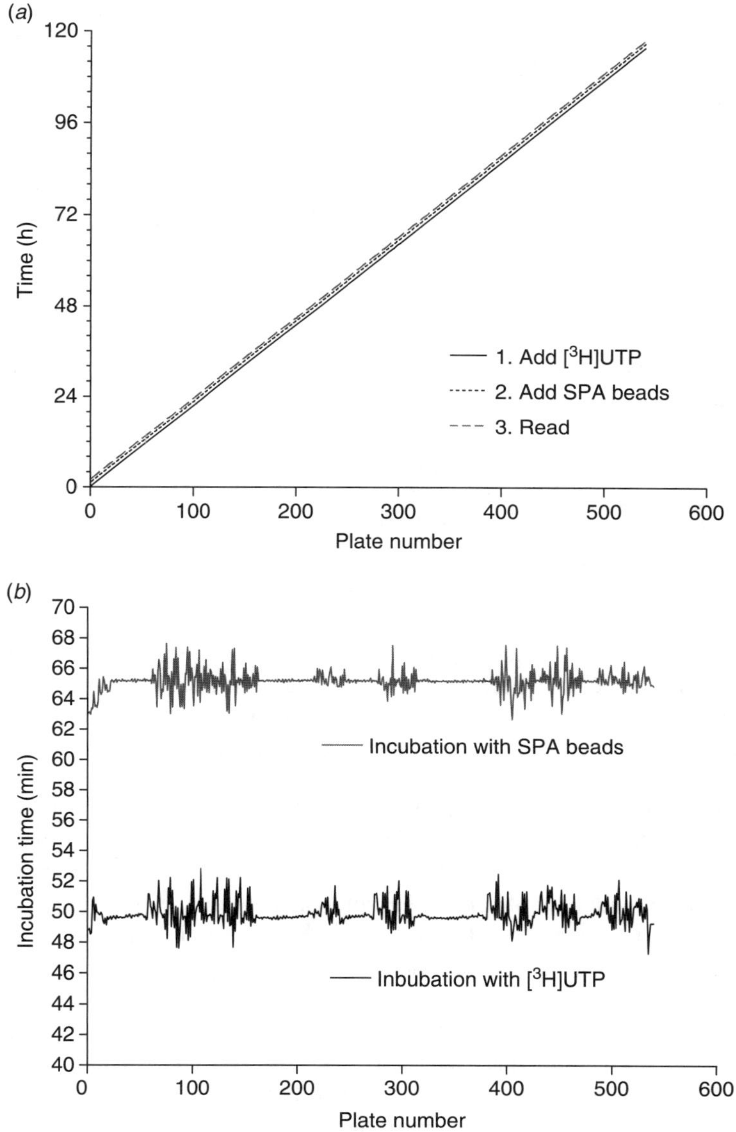

Figure 13.5 Fully automated continuous screening with 540 AssayPlates over 5 days. (*a*) Logged execution time at three critical assay steps. (*b*) Subtracting the three times in (*a*) yields the reaction time for two steps: enzymatic reaction time and the SPA beads incubation time. (See color insert.)

using a fully integrated robotic system with the enzymatic assay discussed in Figure 13.3. This assay was performed in 1998 with a fully integrated robotic system built by Robocon (the company was later purchased by Thermo CRS). It is scheduled to run continuously from Monday afternoon till Saturday morning. The reagents, the AssayPlate, and the CompoundPlate were loaded into the system manually once everyday, which took about 1 h. The rest of my time was spent on quality control of the screening data and on other tasks. On Friday evening before leaving the lab, I made the final loading of reagents and microplates. On Saturday, the robot shut itself down after finishing the remaining work. On the following Monday morning, all the leftover reagents were replaced with new reagents and the screening cycle started again. The execution time for the three critical steps in the assay for each AssayPlate is shown in Figure 13.5*a*. After subtracting the time from (*a*), two critical incubation times for all the plates are shown in Figure 13.5*b*. During the 5-day screening, 540 microplates was screened continuously. The system was capable to schedule up to 1000 AssayPlate at that time, but we chose to run 540 plates per run, which is about 5 days of continuous operation. Figure 13.5 shows that the critical incubation time in the assay only has a maximum variation of about 2 min in both the 50-min and the 65-min incubation steps. The operation was presented to the Robocon user group meeting. According to Robocon, this was the longest unattended operation among its installed instruments by that time. Such an operation requires that the reagents are stable for at least 16 h during nonworking hours after they are placed in the reagent reservoirs in the system. The online reagent reservoirs can be cooled to 4°C if needed by circulating coolant. Continuous parallel screening by feeding microplates in fixed intervals may expose other problems that are not present in batch processing. Here is an example involving dispensing SPA beads with Multidrop. The SPA beads are pumped from a reagent bottle. Because the SPA beads are lighter than the buffer, they will slowly aggregate together on the surface of the buffer. A stirrer is used to mix the SPA beads. When doing batch processing of many plates, the SPA beads can be continuously pumped from the reservoir to all the microplates without a problem. However, there is a problem for fully automated operation when it takes ~10 min to process one microplate (such as the screen described in Fig. 13.3). After dispensing SPA beads to one plate, the Multidrop is inactive before the next plate comes in for processing. During this period, the SPA beads in the reservoir are fine because they are stirred. However, the SPA beads in the tube between the inlet and outlet will separate from the solution and stick to the tube. Thus, the Multidrop tube must be purged before dispensing to each plate. Fortunately, the purged SPA beads can be recovered and reused, allowing fully automated operation to proceed.

When purchasing workstations or individual instruments, HTS managers only deal with one party and the party has full control of its product. After careful evaluation of many instruments based on functionality and reliability, the chosen mass-produced workstation or individual instrument almost always performs as expected. In contrast, a fully integrated HTS system is usually uniquely designed by the HTS manager who decides what kind of instruments are needed in the system and which individual instrument in the market is the best to carry the required functions. The integrator's function is usually to provide the scheduling software and instrument-specific adaptors so that all the instruments in the system communicate with the scheduling software, though some integrators may suggest design ideas as well. Most of problems that happened during building the fully integrated system are the software communication between the adaptors and the ICPs. The instrument manufacturer may make small modifications to ICPs between different versions that the integrator may not know. This can cause infrequent glitches that sometimes can only be caught during extended testing. Thus, it is very important to specify several testing protocols mimicking the actual assays for a longer period of continuous operation as acceptance test criteria. The system is usually built at the integrator's facility. After the system is built, several predefined assay protocols are carried out without using actual biological reagents. This is referred to as a factory acceptance test (FAT). After passing the FAT, the integrated system is dissembled and the components are shipped to the user's site. After reassembling the system at the user's site, another series of tests are carried out. This is referred to as a site acceptance test (SAT). After passing the SAT, the project is concluded. When evaluating potential system integrators, it is very important to evaluate their supporting service in addition to evaluating their technology. Fully integrated systems are not just built and delivered. The system requires maintenance, repair, and modification/upgrading in the future. It is a long-term relationship between the user and the integrator. Close proximity to the integrator or to their local support office is of advantage for promptly solving problems. In addition, the cost of service will be lower because no expense for air travel and overnight hotel stay will occur. Another very important criterion that was often overlooked is the integrator's financial viability. Sadly, the business of integrating a fully automated screening system is not a very profitable business, and there is a chance that an integrator may not exist a few years down the road if it does not have strong financial backing. It takes the HTS manager's experience, self-confidence, and gut to implement HTS operation with a fully automated system. If the implemented system did not work after spending millions of dollars, the company would end up with a bunch of expensive metals occupying a big room and the HTS manager might lose his/her job.

As HTS moves more and more toward cell-based screening, the production of enough cells to feed the screening operation becomes a bottleneck. Preparing a large quantity of mammalian cells is a slow and tedious process. Through collaboration with a consortium of users, The Automation Partnership (TAP) developed SelecT, which became the leading automated cell culture system for multiple cell lines and assay-ready plate production. In this system, cell lines are maintained and expanded in T-175 format flasks. Many cell lines can be cultured in parallel. Cells can be harvested, counted, and seeded without operator intervention. The cells can be made into 96-, 384-, and 1536-well microplates ready for screening. In addition to the large expensive SelecT, TAP also offers many other automated cell culture systems that fit different needs. Several other companies also offer fully automated cell culture systems, such as the Cellerity from Tecan. Traditional cell-based screening uses fresh cells. This imposes a lot of operation constraint since the cells must be produced continually at the same speed as the screening speed. Overproduction of cells will lead to a waste of resources because the cells cannot be stored for long. Underproduction of cells will slow down or even stop the screening. To solve this problem, scientists in several companies investigated the feasibility of using freshly thawed frozen cells in screening. In most cases, it was found that screening with freshly thawed frozen cells produced similar results as compared with screening using fresh cells. With this approach, the cells can be treated the same way as other assay reagents in the screening. They can be mass-produced ahead of time, stored frozen, and thawed before screening. In either mode of operation, automated cell production systems are an indispensable part of the infrastructure in HTS operations.

13.6.2 HTS Operation Management and Control

After setting up the necessary infrastructure, the HTS operation is ready to carry out screening. The assay developers will first develop assays and generate assay protocols. Because the screening is expensive and there is little chance that the same biological target will be screened again, the developed assay must be well thought, be agreed upon by the project team, and be approved by the appropriate authority within the organization. Once the protocol is approved, it should not be changed in the middle of screening without proper authorization. Conflicts among involved parties may arise when the screen produces unexpected results. Clear communication and adhering to proper procedure can avoid finger pointing in this situation. Project management tools, such as MS Project, can help manage the screening process and improve communications when several parties are involved in screening. People who are already comfortable with what they used to do may strongly resist changes when a new system is introduced into an organization. Support from management is very important to implement the new technology. When I introduced MS Project to five prime therapeutics, there was strong resistance initially. After a few years, MS Project was well adopted in the company, and it became an indispensable tool not only for screening operation but also for other projects in the company.

Depending on the setup of individual companies, the screening is executed either by the assay developer or by a screening operator. My preference is to have the assay developer in the screening group, who is well trained in all the HTS instruments, to execute the screening with the help from scientists in the therapeutic group. An alternative setup is to establish a central screening lab with only resources to maintain the lab and to train scientists from the therapeutic group to use the facility. This mode of operation produces most failures because scientists in therapeutic departments usually are not familiar with modern HTS assay technology, instrumentation, and the limitation of HTS. The developed assays usually do not run well in HTS operation. Another setup is to have a relatively large HTS department with many low-level screening operators to execute the assay with the help of scientists from the therapeutic department. The problem with this setup is that some of the low-level scientists just execute operations day-in and day-out without spending time to study either the robotic or the biology behind the assay. Thus, they cannot offer much insightful information on modern assay technology to the scientists in therapeutic group. No matter what kind of setup in a particular organization, the person responsible for executing the screening will calculate the quantity of all required reagents and consumables and obtain all these items in-house before the start of screening. It is not a good practice to order portions of reagents and plan to order more in the middle of the screening. Sometimes the vendor may run out of stock of the reagents when you need them in the middle of screening. In addition, the newly obtained reagents may be from a different batch from the original order and may produce different assay results. The best practice is to obtain all the reagents and pool them together as one batch and then make aliquots of them for daily use. This will ensure consistency of the screening.

The screening quality should be placed at a higher priority than screening throughput. Early screening operation placed too much emphasis on throughput. Throughput is a number that can be misleading. For example, a high-throughput screening operation can claim screening throughput of 1 million compounds a day when the screening was performed by pooling 10 compounds in one well. The actual throughput is reduced by a factor of 10 if the number of wells was counted. On the other hand, if the screening operation is not flexible to take diverse assays but can only take a few simple assays, the operations throughput can be very high too. For example, when several dozen cancer cell lines are screened for inhibitors that can stop cancer cell proliferation, the throughput can be extremely high because several dozen assays are essentially going through the same simple cell proliferation assays. In contrast, a screening operation that is flexible to take all kinds of different screening will have lower throughput because of different and complicated screening procedures for every screening. Recent effort in the industry to clean up compound libraries and to use only single compounds with known identify in screening shows the shifted trend toward quality from throughput. Scientists in the National Institute of Health (NIH) screening center even go as far as to screen compounds at multiple concentrations to obtain dose–response curves in primary screening. In addition to emphasis in screening quality, screening operation should be made flexible to carry a variety of assays either in a fully automated parallel-plate processing mode or in a semiautomated sequential batch-plate processing mode.

The ultimate success of screening is judged by how many good leads are advanced to the next stage or on how much cost reduction is generated by helping to terminate a drug discovery project. Many projects in pharmaceutical companies can drag on for a long time with no clear direction because the hypothesis cannot be tested without small-molecule inhibitors. HTS can rapidly provide the inhibitors to test the hypothesis and terminate the project if the test results are unfavorable. Being able to stop a project early is very important for pharmaceutical companies so that resources can be redirected to more promising projects. The terms "hit rate" and "confirmation rate" are commonly cited as criteria for screening success. The hit rate refers to the ratio of the number of compounds picked as hit based on the primary screen data to the total number of compounds present in a primary screen. The confirmed hits are the hit compounds picked in primary screen that are confirmed in subsequent testing. The confirmation rate is the ratio of the number of confirmed hits to the total number of hit compounds. Both hit rate and confirmation rate are relative numbers that depend on the definition of primary hits (or the criteria to pick primary hits). These terms may be misleading when detailed information is not given. The primary screening data shown in Figure 13.11 is a good example to demonstrate the above argument. Most of the data in the screen clustered at the baseline ($0 \pm 20\%$). The hit rate varies when different hit picking criteria are used: by percentage of inhibition or by multiples of standard deviation. Both of them are arbitrary. For example: defining hits as the compounds that inhibit, more than 50% inhibition will result in lower hit rate than defining the hits as 30% inhibition or 3 standard deviations. One the other hand, defining hits as the compounds that give an activity value beyond

5 standard deviations will result in a lower hit rate than defining hits as the compounds that give an activity value beyond 3 standard deviations. The hit picking criteria also affect the hit confirmation rate. If the hit picking criterion is set at 3 standard deviations or 30% inhibition, many inactive compounds close to baseline will be picked as hits, resulting in a lower confirmation rate in the subsequent testing. When 50% inhibition was chosen as the hit criterion, more than 90% of the hits produced were confirmed in subsequent testing.

13.6.3 Building an HTS Team

"People are not the most important asset of your organization. The right people are," claimed by Jim Collins in his 2001 best seller *Good to Great*. It is true for HTS operation as well that the right people that fit the job should be recruited. To setup a team, the first thing is to define the jobs and then find the people who fit the jobs. HTS operation requires the seamless integration of expertise in therapeutics, assay development, software development, engineering, and project management. The composition of personnel varies depending on the size of the screening operation. For small to mid-sized companies, the majority of the personnel are assay development scientists who also carry out screening functions. A dedicated engineer may not be required because most devices are manufactured by other companies and there is hardly a need to build a device in-house. When purchased instruments break down, it is usually not possible for in-house engineers to service the instrument anyway because the warrantee may be voided by doing so and also because of the lack of special parts. However, there is a need for a person who is familiar with all the instruments and is savvy with mechanics and electronics. This person should be able to identify what is wrong with a failed instrument and facilitate the repair by engineers from the instrument vendor. This function can be filled on a part-time bases by a regular scientist in the screening group who is interested in mechanics. If a commercial HTS database is used in the operation, a full-time DBA is usually required. If the HTS database is developed in-house, a good Oracle developer who helped coding the software can gradually transition from developing the database to maintaining the database and serve as the DBA. HTS scientists should possess broad knowledge in biology and be able to rapidly gain specific knowledge in specific therapeutic areas in a short time because they may be required to developed at least four assays annually in different therapeutic areas. In contrast, scientists in the therapeutic group may stick with one project for several years. In addition to being able to change rapidly within biological disciplines, HTS scientists must be able to interface with people in other disciplines and be able to manage projects. Different from scientific research, HTS is a process. While process excellence demands consistency, precision, and repetition, scientific research calls for variation, failure, and serendipity. When performing assay development, the mindset should be in the research mode. When performing screening, the mindset should be in the process mode. HTS scientists are engaged in both modes of action and must be able to rapidly switch the mindset between the two. Most importantly, an HTS operation must develop clear standard operation procedure (SOP) documents and the team must follow the SOP during operation.

13.7 BUILDING AN HTS OPERATION FOR BIOPHARMACEUTICAL DISCOVERY

13.7.1 Introduction to Biopharmaceuticals

Two decades after the launch of the first human recombinant protein (Humulin), protein biopharmaceuticals now account for a significant portion of total FDA approvals. The major "blockbuster" biopharmaceuticals fall into two categories: secreted proteins normally encoded by the genome, including cytokines, growth factors, and hormones that act by activating cellular receptors; and monoclonal antibodies or soluble receptors capable of blocking the activity of cell surface receptors, hence preventing activation by their native ligands. There is a trend for major pharmaceutical companies to diversify their product mix by increasing more biopharmaceuticals to fill the void left by the expiration of blockbuster small-molecule drugs. It is now possible to comprehensively and systematically enumerate, clone, produce, and screen all secreted proteins by building upon knowledge accumulated in the past two decades in high-throughput screening, genomics, and parallel protein expression technologies. In particular, the concepts and philosophies of high-throughput screening developed in discovering small-molecule drugs can be productively applied to the discovery of protein biopharmaceuticals.

It is estimated that there are roughly 3500 to 4000 gene loci encoding secreted proteins and single-pass cell surface receptors in the human genome. These proteins represent the fundamental set of secreted protein therapeutic candidates and a large proportion of potential antibody targets. Although advances in genomics have had considerable impact on drug discovery, functional understanding of gene products has lagged behind the sequencing effort, and only a small fraction of proteins encoded by the human genome have been assigned functions. For example, of the roughly 4000 proteins in the secreted and single-pass transmembrane protein (STM) classes, we fully understand the function and pharmacology of perhaps 15%, and nearly all existing protein therapeutics emerge from that fraction. It is not heretical to expect that more secreted proteins and antibody targets of therapeutic value are contained in the remaining, not fully characterized, fraction. While the application of genomics, primarily transcriptional profiling, to secreted proteins and cell surface receptors has attracted much attention, there remains the need to determine which of the trove of uncharacterized secreted proteins have medically relevant pharmacology.

While the universe of small molecules is extremely large, the total number of native, nonantibody secreted proteins and cell surface receptors is modest by HTS standards. The combination of new molecular biology and protein expression technologies and the judicious application of high-throughput and high-content screening tools now make it possible to comprehensively screen the entire set of candidates for biotherapeutic lead molecules. Every protein screening effort has three main stages: the selection and establishment of a cDNA library encoding the individual proteins to be screened, the conversion of the cDNA library into a functional protein library, and the screening of the protein library for therapeutically relevant proteins. The multistep nature of the process means that the failure rate in each step multiplies, that is, a 50% success rate in cloning and a 50% success rate in expression yields only

25% of the total possible proteins in the screening step. It is critical to maximize yield at each step while developing the library and expression system, and constantly to emphasize quality and attention to detail during screening operations.

13.7.2 Construction of Screening Library

The first step of the screening operation is the establishment of a library of cDNA clones encoding the proteins to be assayed. Early efforts to catalog the proteins encoded in the genome were based on expressed sequence tags (ESTs), which are useful for tagging the location of expressed sequences in the genome. However, establishment of a robust screening library requires accurate, complete knowledge of the coding sequence of each gene, and EST-based prediction remains subject to the limitations of gene prediction. Computational gene predictions from human genomic sequences, even with the aid of ESTs, are error prone due to long introns and short exons. The most accurate way to establish a comprehensive cDNA collection for screening purposes remains careful construction of a well-curated full-length cDNA library from a large number of tissue sources coupled with the most advanced full-length cDNA cloning technologies. The next task is accurate curation of the cDNA collection to select the components of the library. Not all proteins represent candidate biotherapeutics and as druglike small molecules can be selected using computational chemistry, so too can proteins be classified in order to focus on the most likely classes for discovery research. Secreted proteins and cell surface STM proteins are the two categories most relevant to biopharmaceutical development.

The next step is to convert the cDNA collections to protein screening libraries. While the construction and maintenance of a protein library superficially resembles the corresponding process for a small organic molecule library, there are important differences that dictate operational details for working with protein libraries. The first difference is scale. As noted above, while small-molecule structural space is almost infinite, there are a relatively modest number of proteins in humans. The primary components of a protein screening library, the canonically secreted proteins, are encoded by fewer than 4000 gene clusters. The total number of protein variants and isoforms will be slightly higher when one considers splice variation and posttranslational modifications, such as enzymatic cleavage, phosphorylation, sulfation, lipidation, and glycosylation. Nonetheless, it remains possible to screen all of the secreted and STM proteins in cell-based assays in a library that is quite modest by HTS standards. Another major difference is the protein stability that imposes profound operational implications. Small molecules are generally stable both in dry form and in DMSO, a near universal solvent for small organic molecules. There is no proven medium that can store all proteins for lengthy periods of time as would be necessary for the development of a permanent protein library, ready to be dispensed before screening. Degradation and precipitation are also major problems facing long-term storage of proteins and it is, therefore, necessary to produce the proteins fresh from a cDNA source collection every time before screening for robust and comprehensive screening results. Thus, a cDNA library became the primary long-term screening library, similar to a small molecular library in small-molecule discovery, which must be converted into protein libraries just before screening.

Ultimately, one would like to express all the predicted secreted proteins in the native cells where they are made and processed, purify the secreted proteins, and store them in a universal medium that is compatible with downstream assays. Operational and scientific considerations make this unfeasible, and typically a single or at most a small number of complementary expression hosts are used for the conversion of the cDNA collection to proteins for screening. The main goal when designing a protein screening process is to ensure that a sufficient concentration of the proteins to be screened is available in the assay well in a soluble, active form. Automated parallel expression of a large number of proteins has been described for protein structure studies. Techniques for the expression of tagged proteins in a 96-well format have also been described. While epitope tagging aids in the generalization of purification protocols and can be used to increase protein concentrations in the assay plate, it may also render the protein less active or, in extreme cases, ablate activity altogether. Systems that obviate the need for purification, and therefore allow screening of untagged proteins, are preferable. Many proteins have been expressed in an active form in microbial systems. However, on a genome-wide basis, expression of native proteins in cultivated mammalian cells is superior for proper protein folding, assembly, and posttranslational modification. High-throughput transient transfection of each cDNA can be performed in 96- or 384-well microtitre plates, using an automated robotic system such as the one shown in Figure 13.2. The supernatants are then separated from the transfected cells to form the protein library that can be used for assays in diverse formats, using different cell targets.

One major problem with directly using unpurified recombinant proteins in supernatant from protein producing mammalian cell lines is the uncontrollable media effect to the assay. The protein producing cells are usually in cell culture media containing $\sim$5% serum for optimal protein production. The medium can cause artifacts in cell-based assays, causing either high backgrounds or false-negative measurements by the many growth factors present in the serum. It is not possible to reduce this effect by dilution because the expressed protein concentration will be reduced in dilution. It may be necessary to compromise protein production by the use of serum-free media, which typically results in lower protein concentrations but will give much lower assay background. In addition to serum, nonprotein metabolite composition (such as glucose level) in the media may differ from well to well due to idiosyncratic differences in the metabolic rate caused by the transfected genes. These variations in the media where the proteins are produced can also affect the final assay.

13.7.3 Screening Operation for Biopharmaceutical Discovery

Protein screening presents special challenges and opportunities. Because of the modest size of the protein library and the large number of assays that have to be performed for each protein simultaneously due to protein stability, a screening operation for proteins should not be aimed primarily at maximizing throughput, but rather at achieving high-quality data and maximum flexibility in handling a variety of assays. In protein therapeutic discovery, a majority of the assays are based on cellular systemwide changes (phenotypic screening) rather than on direct molecular interactions. Because there is no defined target or pathway, it is essential to perform multiple parallel assays to catch all possible regulators of the target cells. The modest library size allows executing more complex and time-consuming assays. Due to the fact that secreted proteins primarily act at the plasma membrane and their natural targets in the body are native cells, primary human cells should be considered first in assay development. If they are not available, primary mammalian cells with functions known to be preserved in evolution should be the second choice. However, lack of cross-species reactivity can be an issue, particularly for proteins active in the control of the immune system, which are generally poorly conserved between humans and rodents. Another choice are cell lines, however, these are transformed cells with profoundly perturbed biology and, in many cases, lack receptors that may interact with potential secreted proteins. For example, the L6 cell line, widely used to mimic muscle cells, does not express, or expresses at a dramatically reduced level, EGF-family receptors compared with primary muscle cells. In addition, adipocyte cell lines rapidly lose expression of the insulin receptor and other metabolically important receptors. In the absence of a human primary cell source, it is generally useful to screen rodent primary cells as well as human cell lines to maximize the likelihood of success.

The use of primary cells, rather than cell lines, in the screening operation creates challenges throughout assay development. Scarce cell sources, donor variability, cell viability, and stromal environment are all critical variables that must be understood and controlled. For example, only a limited quantity of immune cells (such as natural killer cells) can be obtained from a single donor at one time. Immune cells from different donors will react with each other and cannot be pooled, so primary cells from multiple donors have to be used separately in screening and can result in variation, due to genetic background and health of the donor, which is not an issue when using cell lines. Most primary cells can only be cultured for a limited time. For example, freshly isolated primary adipocytes survive for less than 24 h in culture, requiring an assay format that can be executed within hours rather than days. In such cases, higher sensitivity assay formats will offer advantages. For example, choosing a fluorogenic substrate over a chromogenic substrate will reduce assay time by a factor of between 10 and 100, due to the greater sensitivity in detecting a fluorescence signal compared with an absorbance signal. Some primary cells (such as cardiomyocytes and chondrocytes) will rapidly dedifferentiate into fibroblast-like cells after being plated on a two-dimensional surface. Three-dimensional matrices can be used to maintain cellular phenotypes, but this imposes challenges in liquid handling and cell dispensing. Finally, some cellular activities only occur in intact tissues. For example, pancreatic beta cells will secrete insulin from excised islets, but rapidly lose this characteristic upon disaggregation. Dispensing of aggregated cells and tissue slices into microtiter plates can be challenging, with the risk of large well-to-well variation. Application of new technologies can help to overcome some of these difficulties. For example, high-throughput automated imaging systems enable the study of specific cell types within a mixed population of cells. While it remains difficult to transfect reporter genes into some primary cells without changing the cell characteristics, label-free technologies can play a unique role in primary cell screening since no genetic manipulation of the cells is required.

13.8 QUALITY CONTROL AND DATA ANALYSIS IN PRIMARY SCREENING

13.8.1 General Statistics Applied to HTS

The major goal of analysis of data generated in primary HTS is to separate the actives (or hits) from the inactives. The hits from primary screen are then retested to confirm they are true hits but not from statistics or experimental errors in the screening process. The screening data in a microplate (no matter 96-, 384-, or 1536-well plate) can always be divided into three data sets: the inactives, the hits, and the controls. In a typical AssayPlate design, the controls are placed in the two columns on each side of the AssayPlate and the test compounds are place in the middle of the AssayPlate. The controls are used to monitor performance of the assay with each AssayPlate, to help normalize plate-to-plate variation, and to help transform the raw data into biologically meaningful values. Because the true hit compounds (the compounds that can be confirmed in retest) are rare, the majority of the test data on each AssayPlate belongs to an inactive data set. In statistics, the central limit theorem dictates that the distribution of the means of a large number of independently collected data should follow Gaussian distribution. Because the inactives in an AssayPlate are very large, they usually follow Gausian distribution very well. On the other hand, the controls may not follow Gaussian distribution because of the limited data obtained in the AssayPlate.

When applying statistics to HTS, we assume the distribution of a particular data set always follows Gaussian distribution (e.g., the data set for all inactives obtained in screening and the data set of each control group). Four terms, the total number of data points (N), the mean (m), standard deviation (SD), and standard error of mean (SEM) are used to describe the data set. The mean of the data set is a value that is close to the true value of the measurement. The SD describes the scattering of the collected data that is independent of how much data is in the data set. The SEM describes the accuracy of the mean calculated from the data set that is dependent on the N. The larger the N, the smaller the SEM, which indicates the calculated mean value is more accurate. The uses of SD and SEM are sometimes confusing for biologists. The following example shows how these terms are used in HTS. In the AssayPlate, the measurement of each well corresponds to an individual compound's activity. Because most of the wells are inactive and they should converge to one value (the mean), the scattering of the value measured in each well (SD) should be used to determine whether we could obtain hits that do not belong to the inactive population. The SD does not change when the number of wells increases or decreases. Now suppose the assay is an image-based cellular assay and individual cells are measured in each well. Within each well, hundreds or thousands of cells are measured and they are expected to follow Gaussian distribution. Whether a compound is active or not in each well should be determined by the mean of all the cells inside the well. The scattering of the data (the SD) is less important here, but SEM is more important because we want to know how much uncertainty is associated with the mean value. SEM is dependent on the total number of cells in the well. Figure 13.6 shows a hypothetical image-based cellular assay result. Each data point represents the measured value of one cell. Experiment A and experiment B are identically performed but with experiment B

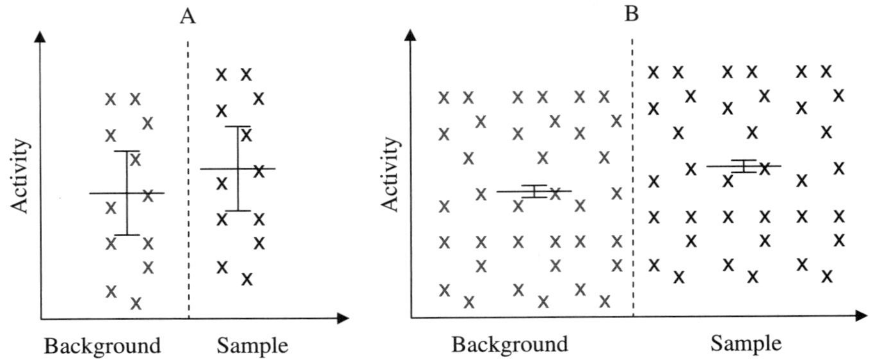

Figure 13.6 Hypothetical image-base cellular assay. Each data point represents the measured value of one cell. Experiment A and experiment B are identically performed but with experiment B having nine times more cells than that in experiment A. The data for both the background (no added compound) and the sample are shown. The mean and the SEM for both the background and the sample are shown in the graph. In experiment A, the SEM is quite large because of a limited number of cells. Thus, it is impossible to determine whether the sample is active or not. In experiment B, the SEM is reduced by a factor of 3 (the square root of 9). Experiment B shows that statistically the sample is active though the scattering of the data (SD) remains the same.

having nine times more cells than that in experiment A. The data for both the background well (no compounds) and the sample well (with compounds) are shown. The mean and the SEM for both the background and the sample are shown in the graph. In experiment A, the SEM is quite large because of a limited number of cells. Thus, it is impossible to determine whether the compound is active or not. In experiment B, the SEM is reduced by a factor of 3 (the square root of 9). Experiment B shows that statistically the compound is active though the scattering of the data (SD) remains the same (not shown) for both experiments.

With Gaussian distribution, the number of data within 1 SD of the mean will account for 68.3% of all the data, within 2 SD will account for 95.4% of all the data, within 3 SD will account for 99.7% of all the data, within 4 SD will account for 99.994% of all the data, and within 5 SD will account for 99.9999% of all the data. These numbers are very important when setting the hit criteria in primary screening. For example, if 1 million inactive compounds are tested in a screen and you choose the hit criteria as any data that are outside of 3 SD from the mean of the background, then 3000 background wells will be picked as hits. That is a lot of false positives and they will be eliminated in the follow-up retesting stage. If 4 SD is used as the criteria to pick hits, only 60 background wells will be picked. However, increasing the SD window to eliminate the false positives may also reject true positives if the SD is relatively large. For example, if the SD is about 15% in an enzyme inhibition assay with the assay window ranging from 0 to 100%, 4 SD cut-off windows will mean a true inhibitor must be able to inhibit more than 60% of the signal in order to be picked as a hit. Any inhibitors that inhibit less than 60% will be included as background population. With this strigent criteria, the confirmation rate of the picked hits from primary screen will be high.

13.8.2 Evaluating Assays During Assay Development

When developing an assay, statistic principles can help to evaluate and validate the assay. In most assay development, a control is required that gives a signal that is different from the background. Figure 13.7 shows the distribution of the measured control value and the measured background value, with each of them forming their own data sets and they both follow Gaussian distribution. The mean of the control signal is m_S and the standard deviation associated with the control dataset is σ_S. The mean of the background signal is m_B with the associated standard deviation of σ_B. Because the commonly used terms of the signal-to-background ratio (S/B, shown in Fig. 13.7 as m_S/m_B) does not take into account the errors associated with either of the data set, and the signal-to-noise ratio (S/N, shown in Fig. 13.7 as m_S/σ_S or m_B/σ_B) only deals with one set of data at a time, none of them individually is useful to evaluate an assay. A new term that combines both S/B and S/N is required to evaluate the performance of an assay. From Figure 13.7, it is clear that a new term, "absolute signal window," or ASW, can be defined as $|(m_S - k\sigma_S) - (m_B + k\sigma_B)|$ to account for the signal, the background, and the standard deviation of the two data sets. The value k is a constant that is arbitrarily decided by the user according to the confidence level the user desires. Sittampalam and colleagues in 1997 proposed this concept, and they used signal window (SW), which I would refer to as relative signal window or (RSW) here. RSW is defined as ASW/σ_S, which reduced the ASW to a simpler dimensionless value. However, few scientists in HTS adopted their term of SW in evaluating assays. Zhang and colleagues in 1999 used the same concept but proposed a new term called the Z' factor by dividing ASW with $|m_s - m_b|$ to reduced ASW to a

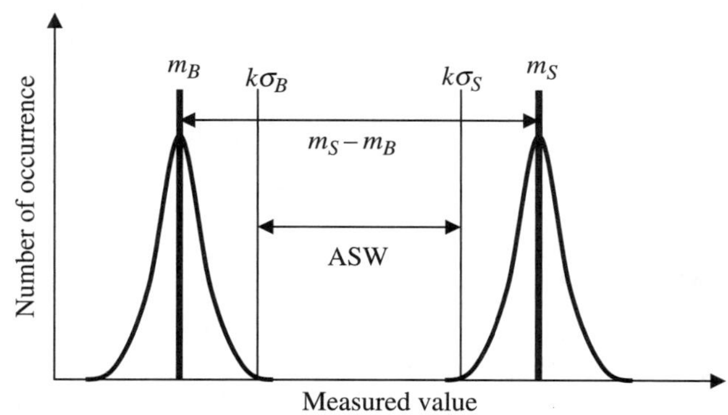

Figure 13.7 Evaluation of assays based on statistics. The distribution of two sets of data commonly encountered in bioassays (background and controls) are shown. The two sets of data follow Gaussian distribution and are separated due to different means. The absolute signal window (ASW), which takes account both the mean and the standard deviation of the two data sets, is a measurement of the assay performance relative to the control employed.

dimensionless value. The only difference between Z' and RSW is which value to be used as a divisor to change ASW to a dimensionless value. However, Z' became the standard term that most scientists in HTS use today. When proposing Z' and Z factors, the k value of 3 (3 times standard deviation) is used. With Gaussian distribution, it means that there is a 0.3% chance for the control values or background values to not be included in their own population. To pick the k value of 3 is truly arbitrary, one can use $k = 1$, 2, 4, or a larger number to obtain a term that may be called A', B', C', or whatever. However, the criteria of evaluating an assay must be adjusted accordingly. As with any industry, a standard is valuable for product evaluation and comparison. That Z' became the standard in HTS is a very positive development.

The biggest issue in applying Z' in evaluating an assay is the selection of the control that defines the ASW. If the ASW cannot be defined, Z' can only be used to validate the consistency of the assay but not be able to qualify an assay. In practice, the Z' can always be applied to qualify assays aimed at finding inhibitors of enzymes or antagonist for a receptor. However, it has limited use for qualifying assays aimed at finding an agonist in a cell-based assay. When designing an assay to find inhibitors for an enzyme, the substrate is allowed to turnover into product. The amount of the product can be detected that form the background signal. A known inhibitor is served as a control. If the inhibitor can inhibit 100% of the enzyme, the detected product will be zero. This forms the assay window ($m_s - m_b$), where m_b, is zero or close to zero. If no known inhibitor can inhibit the enzyme, one can always find a way, such as denaturing the enzyme, to obtain the value of 100% inhibition of the enzyme. In this case, the bigger the Z', the better the assay can find an inhibitor. In contrast, what does the Z' factor mean when an assay is designed to screen for agonist that elicit a cell response? In such assays, the background is the measured value from cells not exposed to the screening compounds. However, there is no definable value for the control signal. Suppose if there exists only one known agonist that can be used as a control and this agonist only gives a value of 2 in the assay with standard deviation of 0.2 and if the background value is 1 with standard deviation of 0.1. Then the Z' factor will be 0.4. On the other hand, if there exist two agonists and the other agonist give a value of 4 with standard deviation of 0.2 in the same assay, using this agonist as a control will give a Z' factor of 0.7 with an identical assay. Thus, the Z' can be any value for a given assay depending on the potency of the arbitrarily chosen agonist used as a control. In this case, the Z' cannot be used to evaluate the assay.

13.8.3 Quality Control During Screening

Before attempting to identify actives out of a large number of inactives, proper quality control (QC) should be applied to each AssayPlate. Statistical analysis can only help interpreting the data and facilitate QC, but it does not decide whether the data is properly collected or not. Applying statistics to improperly collected data will lead to wrong conclusions, as the saying goes: garbage in, garbage out. Because each screen is different, proper QC procedures must be developed for each screen. However, there are some general observed phenomena indicating something may be wrong. For example, if there is a pattern of measurement that gradually increases or decreases from the beginning of the column to the end of the column, it may hint that the reagents may not be properly purged and its concentration is increasing with more reagent filling the tubes in the reagent dispenser. If data in a row is dramatically different from the other rows, it may indicate that the nozzle responsible for dispensing the reagent to the row may be partially clogged. If a fixed well gives a value that is dramatically different from the other wells in several consecutive plates, it may indicate that the tip responsible for delivery of reagent to that well may be loosely attached to the multichannel pipetting head. After rejecting the AssayPlate with obvious issues, statistics can be applied to each plate to detect drifting in screening quality. The mean, SD, and CV (coefficient of variation) for all the noncontrol wells in each AssayPlate are usually calculated and compared to detect a trend in the screen. The results obtained by this method can be skewed if a plate contains a few potent actives that will change the mean, SD, and CV wildly. There is reason to use the median instead of mean. Median is insensitive to the few potent actives because of the presence of a large number of inactives. However, the SD and CV will still be skewed to a larger number. Ideally, only the data from the inactive population should be analyzed instead of analyzing the mixed population containing both actives and inactives. Separating the inactives from the actives can be achieved by treating the actives as outliers and reject them. There are many ways to reject outliers. In this situation, the approach according to the "rule of huge error" works well and is simple to implement. With this method, if a suspected point deviates from the mean by a multiple of 4 or more of SD, it is treated as an outlier. The problem is that the SD of the population is not known. It can be estimated by using all the data (including outliers) or by first rejecting a small percentage of the extreme data in the data set and then compute the SD. In the first method, the outliers are rejected using the 4 SD criterion. The SD of the remaining data is computed again and then 4 SD is used to reject any remaining outliers. In the second method, a predefined percentage of the extreme points in the AssayPlate (e.g., 10%) is first rejected (excluding controls). The remaining data is used to compute the SD. The 4 SD criterion is then applied to all the data (including the rejected 10% of the data) and outliers are rejected. The SD of the remaining points is calculated. This new SD is applied to all the data to reject outliers using the 4 SD criterion. Some people prefer to use CV instead of SD to describe the data variation in a plate. The advantage of using CV is that it is a dimensionless relative number, and it can be used to compare not only between plates in an assay but can also compare the quality between assays. However, CV is only meaningful when the underlying data is an absolute measurement but not relative measurement. For example, the mass of a substance, the concentration of a substance in solution, and 100% inhibition are absolute measurements. However, temperature reading in either Celsius or Fahrenheit and raw fluorescent count without background subtraction from fluorescence readers are not absolute numbers. When applying CVs to relative entities, the result is meaningless. For example, when measuring a freezing water solution (with some salt) at 0.5°C with 1°C error, the CV will be 200% when calculating using the Celsius unit. If using Fahrenheit to calculate the same measurement, the CV will be $1.8/32.5 = 5.5\%$. The absolute value of temperature should be used in this case and the meaningful CV is $1/273.5 = 0.37\%$. When directly calculating CV from the fluorescent read out of a AssayPlate can get different CVs as well depending on the readers.

Sometimes the CVs can be extremely large because the measured relative value is close to zero. Thus, the fluorescent reading must be first converted into a meaningful absolute unit before CV is calculated.

The Z factor can be used as a parameter for QC of screening by monitoring Z factor changes between different plates. A consistent assay performance should yield the same Z value with the same control. A reduction in Z may indicate the deterioration of the assay. However, there is a caveat when using Z factor to QC screening data because only a limited number of controls can be placed in the AssayPlate, causing the large variations of the control values in each AssayPlate and hence the large variation in Z factors. For example, only 4 to 8 wells in a 96-well AssayPlate can be used for one set of controls. This small number of data points can lead to large uncertainty of the mean, especially when one of the points is an outlier. It is very important to reject the outlier within this small set of data. The most widely used method to reject outlier in this situation is the Grubbs test. In this method, the data points are first ranked according their values. The smallest value ($x_{\min}$) and the largest value ($x_{\max}$) may be outliers and are tested. All the data are used to calculate the mean (x_m) and the SD. The $x_{\min}$ or $x_{\max}$ is tested by calculating T, which is equal to $(x_m - x_{\min})/\mathrm{SD}$ or $(x_{\max} - x_m)/\mathrm{SD}$. The value of T is then compared with the critical number with a certain confidence level to decide whether it is an outlier or not. The rejection of outliers not only helps obtain correct Z values but also helps to normalize data from plate to plate.

13.8.4 Identifying Actives in Primary Screen

After rejecting unacceptable plate in QC, the test results in the remaining plate are used to identify the hits (or the outlier) in the noncontrol wells. Several transformations are usually performed with the primary data to aid the hit picking process. The "edge effect" is a well-known phenomenon that the values at the edge of the plate are systematically different from the values in the center of the plate. The edge effect can be clearly observed when the screening data is of high quality. Figure 13.8 shows the primary screening results in the polymerase assay described in Figure 13.3. Four wells in the 96-well AssayPlate, one center well (E7) and three edge wells (A10, H3, and H10), are selected to show the positional effect. The results in the edge wells shifted about 10% above the center well. Since this is a systematic error, it can be corrected mathematically. This can be performed by obtaining the mean value in the edge wells and then normalizing it to the mean of the center wells. The mean of the inactives among different AssayPlates may be different due to experimental variations, and plate mean alone do not allow normalization across AssayPlates. Controls and plate mean together can be used to normalize the screening data across AssayPlates by assigning a value in each well using the following equation:

$$\%\text{Inhibition} = \frac{x - x_m}{x_{\max} - x_m}$$

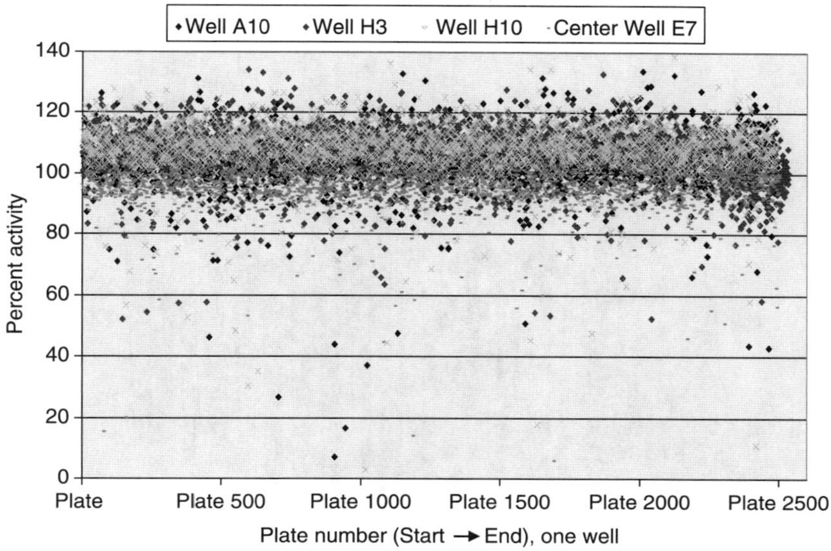

Figure 13.8 Primary screening results in the polymerase assay described in Figure 13.3. Four wells in the 96-well AssayPlate, one center well (E7) and three edge wells (A10, H3, and H10), are selected to show the positional effect. The results in the edge wells shifted about 10% above the center well. (See color Insert.)

where x_m is the mean of the value of inactives in the whole plate, x is the value of any well in the plate, and x_{max} is the control that gives 100% inhibition (or activation). Figure 13.9 shows the normalization of the primary screening data in a proliferation assay for CHO-CD40L activated B cells. Recombinant CD40 is used in the AssayPlate as a control for activation and TGF_{β} is used in the AssayPlate as a control for inhibition. The activated B cells are incubated with the secreted protein library and controls for 4 days. At the end of the assay, the amount of ATP in the AssayPlate is measured that is proportional to the total number of cells. Figure 13.9*a* shows the raw data and Figure 14.9*b* shows the normalized data (also corrected for edge effect). The normalized data can be used for final hit picking across the plate in the whole screen. When the library is small (as in the case of secreted protein library), the library can be screened in multiplicate to increase the screening confidence without incurring operational difficulties if the reagent cost is reasonable. In the previously described screening with B cell proliferation assay, the library was screened in duplicate. The results for one pair of the duplicate AssayPlate is shown in Figure 13.10. The control for activation (recombinant IL4) is used as the normalization control value and is arbitrarily set at 100% activation. The data are also ranked using the criteria of multiples of standard deviation from the mean. The data are clustered at the diagonal line meaning the consistency of the data in the two AssayPlates.

There are two major methods to pick hits and each of them has its advantage in unique situations. The two methods are illustrated in Figure 13.11, using the screening

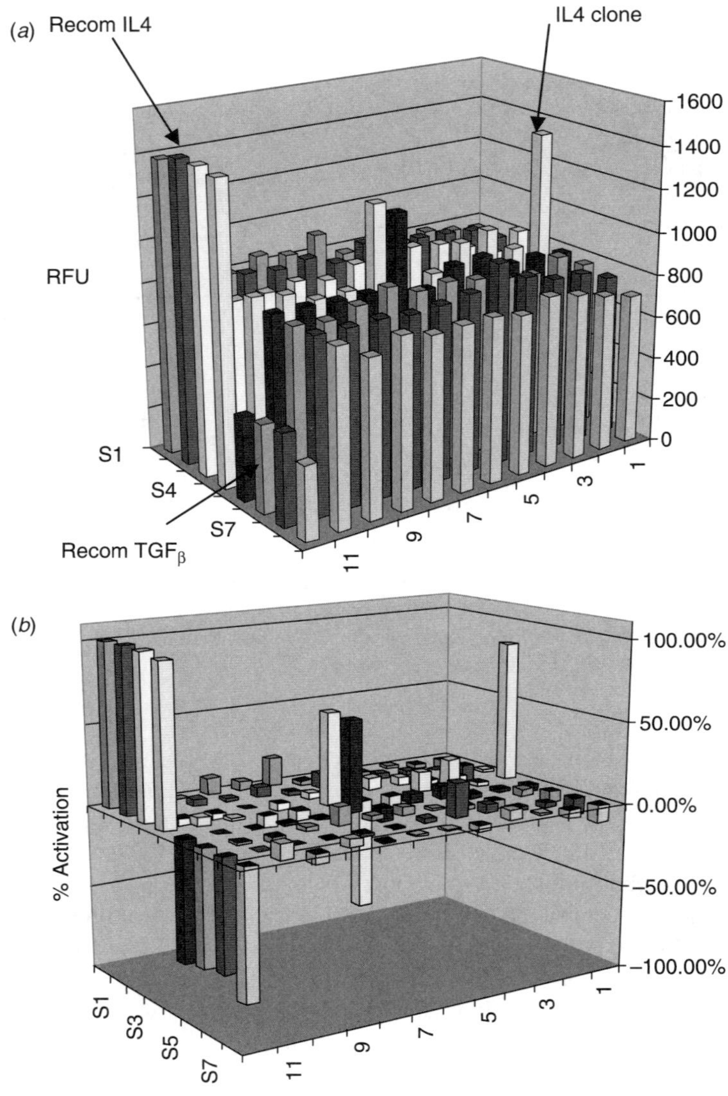

Figure 13.9 Normalization of the primary screening data in a proliferation assay for CHO-CD40L activated B cells. Recombinant CD40 is used in the AssayPlate as a control for activation and TGF_β is used in the AssayPlate as a control for inhibition. The activated B cells are incubated with secreted protein library for 4 days. At the end of the assay, the amount of ATP in the AssayPlate is measured, which is proportional to the total number of cells. (*a*) The raw data and (*b*) The normalized data (also corrected for edge effect).

data obtained in the previously discussed polymerase assay. The first method is based on the confidence level to exclude the hits from the background population. For example, the hit criteria can be set to any wells that give a value more than 3 SD, 4 SD, or 5 SD from the mean. If 1 million wells are screened, the hit-picking criteria with 3 SD from the mean will cause 3000 inactive wells being picked as

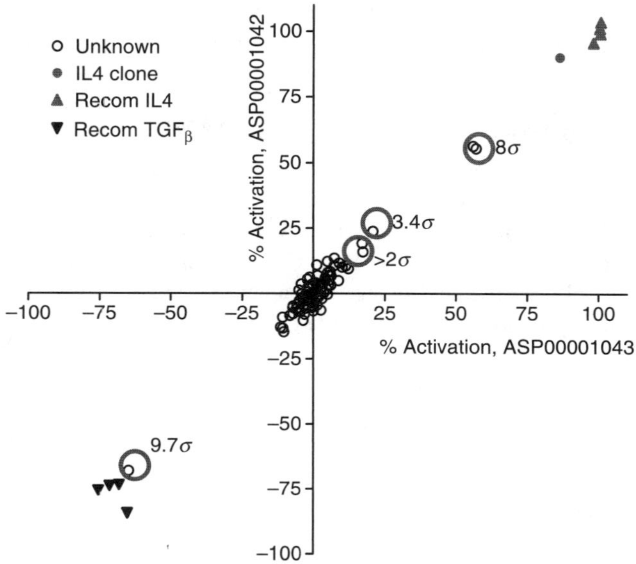

Figure 13.10 Normalized screening data from a pair of duplicate AssayPlate in the proliferation assay for CHO-CD40L activated B cells. The control for activation (recombinant IL4) is used as the normalization control value and is arbitrarily set at 100% activation. The data are also ranked using the criteria of multiples of standard deviation from the mean. The data are clustered at the diagonal line meaning the consistency of the data in the two AssayPlates.

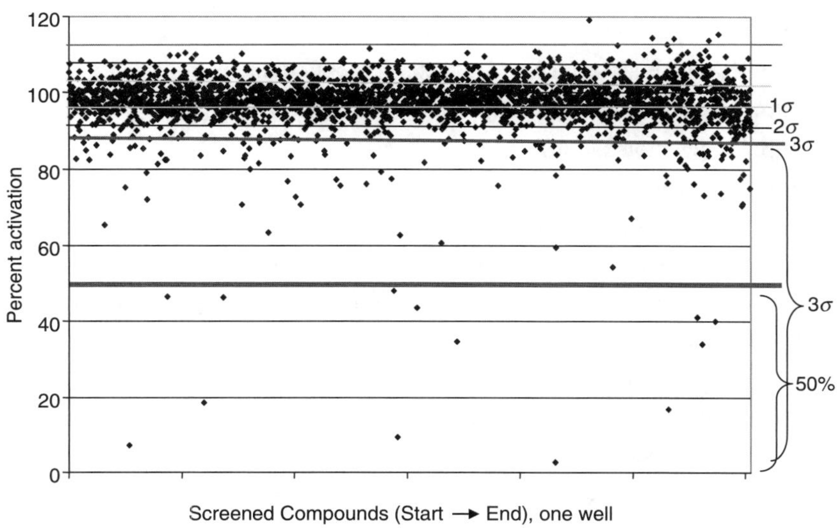

Figure 13.11 Hit picking criteria based on either arbitrarily-set confidence level or arbitrarily-set fixed percentage of inhibition. The screen in this case is performed very well and the SD is only about 5% of inhibition. Hit picking with 50% inhibition criteria will result in a high confirmation rate.

hits. These wells will be eliminated in the retesting stage, resulting in lower hit confirmation rate. This method is purely statistical without regard to the activity. The advantage of this method is that the controls do not have an effect on hit picking. Thus, it is very useful when trying to identify agonists in a cell based assay for which there is no known active compound to serve as a control. The disadvantage is that the SD value may change among AssayPlates. A weak active compound may be included in the inactive population when the SD is very large for an AssayPlate. Another method is to arbitrarily set certain inhibition value as the criteria. For example, 50% inhibition of enzyme can be used as hit picking criteria. With this method, all the AssayPlates must be normalized using the values in the control wells. If the SD of the background population is low as in the case shown in Figure 13.11, hit-picking with 50% inhibition criterion will result in a high confirmation rate. The screen result shown in Figure 13.11 has performed very well and the SD is only about 5% of inhibition. Sometimes the screening data is not as good and the SD can be more than 20% of inhibition. In this case, 50% inhibition will still be inside the 3 SD window ($20\% \times 3 = 60\%$) and the picked hits at 50% inhibition will have a low confirmation rate.

Useful Websites

http://www.cytomat.com/

http://www.liconic.com/products/products.html

http://www.tecan.com/

http://www.matrical.com/MatriStore.php

http://www.labcyte.com/Echo_Liquid_Handling_Systems/Default.81.html

http://www.edcbiosystems.com/bio/

http://www.ttplabtech.com/products/compound/index.html

http://www.bmglabtech.com/

http://www.biotek.com/

http://www.moleculardevices.com/

http://www.bertholdtech.com/ww/en/pub/bioanalytik/produkte/lb940.cfm

http://www5.gelifesciences.com

http://las.perkinelmer.com/Catalog/ProductInfoPage.htm?ProductID=MiniTrak

http://www.staubli.com/en/robotics/products/6-axis-scara-industrial-robot/

http://www.motoman.com/products/default.php

http://www.thermo.com/com/cda/landingpage/0,10255,807,00.html

http://www.mitsubishi-automation.com/products/robots_content.html

http://www.kalypsys.com/systems/index.shtml

http://www.rtslifescience.com/

http://www.cybio-ag.com/

http://www.idbs.com/activitybase/

http://www.genedata.com/products/screener

http://www.symyx.com/products/software/lab-execution/assay-explorer/index.jsp

http://www.automationpartnership.com/

BIBLIOGRAPHY

Bajorath, J. (2002) Integration of virtual and high-throughput screening. *Nat. Rev. Drug Discov.* **1**, 882–894.

Baurin, N., et al. (2004) Drug-like annotation and duplicate analysis of a 23-supplier chemical database totalling 2.7 million compounds. *J. Chem. Inform. Computer Sci.* **44**, 643–651.

Bender, A., et al. (2008) Which aspects of HTS are empirically correlated with downstream success? *Curr. Opin. Drug Discov. Devel.* **11**, 327–337.

Brideau, C., Gunter, B., Pikounis, B., and Liaw, A. (2003) Improved statistical methods for hit selection in high-throughput screening. *J. Biomol. Screen.* **8**, 634–647.

Chung, T. D. Y. (1998) Screening of compound singly: Why muck it up. *J. Biomol. Screen.* **3**, 171–173.

Collins, J. (2001) *Good to Great*. Harper Collins, New York.

Devlin, J. P. (ed.) (1997) *High Throughput Screening*. Marcel Dekker, New York.

Eastwood, B. J., et al. (2006) The minimum significant ratio: A statistical parameter to characterize the reproducibility of potency estimates from concentration-response assays and estimation by replicate-experiment studies. *J. Biomol. Screen.* **11**, 253–261.

Fox, S., Farr-Jones, S., Sopchak, L., Boggs, A., and Comley, J. (2004) High-throughput screening: Searching for higher productivity. *J. Biomol. Screen.* **9**, 354–358.

Fox, S., et al. (2006) High-throughput screening: Update on practices and success. *J. Biomol. Screen.* **11**, 864–869.

Gagarin, A., Makarenkov, V., and Zentilli, P. (2006) Using clustering techniques to improve hit selection in high-throughput screening. *J. Biomol. Screen.* **11**, 903–914.

Gillet, V. J. (2008) New directions in library design and analysis. *Curr. Opin. Chem. Biol.* **12**, 372–378.

Gunter, B., Brideau, C., Pikounis, B., and Liaw, A. (2003) Statistical and graphical methods for quality control determination of high-throughput screening data. *J. Biomol. Screen.* **8**, 624–633.

Hajduk, P. J. and Greer, J. (2007) A decade of fragment-based drug design: Strategic advances and lessons learned. *Nat. Rev. Drug Discov.* **6**, 211–219.

Harper, G., Pickett, S. D., and Green, D. V. (2004) Design of a compound screening collection for use in high throughput screening. *Comb. Chem. High Throughput Screen.* **7**, 63–70.

Hesterkamp, T. and Whittaker, M. (2008) Fragment-based activity space: Smaller is better. *Curr. Opin. Chem. Biol.* **12**, 260–268.

Hoever, M. and Zbinden, P. (2004) The evolution of microarrayed compound screening. *Drug Discov. Today* **9**, 358–365.

Houston, J. G., et al. (2008) Case study: Impact of technology investment on lead discovery at Bristol-Myers Squibb, 1998–2006. *Drug Discov. Today* **13**, 44–51.

Hüser, J. (ed.) (2006) *High-Throughput Screening in Drug Discovery*. Wiley-VCH, Weinheim.

Iversen, P. W., Eastwood, B. J., Sittampalam, G. S., and Cox, K. L. (2006) A comparison of assay performance measures in screening assays: Signal window, *Z′* factor, and assay variability ratio. *J. Biomol. Screen.* **11**, 247–252.

Janzen, W. P. (ed.) (2002) *High Throughput Screening: Methods and Protocols*. Humana, Towata, NJ.

Kempner, M. E. and Felder, R. A. (2002) A review of cell culture automation. *JALA* **7**, 56–62.

Kevorkov, D. and Makarenkov, V. (2005) Statistical analysis of systematic errors in high-throughput screening. *J. Biomol. Screen.* **10**, 557–567.

Kitchen, D. B., Decornez, H., Furr, J. R., and Bajorath, J. (2004) Docking and scoring in virtual screening for drug discovery: Methods and applications. *Nat. Rev. Drug Discov.* **3**, 935–949.

Lin, H., et al. (2008) Discovery of a cytokine and its receptor by functional screening of the extracellular proteome. *Science* **320**, 807–811.

Lindsay, M. A. (2005) Finding new drug targets in the 21st century. *Drug Discov. Today* **10**, 1684–1687.

Ling, X. B. (2008) High throughput screening informatics. *Comb. Chem. High Throughput Screen.* **11**, 249–257.

Lipinski, C. A. (2000) Drug-like properties and the causes of poor solubility and poor permeability. *J. Pharmacol. Toxicol. Methods* **44**, 235–249.

Makarenkov, V., et al. (2007) An efficient method for the detection and elimination of systematic error in high-throughput screening. *Bioinformatics* **23**, 1648–1657.

Mayr, L. M. and Fuerst, P. (2008) The future of high-throughput screening. *J. Biomol. Screen.* **13**, 443–448.

Pereira, D. A. and Williams, J. A. (2007) Origin and evolution of high throughput screening. *Br. J. Pharmacol.* **152**, 53–61.

Rishton, G. M. (2003) Nonleadlikeness and leadlikeness in biochemical screening. *Drug Discov. Today* **8**, 86–96.

Roddy, T. P., et al. (2007) Mass spectrometric techniques for label-free high-throughput screening in drug discovery. *Anal. Chem.* **79**, 8207–8213.

Scheer, A. (2006) Future trends in screening technology for drug discovery. *Expert Opin. Drug Discov.* **1**, 195–198.

Seethala, R. and Fernandes, P. (eds.) (2001) *Handbook of Drug Screening*. Informa HealthCare, New York.

Shelat, A. A. and Guy, R. K. (2007) The interdependence between screening methods and screening libraries. *Curr. Opin. Chem. Biol.* **11**, 244–251.

Sittampalam, G. S., et al. (1997) Design of signal windows in high throughput screening assays for drug discovery. *J. Biomol. Screen.* **2**, 159–169.

Snider, M. (1998) Screening of compound library . . . consomme or gumbo. *J. Biomol. Screen.* **3**, 169–170.

Snowden, M. and Green, D. V. (2008) The impact of diversity-based, high-throughput screening on drug discovery: "Chance favours the prepared mind." *Curr. Opin. Drug Discov. Dev.* **11**, 553–558.

Verheij, H. (2006) Leadlikeness and structural diversity of synthetic screening libraries. *Mol. Diversity* **10**, 377–388.

Walters, W. P. and Namchuk, M. (2003) Designing screens: How to make your hits a hit. *Nat. Rev. Drug Discov.* **2**, 259–266.

Williams, A. J. (2008) A perspective of publicly accessible/open-access chemistry databases. *Drug Discov. Today* **13**, 495–501.

Wu, G. and Doberstein, S. K. (2006) HTS technologies in biopharmaceutical discovery. *Drug Discov. Today* **11**, 718–724.

Wunder, F., Kalthof, B., Müller, T., and Hüser, J. (2008) Functional cell-based assays in microliter volumes for ultra-high throughput screening. *Comb. Chem. High Throughput Screen.* **11**, 495–504.

Zhang, J.-H., Chung, T. D. Y., and Oldenburg, K. R. (1999) A simple statistical parameter for use in evaluation and validation of high throughput screening assays. *J. Biomol. Screen.* **4**, 67–73.

Zhou, H. (2007) Biologics in the pipeline: Large molecules with high hopes or bigger risks? *J. Clin. Pharmacol.* **47**, 550–552.

CHAPTER 14

CASE STUDY: DEVELOPMENT OF A MICROFLUIDIC-BASED KINASE ASSAY PLATFORM

IN THIS chapter, I will use a real-world assay development situation to illustrate many of the assay development principles discussed so far. Many of the mistakes and solutions (if there are any) are presented so that the readers can have a feel about how a new technology is developed. Most end users only used the final finished working product but never knew the many failures and triumphs behind the product they were using. Because I conceived the idea of Caliper's off-chip kinase assays and was deeply involved in the development of the assay, I can provide a rare opportunity for the reader to peek into the assay development process. Both failures and successes are presented. I also included the exact time of some key experiments and some of the emails so that the reader can have a real feel about the assay development process with time stamps, and it is also easier to follow the flow of events. Because the aim here is purely scientific discussion, I avoided the names of key players and also deleted information that may involve Caliper's trade secrets, although all the events discussed here happened in the year 2000 to the beginning of 2001. All the scientific discussions here are either public knowledge (published in journals, presentations in conferences, product manuals) or can be derived by skilled scientists in the field. I hope this chapter can also help startup companies to avoid many of the mistakes and the big companies to avoid potential problems when purchasing new technologies.

14.1 BACKGROUND OF MICROFLUIDIC TECHNOLOGY AND ITS APPLICATION IN BIOASSAYS

All the assays discussed so far are performed by manually or automatically moving the microplate from one liquid dispenser to the other and finally to the reader to obtain the data. In most cases, these liquid-handling devices and the readers are physically separated in different locations. Sometimes limited functions are built

Assay Development: Fundamentals and Practices. By Ge Wu

into one instrument as separate modules (e.g., the well-known FLIPR instruments with both liquid-handling and signal detection built into one instrument). Before the 1990s, the major advances in high-throughput assay technologies have been the efficient integration of instruments and the miniaturization to increase the throughput while reducing the reagent usage. There was no major paradigm shift in the way the assays were performed. Miniaturization down to 1536-well microplates (with total assay volume about 8 μL) and 3456-well microplates (with total assay volume about 2 μL) seemed reaching the limits of the existing assay paradigm due to the physical issues (surface-to-volume ratio) and economic issues (expensive instruments to handle the small volumes of reagents) encountered. In the early 1990s, microfabrication techniques from the electronic industry were adopted to create micrometer-scaled channels in rigid materials, such as glass and silicon. These channels connect the reagent reservoirs on the chip and form interconnected liquid paths so that the reagents from the reservoirs can be selectively mixed at different ratios and the final mixed reagents can be drawn to the area on the chip for detection. For example, Harrison and colleagues in 1993 demonstrated using micromachining technology to build chemical analysis systems on glass chips with the size about 1 cm by 2 cm. Capillaries with diameters about 20 μm and with lengths of about 1 to 10 cm long were etched on the chip. Electroosmotic flow was employed to drive the flow of fluid and electrophoresis is used to separate components in the sample. This system allowed for capillary electrophoresis-based separations of amino acids with up to 75,000 theoretical plates in about 15 seconds, and separations of about 600 plates can be achieved within 4 seconds. They also demonstrated that the directions of the fluidic flow on the chip could be manipulated by controlling the applied voltages. The system had no moving parts because the fluidic flow was driven by EOF, which made the system robust without concern about wearing down after long term repeated use. For each analysis, the sample volume was in the picoliter scale. Thus, this system achieved both miniaturization and integration of the whole analytical system in a chip with a small footprint. In comparison, conventional operation paradigm in bioassays performed in microplate can only reduce the assay volume down to 1 to 2 μL and requires a large footprint of lab space to host all the disparate instruments. In addition to miniaturization and integration, the analysis time with assays performed on the chip is usually shorter because all the sample manipulation and detection steps happen in a small footprint. The total time to analyze one sample is usually less than 2 min (the time for the sample to flow from its reservoir to the detector). This new paradigm for chemical analysis is commonly referred to "labs-on-a-chip" or micro total analysis systems (μTAS).

To capitalize on this new chemical analysis paradigm, several companies were formed to bring products based on this idea to the market. Caliper Technologies was formed in 1995 and was a leader in the field. Remarkably, the first microfluidic-based product for DNA sizing was marketed in less than 3 years in 1998 through collaboration with Hewlett-Packard (currently Agilent). The triumph was the result of a combination of good luck in picking the right application and great engineering. Figure 14.1 shows the DNA sizing chip manufactured by Caliper and marketed by Agilent. It was used together with the Agilent 2100 Bioanalyzer instrument. The size of the whole glass chip is about the size of a penny. It is a planar chip with integrated microfluidic paths that can be seen from the back of the chip shown in

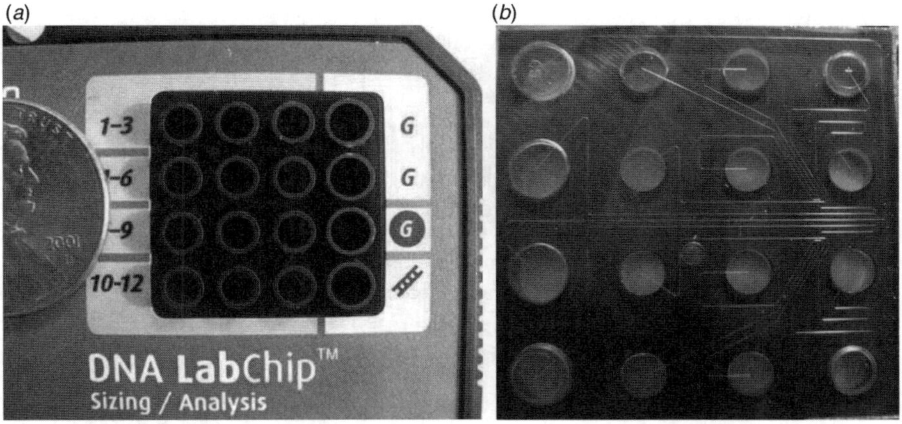

Figure 14.1 Caliper's DNA sizing chip. (*a*) Top, (*b*) bottom (microfluidic channels are visible connecting the 12 sample wells). (See color insert.)

Figure 14.1*b*. Twelve DNA samples can be analyzed with one chip. The DNA sizing is based on traditional gel electrophoresis principles that have been transferred to a chip format. The sample wells (labeled 1 to 12), gel wells (labeled g), and a well for an external standard (labeled with a ladder symbol) are placed on the chip. These wells are connected by microchannels fabricated on the chip. The microchannels are filled with a sieving polymer and fluorescence dye. The 16-pin electrodes of the cartridge in the Bioanalyzer 2100 are arranged so that they fit into the wells of the chip. Each electrode is connected to an independent power supply that controls the flow of the samples in the chip and also provides the voltage gradient for electrophoresis separation. DNA fragments migrate in the gel matrix driven by a voltage gradient and are separated by their size, similar to slab gel electrophoresis. These DNA fragments are labeled by the fluorescent dye initially loaded onto the chip and are detected by laser-induced fluorescence. Because the microchip is much shorter than normal gel, the time it takes to separate the DNA fragments is much shorter than the time it take to separate them with slab gel. In addition, all the procedures involved in a normal DNA sizing experiment are integrated in one chip. Thus, the chip format dramatically reduces the time for the whole DNA sizing experiment to about half an hour. The DNA samples and reagent consumption are dramatically reduced too. In addition, the information collected in chip format is already digitized and the quantity information of each DNA fragment is readily available. To obtain such information, the gel in normal DNA sizing and quantitation experiments must be scanned with a densitometer. The Agilent Bioanalyzer 2100 won the Pittcon gold award in 2000 because of the revolutionary technology.

After the initial success, Caliper wanted to expand the application of microfluidic technology to the then red-hot high-throughput screening applications to realize the dream of performing a whole bioassay (not just separation) on a chip. The idea is to have all the assay reagents placed on the reservoirs on a chip with interconnected channels connecting them. For enzymatic assays, the enzyme and the substrate are placed in separate reservoirs on the chip. The enzyme and substrate are then drawn

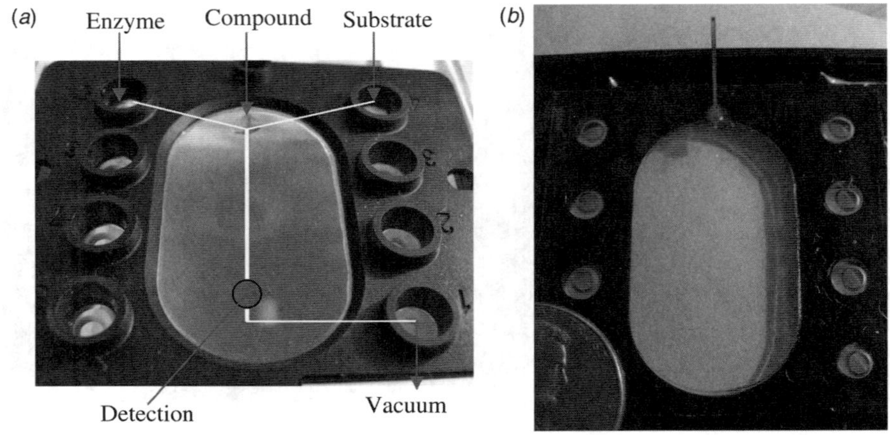

Figure 14.2 First-generation Caliper chip with a single sipper. (*a*) Top of the chip. There are a total of eight wells on the chip. The number 1 well is reserved for vacuum. The other wells may or may not be connected to the main channel depend on the assay. For example, only the enzyme and substrate reservoirs are connected to the main channel in fluorogenic assays. (*b*) Bottom of the chip. The sipper can be clearly seen in this picture. A penny is included in the picture to show the actual size of the chip.

to the main reaction channel where the product is generated. To access the compounds in a microplate, a capillary (called sipper) is perpendicularly attached to the chip and it can insert inside the wells in a microplate. The compounds in the microplate are introduced sequentially to the main channel on the chip. If the test compound is an inhibitor, the enzymatic reaction in the main channel will be stopped. Figure 14.2 shows the first generation Caliper chip with a single sipper. Figure 14.2*a* shows the top of the chip. There are a total of eight wells on the chip. The well 1 is reserved for vacuum. The other wells may or may not be connected to the main channel depending on the assay. For example, only the enzyme and substrate reservoirs are connected to the main channel in fluorogenic assays, which will be discussed below. Figure 14.2*b* shows the bottom (backside) of the chip. The sipper can be clearly seen in this picture. A penny is included in the picture to show the actual size of the chip. The addition of the sipper changed the microfluidic networks from two-dimensional planar interconnected channels to three-dimensional interconnected channels. Many difficulties were encountered to attach the sipper to the chip, which is one of the most labor-intensive procedures in making the three-dimensional chips. With this new microfluidic-based assay format, each assay data requires only picoliters to nanoliters of reagents. Thus, the system in theory will leapfrog from the existing assay paradigm.

The same fluorogenic assay for caspase-3 (discussed in Chapter 6) is used here as an example to show how the fluorogenic enzyme assay works in the microfluidic format. Figure 14.3 shows the fluidic flow diagram and how the fluorogenic assay is configured. When a vacuum is applied to the waste reservoir (well 1), enzyme, substrate, and compounds/buffer are drawn to the main channel from their reservoirs and microplate wells. The stream of different reagents coming from their respective

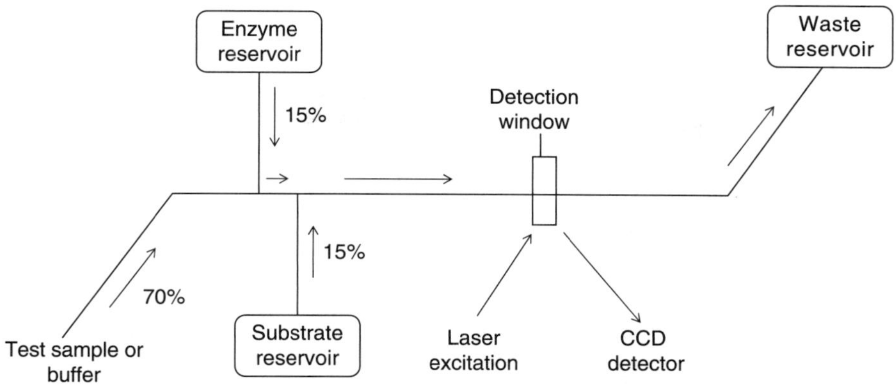

Figure 14.3 Schematic illustration of fluorogenic enzymatic assays in microfluidic format. The enzyme to be assayed is placed in the enzyme reservoir and the substrate of the enzyme is placed in the substrate reservoir. Test compounds in a microplate and buffer in a reagent trough that is attached to the plate stage on the instrument are introduced to the chip through a sipper. When a vacuum is applied at the waste reservoir, enzyme, substrate, and compounds/buffer are drawn to the main channel passing the fluorescence detector located at the end of the main channel. Percent values (%) shown on the figure indicate the final percentage of main channel flow contributed by each reagent channel. The compounds, enzyme, and substrate are incubated on the main channel for ~60 s before reaching the detector.

reservoirs meet at the main channel and the reaction takes place. The reaction mixtures in the main channel are continuously drawn to the end of the main channel to the waste reservoir where the vacuum is applied. A laser excitation source and a fluorescence detector are placed at the end of the main channel and the fluorescence signal from the passing fluid is detected. The percentage of contribution to the flow in the main channel from each branched channel is dependent on the resistance to the flow from it and can be manipulated by varying the length and diameter of each branched channel. In this particular chip design, 70% of the volume in the main channel are contributed from the sipper, 30% of the volume in the main channel are contributed from enzyme channel (15%) and substrate channel (15%). The total flow rate in the vacuum-driven flow is dependent on the vacuum applied and also on the resistance to the flow. In this particular chip, compounds, enzyme, and substrate are incubated on the main channel for ~60 s before reaching the detector. The procedure to develop a fluorogenic assay in microfluidic format is shown in Figure 14.4. Initially, fluorogenic substrate was placed in the substrate well on the chip and all other wells on the chip were filled with assay buffer. The sipper was place in a trough containing the same assay buffer. A vacuum was applied and the fluorogenic substrate filled the main channel. The fluorescence (fluorogenic substrate still can emit fluorescence though at a much lower level) from the continuous flow of substrate was used to focus the laser and was used for the adjustment of the fluorescence detection. After this adjustment, ~500 relative fluorenscene unit (RFU) initial signal from continuous flow of substrate was obtained (zone a in Fig. 14.4). The buffer in the enzyme well was then replaced by caspase-3. After applying the vacuum, caspase-3 met the

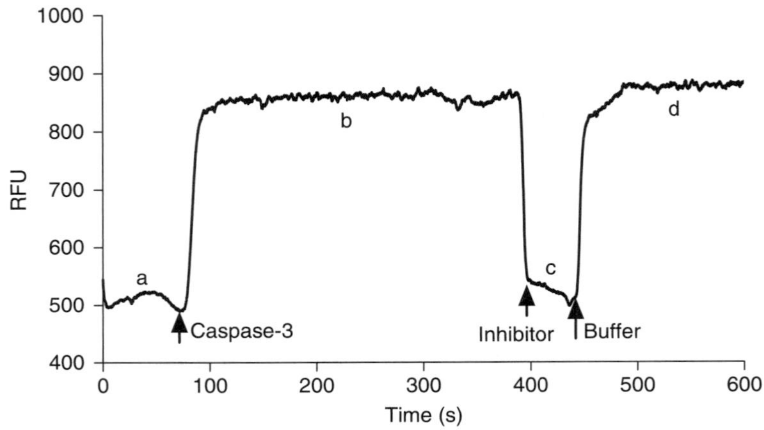

Figure 14.4 Caspase-3 assay development with microfluidic format. Fluorogenic substrate was placed in the substrate well on the chip and a vacuum is applied to fill the main channel with the fluorogenic substrate. The initial 500 RFU signal in zone a was from substrate. Caspase-3 was then placed in the enzyme well. After applying the vacuum, the enzyme-associated increase in fluorescent signal appears at 80 s in zone b. The caspase-3 concentration in the assay channel was 0.9 nM and the substrate concentration was 15 μM. At 400 s, an inhibitor (Ac-DEVD-CHO, 1.4 μM in assay channel) was sipped continuously into the chip through the sipper and the signals decreased close to baseline in zone c. At 450 s, the inhibitor is replaced by the buffer that was sipped through the sipper. All the enzyme associated signals were recovered in zone d.

fluorogenic substrate in the main channel and cleaved the substrate to generate a strong fluorescence signal. The enzyme-associated increase in fluorescent signal appears at 80 s (zone b). The caspase-3 concentration in the main channel was 0.9 nM and the substrate concentration was 15 μM. At 400 s, an inhibitor (Ac-DEVD-CHO, 1.4 μM in the assay channel) was sipped continuously into the chip through the sipper and the signals decreased close to the baseline (zone c). At 450 s, the buffer was sipped through the sipper, replacing the inhibitor. All the enzyme-associated signals were recovered (zone d). Using this assay format, a test screening was performed. The assay compounds in the microplate were introduced to the chip sequentially through the sippers. The injection time for each compound was 1 s, which was followed by injection of the buffer for 5 s. The assay results are shown in Figure 14.5. The large dip at the end of the trace was from the 100% inhibition control. The two smaller dips in the trace were from two 30% inhibition controls. The baseline was affected by the movement of the sipper from the sample wells to the buffer trough repeatedly (shown in the insert). It was about 3% of the total signal in Caliper's assay format depending on the chip. This was caused by the change in the pressure from the end of the sipper when it was inserted inside a solution and when it was taken out of a solution. The sipper movements were associated with constant changing in surface tension and caused the perturbation to the fluorescence signal at the detector. Thus, there was a periodic perturbation every 6 s that equal to the total time to analyze each sample.

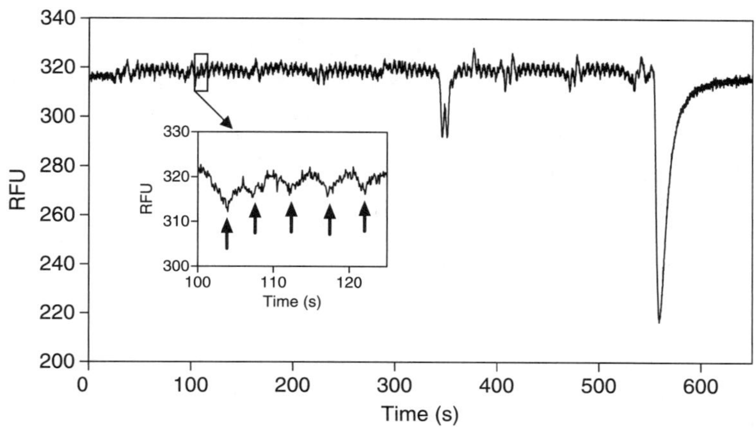

Figure 14.5 Screening caspase-3 in microfluidic format. The assay compounds in the microplate were introduced to the chip sequentially through the sippers. The injection time for each compound is 1 s, which is followed by injection of buffer for 5 s. The large dip at the end of the trace is from the 100% inhibition control. The two smaller dips in the trace were from two 30% inhibition controls. The baseline is affected by the movement of sipper from sample wells to buffer trough (shown in the insert). It is always about 3% of the total signal in Caliper's assay format.

In addition to the simple fluorogenic protease assays, Caliper also demonstrated the feasibility to assay kinases by combining the hydrodynamic flow and electrophoresis on the same chip. This assay was called "mobility shift assay" because it is based on the mobility difference between the kinase substrate and the products. Because there is a -2 charge difference between kinase product and substrate, the kinase substrate and product will move at different speeds in a voltage gradient and they should be separable if the substrate is relatively small, such as a peptide. A technology access program (TAP) was set up at Caliper to apply the technology to practical uses. Several large companies, from both pharmaceutical and biotechnology industries, signed up for the program with most of them interested in applying the technology to HTS. After signing up these big companies, scientists having expertise in protease assays and kinase assays were recruited at Caliper to deliver the assays to the partners. After spending more than 2 years at Merck doing HTS, I decided to move on, and Caliper was one of the companies that attracted my attention. I joined Caliper's TAP department in April 2000 responsible for delivering the mobility shift kinase assays to the partners.

14.2 THE ORIGINAL MOBILITY SHIFT KINASE ASSAY FORMAT

A kinase assay using the catalytic domain of PKA and a fluorescently labeled neutral Kemptide substrate (Fl-Kemptide) was regularly used to demonstrate the technology both internally and externally. The PKA-catalyzed reaction product is negatively charged Fl-Kemptide-PO_4^{2-}. The assay scheme and the liquid flow on the chip (code named the NS75 chip) are shown in Figure 14.6. The contribution to the total liquid flow at the detection points are: 8% from well 4 and well 5, respectively, 6% from well 7, and the remaining 76% from the sipper. PKA was placed in well 4 on the chip at a concentration of 1.25 μM (100 nM in the main channel because of dilution). The Fl-Kemptide substrate was placed in well 5 on the chip at a concentration of 125 μM (10 μM in the main channel because of dilution). The other wells on the chip were filled with assay buffer, which contained 100 mM HEPES, pH 7.5, 10 μM ATP, 5 mM $MgCl_2$, and some other components. Compounds to be assayed were dissolved in the same assay buffer and placed in a microplate. The compounds were sequentially drawn through the sipper to the chip by a vacuum (usually −1 to −2 psi) that was applied at well 1 on the chip. In the main channel, the compounds or assay buffer sipped from the microplate was first mixed with the enzyme for approximately 2 s. The mixture was then mixed with the Fl-Kemptide substrate in the main channel to initiate the kinase-catalyzed reaction. The reaction continued in the main channel while the fluid flowed downstream. By the time the reaction mixture reached the starting point of the separation channel on the main channel, a certain percentage of the Fl-Kemptide is converted into Fl-Kemptide-PO_4^{2-} product. A voltage (usually about 2000 V) is applied between well 6 and well 8, defining the separation channel. The kinase substrate and the product are separated in the voltage gradient as they move along the main channel toward well 1. A detector is placed at the end of the separation

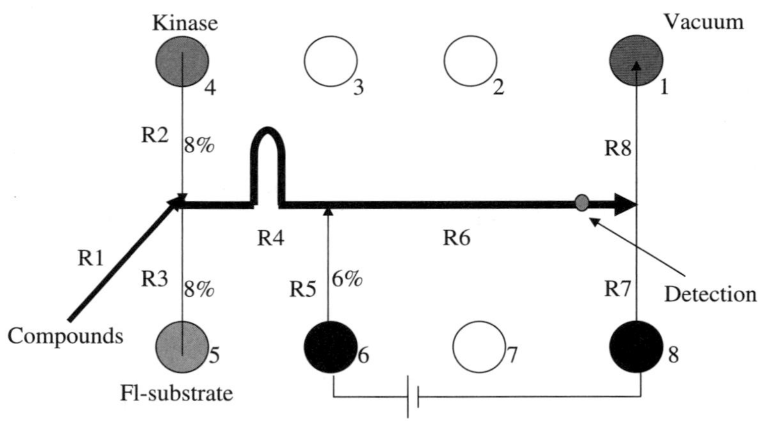

Figure 14.6 Graphical representation of how Caliper's original kinase mobility shift assay works. Kinase is placed in well 4 on the chip. Substrate is placed in well 5 on the chip. Compounds to be screened are placed in a microplate and sequentially drawn through the sipper to the chip by vacuum applied at well 1 on the chip. There is a small shift where the channel R2 and R3 meet the main channel. Thus, the compounds first mix with the kinase. After approximately 2 s incubation between the two, the mixture meets the substrate from well 5. The kinase-catalyzed substrate turnover takes place continuously in the assay channel while the compound plug flows further downstream until reaching the side channel (R5) coming from well 6. A voltage gradient is applied between well 6 and well 8, defining the separation channel. The substrate and the product are separated in the voltage gradient as they move along the channel toward well 1. A detector is placed at the end of the separation channel to detect the fluorescence signal from the product and substrate.

channel to detect the fluorescence signal from both the product and substrate. The detected fluorescence signal should remain the same because the amount of kinase product generated is the same as the amount of substrate depletion. This conclusion is based on the assumption that the extra phosphate group on the kinase product does not affect the fluorophore attached to the peptide. Thus, a flat line of fluorescence signal should be observed over time. However, when a plug of PKI (inhibitor of PKA) at a low concentration (e.g., 100 nM) is introduced into the channel, somehow a "peak-and-dip" pattern was observed (Fig. 14.7). The amplitude of the peak or the dip appeared to be proportional to the concentration of the inhibitor under certain conditions.

The microfluidic-based mobility shift kinase assay eliminated the need for all the liquid pipettors and dispensers. All the assay steps were performed on a small chip. If this technology could be generalized to all enzymatic assays, it would be a true revolution. I did not see any major issues with the protocol except that the emission from fluorescein should be kept in a more basic condition for the maximum signal. This is a minor issue here because we had plenty of signals due to the high concentrations of the substrate in the micromolar range. Another potential issue to apply this technology to HTS was the requirement that the composition and concentration of the buffer in the microplate where the test compounds were sipped from should closely match the buffer on the chip. A mismatch would result in a disturbance to the fluorescence signal whenever a sample is sipped. When just starting a new experiment, all the solutions are freshly made from the same stock buffer. After a long period of screening, the different evaporation rates between the buffer on the chip

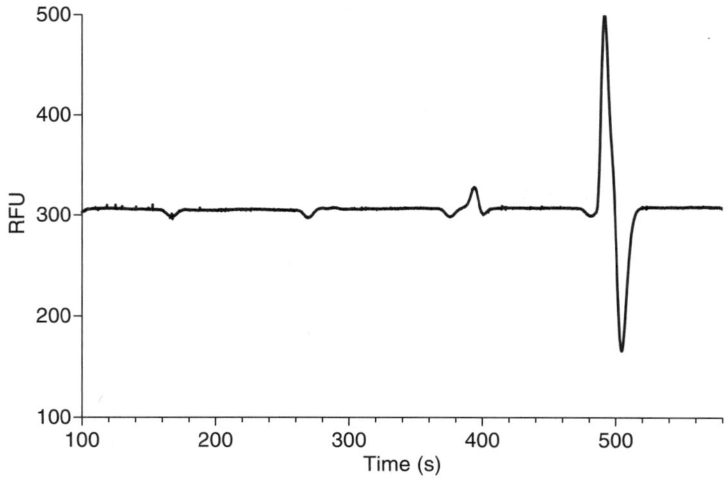

Figure 14.7 Typical peak-and-dip pattern in Caliper's mobility shift kinase assay. The substrate is fluorescently labeled. When a phosphate group is added to the substrate by the kinase, the fluorescence intensity should be about the same in the product as in the substrate. So a flat line of fluorescent signal should be observed over time. Even when there is a voltage gradient in the separation channel, a flat line is still observed. However, if an inhibitor plug is introduced into the channel, somehow a peak and a dip pattern is observed. The amplitude of the peak or the dip is proportional to the inhibitor concentration under certain conditions.

and the buffer in the microplate will cause a mismatch. Since the objectives for the assay only required it to run continuously for less than 8 h, this issue could be resolved by installing a humidity chamber to minimize the evaporation.

The separation power of the NS75 chip can be demonstrated by injecting a plug of dye mixtures containing two dyes with different charges when all the wells on the chip are filled with buffer. Figure 14.8 shows the separation of a mixture of dyes containing BODIPY (no charge) and fluorescein (−2 charge) with the NS75 chip. The vacuum of −1 psi was applied at well 1. A plug of the dye mixture is sippered into the chip for 2 s through the sipper. The main channel and the channels on the branch leading to well 6 and well 8 on the chip were coated with a polymer to eliminate EOF. In the absence of applied voltage, fluorescein and BODIPY co-eluted with a peak at 188 s. When a positive voltage gradient of 1435 V was applied from well 6 to well 8, the negatively charged fluorescein was eluted much slower with a peak at 215 s. When the voltage gradient was reversed to −2335 V, fluorescein was eluted out first with a peak at 175 s. The elution time for the neutral BODIPY does not change with the voltage (the small changes observed were due to residue EOF). The same experiment can be performed to demonstrate the separation between the neutral PKA substrate (Fl-Kemptide) and the −2 charged product (Fl-Kemptide-PO_4^{2-}), as shown in Figure 14.9. The vacuum applied here was −1 psi. A plug of the substrate and product mixture was sippered into the chip for 2 s through the sipper. The same channels in the chip were coated with a polymer to eliminate

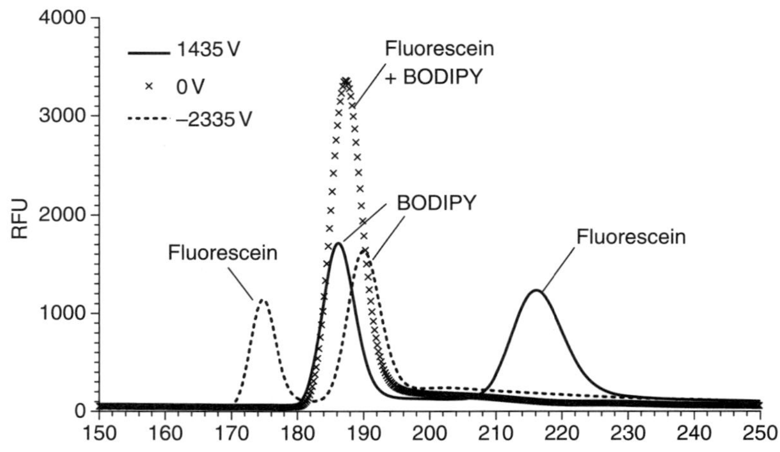

Figure 14.8 Separation of a mixture of dyes containing BODIPY (no charge) and fluorescein (−2 charge) with the NS75 chip. The vacuum applied is −1 psi. A plug of the dye mixture is sippered into the chip for 2 s through the sipper. The channels in the chip are coated with a polymer to eliminate EOF. In the absence of applied voltage, fluorescein and BODIPY co-elute with a peak at 188 s. When positive voltage gradient of 1435 V is applied from well 6 to well 8, the negatively charged fluorescein is eluted much slower at 215 s. When the voltage gradient is reversed to −2335 V, fluorescein is eluted out first at 175 s. The elution time for the neutral BODIPY does not change with the voltage (the small change observed is due to residue EOF).

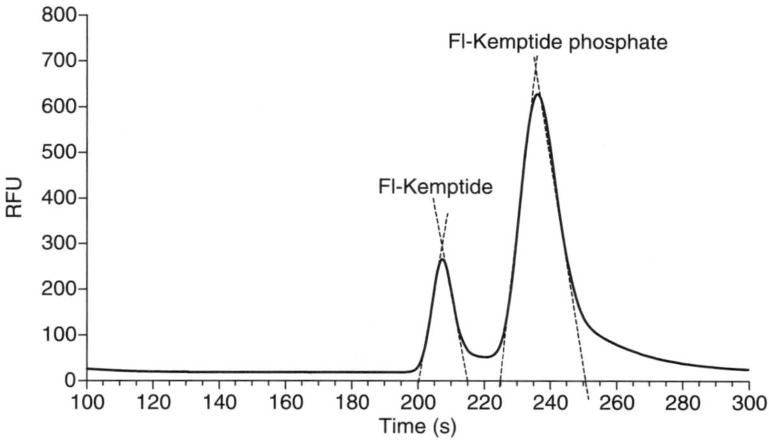

Figure 14.9 Separation of a mixture of neutral Fl-Kemptide and negatively 2 charged Fl-Kemptide phosphate with the NS75 chip. The vacuum applied is −1 psi. A plug of the mixture is sippered into the chip for 2 s through the sipper. The channels in the chip are coated with a polymer to eliminate EOF. A positive voltage gradient of 2335 V is applied from well 6 to well 8, the negatively charged Fl-Kemptide phosphate is eluted slower than the neutral Fl-Kemptide. The first peak width at the base is about 15 s and the second peak width is about 25 s. The total time for both peaks to elute is about 50 s.

EOF. A positive voltage gradient of 2335 V was applied from well 6 to well 8. The negatively charged Fl-Kemptide phosphate was eluted slower than neutral Fl-Kemptide. The width of first peak (substrate) at the base was about 15 s and the width of the second peak (product) was about 25 s. The total time to elute both product and substrate was about 50 s. It should be noted that the fluorescence signal between the two peaks did not go back to the baseline even though the two peaks were well separated. This was due to the broadening of the peak with a pressure-driven flow, and there was always a relatively long tail for any peak with the technology. This feature helped to explain the peak-and-dip phenomenon as discussed in Section 14.7.

After I was involved deeper into the microfluidic technology, I learned that several technical issues should be addressed in the chip-based assay format before it can be implemented in the screening mode. Thus, I had to help to develop the technology before I could implement it in the partner's kinase assays. One issue was how to identify the location of the sample in the microplate that gave an inhibition signal in the time-based experiment results obtained by the microfluidic assay format. When assays were performed in a microplate, the assay results could be easily assigned to the wells on the microplate because the samples were separated spatially. However, when an unknown compound in each well in a microplate is sequentially injected into the chip and there is a delay between sample introduction and detection, the situation has changed from the assay with spatial resolution to the assay with temporal resolution. When samples are introduced into the chip with large time intervals between samples, it is easier to identify the time region that defines the sample. However, when samples are introduced adjacent to each other, as required for HTS applications to have adequate throughput, it is difficult to find the time boundaries for each sample. The problem exists in both the kinase mobility assay format and the protease fluorogenic assay format. In these two assays, samples were sequentially sipped into the chip at a time interval of less than 10 s between samples. Thus, each sample occupied the same time slots (<10 s) in the time-based assay trace. Because most of the samples were inactive in real screening, a flat line was observed for most of the regions in the time-based assay trace as seen in Figure 14.5. It is difficult to pinpoint which of the 10-s regions correspond to which well in the microplate. To help assign the region in the time-based trace to a specific well location on the microplate, a second dye (marker dye) with a different fluorescence wavelength from the substrate fluorescence was employed. The two independent fluorescence traces must be re-aligned because the marker dye usually had different mobility from the fluorescent substrate or product. The alignment was done by placing the marker dye and a known inhibitor in the same well in the microplate. When a plug of the marker dye and the inhibitor was introduced into the chip from the control well, a pattern of inhibition will appear in the main fluorescent channel and a peak will appear in the second (marker) fluorescent channel. Because both of them were from the same well in a microplate, these peaks were used to align the two traces from the two fluorescence channels. A series of marker dyes were placed on the microplate (usually in one column of the microplate) to increase the time resolution of assigning the region of the time-based trace to the exact location on the microplate. Due to the requirement for a marker dye, a minimum of two laser sources were installed in early Caliper HTS instruments to carry the microfluidic-based assays.

Here I will use the actual experimental data from a PKA assay to illustrate how to assign a time region to a specific well in a microplate with the help of a marker dye. The same NS75 chip was used in the experiment. The concentration of PKA in the main channel was 50 nM and the concentration of the substrate was 5 μM. In a 96-well microplate, every well in column 1 was filled with buffer, which was used to elute the samples after each sample injection. The marker dye was placed in every well in column 2. The first well in column 2 (A2) also contains PKI at 1.75 μM. PKI at different concentrations was placed in several wells in the microplate to obtain the peak-and-dip pattern of inhibition. The experiment was performed with 100 s initial delay, followed by 1 s sample injection. After each sample injection, the buffer was sipped for 10 s to elute the sample. After injection of the last well, 50 s of final delay was applied. A vacuum of −1 psi was applied to well 1 on the chip. A snapshot of the assay results with several continuous runs of a 96-well microplate is shown in Figure 14.10. The top trace was the signal from the main channel with the fluorescence from the Fl-Kemptide and Fl-Kemptide phosphate. The bottom trace was the signal from the marker dye channel, which had a different wavelength from the fluorescence in the main channel. The marker dye was injected periodically (at spacing of every 10 samples in a row) into the chip. The first wells containing both the inhibitor and the marker dye was used to align the two traces and to help assign each region of the time-based trace to the physical location of the well in the microplate. The regions on the trace between two marker dyes were assigned to 10 samples that were introduced into the chip between the two marker dyes. Each region containing 10 samples was further divided into 10 smaller regions by equally dividing the

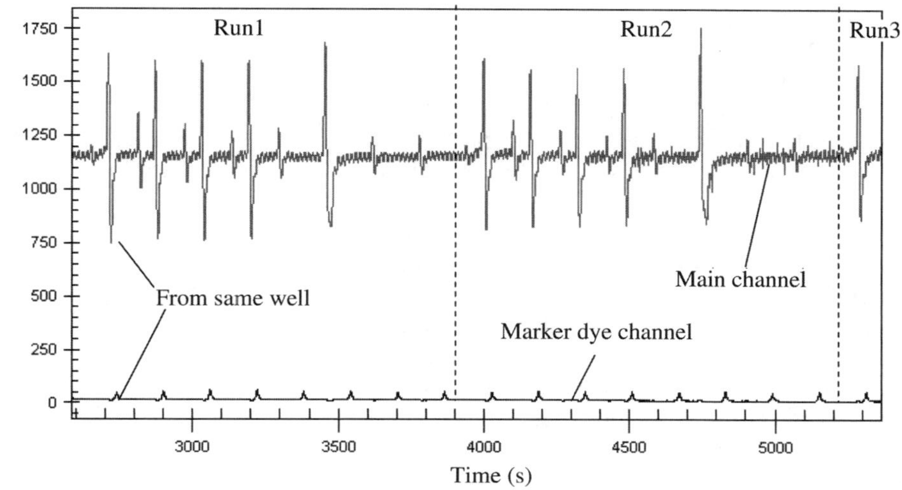

Figure 14.10 Assigning the region of the time-based assay trace to the position in a 96-well microplate with the help of a marker dye. This is a snapshot of assay results when one 96-well microplate is tested repeatedly. A marker dye plug is introduced into the chip between every 10 samples. The experiment was performed with 100 s initial delay, followed by 1 s sample injection. After each sample injection, buffer is sipped for 10 s. After injection of the last well, 50 s final delay is applied. A vacuum of −1 psi is applied on the chip. The top trace is the channel with fluorescence signal from the Fl-Kemptide and Fl-Kemptide phosphate. The bottom trace is the marker dye channel with different wavelength from fluorescein.

total time in the region by 10. Each of the 10 regions was then assigned to the well locations in the microplate according to the sample injection sequence.

Another issue in the kinase mobility shift assays was how to analyze the data to quantify the level of inhibition. By the time I arrived at Caliper, a consultant was already hired to develop an algorithm to calculate the extent of the inhibition. The consultant presented an algorithm that was implemented in the first draft of Caliper's data analysis software (alpha version) that was bundled together with the instrument control software. In this method, the distance between the peak and dip in kinase assay was used as a measurement to quantify the extent of kinase inhibition. This algorithm was used because it was simple to implement (no need to calculate the baseline). There was another proposal to calculate the extent of inhibition by double integrating the peak-and-dip curve. This thinking was based on the fact that the peak-and-dip pattern looked like the first derivatives of a normal Gaussian-shaped curve. Thus, the first integration would obtain the underlying Gaussian-shaped curve and the second integration would obtain the area under the Gaussian-shaped curve. However, this proposal was ruled out because finding the baseline was required for integration, and according to the consultant there was no baseline-finding algorithm at Caliper at that time. In addition to implementation challenges, double integration would require experimental data with good baseline. Otherwise, the error associated with the measurement would be significantly amplified. Thus, the algorithm for measuring the amplitude between the top of the peak and the bottom of the deep was adopted.

The consistency of the surface on the channel on each batch of chips was a major problem. The variation in the surface on the channel could affect the interactions between the analytes and the channel surface leading to inconsistency in the final assay results. When the surface on the channel was chemically modified to suppress the EOF, the inconsistency between different batches of chips was reduced. However, EOF might not be completely eliminated, and the variations on the surface of the channel could still affect the separation of analytes. To reduce the variations in the surface on the channel between the batches of chips, high-quality fused silica was used as the substrate to make the chip, which increased the cost of the chip. The high-quality silica is harder to machine (drilling holes), which increased the cost further. Another difficulty in manufacturing the chip was the insertion of the sipper in the holes on the chip. This was done manually under a microscope to position the capillary in the hole and to glue the sipper to the chip. Misplaced sipper could cause problems in the flow pattern and result in failure of the chip. For example, a sharp peak followed by a broader peak with a long tail was often observed when a plug of dye was injected into a chip with the sipper poorly inserted. When an air bubble was trapped inside the channel, the channel would most likely appear as blocked. To prevent this from happening in customer's hand, the chips were filled with water and they were shipped to customers in a container containing water. This scheme had caused other problems. The water weakened the adhesives used to bind the plastic caddy and the fused silica chip together. After a long period of exposure to water, delamination occurred due to the weakened adhesives, which caused leakage of current between the wells on the chip when a voltage was applied between the wells. The short circuit was a serious problem and led to the failure of the assay.

14.3 REALIZING THE FLAWS IN THE ORIGINAL KINASE ASSAY FORMAT

14.3.1 Understanding the Peak-and-Dip Pattern

After learning how to perform the mobility shift kinase assays, I decided to obtain the PKI's concentration–response curve and extract its IC_{50} value in PKA assays. This information would help to compare the sensitivity of the mobility shift kinase assay with existing kinase technologies. About one month after I joined Caliper (on May 16, 2000), I performed a critical experiment to assay PKA, following the same procedure I was taught. In this experiment, the concentrations of all reagents used were the same except that the concentrations of the PKI were varied. I made a serial 1 : 3 dilution of PKI in a microplate. Out of curiosity, I started at very high concentration of PKI (100 μM) and diluted down. Since the serial dilution scheme was always from high concentrations to low concentrations, I started introducing the PKI into the chip in the same sequence. To make sure that the high concentrations of PKI were fully eluted from the chip after they were sipped into the chip, I doubled the buffer elution time to 200 s from the 100 s elution time that I normally used. After

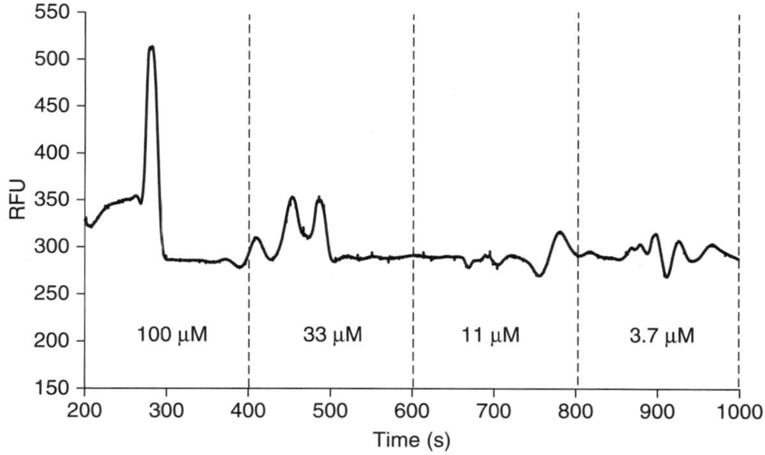

Figure 14.11 Mobility shift kinase assay at high concentrations of PKI. The PKA concentration is 100 nM in the main channel. The PKI concentrations for four samples are labeled in the graph. At 100 μM PKI, a peak was observed but no dip was observed. The following PKI samples introduced to the chip produced unfamiliar patterns.

injecting four of the highest concentrations of PKI with 2 s of sample injection time, I realized that something was wrong, as seen in Figure 14.11. There was no peak-and-dip pattern for any of the samples. The first sample only had a peak. I thought that the batch of PKA used might lose its function since it was an old batch. I added another 100 nM new batch of PKA to well 4 on the chip and repeated the experiment. I obtained the same results. After considering various causes, I concluded that PKI at high concentrations might stick to the surface of the channel and fully inhibited the conversion of the substrate to the product. If this happened, the observed fluorescence should be from the substrate alone and freshly injected PKI would not change anything. Thus, there would be no peak-and-dip pattern since there was no product at all. After eluting the chip for a long time with buffer, I reversed the order of the PKI injection to the chip by sipping PKI from low concentration to high concentration with the normal 100 s of buffer elution time after 2 s injection of PKI samples. The results are shown in Figure 14.12. Now I have the expected peak-and-dip pattern at low concentrations of PKI. However, the dip disappeared at high concentrations of the PKI. I realized that if this were true, our algorithms implemented in the analysis software would be wrong. A bell-shaped concentration response curve would be obtained if the amplitude from the peak to the dip was used as a measure of inhibition. I repeated this experiment several times to make sure this experiment was correct.

I communicated the data in meetings and to several individuals. However, the new finding was not well-perceived, because poor understanding the of peak-and-dip phenomenon within the company. A few people even suggested I might have done something wrong or the chips I used might have a problem. In the early days, chip manufacturing was not very consistent and many inferior chips were used internally. There is a tendency in the lab that people often attributed unexpected experimental results to chip failures though that might not be true. I spent a couple of weeks

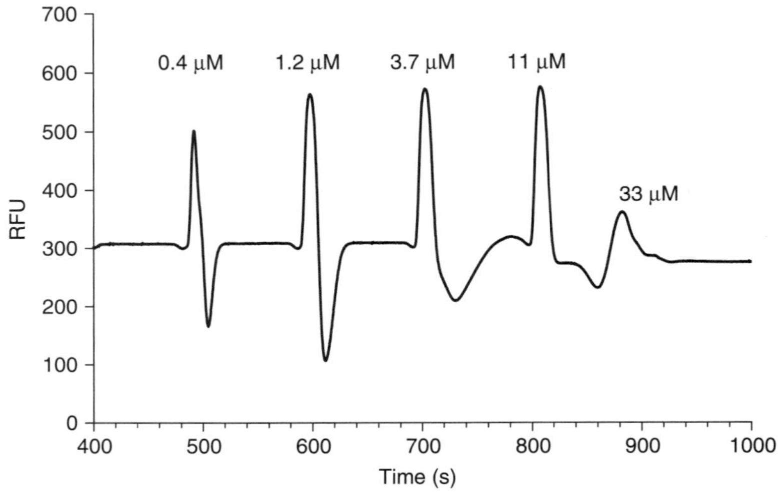

Figure 14.12 Inhibition of PKA by PKI. The concentrations of PKI were labeled on the graph. When injecting PKI from lower concentration to higher concentration, the peak and dip was seen at lower PKI concentrations. At higher concentrations of PKI, the dip is distorted or disappeared.

thinking about the data and finally figured out how the pattern was formed and thought of a simply way to explain this pattern graphically. (The original graphical presentation was made with straight lines only in PowerPoint in green and red color. It appeared on most of the training manuals in Caliper's on-chip kinase assay format.) With the graphical illustration, it is much easier to communicate with other scientists. After demonstrating the new analysis to several scientists, some of them (including the consultant who wrote the original algorithm to calculate the extent of inhibition) agreed with the analysis.

I will not repeat the analysis in this section. Instead, a poster containing the detailed analysis that I presented in the 2002 annual Society for Biomolecular Screening meeting is attached in the appendix to this chapter (Section 14.7). The basic conclusions are: (1) the peak-and-dip pattern is formed by the overlap of a peak (increased substrate concentration in the presence of an inhibitor) and a dip (decreased product concentration in the presence of an inhibitor) and the peak and the dip are separated by the voltage gradient. (2) The widths of the peak and the dip do not depend on the concentration profile of the inhibitor (Gaussian-shaped curve with no cap at the top). Instead, they depend on the inhibition profile (with the top capped at 100% inhibition). Thus, a potent inhibitor will always give the shape with a flat top in the leading peak (or dip) even if there is a good separation between the product and the substrate. (3) It is close to impossible to obtain complete separation of the peak and the dip when a strong inhibitor is introduced (e.g., even 99.9% of the inhibitor is eluted after a long elution, the 0.1% remaining inhibitor can still cause 100% inhibition). However, the amplitude of the leading peak (or dip) will not be affected by this phenomenon. Thus, the extent of inhibition can be measured correctly by measuring the amplitude from the baseline to the leading peak (or dip). (4) At 100% inhibition, the ratio of the fluorescence reading at the top of the leading peak to the baseline fluorescence reading is approximately equal to the substrate turnover ratio if the fluorescence reading is correctly calibrated. Thus, the substrate turnover in the Caliper's model assay is about 80 to 90%, which is far more than the acceptable turnover rate according to traditional enzyme kinetic theory, which requires all the experimental data being collected at an initial velocity (or less than 10% substrate turnover). This extreme substrate turnover condition in the Caliper's model assay was caused by the lack of understanding of why the peak-and-dip pattern appeared. The data obtained from the model assay were presented numerous times internally and externally and nobody realized that the substrate turnover was about 80 to 90% until this analysis.

14.3.2 Can the Chip-Based Assay Be Performed Consistently over Time for HTS?

Though we were able to correctly analyze the data by then, we did not know whether the assay could hold over an adequate period of time that was required for HTS applications. As seen from the above analysis, the kinase assay depended on the consistent separation between the kinase's substrate and product. The separation in turn depends on the stability of the surface of the microfluidic channel. Because it was desirable to eliminate EOF, the channels on the chip had to be modified to suppress EOF. There were several polymers known to suppress the EOF. Before I joined Caliper, a method to activate one of the EOF-suppressing polymers was already developed at Caliper that enables to covalently link the polymer to the silanol group on the surface of the channel. This modified polymer was used to modify the channels in all the kinase assay chips we discussed so far. It had been shown to work well in a short assay period and good data could be generated. However, it was not demonstrated whether the polymer could sustain repeated long run in typical HTS operations or not. To deliver a working assay to partners, we had to demonstrate that this assay could be carried out for the duration of at least a typical working day. In the mobility shift kinases assay, the flow rate in the main channel (determined by the applied vacuum if there is no EOF) and the voltage gradient determined the extent of the separation between the peak and the dip. A separation condition was predefined in the assay development stage and it depended on the surface condition when the assay development was performed with a particular chip. The separation condition was only good if the surface on the separation channel could remain the same during the screening period. On July 17, 2000, I performed an experiment to test the duration of the standard mobility shift assay with PKA. One microplate containing PKI in some wells were repeatedly assayed for about 6.7 h. The assay results changed continuously after running about 2 to 3 h. The early data (2600 to 5400 s) is shown in Figure 14.10 and the later data (after 6 h) is shown in Figure 14.13. By comparing the two figures, it is clear that there was a dramatic change in the pattern of the inhibition.

Because a voltage was applied between well 6 and well 8 on the chip and there was a current flowing between the two wells, electrochemical reactions should occur in them, leading to pH changes in the solution in the two wells. Initially, I suspected

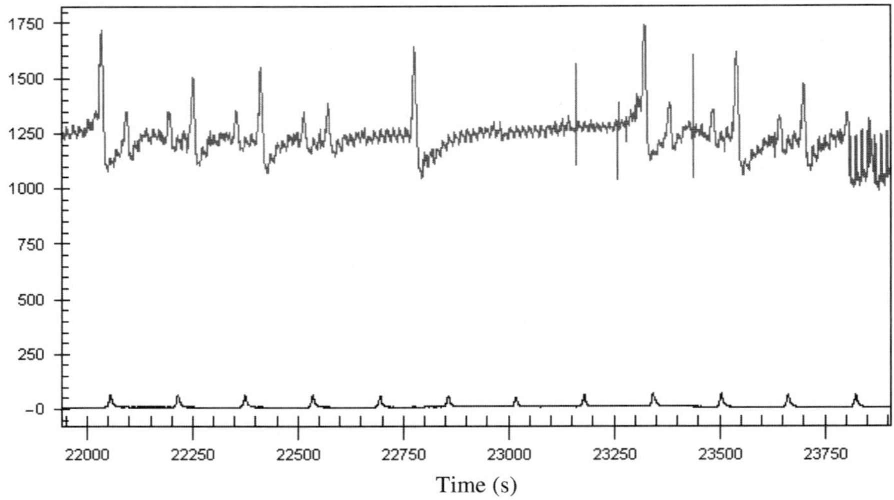

Figure 14.13 Distortion of the peak-and-dip pattern after continuous running of the mobility assays for more than 6 h as compared with the initial pattern shown in Figure 14.10.

that the pH in the wells containing electrodes might have changed dramatically and the solution with low pH from well 6 may facilitate the breaking of the covalent bond between the polymer and the silanol group on the surface of the channel. If the pH changed dramatically, the separation might be affected as well. After finishing this long run (close to 7 h) with constant electrochemical reactions taking place in the two wells, I tested the pH in every well on the chip. Due to the small volume of solutions in the wells (~50 μL at the start), narrow-range pH papers were used for the test. As expected, all the wells without electrodes inserted had the same pH (8) as they did at the start. However, the pH in well 6 changed from 8 to 7.5 while the pH in well 8 changed from 8 to 8.5. While the change was not dramatic and might not affect the assay results, this experiment indicated that the wells inserted with electrodes should have a large buffer capacity to reduce the change in the pH caused by electrochemical reactions. However, it was not realized in the early version of the chip design and all the wells were the same on the chip (see Fig. 14.2). In newer chip designs, the wells that take electrodes are much larger than other wells. After repeating several times the long-period assays, I confirmed this observation was true. However, I did not have a solution to this problem. I wrote on July 28 to the group the following (the identity of the polymer is changed to X to protect trade secret):

> Hi everyone:
>
> We have quite good understanding now with most of the puzzles presented in the last kinase meeting. The Major Problem Indeed Comes from unstable X coating. We also identified that the inhibitor PKI stick to the channel that have complicated the situation. This is what happened:
>
> X prevents both EO flow and PKI sticking to the wall. We based on this condition to adjust the pressure and the voltage to give us the best separation of product and substrate

> to obtain good signal shape and intensity. The X coating gradually getting eluted from the surface. Within 2 to 3 hours, there is still enough X to coat the surface and we obtain stable signals (intensity and shape). After 2 to 3 hours, significant amount of X has been eluted out. PKI started to stick to the wall that results in the reduced signal intensity. EO flow also takes effect now. The reaction time becomes shorter that results in decreased signal intensity too. The original condition that has been optimized to separate product and substrate becomes a bad condition now with the new surface. This reduced the signal and also changed the shape of the signal.

I also performed long-term studies of the PKA assay with chips with no coating on the surface of the channel. In this case, significant EOF was present and the vacuum and the voltage gradient had to be adjusted to find the initial separation conditions. The assay did not last much longer than the coated one. After running the assay for about 5 h, the data is not acceptable. I postulated that the bare surface of the channel might be continuously and slowly coated by some components present in the assay reagents (such as PKI, the PKA, and the substrate) that were absorbed to the surface of the channel. Indeed, a slow rising of the baseline was observed, indicating that Fl-Kemptide and PKA (turnover more substrate) might have been absorbed at the surface. I did not spend any further time on this problem because several other newly discovered issues hinted to me that the mobility shift assay might not work in the real world at all.

14.3.3 Can the Microfluidic-Based Assays Save Reagent More Than 1000-fold?

Microfluidic-based assays were often touted as the promising technology that could save the assay reagents on the order of 1000- to 10,000-fold. This claim was based on the fact that each compound is tested in a plug containing all the assay components and the volume of the plug is in the picoliter scale. In reality, the consumption of reagents in a microfluidic-based assay should be calculated based on the total reagents to be placed on the chip in the beginning of the screening minus any recoverable reagents at the end of the screening. A total number of the compounds assayed with these reagents are calculated and then how much the reagents will be used in a normal assay to screen the same number of compounds are calculated. The use of each reagent in the two-assay formats are then compared. Based on this method of calculation, there is substantial savings of reagents that are placed on the wells in the chip (it could be several hundred fold depending on the assay, which might not be more than 1000-fold) with nonenzymatic assays. There is no reagent savings for the compounds placed in the microplate because they are in matching buffer and are not recoverable after the assay though only minute amounts of compounds are sipped into the channel. For enzymatic assays, I found that there might not be any savings in enzyme in some situations (e.g., kinases). After a few months on the job, I was assigned to study whether we could develop an assay for an enzyme containing more than nine subunits. This project was from a partner company. The scientists there reconstituted many component proteins to obtain the functional enzyme in small quantities. The enzyme complex was very expensive to make. Thus, it is very difficult to run HTS with the enzyme using traditional microplate-based assays. The chip assay would make it possible because of the promised more than 1000-fold savings in reagents. Since I was busy with other projects, I asked my associate to give it a try first. The next day, he told me that he used up all the enzymes sent from the partner and was not able to continue developing the assay. I was shocked and took a closer look at what happened. After this investigation, I found that there might not be any savings in enzymes with the microfluidic-based assays. It might take more enzymes to develop an assay in microfluidic-based assay format than in the traditional assay format. In addition to the enzyme saving (wasting) problem, a few other fundamental problems with the chip-based enzymatic assays became obvious. Below I will use the kinase assay format to illustrate why there is hardly any savings in kinase when performing screening in microfluidic-based assay formats.

One of the most important properties of fluidic flow in micrometer-sized channels is the laminar flow. Because of laminar flow, there is no active mixing between streams of fluids flowing inside the channel. However, there is passive diffusion of the molecules in all directions. The diffusion can be desirable in some situations and undesirable in other situations. For example, the enzyme stream from well 4 and the substrate stream from well 5 in the chip for kinase assay (see Fig. 14.6) must be mixed for enzymatic reaction to take place. Because there is no active mixing, the interaction between the enzyme and the substrate depends solely on the diffusion. The time (t) for a molecule to diffuse distance (x) is approximately equals to $x^2/2D$ where D is the diffusion coefficient that depends on the viscosity of the solution and the size of the diffusing particle. The larger enzyme will not diffuse much across the channel during the assay. Thus, the enzymatic reaction depends on the diffusion of the smaller peptide substrate across the channel. This passive diffusion also makes it possible for the small molecule introduced through the sipper stream to rapidly diffuse across the channel to interact with the enzyme (within 2 s). However, diffusion also happens in the longitude. If a plug of a compound is sipped into the chip and parked there without fluidic flow, the plug will diffuse and disappear over time. Thus, the peak broadening in microfluidic-based assays is caused by both passive diffusion and Taylor dispersion (discussed in Chapter 4). Thus, the plug containing all the assay components, where the chemical reaction takes place, must rapidly flow through the channel for the reaction signal to be detectable. Otherwise, a broad smear of signal barely above baseline will show up at the detector. Thus, the NS75 chip designed for kinase assay has the total time of about 90 s for the plug moving from the tip of the sipper to the detector (~20 s to travel the length of the sipper, ~30 s to travel from the point that the enzyme channel intersects with the main channel to the start of the separation channel, and ~40 s to travel the length of the separation channel). This means only ~30 s of enzyme-catalyzed substrate turnover is allowed in the microfluidic-based kinase assay.

When discussing fundamental principles of assay development with isolated proteins in Chapter 3, I pointed out that one of the great advantages of enzymatic assay over a protein binding assay was the amplification of the signal through repeated turning over of the substrate to a large number of product by the action of enzyme. The total amount of final product generated (or the ratio between the final product and substrate in some assays) determines the strength of the final detectable signal, which is linearly proportional to the concentration of the enzyme and the time of incubation (assuming substrate is in large excess). Most enzymatic assays are performed at a low enzyme concentration (commonly in the nanomolar range). To obtain a robust signal, the limited number of enzymes must work more rounds to turn over more substrate. This requires a long incubation time between the substrate and the enzyme to allow more round of substrate turn over to compensate for the reduced enzyme concentrations. Using a low concentration of enzyme in an assay not only reduced the enzyme consumption but also allowed the enzyme to stay at a concentration lower than the concentration of substrate and inhibitor. This is an ideal condition that allows better interpretation of the data with existing enzyme kinetic theories. Thus, it is common for the enzymatic reaction to be incubated for many hours and even over night (about 15 h from 6 pm to 9 am) when low concentrations of enzymes are used in traditional microplate-based assays. However, only 30 s were allowed for the enzyme to work on the substrate in the mobility shift kinase assay in the microfluidic format. To compensate for the short incubation time to achieve the same amount of substrate turnover (e.g., 10%) as the normal kinase assay, the concentration of the kinase in the main channel in the chip with the microfluidic-based kinase assay must be increased accordingly. For example, if the incubation time between substrate and kinase in normal kinase assay is 5 h, $5 \times 3600/30 = 600$-fold higher concentration of the kinase must be present in the main channel to achieve the same amount of substrate turnover when all other assay conditions are the same. Looking back at the standard mobility shift assay protocol, it was clear why the original microfluidic-based kinase assay protocol used 100 nM PKA in the main channel while only subnanomolar of PKA was used in normal microplate-based PKA assays. Because there is an 8% dilution from the enzyme storage well on the chip to the main channel, the concentration of the kinase in the well must be $600/0.08 = 7500$-fold more concentrated than the kinase in a normal assay. On top of this, the microfluidic-based kinase assay requires a higher substrate turnover (~50%) to obtain a reliable signal, which is another factor of 5 higher than the common substrate turnover (<10%) in normal kinase assays. Thus, the concentration of kinase placed in the enzyme well on the chip should be about 37,500-fold higher than the concentration of kinase in a normal kinase assay with 5 h incubation to obtain comparable signals. The wells that store the enzyme on the chip have a capacity of about 50 μL. To run a screening, this volume must be filled so that the evaporation does not result in a significant change in the concentrations of the components in the enzyme solution. The microfluidic-based assay can usually sustain a continuous run within ~8 h in the simple fluorogenic assay. At the end of the 8-h screening, less than half of the initial volume in the enzyme well remains. The remaining enzyme solution is usually discarded because there is always change in the concentrations of the reagents in the solution due to evaporation and the enzyme's integrity may be compromised after sitting at room temperature for 8 h. Thus, I can assume the best scenario for the kinase assay will be a continuous run for 8 h if the kinase assay can be successfully developed. Based on these facts and assumptions, let's compare the overall kinase usage between the microfluidic-based assay and the traditional microplate-based kinase assay.

A hypothetical kinase is assayed with a final concentration of 1 nM in the traditional kinase assay format and the assay is performed in a 384-well microplate with an assay volume of 50 μL per well. It takes 5 h incubation time to obtain 10% substrate conversion to product. To assay the same kinase in a microfluidic format, the final concentration of the kinase in the wells on the chip must be 7.5 μM (7500-

fold higher based on the calculation in the paragraph above) to obtain the same amount of substrate turnover in 30 s. (I did not account for the fact that microfluidic-based kinase assay requires even more substrate turnover to make the argument more general since fluorogenic assays in microfluidic format do not require more substrate turnover.) Thus, a full kinase screening lasting 8 h in the microfluidic-based assay needs 50 μL × 7.5 μM = 0.375 nmol of kinase. We will calculate next how many samples can be tested in the microfluidic-based kinase screening and then calculate how much kinase is needed to screen the same number of samples in a normal kinase assay.

The maximum throughput of screening kinases in microfluidic format is dictated by the separation of the peak and the dip. Figure 14.12 showed that it took about 30 s (with the sample producing normal shaped peak and dip) for both the peak and the dip to pass the detector. Thus, 30 s per sample is the minimum assay time to avoid overlapping of signals from adjacent wells. However, 30 s per sample (equivalent to a throughput of 48 min per 96-well microplate) is too slow to carry a screen. With lower throughput, only a small number of samples can be assayed in 8 h, resulting in more usage of reagents (including kinase) per sample because whatever is placed on the chip wells are not recovered after the 8-h run. To increase the throughput, a speed of 10 s per sample was adopted in the microfluidic-based kinase assay format at Caliper. With 10 s assay time per sample, one nonsticky inhibitor will wipe out the signals from the next 2 samples. A sticky inhibitor will wipe out even more wells and those wells must be retested. Even if we allow this to happen, the total number of samples that can be processed in 8 h is only about $8 \times 3600/10 = 2880$. To assay the same 2880 samples in a normal 50-μL per sample assay in 384-well format with 1 nM final kinase concentration, it will require only 2880 × 50 μL × 1 nM = 0.144 nmol of kinases. If the normal assay only incubates the enzyme and substrate for 1.5 h instead of 5 h, the kinase usages between the two formats is about equal. Thus, there is hardly any savings in kinase in microfluidic-based assays. This argument can be applied to any enzymatic assays. The absolute amounts of enzymes required for the two assay formats are not important here. The major purpose of this calculation is to show a general way to calculate the enzyme usage based on what happens in the true operation mode. It is clear that the hypothetical 1000- to 10,000-fold of reagent saving does not apply for the enzyme because of the time scale mismatch between the two assay formats.

Now we can go back to see why all the enzymes in the partner's complex enzyme assay were used up before we even have an assay. The partner's microplate-based assay was done in 50 μL volume in 384-well microplate with the reaction time of ~3 h for substrate turnover. They gave us enough enzymes based on their microplate-based assay. In microfluidic-based assay, my associate had to raise the enzyme concentration until he could observe a signal. The chip he used allowed a longer incubation time between the substrate and enzyme (~60 s) and the dilution from the enzyme well to the main channel is about 15%. Based on my analysis, he should have used $3 \times 3600/60 = 180$ times higher concentration of the enzyme in the main channel to achieve the same amount of substrate turnover as that of the microplate-based assay to observe a comparable signal. This translates into $180/0.15 = 1200$ times higher concentration of the enzyme that must be placed in the enzyme well on the chip. In the assay development mode, there is no need to fill the whole enzyme well on the chip with 50 μL of enzyme. However, at least 10 μL of enzyme should be placed in the enzyme well to cover the hole leading to the channel to prevent air bubbles from entering the channel. Assuming the enzyme concentration is 1 nM in the partner's microplate-based assays, one test in the microfluidic-based assay will use 1200 nM × 10 μL = 12 pmol of enzyme. Further we assume that we can collect 20 data points in assay development with one concentration of enzyme on the chip before changing to next assay conditions. To test the same 20 samples when developing the normal enzymatic assay in microplate format with a final assay volume of 50 μL per sample, only 20 × 1 nM × 50 μL = 1 pmol of enzyme is required. Thus, a lot more enzymes were consumed in the assay development phase in the microfluidic format because of the requirement of the high concentration of the enzyme being placed in the enzyme well.

The more than 1000-fold increase in enzyme concentration in the microfluidic assay format have several negative side effects that made the assay results hard to interpret because the assay condition is out of range of the boundary conditions when the common enzyme kinetic equations were derived. As discussed in Chapter 3, the enzyme kinetic equations were derived based on the assumption that the enzyme concentration is much lower than the substrate concentration. While PKA is the most active kinase and the assay concentration for PKA was already at 0.1 μM in the microfluidic-based assay, the other less active kinase will be required at even higher concentrations (more than 1 μM) if they can be studied in this assay format. The substrate concentration cannot increase by the same magnitude to balance the increased kinase concentration because the assay signal depends on the percentage but not the absolute number of substrate turnover in microfluidic format. Higher substrate concentration will in turn require even more kinase to be able to convert 50% of substrate in 30 s. Thus, the high concentration of kinases may be close to the concentration of the substrate in mobility shift kinase assay. In addition to the fact that the substrate concentration may be close to the kinase concentration, the high kinase concentration may well exceed the K_i value of an inhibitor causing increases in apparent IC_{50} for the inhibitor. Since only one concentration of inhibitor is tested in normal screening, the shift of the apparent IC_{50} will lead to less inhibition when the inhibitor is fixed at the screening concentration. Furthermore, the screening concentrations of inhibitor are usually between 1 and 10 μM, which will be very close to the concentrations of less active kinases in microfluidic-based kinase assays. In discussing enzyme inhibition in Chapter 3, we set the boundary conditions that the enzyme concentration must be less than the inhibitor concentration and the K_i value. When these conditions are met, the apparent IC_{50} value should not change with different enzyme concentration (but will change with different substrate concentration with competitive and uncompetitive inhibitors). When the IC_{50} value changes with enzyme concentrations, the normal enzyme inhibition treatment will not work. The situation has to be treated as a tight binding situation with very potent inhibitors. In this situation, the apparent IC_{50} value will be simply close to half of the concentration of the enzyme. Thus, it becomes a less meaningful parameter, and it is hard to obtain K_i values from the apparent IC_{50} value obtained experimentally. This phenomenon can be illustrated when using the chip-based kinase assay to determine the IC_{50} values of PKI at three different PKA concentrations. The assay results are shown in Figure 14.14 where the IC_{50} values change with the PKA concentrations.

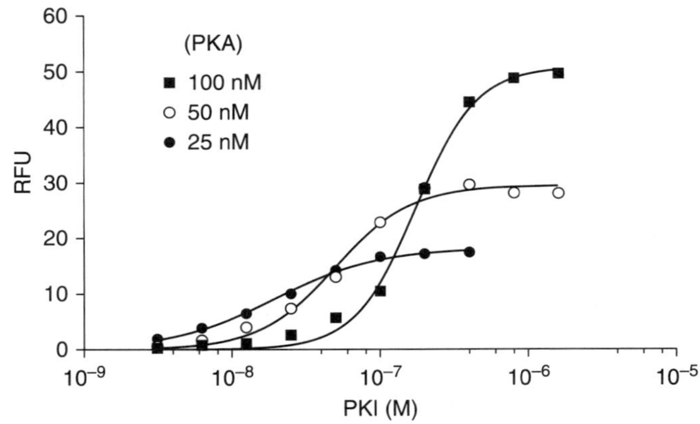

Figure 14.14 Determine the apparent IC_{50} values of PKI at different concentrations of PKA in chip-based assay. The concentration of PKA is 100 nM in the original PKA assay protocol to achieve the maximum signal. However, the apparent IC_{50} value increased to 174 nM. When the PKA concentration is reduced to 25 nM, the apparent IC_{50} value is reduced to 19 nM, which is still far away from the ideal situation because the K_i for PKI is about 5 nM. However, it is not possible to reduce the concentration of PKA further because the signal will be too weak to be reliably determined.

The above analysis of reagent usage in the microfluidic-based format is only applicable to the enzymes because the assay format takes away the advantage of a substantial amount of substrate turnover due to the short reaction time. There is substantial savings in the substrate in the microfluidic-based assay format though the savings may not be as high as the claimed 1000- to 10,000-fold savings. This is because the concentration of the substrate is the same (usually fixed at its K_m) in enzyme-catalyzed reactions no matter what assay format was used. The final savings in substrate in the microfluidic-based assay is on the order of 100-fold compared with a normal assay. Unfortunately, the savings in peptide substrates does not translate to meaningful overall savings in the assay because peptide substrates are cheap. The test sample usage is the same between the two assay formats. Though only picoliters of samples are actually introduced into the chip, at least 10 to 20 μL of samples must be present in the microplate to feed the sipper. Thus, there is no saving for the test sample either. The true reagent savings in microfluidic-based kinase assays is indirectly coming from the way the kinase was performed. In this rapid (30 s) separation-based assay, the only extra reagent needed to perform the assay is the labeled substrate. There is no need to use any other expensive reagents, such as antibodies, SPA beads, TRF reagents, and the like. After realizing this true advantage of the technology, a much better assay format without the many inherent problems discussed so far can be developed, which will be discussed later.

14.3.4 Is the Microfluidic Kinase Assay Applicable to Other Kinases?

Though the feasibility of the Microfluidic-based kinase assay was demonstrated to work with PKA and the performance of the assay could sustain in a short period (less than 2 h), I had doubts that the assay would be applicable broadly to other kinases besides PKA. After all, PKA is one of the most active kinases. Even with kinase as active as PKA, the original assay protocol used 100 nM PKA in the main channel, which is very high for an enzymatic assay. I have demonstrated that PKA can be reduced to 25 nM in the assay and a good dose–response curve can still be obtained. However, the signal is not very robust and in this situation the concentration of PKA cannot be reduced further. For less active kinases, it would not be wise to further increase the kinase concentration above 100 nM because of the issues of high concentration of enzyme that was discussed previously. Thus, if we limit the assay concentration of kinase to 100 nM, the kinases that possess less than one-fourth of the PKA's activity cannot be assayed with this assay format. I decided to search for available kinetic parameters of all kinases and to calculate how many of them can be assayed in the microfluific kinase assay format. In 2000, Upstate (now part of Millipore) had the most kinases in its collection. I searched its catalog and found 28 kinases with specific activity information (Table 14.1). However, there was no information about the K_m value for the substrate listed that was required for my calculation. Since the K_m value for all the kinase substrates (except the Kemptide substrate for PKA for which I knew the K_m to be $\sim$10 μM) are unknown, two scenarios are tested based on K_m of 50 and 10 μM. These two values were chosen based on my past experience that the K_m for a peptide substrate usually is in the range between 10 and 200 μM in kinase assays. The higher the K_m, the less favorable for the microfluidic-based kinase assay format. Thus, I picked the low end of the K_m value to give the assay a better chance for the broader applicability. I then calculated the maximum percentage of substrate conversion to product within 30 s of reaction time based on the reported specific activity. The concentration of the kinase was set at 100 nM and the substrate concentration was set at K_m. The 25% conversion is used as criteria to decide whether the kinase could be assayed or not. When assuming K_m equals to 50 μM for all the kinases, none of them can be assayed in the microfluidic assay format except PKA. When assuming K_m equals to 10 μM, only 5 out of the 28 kinases could be assayed. However, it is very unlikely for a peptide substrate to have such a low K_m value with these kinases. Thus, the chance for the microfluidic-based kinase assay to be applicable to other kinases is very slim. In addition, there is always an initial delay to generate product for many catalytic domains of receptor Tyr kinase after the kinases are mixed with their substrate (see Fig. 7.12). Those kinases cannot be assayed with the microfluidic-based kinase assay at all. I concluded that it would be very difficult to market the product with only limited use. Thus, it would not be wise to spend a lot of effort to further develop the existing kinase assay. An alternative strategy had to be developed to fulfill the partner's needs.

14.3.5 Can the Microfluidic Kinase Assay Detect the Same Number of Inhibitors in a Library?

There are many studies to compare different microplate-based assay formats for their ability to detect inhibitors from the same compound library. In general, there are always discrepancies between different assay formats and the reasons for the discrepancies are not fully understood. Thus, experimentally comparing the microfluidic-based kinase assay format with any of the many microplate-based kinase assay formats

TABLE 14.1 Calculation of Maximum Percentage of Substrate Conversion Based on Reported Specific Activity of Kinases[a]

Enzyme	MW	Substrate	Specific Activity (nmole/min/mg)	Turnover $K_{cat}(s^{-1})$	Percent Conversion ($K_m = 50\ \mu M$)	Percent Conversion ($K_m = 10\ \mu M$)
ATK1	60000	RPRAATF	30.75	3.08E−02	0.15	0.76
AMP-PK	136000	SAMS peptide, HMRSAMSGLHLVKRR	337	7.64E−01	3.68	16.03
CaM Kinase II[b]	50000	Auto Camtide II	370	3.08E−01	1.52	7.16
CaM Kinase IV	50000	Auto Camtide II			0.00	0.00
CHK1	80000	KKKVSRSGLYRSPSMPENLNRPR	155	2.07E−01	1.02	4.91
CK2	148000	RRRDDDSDDD	2578	6.36E+00	**24.12**	**61.39**
Fes (Tyr)	93000	Poly(GluTyr) 4:1	16	2.48E−02	0.12	0.62
Fyn (Tyr)[b]	50000	KVEKIGEGTYGVVYK	0.001	8.33E−07	0.00	0.00
GSK3 beta	50000	YRRAAVPPSPSLSRHSSP-HQ(pS)EDEEE	754	6.28E−01	3.05	13.58
Lck (Tyr)	56000	KVEKIGEGTYGVVYK	11.5	1.07E−02	0.05	0.27
Lyn (Tyr)	56000	KVEKIGEGTYGVVYK	0.12	1.12E−04	0.00	0.00
Erk1	44000	MBP	100	7.33E−02	0.37	1.80
Erk2	50000	MBP	1231	1.03E+00	4.88	20.41
MAPKAPK2	66000	KKLNRTLSVA	2853	3.14E+00	13.56	**43.96**
MSK1	95000	RPRAATF	150	2.38E−01	1.17	5.60
PKA[c]	40000	LRRASLG	12000	8.00E+00	**66.67**	**66.67**

PKC delta	79000	2066 dalton peptide	2300	3.03E+00	13.15	**43.09**
PRAK	54000	KKLRRTLSVA	207	1.86E−01	0.92	4.45
ROK alpha	66000	AKRRRLSSLRA	1867	2.05E+00	9.31	**33.92**
P70 S6 kinase	55000	AKRRRLSSLRA	48	4.40E−02	0.22	1.09
JNK2alpha2/ SAPK1a	51000	GST-AFT2(19-96)	97	8.25E−02	0.41	2.02
JNK1alpha1/ SAPK1c	43000	GST-AFT2(19-96)	99	7.10E−02	0.35	1.74
P38alpha/ SAPK2a	64000	MBP	38.4	4.10E−02	0.20	1.01
p38beta/ SAPK2b	65000	MBP	56	6.07E−02	0.30	1.49
p38gamma/ SAPK3	71000	MBP	116	1.37E−01	0.68	3.32
p38delta/ SAPK4	71000	MBP	31.3	3.70E−02	0.18	0.92
SGK	48000	GRPRTSSFAEG	833	6.66E−01	3.22	14.28
SRC	60000	KVEKIGEGTYGVVYK	0.6	6.00E−04	0.00	0.01

[a]The concentration of the kinase is set at 100 nM and the substrate concentration is set at K_m. Since the K_m value for all the kinases are unknown except the Kemptide for PKA, two scenarios are tested based on K_m of 50 μM (more realistic) and 10 μM (rare). 25% conversion is used as criterion to decide whether the kinase can be assayed or not. The kinases that fit the criterion are in bold face.

[b]MW is estimated.

[c]Known value of K_m for PKA of 10 μM is used.

will definitely lead to discrepancies. However, based on my understanding of the microfluidic assay format in late 2000, several shortcomings in the microfluidic-based kinase assay were obvious, and I concluded that the microfluidic-based assays would miss inhibitors in several ways.

One class of inhibitors—slow inhibitors—have a very slow rate to exert their effect on the enzyme. In the microfluidic assay format, there are only 2 s for the interaction between the liquid stream containing the inhibitor and the liquid stream containing the enzymes. The 2 s are allocated for both the diffusion of the inhibitor to the enzyme (about 1 to 2 s) and the onset of inhibition. Thus, there is little time left for the inhibitor to exert its function on the enzyme before the enzyme meets with the substrate to start enzymatic reactions. Though the inhibitor is still present and can still exert inhibition on the enzyme in the 30-s time for the enzymatic reaction, the 30-s time period is still relatively short for slow inhibitors. If the onset inhibition rate matches the 30-s scale, the inhibitor still will behave as less potent because of substrate turnover at the same time. Thus, the IC_{50} value for these inhibitors will increase. On top of the time constraint, the IC_{50} values of potential inhibitors are further increased because the kinase concentrations in microfluidic kinase assay formats are more than 1000 times higher than that in normal microplate-based assays. Thus, most inhibitors will appear less potent in microfluidic-based kinase assays. For example, the IC_{50} value was 174 nM for PKI when 100 nM of PKA was used in the microfluidic-based kinase assay (see Fig. 14.14) while the actual value should be $\sim$1 nM in the normal microplate-based assays. This is a more than 100-fold shift in IC_{50} values. Because PKI is very potent, its inhibition can still be detected in microfluidic-based assays though the IC_{50} value already shifted to 174 nM. For weak inhibitors with IC_{50} values in the submicromolar range, the microfluidic assay system will have a hard time to detect them when the IC_{50} values shifted more than 10-fold.

14.4 SEARCHING FOR ALTERNATIVE KINASE ASSAY METHODS

By the end of July 2000, I reached the conclusion that the microfluidic-based kinase assay would not work based on the experimental results and theoretical calculations. These issues are summarized here:

1. There is little, if any, savings in enzymes.
2. The channel surface stability issue prevents the assay from a continuous run longer than a few hours, and there was no solution to the problem at that time.
3. The assay is only applicable to a small number of kinases even if the assay can be developed.
4. Many false negatives will occur when the assay is applied to HTS.

The fundamental issue with the original microfluidic-based kinase assay is the mismatch of the assay time scale between the new assay and the assays it tries to replace. This mismatch is caused by the limitation of Caliper's technology that no valves can be placed on the chip to prevent fluidic plug diffusion and allow suitable time for incubation. The dramatically reduced incubation time, on the order of a few 1000-fold, in microfluidic-based assays forced the same magnitude of increase in enzyme concentration to obtain the same level of signal. This scheme takes away the key advantage of the normal enzymatic assays, that is, high substrate turnover per enzyme (or signal amplification). When discussing measurement in Chapter 2, I pointed out that when deciding a measurement method in a particular situation, two important aspects of the potential measurement must be considered: spatial resolution (the size of the probe or the resolution of the detection should not be close to or exceed the size of the object) and temporal resolution (the speed of the assay should match the speed of the object to be measured). The microfluidic-based assay was designed to solve a problem that happens in the scale of hours with a solution with the temporal resolution on the order of seconds. Thus, for the microfluidic assay to work, the long incubation time must be restored. A technology that can park the mixtures containing enzyme, substrate, and inhibitor for a controllable time without diffusion is required, which could not be done with Caliper's technology.

Is there something useful left in the microfluidic-based kinase assay? Though the original kinase assay that I inherited does not work, as a project manager I had to deliver some kind of kinase assays that can take advantage of microfluidic technology to Caliper's partners. After eliminating the perceived advantages of enzyme savings and ruling out the feasibility for the general use of the original kinases assay, I concluded that there was still a distinct feature of the technology: the rapid separation of the kinase product and substrate within 1 min after sample injection. This was the key competitive advantage of the technology compared with other existing kinase assays discussed in Chapter 7. All the existing homogenous kinase assays required special costly reagents, such as SPA beads, IMAP beads, antibodies, and ATP/ADP measurement reagents. These assays were either indirect or they only measured the product or the substrate. I reasoned that the Caliper chip could be used to separate the fluorescently labeled peptide substrate and product if the kinase assay was performed in normal microplate-based assays with no incubation time constraint. Though this scheme did not realize the full potential of the lab-on-a-chip scheme that was promised to the partners, it still had many advantages, especially that the product and the substrate can be observed directly.

There was little scientific risk in developing this assay. Separation of peptides electrophoretically in glass capillary was a well-known scientific practice. In fact, Cetek (Marlborough, MA) regularly performed kinase assays for customers by separating kinases product and substrate with capillary electrophoresis. In addition, we routinely separated Flu-Kemptide substrate and the product with the NS75 chip to determine the separation conditions before we proceeded to develop the chip-based assay (see Fig. 14.9). However, there are implementation risks to developing any marketable product. There is a huge gap between scientific feasibility and a marketed product. A marketed product has to meet stringent criteria, including broad applicability, reliability, and robustness. In comparison, only a few special situations were picked to show the feasibility in scientific feasibility studies. The original microfluidic-based kinase assay is a good example to pick a "one of the kind" kinase (PKA) to show the feasibility. Many new technology companies trying to move new scientific findings into marketable products failed because of the same reasons. At Caliper, there was a great resistance to the proposed change because it did not fit the original idea when the company was found. However, there was nothing else that could be done in order to deliver the kinase assays except the one I proposed. Thus, I was finally allowed to develop the new assay but with limited resources. The major resources at Caliper were still devoted to the assays that supposedly could perform in the lab-on-a-chip. Because the original kinase assay was called a "mobility shift assay," the new assay I proposed needed a different name to distinguish it from the original assay. Though the new assay was also based on mobility shift, I could not use the same assay name. Thus, I started to call the original kinase assay as "on-chip kinase assay" and the new assay as "off-chip kinase assay." The somewhat awkward "off-chip kinase assay" name thus continues today. One regret I have now is that I should have given it a better name.

Though the microfluidic-based enzymatic assay for kinases encountered serious problems, the technology still has advantages over traditional assays in special situations. For example, the caspase-3 fluorogenic assay (see Chapter 6) is a good example that the technology can be used successfully. The microfluidic-based assay worked well in this case because and fluorogenic assay format and the fast kinetics of caspase-3. Fluorogenic assays do not require a certain percentage of substrate conversion to product. As long as enough absolute amounts of fluorescent product are generated (above certain levels of the detector's detecting limits), the assay will work fine. For example, 20 nM of 7-amino-4-methyl coumarin (AMC) can be easily detected with modern fluorimeter. Thus, when 4 μM fluorogenic substrate is used, substrate conversion of only 0.5% will generate 20 nM AMC to give a strong signal. In comparison, the microfluidic-based kinase assay required at least 50% substrate conversion to the product that is 100-fold higher than the example in the fluorogenic assay. In addition, caspase-3 is so active that only 3 nM caspase-3 converted almost all 4 μM fluorogenic substrate into fluorescent product in 10 min. Thus, the microfluidic-based fluorogenic assay only needs to use 1 nM caspase-3 in the main channel (see Fig. 14.5). In this special situation, the microfluidic-based assay works well and realizes the dream of lab-on-a-chip without the need to use any liquid transfer instruments in the assay (except compound plate preparation). However, the fluorogenic assays can be conveniently performed in microplate-based assay format, and there are not many incentives for the user to switch to microfluidic-based assays just for fluorogenic assays. The cost of Caliper's instruments and the steep learning curves worked against the wide adoption of the technology unless it can handle some special assays (such as kinase assays) that could not be done easily with conventional methods.

14.5 DEVELOPMENT OF THE OFF-CHIP KINASE ASSAY FORMAT

After analyzing the situations to find critical issues that should be resolved to deliver the final product, I narrowed them down to the following critical items:

1. Spectra separation of the substrate and product
2. How to calculate the kinase inhibition

3. How to assign the region of the time-based signal trace to the physical position of the test sample in the microplate
4. Stability of the surface on the separation channel
5. Extent of substrate turnover.

14.5.1 Separation of the Substrate and Product and the Calculation of Inhibition

I decided first to run PKA assays in 96-well microplates with a few sample wells containing inhibitors to get a feel of what the data would look like in off-chip kinase assays. The concentration of Fl-Kemptide was set at between 1 and 5 μM. The concentration of PKA was varied to obtain substrate conversion at about 50% in 1 h. In general, the final PKA concentration was in the subnanomolar range. Figure 14.15 shows the off-chip kinase assay data of 10 samples in the first row in a 96-well microplate. Three of the samples contained inhibitors (83 μM staurosporine, 166 μM staurosporine, and 10 mM EDTA). After reaction, the 10 samples were sequentially introduced into the chip where the substrate and the product were separated. A new chip (SP216) specifically designed for off-chip kinase assay was used. The data were collected on August 25, 2000. In this experiment, the polarity of the voltage gradient was reversed from the experiments discussed before, such as the one shown in Figure 14.9. Thus, the product was eluted to the detector before

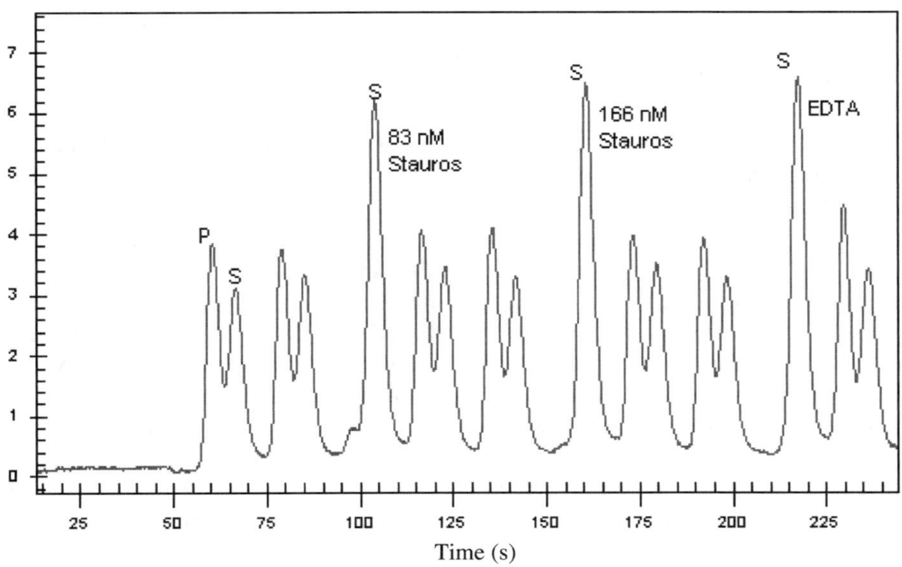

Figure 14.15 Pattern of 10 PKA reaction mixtures separated with off-chip kinase assay. The same concentration of PKA substrate (Fl-Kemptide) was mixed with the same concentrations of PKA in 10 sample wells in a row in a 96-well microplate. Two of the wells contained staurosporine at 83 nM and 166 nM and one of the wells contained 10 mM EDTA. The reaction is allowed to proceed until slightly more than 50% of substrate is converted into product. The 10 reaction mixtures were then sipped into the microfluidic chip and separated.

the substrate. The experiment was designed so that the substrate and the product peaks from each well were only partially separated to form doublet peaks, and the doublet peaks from each well were separated from adjacent wells with a relatively long buffer injection time between samples. I decided to use this condition as the starting point to establish the assay. I reasoned that the separation between the substrate and product peaks should be small to increase the throughput. Otherwise, longer buffer time will be required to fully elute the lagging peaks before the next sample can be injected and decrease the throughput. Longer elution time would also cause broadening of the lagging peak and affecting the accurate quantitation of substrate or product in the assay. Allowing overlapping of substrate and product peak would increase the robustness of the assay because some peptide substrates and their products may not be fully separated at all in the chip. The disadvantage of allowing overlapping peaks was that one huge peak would affect the adjacent small peak, resulting in high uncertainty in the quantitation of the small peak. Thus, this strategy required the two peaks to be separated at about similar height (i.e., ~50% substrate turnover).

Even with significant overlapping of substrate and product peaks as shown in Figure 14.15, it still took 20 s for the two peaks to elute. Without a good algorithm to define a time region to a well on the microplate, I had to rely on a relatively long buffer time between samples so that each doublet peak is clearly separated from the others. Otherwise, continuous peaks one after another would appear in the trace. If samples were injected every 20 s, the maximum throughput of the assay with this scheme would be $20 \times 96/60 = 32$ min per 96-well microplate. By that time, a chip having 4-sippers was already in development at Caliper. With 4-sipper chips, the throughput of the assay could reach a maximum of 8 min per 96-well microplate, which is acceptable. With this assay scheme, the baseline between the substrate peak and the product peak within a sample well cannot be determined, and thus it would not be possible for this assay to accurately quantify the area under the peak. Thus, I decided to use the peak height to estimate the quantity of each substance. The peak height is measured at the middle of the peak area and is less affected by adjacent overlapping peaks. The measurement of peak height also allows the maximum reduction in the buffer time between samples and allows further increases in the throughput of the assay. I also decided to use the sum of the peak heights of the substrate and product as the total fluorescence signal so that the percentage of the product can be calculated by dividing the height of the product by the sum of the height of the substrate and the product. The use of percentage instead of just the height of the product peak (or substrate peak) increased the robustness of the assay. However, this treatment required that the product peak and substrate peak should be close to each other. Otherwise, substrate conversion would be significantly underestimated when the large separation of the product and substrate peaks causes significant broadening of the lagging peak and reduction in its height. The following email sent to the software programmer working with me on the project on September 1, 2000, summarized my thoughts.

The data processing requirement:

1. No marker dye is needed.

2. Search from the beginning of the file and use the buffer time period as the baseline.

3. Each well contains two peaks. The first is the product. The second is the substrate.
4. The time for each well is about 20 s for this 216 chip. In the future with better chip design, we hope to lower this to 10 s.
5. Compute two values for each well: sum of the product peak + substrate peak and the ratio of product peak/sum.
6. Take the median of the ratio for each plate (or cycle in this case) as 100% control.
7. Divide the ratio from each well by the median and report this value for each well as % reduction of activity.
8. Compare the sum of each well with the median. If the sum of a well is >2 times the median, mark that well as possible fluorescence compound interference.

14.5.2 Assigning the Region in the Time-Based Trace to the Physical Location of the Sample

In the above email, I decided to eliminate the marker dye in this assay. There was a history at Caliper to include a marker dye in all the assays because the requirement of the on-chip assay format and many people demanded to use the marker dye in the off-chip assay. However, I did not see a need for a marker at all because every assay well contains at least one peak, and this information can be used to define the boundary region of the well on the time-based trace. Since the signal is from the assay well directly, it provides more accurate information than using a reference marker dye that migrates differently from the samples in the assay well. In addition, there are several issues with using marker dyes at Caliper that I discussed before. The marker will require another fluorescence channel in the instruments and increases the cost. Caliper's instrument with three lasers was selling at about $300,000. The major reason for three lasers was that the on-chip assays (fluorogenic, kinase, and cell-based) required them. Based on my analysis, those assays have a slim chance to become acceptable products in the market. Thus, I proposed to build a new instrument to handle off-chip kinase assays with only one laser. I even thought of a name, "kinase analyzer," for this instrument. This instrument could be sold at a low price of under $100,000 because only one laser is required. The cost of the instrument could be further reduced if we use the red pointer laser, which cost about $100 then and only costs a few dollars now. To use this instrument, customers should switch from fluorescein-labeled peptide to a red-dye-labeled peptide. There was no particular reason for scientists to label a peptide with fluorescein. Red dye will be as good as, if not better than, the fluorescein as a label. Unfortunately, this suggestion was not adopted at Caliper and the majority of the effort in the company was spent with the on-chip assay concept.

In my initially proposal outlined in the preceeding email, I was thinking to use a combination of the shape (duplicates in the absence of inhibitor) and the total time for each sample (sample injection time + buffer injection time) to define the region in the time-based trace to specific wells in the microplate. This algorithm required relatively long buffer time between samples so that the doublet peaks can be clearly identified. However, the throughput would be low when long buffer time between samples was used. If the buffer time was reduced, we would see a series of peaks one next to another all over the graph and would not be able to visually identify which time region is from which well in the microplate. I finally thought of a new robust algorithm that used the peak of the substrate alone as the marker and the marker was identified not based on the wavelength of fluorescence but on the amplitude of the peak by injecting the substrate at a concentration that was twofold more concentrated than the substrate concentration used in the assay. Thus, the real substrate served the same function as the marker dye in the on-chip assay format. Using the amplitude to identify the marker dye instead of using a separate fluorescence channel with a foreign marker dye had many advantages. Because the marker dye was the same molecule as the molecules to be separated and they are in the same fluorescence channel, there was no need for two expensive fluorescence channels and no need to align the marker dye. In this well assignment scheme, substrates alone at high concentration were injected periodically to the chip between assay samples. The region defined by the two high concentration substrate injections was equally divided by the total number of samples. The location of each sample on the trace was then assigned by their sequence of injection to the chip. This algorithm was adopted in the final off-chip kinase product.

14.5.3 Stabilizing the Surface on the Channel During the Assay Period

One of the major obstacles in implementing the original on-chip kinase assay was the instability of the coating of the surface on the channel. The covalent modification of the channel by a polymer could only maintain the stability of the channel for a few hours. Beyond that period, loss of coating caused the collapse of the separation between product and substrate. We had tried several different polymers with noncovalent coating too. None of them worked beyond a few hours. Finally, I thought it might work if the nonactivated polymer is included in all the assay wells. I did not think in this direction in on-chip assays because the constant presence of the polymer in the main channel might affect the enzymatic reaction that took place in the main channel. The situation was different with off-chip kinase assays because all the reactions were stopped already before the samples were introduced to the chip for analysis. It would be fine to add other reagents to the stop solution to help analyze the final product. To distinguish this coating scheme from the old covalent coating, this scheme is called "dynamic coating," because the polymer was continuously coated onto the surface of the channel and was also continuously eluted out from the surface. For initial proof of the concept studies, I used high concentrations of the polymer in a microplate. After about 7 h of continuous run, the separation times between the substrate and product were quite stable. However, the separation still changes gradually at a much a slower rate. Since we know that the pH in the two wells on the chip where electrodes were inserted changes after long electrophoresis, the instability of the separation between product and substrate could be attributed to the pH effect. By October 27, 2000, we obtained relatively stable off-chip kinase assay conditions. One experimental data is shown in Figure 14.16. In this experiment, the separation between the substrate peak and the product peak over time is measured. The separation time is relatively stable at $\sim$8 s. There was some drifting but it was within 1 s. The current flowing between the two electrodes was monitored as well, which continuously drifted downward. Since the voltage applied between the two wells was constant throughout the experiment, the electric resistance between the two wells should have changed. Thus, the drifting in separation

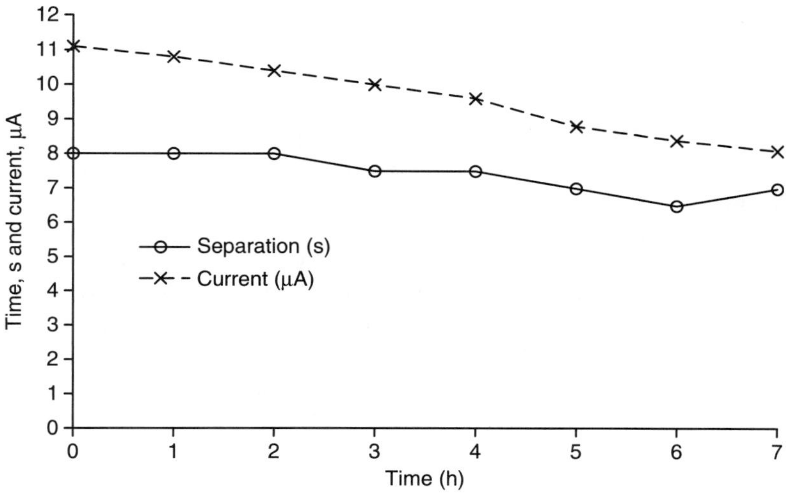

Figure 14.16 Stability of the off-chip kinase assay over time. The separation between the substrate peak and product peak is measured. The separation time is relatively stable at ~8 s. There is some drifting but was within 1 s. The current flowing between the two electrodes are monitored as well, which continuously drifted downward. The voltage applied between the two wells is constant throughout the experiment.

might be caused by the change in the composition of the chemicals in the two wells. Most likely the pH changes in the two wells caused the changes in the resistance. Changing the buffer in the two wells after an extended run could restore the separation of the two peaks. In the newer generation of chip design, relatively larger wells with more capacity to buffer the pH changes were placed on the chip to receive electrodes.

14.5.4 Dealing with High Substrate Turnover

Traditional enzymology theory was derived based on the initial velocity that dictates that the substrate turnover cannot be over 10% when performing enzyme kinetic studies to extract correct kinetic data. If we followed this rule, the substrate peak would be 9 times higher than the product peak. Because we had to allow overlapping between the substrate and product peaks to maintain adequate throughput, the large substrate peak would overwhelm the much smaller product peak, which was not acceptable. For the off-chip kinase assay to work, the substrate turnover should be at about 50% so that the two peaks were about equal. I reasoned that the unusually high substrate turnover might be fine for high-throughput screening since the exact kinetic parameter was not important in the screening mode. Instead, the ability to find any inhibitor was more important. An assay with 50% or more substrate turnover will have the same ability to find an inhibitor if the IC_{50} value of an inhibitor obtained in the assay is similar to the standard assay with substrate turnover at less than 10%. From my past experience, I did not see much change in the IC_{50} value when the substrate turnover exceeds 10% but below 50%. In fact, many commercial assays must be performed at substrate turnover well over 50% to obtain an acceptable signal (e.g., IMAP kinase assays discussed in Chapter 7). However, it was not obvious to the end users that many assays they were using had more than 50% substrate turnover. The same is true for the on-chip kinase assay in which the substrate turnover was well over 50%. However, end users and even the people who developed the on-chip kinase assay might not be aware of the fact. The situation would be different for the proposed off-chip kinase assay because the peaks of both substrate and product were clearly displayed. It would be difficult to promote the assay to people who had basic knowledge of enzymology. Thus, I had to develop a new theory that clearly demonstrated the effect on an inhibitor's IC_{50} value when the substrate turnover was higher than 10%.

Since most enzymatic assays were performed in first-order or pseudo-first-order kinetic conditions, I derived the equations based on first-order kinetics and with the condition that there was only one inhibitor's binding site on the enzyme. In Chapter 3, we discussed that the first-order reaction progressive curves could be described as

$$[\mathrm{P}] = [\mathrm{S}]_0(1 - e^{-kt}) \tag{3.18}$$

Thus, the fraction of substrate turnover (x) at time t with different value of k_i (represented different levels of inhibition) could be obtained by

$$x = \frac{[\mathrm{P}]}{[\mathrm{S}]_0} = 1 - e^{-k_i t} \tag{14.1}$$

At different inhibition levels, a series of reaction progressive curves could be drawn based on Eq. (14.1) with different k_i values as shown in Figure 14.17. When there was no inhibition, $k_i = k_0$.

The fraction of enzyme activity at time zero (no substrate turnover) was defined as

$$f_b = \frac{k_i}{k_0} \tag{14.2}$$

The fraction of enzyme activity when the enzyme was inhibited by an inhibitor with only one binding site on the enzyme was also defined by Eq. (3.39) in Chapter 3:

$$f_b = \frac{\mathrm{IC}_{50}}{\mathrm{IC}_{50} + [\mathrm{I}]} \tag{3.39}$$

Combining Eqs. (14.2) and (3.40), we obtained Eq. (14.3):

$$\frac{k_i}{k_0} = \frac{\mathrm{IC}_{50}}{[\mathrm{I}] + \mathrm{IC}_{50}} \tag{14.3}$$

Similarly, we could obtain the same equation as Eq. (14.3) for a situation at other time (t) in the reaction progressive curve when there was substrate turnover as

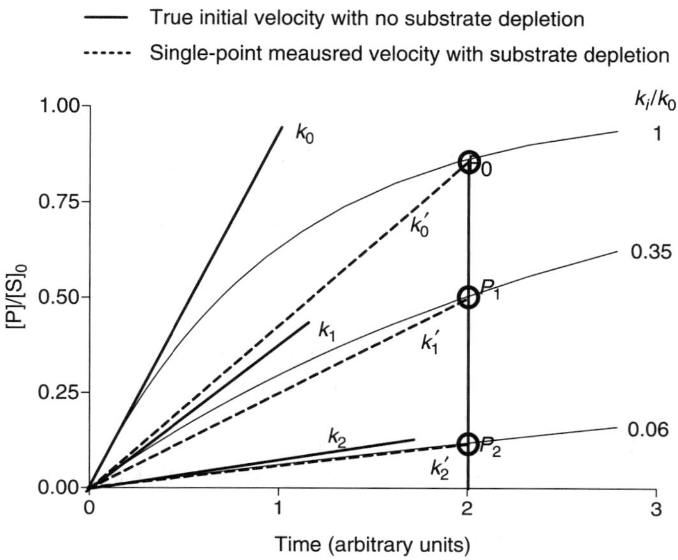

Figure 14.17 Illustration of the deviation of measured velocity from true initial velocity at different conversion levels. Simulated reaction progression curves in the absence of inhibitor and at two different inhibition levels are shown as the curved lines. The true rate constants obtained from initial velocity measurement are represented as the slope of the solid lines, which are labeled as k_0 (in the absence of inhibitor), k_1 (65% inhibition), and k_2 (94% inhibition). k_0, k_1, and k_2 can be obtained by measuring the reaction progression curve and then fitting the curve to Eq. (3.13). With single-point measurement at time $t = 2$ when significant substrate depletion occurs, the measured rate constants are represented by the slope of the dashed lines which are labeled as k_0' (in the absence of inhibitor), k_1' (65% inhibition), and k_2' (94% inhibition). The fraction of converted product at this time with three different inhibitions at labeled P_0, P_1, and P_2.

shown in Eq. (14.4):

$$\frac{k_i'}{k_0'} = \frac{\mathrm{IC}_{50}'}{[\mathrm{I}] + \mathrm{IC}_{50}'} \tag{14.4}$$

From Eq. (14.1), time t could be replaced by a fraction of the substrate turnover x in the absence of inhibition as shown in Eq. (14.5):

$$t = \frac{\ln(1 - x)}{-k_0} \tag{14.5}$$

When there was no inhibition,

$$k_0' = \frac{[\mathrm{P}]_0/[\mathrm{S}]_0}{t} = \frac{x}{t} = \frac{-xk_0}{\ln(1 - x)} \tag{14.6}$$

When there was inhibition,

$$k_i' = \frac{[\mathrm{P}]_i/[\mathrm{S}]_0}{t} = \frac{1 - e^{-k_i t}}{t} = \frac{1 - e^{(k_i/k_0)\ln(1-x)}}{\frac{\ln(1-x)}{-k_0}} \tag{14.7}$$

Substituting Eqs. (14.6) and (14.7) into Eq. (14.4), we obtained

$$\frac{1 - e^{(k_i/k_0)\ln(1-x)}}{x} = \frac{\mathrm{IC}_{50}'}{[\mathrm{I}] + \mathrm{IC}_{50}'} \tag{14.8}$$

Because the total number of independent equations were less than the total number of variables, there was no solution to this problem. I had to rely on numerical solutions to obtain the ratio of $\mathrm{IC}_{50}'/\mathrm{IC}_{50}$. However, all my simulated data and experimental data suggested that there should be a solution to the problem. I asked several friends majoring in mathematics and physics and they could not solve the problem either. After struggling for half a year, one day I suddenly realized that the problem could be solved by applying the known condition, that is, when $[\mathrm{I}] = \mathrm{IC}_{50}$, $k_i/k_0 = 0.5$. Substituting these two values into Eq. (14.8), I obtained

$$\frac{1 - e^{0.5\ln(1-x)}}{x} = \frac{\mathrm{IC}_{50}'}{\mathrm{IC}_{50} + \mathrm{IC}_{50}'} \tag{14.9}$$

Rearranging Eq. (14.9), I obtained

$$\frac{\mathrm{IC}_{50}}{\mathrm{IC}_{50}'} = \frac{x}{1 - e^{0.5\ln(1-x)}} - 1 \tag{14.10}$$

Equation (14.10) clearly demonstrated the relationship between the IC_{50}' value (single point measured along the reaction progressive curve with a certain level of substrate conversion) and the IC_{50} value (measured by initial velocity in the absence of substrate conversion) under the same assay conditions. This relationship was established based on the assumption that the substrate depletion follows first-order kinetics. Equation (14.10) could be graphically represented as shown in Figure 14.18. When keeping the substrate conversion at less than 10%, the IC_{50}' value was only expected to be shifted to $1.05 \times \mathrm{IC}_{50}$, which was close enough to the IC_{50} value. At 50% substrate conversion, the IC_{50}' value was expected to be shifted to $1.4 \times \mathrm{IC}_{50}$. Using Eq. (14.4), we could transform the effect of the IC_{50} change to the percentage of inhibition changes for an inhibitor that gave 50% inhibition in initial velocity studies. The results were summarized in Table 14.2. At low substrate conversion, the measured percentage of inhibition became lower gradually as the substrate conversion increased. At higher than 80% substrate conversion, the IC_{50}' deviated significantly from the true IC_{50} value, and the percentage of inhibition for a 50% inhibitor would be reduced to below 31%.

The above model was developed based on an enzyme reaction that had a reaction progressive curve following first-order kinetics. With enzyme reactions that had substrate concentration less than the K_m, the reaction progressive curve should be close to first-order kinetics. Inhibition of PKA by PKI using 5′-FAM-Kemptide as a substrate

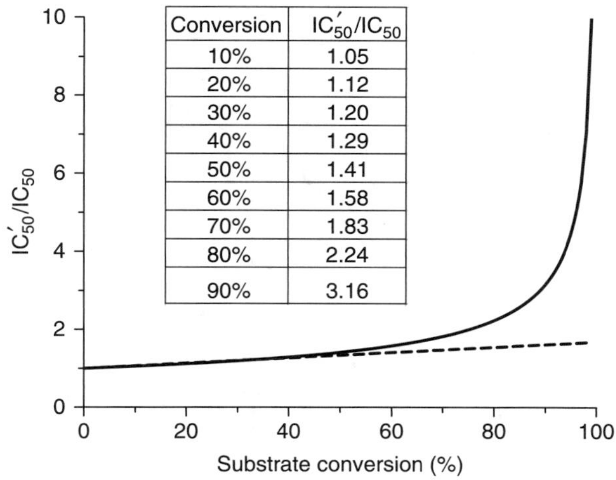

Figure 14.18 Graphical representation of the relationship between single-point measured IC'_{50} value and the percent of substrate conversion. This graph is based on Eq. (14.10). The IC_{50} is the inhibitor concentration at which 50% of reduction in rate constant is observed when the rate constant is measured by initial velocity with no substrate conversion to product. The inserted table lists theoretical ratio of IC'_{50}/IC_{50} at different substrate conversions.

TABLE 14.2 Calculated Percent Inhibition Based on Eq. (14.4) at Different Substrate Conversion Levels for Inhibitor That Gives 50% Inhibition in Initial Velocity Studies

Substrate conversion (%)	0	10	20	30	40	50	60	70	80
IC'_{50}/IC_{50}	1	1.05	1.12	1.2	1.29	1.41	1.58	1.83	2.24
% Inhibition	50	49	47	45	44	41	39	35	31

at a concentration less than its K_m was used as an example to examine the above analysis. The assay was performed at substrate concentration of 1 μM, less than its K_m of ~10 μM. The ATP concentration was set at 10 μM, which is 10 times higher than the substrate concentration, to obtain a pseudo-first-order reaction condition. The K_m for ATP is ~3 μM. The phosphorylation reaction at different PKI concentrations and different times was measured and plotted as shown in Figure 14.19. The 12 data points at different reaction times and at 8 different PKI concentrations were fit to Eq. (3.18) to obtain the reaction progressive curves. To obtain the IC_{50} values of PKI at different substrate conversions, the 8 fitted reaction progressive curves at different PKI concentrations obtained in Figure 14.19 were used to calculate the product generated at different substrate conversions. The details are not presented here (interested readers can find the information in the Bibliography) but the results are summarized in Table 14.3. The experimental results for PKI inhibition of PKA agreed very well with the results calculated from Eq. (14.10). This theory could guide both primary screen design and follow up IC_{50} studies when the system to be

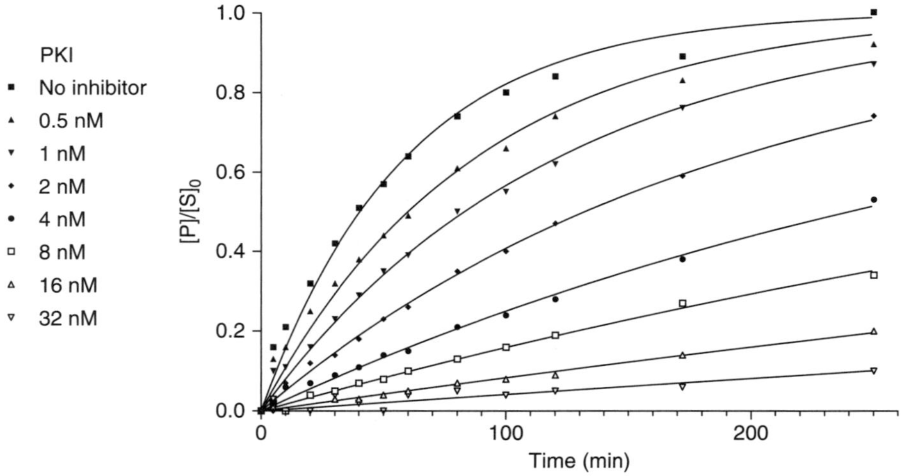

Figure 14.19 Measured reaction progression curves for PKA-catalyzed phosphorylation of fluorescently labeled substrate at different level of inhibition by PKI. The reaction was carried out with 0.2 nM PKA, 1 μM 5′-FAM-Kemptide, and 10 μM ATP in the presence of 5 mM $MgCl_2$ at pH 7.5. PKI at concentrations of 0, 0.5, 1, 2, 4, 8, 16, and 32 nM were present in different wells in the assay. The reaction progress was measured at different time and plotted. The data points were fitted to Eq. (3.13) to generate the reaction progression curve.

TABLE 14.3 Comparison of the IC'_{50}/IC_{50} Ratio Obtained Experimentally from the Inhibition of PKA by PKI at Different Substrate Conversion with Ratio Calculated from Eq. (14.10)

Substrate conversion (%)	0	10	20	30	40	50	60	70	80
IC_{50} (nM) for PKI	0.81	0.83	0.88	0.95	1.03	1.12	1.25	1.43	1.71
Measured IC'_{50}/IC_{50}	1	1.02	1.09	1.17	1.26	1.38	1.54	1.76	2.11
Predicted IC'_{50}/IC_{50}	1	1.05	1.12	1.2	1.29	1.41	1.58	1.83	2.24

studied met the conditions on which the equations were derived. For example, in typical primary screens with compound screening concentration at 10 μM, 50% substrate turnover will have little effect on the ability to detect weak inhibitors (with IC_{50} close to 10 μM) if the hit identification criterion is set at >50% inhibition. However, if substrate turnover is at 90%, the same inhibitor's apparent IC'_{50} would change from 10 to 30 μM, and this inhibitor would not be picked as a hit. For IC_{50} studies, in order to obtain robust signals, almost any substrate turnover (<90%) could be used in the experiment as long as the level of the substrate conversion is known and the reaction progressive curve had been demonstrated to follow first-order kinetics. The true IC_{50} value could be calculated from the measured IC'_{50} value using Eq. (14.10).

14.6 CURRENT STAGE OF MICROFLUIDIC TECHNOLOGY IN BIOASSAYS

After successfully solving many problems encountered, the off-chip kinase assay was in good shape to become a commercial product. In February 2001, we made a presentation to a large audience in the company. Here I attach the first and the last of my PowerPoint presentation (with sensitive information deleted in parenthesis).

First Slide:

HISTORY OF THE OFF-CHIP MOBILITY SHIFT ASSAY

- Contract with (CompanyX) to evaluate (x) kinase targets with on-chip mobility shift method for high throughput screening.
- However, peptides derived from kinase's target proteins usually are not good substrate for kinase (low k_{cat} and high K_a).
- Therefore, there are no kinases out of (CompanyX)'s potential kinase targets can be assayed with on-chip mobility shift method.
- After investigating several alternative methods, one approach works the best: Do the assay on plate and use chip as a device serving only for separation of product and substrate. Though this method does not take the full advantage of LabChip technology, it still has several advantages over existing kinase assay format.
- Project planning started in July, 2000 (Proving principals, throughput, data handling, persuade (CompanyX) to accept this alternative assay, etc.).

Last slide:

SUMMARY AND FUTURE DIRECTIONS

- Off-chip mobility shift assay offers a very unique method to assay kinases.
- Due to the simplicity, robustness, and general application to both Ser/Thr and Tyr kinase, it can be used as a gold standard assay for all the kinases with known peptide substrate.
- Should aggressively market this assay to all the partners and other potential users.
- Sell stand alone kinase analyzer at ~\$60K?

Due to internal management changes, I handed over this project to a new project manager a few weeks after this presentation. My supervisor spun off a new company from Caliper a few months later, and I went with him to the new company. At the time I was busy developing the off-chip kinase product, I had several concerns. One concern was the existence of technologies that could do the same thing cheaper than Caliper. Another concern was the high cost associated with both the Caliper instrument and the chip. While there was no solution to reduce the cost of manufacturing the chip (with my limited knowledge in this area), I really wanted to develop a cheaper version of the instrument, the kinase analyzer, with a price tag at about $60,000 instead of the $300,000 price tag for the Caliper instrument. This idea was not adopted by the old Caliper management and the off-chip kinase assay was not aggressively marketed either. A lot of effort and money were pouring into the on-chip assay ideas. In 2003, Caliper purchased Zymark, and the Zymark management team was retained to manage the company. A few years ago, a stripped-down version of the original Caliper instrument was marketed that was sold at a much lower price than the original instruments. Fortunately for Caliper, no competing cheaper instrument emerged during the lost time. However, those years are very crucial because kinases proved to be very valuable cancer targets. There was a great demand for kinase assays in those years. The off-chip kinase assays could have grabbed a much higher market share had Caliper developed the $60 k version of the "kinase analyzer."

Caliper was not the only company that faced so many challenges to develop a marketable product based on microfluidic technologies. Since the hype started by Caliper in the late 1990s, many competing companies in the field (e.g., Aclara Biosciences) have disappeared. Caliper can be considered a success story because of the DNA and protein sizing products plus the off-chip kinase product in the market. These successful applications do not depend on long-time incubations on chip, that is, no mismatch in temporal resolution. For the lab-on-a-chip idea to be applicable to bioassays, especially in enzymatic assays, a new technology must be developed to place a valve on the chip to allow long incubation time. A search on the Web can only find a handful of companies that are currently making commercial products based on microfluidic technology for applications in bioassay. Interestingly, all these products contain a valve in the channel to allow incubation. Gyros markets Gyrolab CD products for immunoassays. The technology is based on a compact disc (CD) fluidic platform in which the liquid is driven by centrifugal forces. A Gyrolab CD consists of 96 or 112 individual microstructures, which are processed in parallel, under the control of Gyrolab Workstation. Each microstructure contains a column prepacked with streptavidin-coated beads that are activated with biotinylated capture reagents. Capillary force is used to draw liquids into a distribution channel, filling a volume definition chamber. A hydrophobic break is used to serve as a valve to prevent the liquid from moving further into the microstructure. Spinning the CD generates centrifugal force, causing the distribution channel to empty, leaving behind a defined liquid volume. A second spin, at a higher speed, creates a *g*-force sufficient to drive the liquid over the hydrophobic break and through the capture column. Predetermined spin speeds generate the exact flow rate required for each step to capture the maximum amount of protein. SpinX technologies use the Virtual Laser Valve (VLV) that enables the SpinX gCards to be user programmed to suit a specific applications. A VLV connects two microfluidic structures at a precise location on the gCard, which defines a precise volume of liquid. By applying centrifugal force at a controlled time, liquid moves through this newly created connection to a new chamber, where it can be combined with other liquids to create precise dilutions or biochemical reactions for controlled incubation times. The VLV technology relies on focusing a high-radiance laser beam onto a microscopic area on the surface of a thin film that is specially designed to strongly absorb laser light. The film separates two plastic substrates containing microfluidic structures. A laser shot is used to perforate the film. This perforation creates a channel between microfluidic components (either channels or chambers) facing the film. This closed-to-open transition, when combined with the controlled application of centrifugal force, functions as a valve, allowing precise volumes of liquids to flow at precise times. Fluidigm is another company that is currently marketing products based on microfluidic technologies. The technology was based on the work of Quake and colleagues. A multilayer soft lithography fabrication process was employed to build sophisticated networks of channels with valves, mixers, pump, and the like. A rubber like soft material that deflects under pressure was used to create an effective seal (NanoFlex valve). The structures are very small to enable tens of thousands of them to be integrated into a dense network of channels for regulating aqueous solutions on a nanoliter and picoliter scale. This technology was applied to many applications involving large protein molecules and cells. However, such devices cannot be used in assays involving small molecules because the small molecule can easily pass through the soft material.

14.7 APPENDIX: POSTER PRESENTED AT THE 2002 SOCIETY FOR BIOMOLECULAR SCREENING ANNUAL MEETING: ANALYSIS OF MOBILITY SHIFT DATA OBTAINED FROM LABCHIP IN KINASE ASSAY

Summary Microfluidic technology has found increasing use in high-throughput screening due to its promise of reagent savings, ease of operation, and high-quality data. Kinases are among the first classes of enzymes to which microfluidic technology has been applied, and the microfluidic method compares favorably with alternative approaches using microplates. In an on-chip microfluidic kinase assay using the Caliper microfluidic system, an inhibitor typically produces a peak and a dip. The appearance of this pattern can be explained qualitatively by considering the decreased product concentration and increased substrate concentration, and the separation between the two species. This analysis leads to the expectation that the extent of inhibition can be measured by measuring the amplitude from the peak to the dip. However, with potent inhibitors at relatively high concentration, such treatment produces erroneous results. We analyzed the underlying mechanism of the mobility shift assay and found that the more accurate representation of the extent of inhibition is the measurement of the amplitude from the baseline to the first emerging peak or dip. This analysis also showed that the substrate turnover could be estimated by dividing the first appearing peak or dip amplitude by the baseline amplitude. Updated versions of Caliper data analysis software now incorporate this new analysis method.

Introduction The kinase family is among the most important classes of drug targets. Common kinase assays include (1) filtration assay where radioactive kinase product, formed after ^{32}P-phosphate has been transferred from ATP to kinase substrate, is separated from the reaction mix and counted; (2) Scintillation Proximity

Assay where radioactive product is bound to beads or a scintillant-coated microplate and counted; (3) fluorescence resonance energy transfer or AlphaScreen assay where phosphorylated peptide complexes with high-affinity antibody and where both peptide and antibody have been labeled with secondary reagents; (4) competitive fluorescence polarization assay where fluorescence-labeled phosphorylated peptide product competes with the kinase assay product for binding to a high-molecular-weight antibody; and (5) direct fluorescence polarization assay where substrate is fluorescently labeled and the phosphorylated product selectively bound to a high-molecular-weight molecule or bead.

The above-mentioned methods all require radioactivity and/or additional assay reagents other than enzyme and substrate. The microfluidic On-Chip Mobility Shift assay, where substrate and phosphorylated kinase product are separated directly in the assay channel, offers an alternative (refer to Fig. 14.6). In this assay, a fluorescently labeled peptide substrate and a kinase are placed in two separate reservoirs that are connected to the main assay channel. Compounds to be screened are placed in a microplate and sipped to the chip through a sipper that is also connected to the assay channel. The three components are mixed while flowing through the assay channel and after a short incubation time a voltage is applied to separate product and substrate molecules. An inhibitor produces a distinctive pattern that typically has a peak and a dip. Intuitively, one would think that the extent of inhibition can be measured by measuring the amplitude from peak to dip (Fig. 14A.1; ΔPD). Such measurement is simple to implement since it does not require a baseline determination, and early versions of Caliper software implemented this algorithm. Another proposed analysis is to measure the area under the peak and dip. Experimentally, we found these data treatment produce erroneous results with potent inhibitors at relative high concentration. We reanalyzed the underlying mechanism of the on-chip mobility shift assay and showed that the true and reliable measurement of the extent of enzyme inhibition should be the measurement from the signal baseline to the first emerging peak or dip (Fig. 14A.1; ΔPB or ΔDB).

Assay Principle The chip used in the current study is the Caliper NS75 single sipper chip, and the channel has been pretreated by a polymer that eliminated most electroosmotic flow. Figure 14A.2 (same as Fig. 14.6.) shows how the chip works. In this assay design, a vacuum (typically -1 psi) is continuously applied to well 1 and thus all fluid flow is toward well 1. The substrate and enzyme flow continuously from their reservoir into the main channel and product is generated continuously. Because the substrate is fluorescently labeled, the product is the same molecule plus a phosphate group. The sipper, which draws solutions into the assay channel, alternates between sipping buffer and test compounds dissolved in the same buffer in a microplate. Typically, the sipper dips into a compound well for 1 s. The sipper then dips into a buffer reservoir (usually >6 s) before going to the next compound. In this scheme, a series of plugs of test compounds are introduced to the assay channel, with zones of buffer in between the compound plugs. In the absence of an electrical field, the substrate and product flow at the same rate. The conversion from substrate to product decreases the substrate concentration and at the same time increases the product concentration by the same magnitude, and there is no net change in the

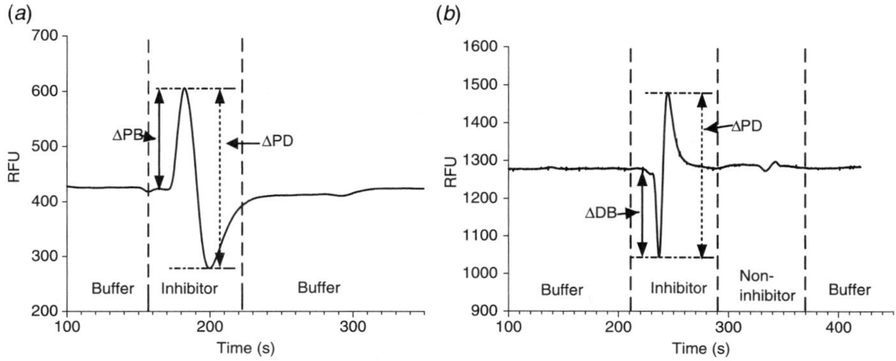

Figure 14A.1 Typical inhibitor-induced mobility shift kinase assay results. Depending on the way the voltage is applied, two patterns of inhibition are observed. (*a*) Peak emerges first when cathode is upstream. (*b*) Dip emerges first when anode is upstream.

fluorescence signal at any point along the main channel. In the presence of an electrical field, however, the substrate and product flow at different speeds due to the -2 charge difference conferred by the phosphate group present only on the product molecule. In the electrical field, substrate and product in the inhibitor plug exchange with neighboring buffer plug, typically producing a peak and dip as shown in Figure 14A.1.

Analysis of the Mobility Shift Data In order to study the mechanism of the on-chip mobility shift assay, we used a fluorescence compound peak to represent an inhibitor zone in the following simulation, illustrating an inhibitor concentration profile. It should be noted that the peak has a lag due to the interaction with the channel surface that resembles the behavior of the majority of test compounds. This is the reason we choose a real experimental compound peak in the current simulation because theoretically generated Gaussian or other type peak cannot accurately describe the real peak. For simple illustration purposes without changing the essence of the analysis, we assumes that (1) the enzyme meets the substrate and test compound in the main channel, (2) 40% substrate has been converted to product when the mixture reaches the separation channel, and (3) substrate and product peaks are separated by the half-peak width (5 s) in the electrical field of the separation channel. Under these conditions, while sipping buffer, there will be a constant fluorescent signal, 60% of it contributed by the substrate and 40% contributed by the product. When a plug of inhibitor at 10 μM is introduced into the assay channel containing enzyme and substrate and product, the inhibitor interaction with enzyme creates different substrate and product response patterns depending on the potency of the inhibitor (see Fig. 14A.2, Inhibition Profile). However, since any change in the number of substrate molecules is balanced by an opposing change in the number of product molecules, the total fluorescence signal is unchanged (Fig. 14A.2, Before Separation). In the electrical field of the separation channel, product peak mobility is reduced while substrate mobility remains unchanged. The distinct signal pattern of inhibition by inhibitors of varying potencies was finally visualized in the detection window (Fig. 14A.2, After Separation, solid trace).

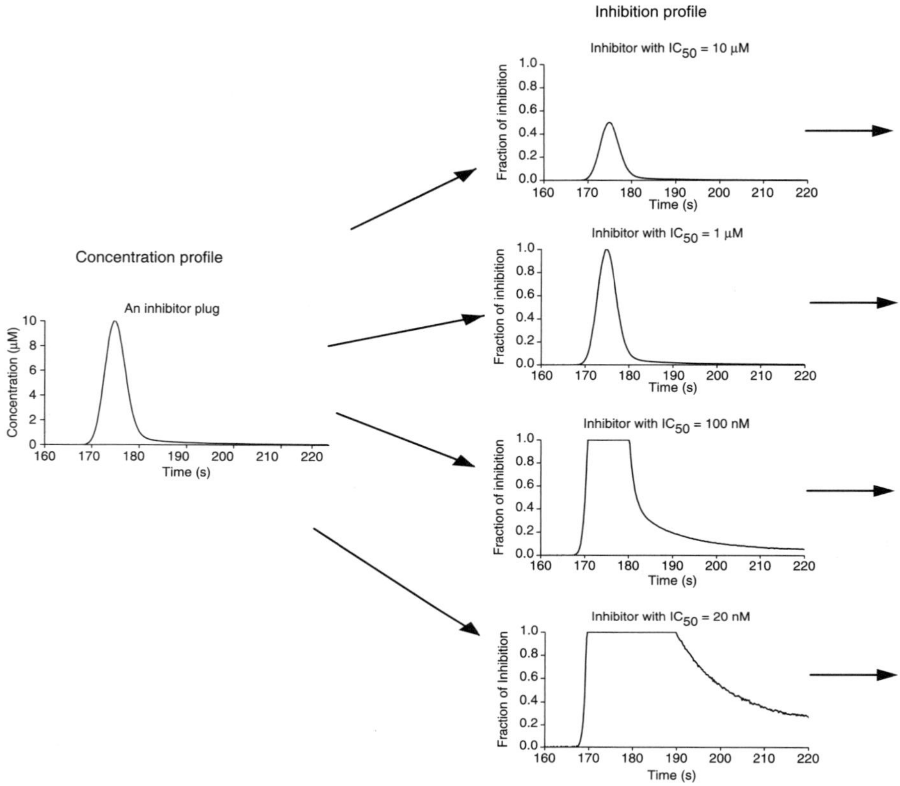

Figure 14A.2 Simulation of the inhibition pattern for inhibitors with different potency (IC_{50} = 10, 1, 0.1, and 0.02 μM) at same concentration of 10 μM. The concentration profile for the four inhibitors are the same. This concentration profile then is transformed to inhibition profile that are dramatically different from the concentration profile. This inhibition profile produces the same pattern of fluorescence signal from the substrate and the product in a compound plug. However, the visualized pattern is a flat line in the absence of voltage due to the two opposite patterns that cancel each other. When the same voltage is applied to the four patterns, the shift in the substrate and product patterns create the final observed mobility shift patterns, which are different from each other depending on the inhibitor potency.

The fluorescence assay signal trace in the presence of an inhibitor shows an initial peak above the baseline that is contributed by increased substrate concentration. The closely following dip below the baseline is due to decreased product concentration. The magnitude of the peak reflects the extent of substrate conversion and the level of inhibition. At maximal inhibition, the ratio of the peak to the baseline is equal to the percent conversion. The key feature of this new analysis is the inclusion of the step that transforms the inhibitor concentration profile to an enzyme inhibition profile. In high-throughput screening, the same concentrations of compounds are tested. Thus, the same concentration profile is expected for every compound. However, different compounds will have different inhibition potency that result in

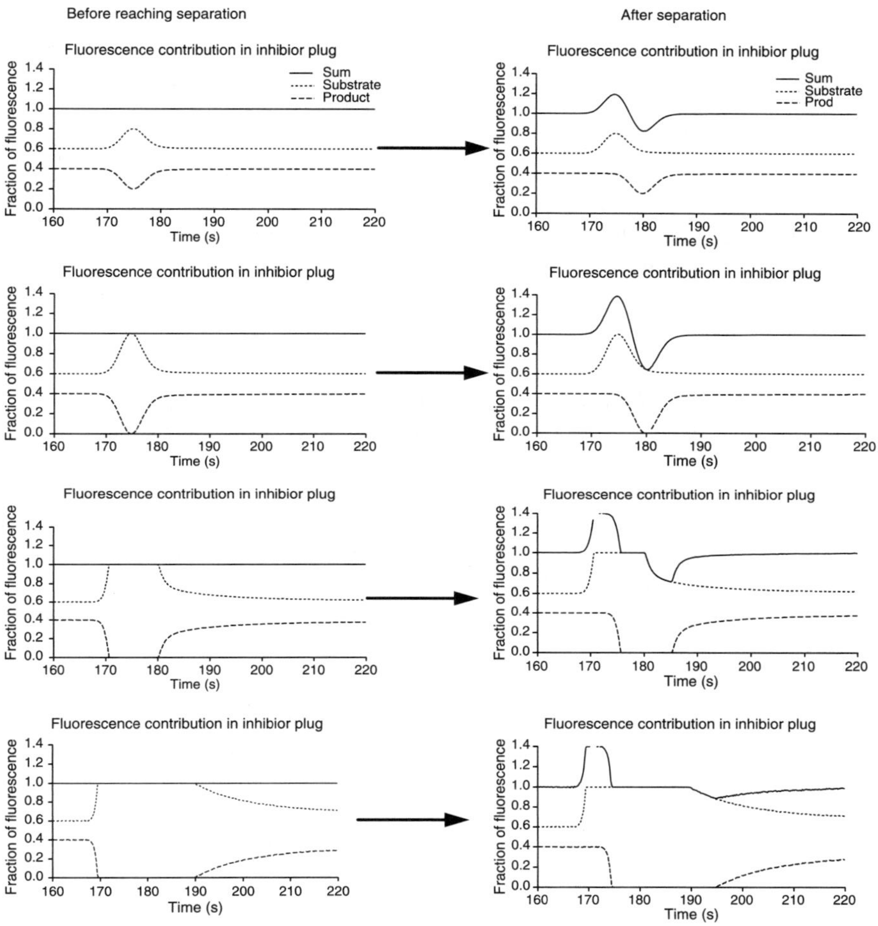

Figure 14A.2 *Continued.*

different inhibition profiles. If directly analyzing the concentration profile without this transformation, the Gaussian-like peak would lead to the conclusion that a symmetric peak and dip would be observed for any inhibitors, and thus ΔPD should be a good measure of the degree of inhibition. This analysis shows that the peak-to-dip measurement would only be a valid measurement of inhibition when inhibition is submaximal. When the potency of an inhibitor reaches maximal inhibition and beyond, the inhibition patterns are distorted and drag a long trace so that the substrate and product can no longer be separated at the preset separation power. This overlap cancels each other's signal that significantly reduced the magnitude of the lagging dip. The peak-to-dip measurement is no longer an accurate measurement of inhibition. Though this overlap can be theoretically resolved by increasing the substrate and the product separation, there is a practical separation limit that can be achieved experimentally. In addition, there is no way to know the potencies of compounds before screening in order to adjust the separation power accordingly. The method of measuring the area under the peak and dip run into the same problem too because the overlapping region reduces each other, and hence the measured area will be significantly less than the total area. In addition, it is very difficult to calculate the area reliably by integration over a broad range due to uncertainty in selecting the range to integrate when strong inhibitors produce a very spread area coupled with baseline fluctuation. Thus, the only reliable measurement of mobility shift data is to measure the amplitude of the first peak (or dip) to the baseline. Though this analysis was demonstrated with inhibitors of different potencies at a fixed concentration, it can be applied to one inhibitor at different concentrations as well.

Testing the Analysis Using Inhibition of PKA by PKI Figure 14A.3 shows the concentration–response curve of PKI on the activity of PKA. In this case, we use different concentrations instead of different IC_{50} to achieve the different inhibition profiles. Increasing concentrations of PKI are sipped through the sipper. The highest concentration of PKI showed here is 1.6 μM. Figure 14A.3*b* shows the analysis of the data with different methods. PKI at higher concentrations would give less inhibition than at lower concentrations when peak-to-dip measurement is used for analysis, which is obviously wrong. The correct result is obtained when the peak to the baseline analysis is used.

Determining Suitable Assay Separation Condition This analysis can also guide us to determine the minimum product and substrate separation needed for an on-chip mobility assay. Due to the nature of the substrate and the limits of the maximum electrical field that can be applied, it is not always possible to separate the product and substrate as desired. We simulated the results of on-chip mobility shift assay

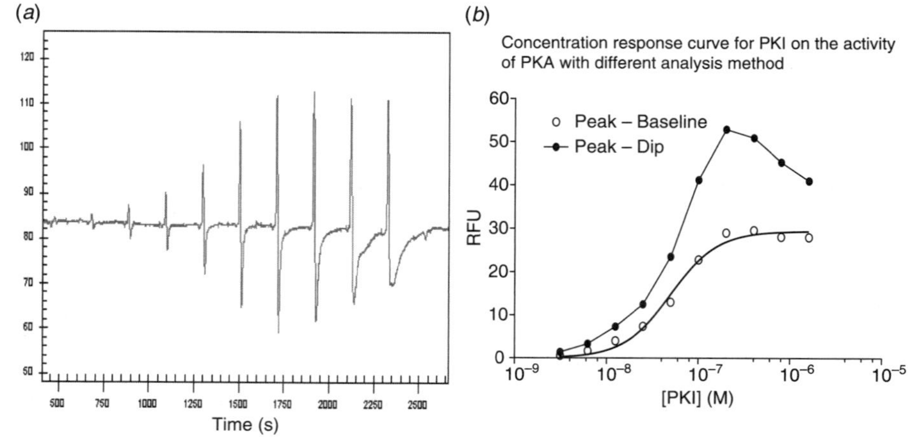

Figure 14A.3 (*a*) PKI concentration–response curve measured by the mobility-shift method. (*b*) Data analysis using pick-to-dip method vs. pick-to-baseline method.

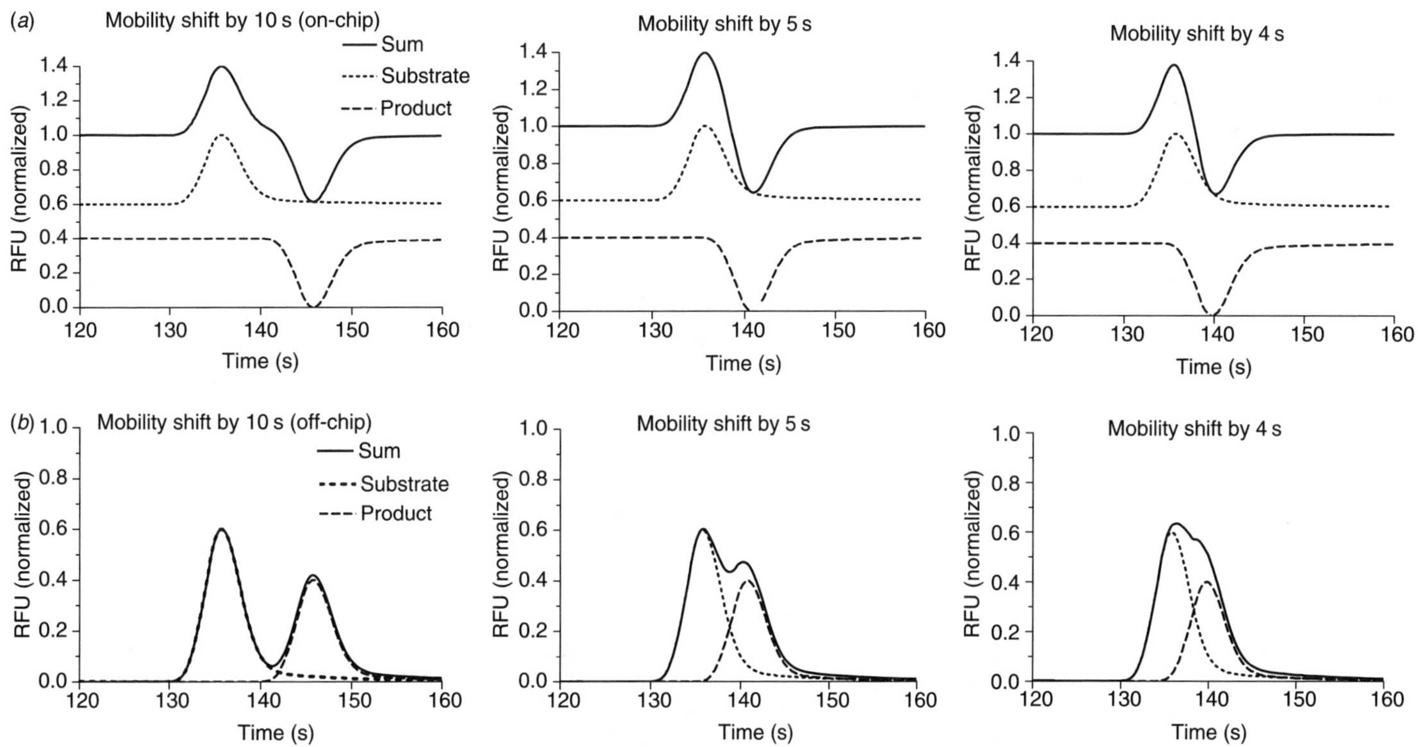

Figure 14A.4 (*a*) Simulation of on-chip mobility assay at different separation condition. (*b*) Simulation of peak overlap when substrate and product are mixed together and separated under the on-chip separation condition.

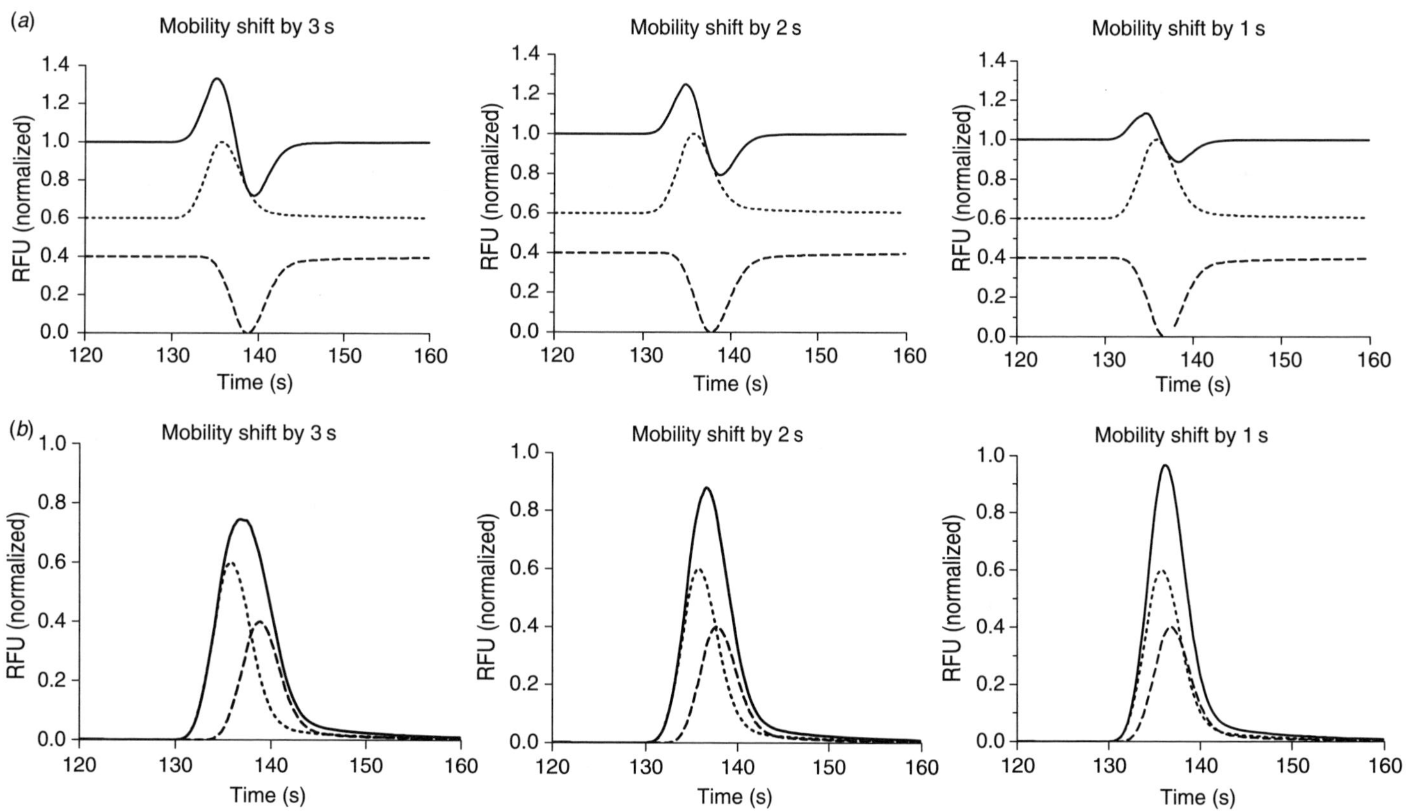

Figure 14A.4 *Continued.*

at different separations shown in Figure 14A.4*a*. It is clear that all the patterns look the same even at poor separation when the total signals are significantly reduced. Thus, the assay separation condition cannot be determined by on-chip mobility shift assay alone. To guide the on-chip assay development, substrate and product should be mixed together at a proportion equal to the conversion with on-chip assay and separated using the same chip and the same buffer. In Figure 14A.4*b*, we simulated the off-chip results for the same condition as shown in Figure 14A.4*a*. A separation condition for on-chip assay should be chosen based on the criteria that the first evolving peak is not affected by the lagging peak (as a rule of thumb, the lagging dip should affect less than 10% of the first peak). In the current demonstrated case, more than 4 s separation is needed for a good on-chip assay.

Discussion Several factors complicate the mobility shift assay, which cannot be easily quantified and thus could not be considered in the current analysis. These factors are: (1) We did not account for that fact that the product is continuously generated in the separation channel at a reaction rate that may be different from the reaction rate in the incubation channel. (2) The inhibitor may flow at a different velocity in the separation channel from the substrate and product patterns created before reaching the separation channel. A new inhibition pattern will be established in the separation channel that will gradually move away from the original pattern. The final pattern is the sum of the two. To minimize the above two complications, the transient time in the separation channel is designed to be shorter than in the incubation channel. (3) We assumed that the labeled substrate and product have the same fluorescence properties (quantum yield, quenching, etc.) in the electrical field. However, our recent experiments have shown that the product gives a relatively higher fluorescence signal than the substrate in electrical field. This will make the dip amplitude bigger than the peak amplitude. (4) We did not account for the fact that the fluorescence signal from the lagging species (product) will have broader peak and lower amplitude than the leading peak. However, these complications should not undermine the major conclusion of the current analysis.

Useful Websites

http://www.caliperls.com

http://www.chem.agilent.com/en-US/products/instruments/lab-on-a-chip/pages/default.aspx

http://www.fluidigm.com/

http://www.gyros.com/

http://www.spinx-technologies.com/

http://www.cetek.com/

BIBLIOGRAPHY

Bilitewski, U., Genrich, M., Kadow, S., and Mersal, G. (2003) Biochemical analysis with microfluidic systems. *Anal. Bioanal. Chem.* **377**, 556–569.

Bousse, L., et al. (2000) Electrokinetically controlled microfluidic analysis systems. *Annu. Rev. Biophys. Biomol. Struct.* **29**, 155–181.

Bruin, G. J. M. (2000) Recent developments in electrokinetically driven analysis on microfabricated devices. *Electrophoresis* **21**, 3931–3951.

Cohen, C. B., Chin-Dixon, E., Jeong, S., and Nikiforov, T. T. (1999) A microchip-based enzyme assay for protein kinase A. *Anal. Biochem.* **273**, 89–97.

Gerber, D., Maerkl, S. J., and Quake, S. R. (2009) An in vitro microfluidic approach to generating protein-interaction networks. *Nat. Methods* **6**, 71–74.

Haeberle, S. and Zengerle, R. (2007) Microfluidic platforms for lab-on-a-chip applications. *Lab-on-a-Chip* **7**, 1094–1110.

Harrison, D. J., et al. (1993) Micromachining a miniaturized capillary electrophoresis-based chemical -analysis system on a chip. *Science* **261**, 895–897.

Hong, J., Edel, J. B., and deMello, A. J. (2009) Micro- and nanofluidic systems for high-throughput biological screening. *Drug Discov. Today* **14**, 134–146.

Huikko, K., Kostiainen, R., and Kotiaho, T. (2003) Introduction to micro-analytical systems: Bioanalytical and pharmaceutical applications. *Eur. J. Pharm. Sci.* **20**, 149–171.

Kang, L., Chung, B. G., Langer, R., and Khademhosseini, A. (2008) Microfluidics for drug discovery and development: From target selection to product lifecycle management. *Drug Discov. Today* **13**, 1–13.

Khaledi, M. G. (ed.) (1998) *High-Performance Capillary Electrophoresis*. Wiley, New York.

Knapp, M. R., Sundberg, S., and Parce, J. W. (2000) Test tube's end. *J. Biomol. Screen.* **5**, 9–12.

Madou, M., et al. (2006) LAB ON A CD. *Annu. Rev. Biomed. Eng.* **8**, 601–628.

Mao, H., Cremer, P. S., and Manson, M. D. (2003) A sensitive, versatile microfluidic assay for bacterial chemotaxis. *Proc. Natl. Acad. Sci. U.S.A.* **100**, 5449–5454.

Melin, J. and Quake, S. R. (2007) Microfluidic large-scale integration: The evolution of design rules for biological automation. *Annu. Rev. Biophys. Biomol. Struct.* **36**, 213–231.

Oosterbroek, R. E. and van den Berg, A. (2003) *Lab-on-a-Chip*. Elsevier, New York.

Perrin, D., Fremaux, C., Besson, D., Sauer, W. H., and Scheer, A. (2006a) A microfluidics-based mobility shift assay to discover new tyrosine phosphatase inhibitors. *J. Biomol. Screen.* **11**, 996–1004.

Perrin, D., Fremaux, C., and Scheer, A. (2006b) Assay development and screening of a serine/threonine kinase in an on-chip mode using caliper nanofluidics technology. *J. Biomol. Screen.* **11**, 359–368.

Quake, S. R. and Scherer, A. (2000) From micro- to nanofabrication with soft materials. *Science* **290**, 1536–1540.

Ramsey, J. M., Jacobson, S. C., and Knapp, M. R. (1995) Microfabricated chemical measurement systems. *Nature Med.* **1**, 1093–1096.

Sia, S. K. and Whitesides, G. M. (2003) Microfluidic devices fabricated in poly(dimethylsiloxane) for biological studies. *Electrophoresis* **24**, 3563–3576.

Sundberg, S. A., Chow, A., Nikiforov, T., and Wada, H. G. (2000) Microchip-based systems for target validation and HTS. *Drug Discov. Today* **1**, 92–103.

Tabeling, P. (2005) *Introduction to Microfluidics*. Oxford University Press, Oxford.

Toriello, N. M., et al. (2008) Integrated microfluidic bioprocessor for single-cell gene expression analysis. *Proc. Natl. Acad. Sci.* **105**, 20173–20178.

Wang, J. (2002) On-chip enzymatic assays. *Electrophoresis* **23**, 713–718.

Wu, G. and Hodge, C. N. (2002) Analysis of mobility shift data obtained from Labchip in kinase assay. *Poster presented at the annual meeting of the Society for Biomolecular Screening.*

Wu, G., et al. (2003a) Assay development and high-throughput screening of caspases in microfluidic format. *Comb. Chem. High Throughput Screen.* **6**, 303–312.

Wu, G., Yuan, Y., and Hodge, C. N. (2003b) Determining appropriate substrate conversion for enzymatic assays in high-throughput screening. *J. Biomol. Screen.* **8**, 694–700.

INDEX

Assay Development: Fundamentals and Practices. By Ge Wu